1101420

4-26

# CIRCUIT THEORY: A COMPUTATIONAL APPROACH

# CIRCUIT THEORY
# A Computational
# Approach

**Stephen W. Director**
University of Florida

JOHN WILEY & SONS, INC.
New York  London  Sydney  Toronto

*Library of Congress Cataloging in Publication Data:*

Director, Stephen W.      1943–
  Circuit theory.

  Bibliography: p.
  Includes index.
  1. Electric networks. 2. Electric circuits.
  3. Electronic data processing—Electric engineering.
  4. Electric engineering—Mathematics. I. Title.

TK454.2.D57      621.319'2      75-2016
ISBN 0-471-21580-5

Printed in the United States of America

10 9 8 7 6 5 4 3 2 1

To Lori

# PREFACE

Because of the continuing need to design larger and more complex circuits and systems, the computer has become an indispensable tool for analysis and design. This is especially true for the design of integrated circuits where advances in technology now allow the realization of circuits with hundreds of transistors on a single silicon chip. This factor has altered the point of view of the first circuits course, which usually introduces the student to a number of analysis techniques that can be used not only for circuit analysis but also in subsequent courses for the analysis of a variety of other systems. A modern circuit theory course, in addition to introducing the basic concepts of analysis, should emphasize those techniques that can be efficiently implemented on the computer. Furthermore, in addition to presenting the usual analytical methods, the modern circuit theory course should introduce those computational methods that are particularly useful for circuit analysis. Finally, in addition to teaching the student which facts of circuit theory apply in a given situation, the modern circuit theory course should demonstrate the use of the computer as an analysis tool, and should teach the student how to interpret the results of the computer correctly.

This textbook is the outgrowth of notes used in the first circuits course at the University of Florida, and it is intended for a two-quarter or three-quarter (or two semester), sophomore or junior level course. A discussion of typical chapter sequences, which may be employed in courses with various emphases, is included in the instructor's manual.

Below are several features of the book that should be mentioned.

*A thorough investigation of resistive circuits is undertaken before the introduction of energy storage elements.* The philosophy here is to present as many of the circuit theoretical concepts as possible before clouding the issues with differential equations. This approach is meaningful when it is realized that, from a computational viewpoint, the dc analysis of linear resistive networks is the basis for all forms of analysis—even the transient analysis of the most complex nonlinear network. Moreover, by initially avoiding the use of differential equations, the material in these chapters (Chapters 1–10) can be taught concurrently with a calculus course.

*A modified nodal analysis formulation is emphasized,* although other formulations are discussed. There are several reasons for choosing this formulation. First, it is particularly easy scheme to implement on the computer because the equations are easily written by inspection. Second, although nodal analysis may not be the most efficient scheme when invoking sparse matrix considerations, only full matrix solution methods are presented in this introductory text so that the fewer, more dense, nodal equations save computer time and storage.

*Computational methods that form the basis for methods employed in modern circuit simulation programs are introduced, as appropriate, and numerous special purpose programs are given.* When numerical techniques are to be introduced, one always faces the question as to how much programming should be required of the student. Clearly, the student can only fully understand the use of a computational method when he himself has used it. On the other hand, the student cannot be expected to write all of his own programs because too much time is usually wasted on debugging them. A compromise is to present to the student a number of example programs that work, and to ask him to modify, or edit, portions of these programs for new situations. Once experience is gained by using these techniques, the student will be in a better position to write programs from the beginning. This approach is used throughout the book. A few general purpose subroutines, such as one for Gaussian elimination, are presented, and numerous special purpose programs that use these subroutines to solve particular network problems are given. These special purpose analysis programs have the same structure as a general purpose analysis program so that the student also gains familiarity with the form of a general purpose analysis program. (A general purpose analysis program that uses the techniques discussed in this text is available and is described in detail in the instructors manual. All programs in the text are available from the author.)

*Many of the examples and problems are aimed at tying circuit theory to modern electronics.* This feature is especially important when the first circuit theory course is taken concurrently with an introductory course.

*There are two chapters that introduce some of the basic ideas behind computer-aided circuit design* (Chapters 6 and 19). Since the real motivation behind studying circuit theory is to be able to design circuits, it is important to introduce design concepts in the first circuits course. In this regard, worst-case analysis is introduced during the discussion of resistive networks, and the computer-aided optimization of circuits operating in the frequency domain is introduced in the final chapter.

Finally, I wish to express my sincere appreciation to my friends and colleagues who taught from various revisions of the notes and made many helpful suggestions for their improvement. I also thank the many students at the University of Florida who good-naturedly suffered through the courses while the notes were being developed and tested. I especially thank Mac Van Valkenburg and Chris Pottle for their detailed reviews and valuable suggestions, and Art Brodersen who provided helpful advice on the physical behavior of electronic circuits. I acknowledge the help of Bill Bristol, Mike Lightner and Ray Majors for providing detailed solutions to the problems, and Mike Thorn who wrote and tested many of the computer programs. Finally, I am indebted to Ron Rohrer who initially motivated me to begin this project.

S. W. Director

# CONTENTS

# CIRCUIT THEORY: A COMPUTATIONAL APPROACH

## CHAPTER 1

# Introduction to Circuit Analysis

During the past decade the digital computer has become an efficient and effective engineering design tool. In fact, it might be argued that in some areas of design the computer is now an *indispensable* design tool. The design of large scale integrated circuits is a case in point. Because of major advances in semiconductor technology and the emergence of powerful circuit analysis computer programs, it is now possible to design and build complex networks consisting of thousands of transistors on a single silicon chip, which has an area of a few square mils. One of the hand-held electronic calculators or, more appropriately, "electronic slide rules," presently on the market, contains three integrated circuit chips that contain a combined total of 50,000 transistors.

Since the computer is playing a larger and larger role in analysis, or *simulation*, and design, it is becoming more and more important for the engineer to be familiar not only with the basic theoretical foundations of analysis and design but also with the basic numerical techniques that can allow him to make effective use of the computer. It is not enough to have a versatile analysis program that can simulate a circuit's behavior and thereby avoid the "breadboard" step in design. The engineer, who uses such a program, but who is not familiar with the basic theoretical and numerical techniques employed by the program, is ill-prepared to assess the validity of the simulation. It is well known among computer users that too often a computer spits out numbers that have no relationship to anything. The engineer who bases design decisions on computer simulations must be sure that the simulation is accurate.

The purpose of this text is to present the theoretical foundation of electrical network, or circuit, analysis and to introduce those elementary numerical techniques used for the computer implementation of circuit analysis. This computational approach to circuit theory, should enable the reader to use the more advanced methods found in modern day circuit analysis programs for the accurate simulation of the large complex circuits of today, as well as to reliably design the even larger and more complex circuits of tomorrow.

In this study of circuit analysis, it is our intention to introduce fundamental concepts only as they are needed. We will try to avoid enumeration of sets of "fundamental" definitions and assumptions that so often obscure the true objectives by emphasizing the general before the particular. The disadvantage to this approach is that concepts may at times seem vague and lacking in exact meaning. However, we believe that the advantages of this approach far outweigh the disadvantages. For this reason, we begin by familiarizing the reader with some of the elementary ideas of which circuit analysis is based, by investigating the behavior of a simple electrical network.


## 1-1  THE FUNDAMENTAL VARIABLES OF CIRCUIT ANALYSIS

Before launching our investigation of circuit theory, it is worthwhile to discuss briefly the variables that are used to describe the behavior of a circuit. The basic electrical quantity is a property of matter known as *charge*. The *MKS* (meter-kilogram-second) units of charge is the *coulomb*—it is the charge contained on $6.24 \times 10^{18}$ electrons. We will use the letter $q$ to denote charge. In particular, $q(t)$ denotes the charge at the instant of time $t$. It is a physical law that charge cannot be created nor destroyed, that is, *charge must be conserved*.

In our study of circuit behavior we will be concerned with the movement of charge or *current*. The basic unit of current is the *ampere* (abbreviated amp or a) that is equal to one coulomb per second. Current is denoted by the symbol $i$ or $i(t)$. In particular, we have

$$i(t) = \frac{d}{dt} q(t)$$

It is now important to point out that there are two kinds of charge: positive and negative. Therefore, current refers to the *net* transfer of charge.

The movement of charge between two points requires *energy* or *work*. We refer to the ratio of the energy required to move charge between two points to the amount of charge as *voltage*. In particular, if $w$ *joules* of work are required to move $q$ *coulombs* between points $a$ and $b$, the voltage, denoted by $v$, is

$$v = \frac{w}{q} \frac{\text{joules}}{\text{coulomb}}$$

Voltage is measured in units of *volts* (abbreviated v). One volt equals one joule per coulomb.

The final physical quantity immediately of interest is *power*. Power is the time rate of change of energy and has units of watts. We use the symbol $p$ to denote power:

$$p(t) = \frac{d}{dt} w(t)$$

Since current is the time rate of change of charge

$$i = \frac{dq}{dt}$$

and voltage is the ratio of work to charge

$$v = \frac{w}{q}$$

we have

$$p(t) = v(t)i(t)$$

that is, power equals the product of voltage and current.

## 1-2  CURRENT SOURCES AND RESISTORS

We begin our study of circuit analysis with the investigation of the behavior of the *current-divider* network of Fig. 1-1. (The reason for this terminology is discussed later.) This electrical network, or circuit, is comprised of two types of *two-terminal circuit elements*, namely *current sources* and *resistors*; the symbols used to represent these elements are shown in Fig. 1-2. A circuit's behavior is characterized in terms of the *voltages* and *currents* presented in the circuit. These voltages and currents are determined by consideration of the

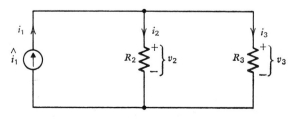

**Fig. 1-1   A current-divider network.**

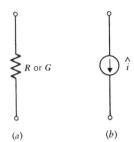

(a)               (b)

**Fig. 1-2  Symbols used to designate (*a*) resistors and conductors, and (*b*) current sources.**

*constraints* imposed by the circuit elements as well as the constraints imposed by the particular *connection* of the circuit elements. First, we will consider the constraints imposed by the circuit elements.

Each two-terminal circuit element imposes one constraint between the *instantaneous branch voltage v(t)* across it and the *instantaneous branch current i(t)* that flows through it.[1] Any two-terminal circuit element for which the constraint imposed on its branch voltage *v(t)* and branch current *i(t)* can be ·written as

$$v(t) = Ri(t)$$

is called a *resistor* where the resistance *R* is measured in ohms. The symbol Ω is used to designate units of ohms. If the branch relationship is of the form

$$i(t) = Gv(t)$$

then the element is termed a *conductor* where the conductance *G* is measured in mhos. The symbol ℧ is used to designate units of mhos. If the resistance *R* is nonzero, then the resistor can be described by the conductor branch relationship with $G = 1/R$. Similarly, if the conductance *G* is nonzero, then the conductor can be described by the resistor branch relationship with $R = 1/G$. In the sequel, both resistors and conductors will be referred to as *resistive elements*.

We should point out here that in the "real world" there is no element that is exactly characterized by either of the above branch relationships. However, these branch relationships do *approximate* the behavior of physical resistors over reasonable ranges of voltages and currents. In particular, observe that if a large enough voltage is put across a physical resistor it will burn up, and the current through it will ultimately go to zero. However, this behavior is not predicted by the *ideal* resistive branch relationship. Over the range of close approximation, we say that the physical resistor is *modeled* by the ideal resistor.

---

[1] Actually, there are two pathological two-terminal lumped elements, the nullator and norator that impose two constraints and no constraints, respectively, on their branch voltages and currents. Pathological elements are not physically realizable and are to be ignored for the present.

During the course of our study we will introduce a number of other ideal elements. Some ideal elements are similar to the ideal resistor because they have physical counterparts, while other ideal elements have no direct physical counterpart but may be used in combinations that mathematically model the behavior of more complex physical devices. Henceforth, we will drop the adjective "ideal"; it will be used when we wish to emphasize the distinction between ideal elements or mathematical models and physical devices.

A *current source*, or more precisely, an *ideal independent current source*, is characterized by the branch relationship

$$i(t) = \hat{\imath}(t)$$

where $\hat{\imath}(t)$ is specified and is measured in amperes (designated by amps or a). The voltage across a current source is arbitrary.

## 1-3    ASSOCIATED REFERENCE DIRECTIONS

A word about the voltage and current reference directions, which are to be used in this text, is in order. Consider the arbitrary two-terminal elements of Fig. 1-3. The term $v(t)$ represents the instantaneous branch voltage and $i(t)$ the instantaneous branch current of the element. The voltage reference plus and minus symbols and the current reference arrow symbol of this element are merely notational conveniences. These symbols do not necessarily represent the actual directions of positive voltage drop (plus to minus) or positive current flow (in the direction of the arrow). Instead, they are convenient reference conventions that may apply arbitrarily to any element. The reference directions apply as follows: if the actual voltage has the same polarity as the voltage reference direction, it is said to be a positive element voltage (with respect to the assigned voltage reference direction), if opposite, negative. Similarly, if the actual direction of current flow is the same as that of the reference current, it is said to be a positive element current (with respect to the assigned reference direction), if opposite, negative. To approach the characterization of an element, we must assign it both a voltage reference direction and a current reference direction. The choices of reference directions are arbitrary, and in various situations one may be more convenient than another.

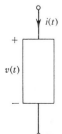

Fig. 1-3    **An arbitrary two-terminal circuit element whose branch voltage is** $v(t)$ **and branch current is** $i(t)$.

Of the four possibilities for labeling an element with voltage and current reference directions shown in Fig. 1-4, the two *associated reference directions* emerge as notationally and mnemonically the more convenient. The reference directions assigned to an element are said to be *associated reference directions if the direction of positive current flow coincides with the direction of positive voltage drop.* Figures 1-4a and 1-4d show associated reference directions, where the current reference direction arrow points from the plus to the minus of the voltage reference.

The associated reference directions have the property that when both the instantaneous voltage and current have the same sign, *power,*

$$p(t) = v(t)i(t)$$

is being instantaneously delivered to the element in question. Similarly, when the instantaneous voltage and current have opposite signs the particular element characterized in terms of associated reference directions is instantaneously delivering power to the remainder of the circuit connect to it.

In the remainder of this study we shall use associated reference directions *exclusively.*

Because the associated reference directions relate the voltage and current polarity conventions, there is redundancy in using both. Given the plus and minus symbols of the voltage reference direction, we recognize that the associated current reference arrow must point from the plus toward the minus symbol. Similarly, given the arrow of the current reference direction, we recognize that the associated voltage reference must have the plus voltage symbol coincide with the arrow's tail and the minus voltage symbol with the arrow's head. Consequently, under the associated reference convention the situations of Figs. 1-5b and 1-5c are both equivalent to that of Fig. 1-5a. In a later section we will find it convenient to abstract the situation still further to that of Fig. 1-5d, where the two-terminal circuit element is represented in

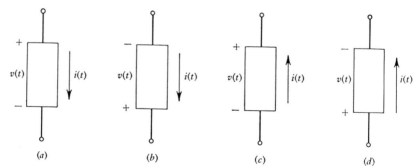

(a)          (b)          (c)          (d)

**Fig. 1-4   The four possible choices for voltage and current reference direction assignment for a given two-terminal lumped circuit element. (a) and (d) show associated reference directions; (b) and (c) are not associated reference directions.**

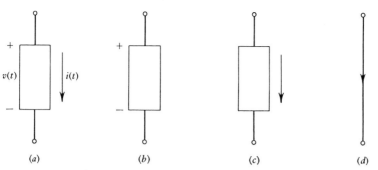

**Fig. 1-5** Under the associated reference direction convention, (*b*) and (*c*) are equivalent to (the more cluttered) (*a*). The ultimate in abstraction is the oriented branch (*d*), which conveys all of the information of any of the others.

terms of its associated reference directions merely by an *oriented branch*, the direction of the arrow indicating both the voltage drop and the current flow reference polarities.

Referring again to Fig. 1-1, and assuming associated reference directions, we see that the constraints imposed by the circuit elements are

$$i_1(t) = \hat{\imath}_1(t)$$
$$v_2(t) = R_2 i_2(t)$$

and

$$v_3(t) = R_3 i_3(t)$$

We have written three equations in terms of six unknowns: $i_1(t)$, $i_2(t)$, $i_3(t)$, $v_1(t)$, $v_2(t)$, and $v_3(t)$. ($R_2$, $R_3$, and $\hat{\imath}_1(t)$ are specified, $v_1(t)$ is implicitly unknown). Let us now turn to the constraints imposed by interconnection.

## 1-4  KIRCHHOFF'S LAWS

The constraints on an element's voltages and currents dictated by its interconnection in a circuit are independent of the type of element involved. *Kirchhoff's voltage law* and *Kirchhoff's current law* give the basic constraints of interconnection for circuits, independently of the two-terminal element types involved.

Before presenting Kirchhoff's laws we need to define a *node* and a *loop*. A *node is a point of connection of two or more branches.* The network of Fig. 1-1, redrawn in Fig. 1-6, has two nodes, labeled ① and ②. A *loop is any closed path among the branches (regardless of their orientation) that begins and ends at the same node.*[2] There are three possible loops for the network of Fig. 1-6 as shown

---

[2] The loop concept is usually more restrictively defined, to insure that no included branch is encountered more than once, for example. For the purposes of our study we need not consider such common sense restrictions, since we will not really require the loop concept once we have interpreted it properly.

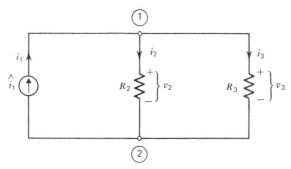

**Fig. 1-6   The current-divider network of Fig. 1-1 has two nodes, labeled 1 and 2.**

in Fig. 1-7. These loops are formed by branches 1 and 2, branches 2 and 3, and branches 1 and 3.

We are now in a position to state *Kirchhoff's Current Law (KCL): the algebraic sum of all of the instantaneous currents leaving (or entering) any node of a circuit is identically zero for all time.* The KCL equations for nodes ① and ② in the network of Fig. 1-6 are:

$$\text{node} ①: \quad -i(t) + i_2(t) + i_3(t) = 0$$

and

$$\text{node} ②: \quad i_1(t) - i_2(t) - i_3(t) = 0$$

(We assume that positive current leaves a node; if a branch current enters a node it is multiplied by $-1$ before adding it to the other currents leaving the node.) Observe that the KCL equation for node ② is the same as the KCL equation for node ①, except that each term has been multiplied by $-1$. We say that the KCL equation for node ② is *linearly dependent* on the KCL equation for node ① since it yields no new information. In Chapter 2 we will discuss the linear dependence of KCL equations in more detail.

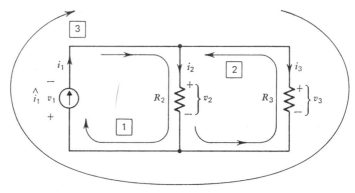

**Fig. 1-7   The network of Fig. 1-1 has three loops, labeled 1, 2, and 3. (*Note.* The direction of the loop is arbitrary.)**

Thus we see that application of KCL to the network of Fig. 1-6 yields one equation in terms of three unknowns: $i_1(t)$, $i_2(t)$, and $i_3(t)$. Together with the branch relationships discussed we now have four equations in terms of six unknowns. In order to solve for the unknowns we need two more equations. These equations come from application of *Kirchhoff's voltage law (KVL)*: *the algebraic sum of all of the instantaneous voltage drops (or rises) encountered in traversing any loop of a circuit, is identically zero for all time.*

The KVL loop equations for the three loops of the networks of Fig. 1-7 are:

$$\text{loop}\;\boxed{1}\qquad v_1(t) + v_2(t) = 0$$

$$\text{loop}\;\boxed{2}\qquad v_2(t) - v_3(t) = 0$$

and

$$\text{loop}\;\boxed{3}\qquad v_1(t) + v_3(t) = 0$$

(*Note:* the voltage drop is taken to be positive if the branch current flows in the same direction as assumed loop direction and negative otherwise.) Observe that the KVL equation for loop $\boxed{3}$ is equal to the KVL equation for loop $\boxed{1}$ minus the KVL equation for loop $\boxed{2}$. Therefore this loop equation is linearly dependent and yields no new information. The KVL equation for loops $\boxed{1}$ and $\boxed{2}$ are *linearly independent* because each contains a term that is not in the other. In particular, the KVL equation for loop $\boxed{1}$ contains $v_1(t)$ that is not in the KVL equation for loop $\boxed{2}$. In Chapter 2 we will study this dependence of KVL equations further. Thus, application of KVL to the network under consideration yields two equations in terms of three unknowns $v_1(t)$, $v_2(t)$, and $v_3(t)$.

## 1-5  SOLUTION OF THE NETWORK EQUATIONS

To summarize our progress to this point, by considering the branch relationships, we obtained three equations that can be written as:

(1.1)
$$i_1(t) = \hat{\imath}_1(t)$$

(1.2)
$$R_2 i_2(t) - v_2(t) = 0$$

and

(1.3)
$$R_3 i_3(t) - v_3(t) = 0$$

by considering Kirchhoff's Current Law we obtained one independent node equation:

(1.4)
$$-i_1(t) + i_2(t) + i_3(t) = 0$$

and finally, by considering Kirchhoff's voltage law we obtained two independent loop equations:

(1.5)
$$v_1(t) + v_2(t) = 0$$

and

$$(1.6) \qquad\qquad v_2(t) - v_3(t) = 0$$

for a total of six equations in terms of six unknowns. This set of six equations could be solved simultaneously to determine values for the six unknowns. Usually we are not concerned with determining values for all the unknowns, but are only interested in some subset of the unknowns. For instance, we might be satisfied with values for the branch voltages or the branch currents. For these situations we can reduce the six equations to a smaller set of equations that involve only the unknowns of interest.

Assume first that we only wish to determine the branch voltages. From the KVL loop equations (1.5) and (1.6) we see that if $v_2(t)$ is known so is $v_1(t)$ and $v_3(t)$ since

$$(1.7) \qquad\qquad v_1(t) = -v_2(t)$$

and

$$(1.8) \qquad\qquad v_3(t) = v_2(t)$$

We can consider $v_2(t)$ to be the independent or *basis* voltage from which all other voltages can be determined. The KCL node equation (1.4) can be written in terms of the branch voltages through the use of the branch relationships. From expressions (1.1) to (1.3) we have

$$(1.9) \qquad\qquad i_1(t) = \hat{\imath}_1(t)$$

$$(1.10) \qquad\qquad i_2(t) = \frac{v_2(t)}{R_2}$$

and

$$(1.11) \qquad\qquad i_3(t) = \frac{v_3(t)}{R_3}$$

Substitution of (1.9)–(1.11) into (1.4) yields

$$(1.12) \qquad\qquad -\hat{\imath}(t) + \frac{1}{R_2} v_2(t) + \frac{1}{R_3} v_3(t) = 0$$

Finally, $v_3(t)$ is related to the basis voltage $v_2(t)$ by (1.8) so that (1.12) becomes

$$-\hat{\imath}_1(t) + \frac{1}{R_2} v_2(t) + \frac{1}{R_3} v_2(t) = 0$$

or

$$(1.13) \qquad\qquad \left( \frac{1}{R_2} + \frac{1}{R_3} \right) v_2(t) = \hat{\imath}_1(t)$$

Expression (1.13) can be solved for $v_2(t)$:

$$v_2(t) = \frac{\hat{i}_1(t)}{1/R_2 + 1/R_3}$$

$$= \frac{R_2 R_3}{R_2 + R_3} \hat{i}_1(t)$$

Once $v_2(t)$ has been found, the voltages and currents in the rest of the circuit are easily determined by use of (1.7), (1.8), (1.10), and (1.11):

$$v_1(t) = \frac{-R_2 R_3}{R_2 + R_3} \hat{i}_1(t)$$

$$v_3(t) = \frac{R_2 R_3}{R_2 + R_3} \hat{i}_1(t)$$

(1.14)
$$i_2(t) = \frac{R_3}{R_2 + R_3} \hat{i}_1(t)$$

and

(1.15)
$$i_3(t) = \frac{R_2}{R_2 + R_3} \hat{i}_1(t)$$

It is now possible to see why we have referred to the network of Figs. 1-1, 1-6, and 1-7 as a current divider. Expressions (1.14) and (1.15) show that the current $\hat{i}_1(t)$ "divides" into two currents $i_2(t)$ and $i_3(t)$ that are proportionate to resistors $R_2$ and $R_3$. Given that $R_2 = 4\,\Omega$, $R_3 = 6\,\Omega$, and $\hat{i}_1(t) = 5$ amp, then

$$i_2(t) = (\tfrac{6}{10})5 = 3 \text{ amp}$$

and

$$i_3(t) = (\tfrac{4}{10})5 = 2 \text{ amp}$$

Observe that

$$i_1(t) = i_2(t) + i_3(t)$$

as required by Kirchhoff's current law.

Let us now consider the possibility of reducing the set of six equations (1.1)–(1.6) to a set of equations that involve only branch currents. First, observe that the KCL node equation (1.4)

(1.16)
$$-i_1(t) + i_2(t) + i_3(t) = 0$$

allows only two branch currents to be independently specifiable. In other words, any two, but *only* two branch currents can be chosen as a basis and the third branch current can be found from (1.16). Since we know that $i_1(t)$ must be equal to the current source current $\hat{i}_1(t)$ that is given, let us choose it along with $i_2(t)$ as our two basis currents, then from (1.16)

(1.17)
$$i_3(t) = i_1(t) - i_2(t)$$

We now use the branch relations (1.1) to (1.3) to express the branch voltages in terms of the branch currents:

(1.18) $$i_1(t) = \hat{i}_1(t)$$

(1.19) $$v_2(t) = R_2 i_2(t)$$

and

(1.20) $$v_3(t) = R_3 i_3(t)$$

Substitution of these expressions into the KVL loop equations (1.5) and (1.6) yields

(1.21) $$v_1(t) + R_2 i_2(t) = 0$$

and

(1.22) $$R_2 i_2(t) - R_3 i_3(t) = 0$$

Finally, expression (1.17) can be substituted into (1.22) to eliminate $i_3(t)$:

$$R_2 i_2(t) - R_3[i_1(t) - i_2(t)] = 0$$

but since $i_1(t) = \hat{i}_1(t)$,

$$i_2(t) = \frac{R_3}{R_2 + R_3} \hat{i}_1(t)$$

which agrees with the results obtained previously. The reason that a voltage still remains in expression (1.21) will become clearer as we proceed.

## 1-6  SUMMARY

In this chapter we introduced the basic variables used to characterize network behavior—voltage and current—and we discussed their relationship to other physical quantities of interest: *charge*, *work*, and *power*. Two *ideal* circuit elements were presented—the *resistor* and the *independent current source*. Then a simple network, the current divider, was used as a vehicle to introduce the basic circuit analysis technique. For each circuit element we write a relationship between the branch voltage and current. In particular, for resistors we have

$$v(t) = Ri(t)$$

and for current sources we have

$$i(t) = \hat{i}(t)$$

where $\hat{i}(t)$ is specified. *Kirchhoff's current law* and *Kirchhoff's voltage law* impose constraints on the branch voltages and currents due to the interconnection of elements. We found that not all of the KCL node equations are independent nor are all of the KVL loop equations independent. The branch relationships, together with independent KCL and KVL equations, form a set of equations that could be solved to yield values for the branch voltages and currents. Alternatively, a set of *basis voltages or currents* could be chosen and by suitable manipulation, this set of equations could be reduced to a smaller set of equations written in terms of the basis variables.

In the next chapter we investigate more fully some of the fundamental concepts introduced here.

## PROBLEMS

**1.1**   Write the KCL node equations for nodes ② and ③ in each of the networks of Fig. P1.1.

**1.2**   Write the KVL loop equations for loops ① and ② in each of the networks of Fig. P1.2.

**1.3**   Determine which of the following sets of KCL node equations or KVL loop equations are linearly dependent or independent.

(a)  $-v_2 + v_3 = 0$
$-v_1 + v_2 + v_4 = 0$
$-v_4 - v_5 = 0$

(b)  $-v_1 - v_2 + v_4 = 0$
$-v_3 + v_4 - v_5 = 0$
$-v_1 - v_2 + v_3 + v_5 = 0$

(c)  $i_1 + i_2 - i_4 = 0$
$-i_1 - i_2 - i_3 = 0$

(d)  $-i_2 - i_4 + i_5 + i_6 = 0$
$i_1 + i_4 = 0$
$-i_3 - i_5 - i_6 = 0$

(e)  $-i_1 - i_2 + i_3 = 0$
$i_2 + i_4 = 0$
$i_1 - i_3 - i_4 = 0$

(f)  $v_1 - v_4 + v_5 = 0$
$v_2 - v_5 + v_6 = 0$
$v_3 - v_6 + v_7 = 0$

**1.4**   Write all the KCL node equations for each of the networks of Fig. P1.4. Show that in each case one of the equations is merely the negative sum of the others. This result confirms the fact that not all of the KCL node equations are linearly independent.

**1.5**   Write the KVL loop equations for the loops shown in each of the networks of the networks of Fig. P1.5. Try to determine if the loop equations written for each network are linearly dependent or independent.

**1.6**   In the circuit of Fig. P1.6 which element (branch) has the associated reference directions indicated?

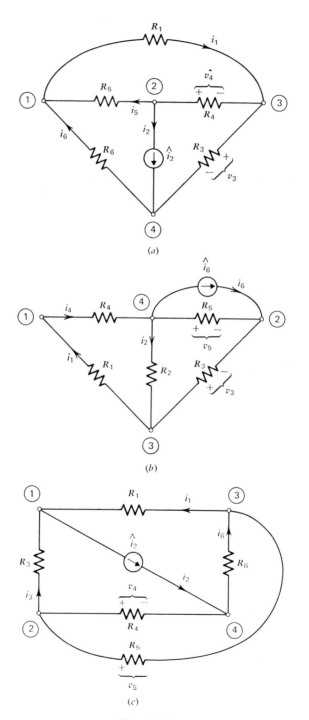

**Fig. P1.1**

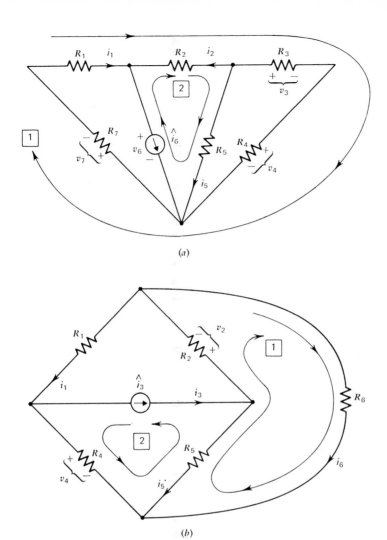

(a)

(b)

**Fig. P1.2**

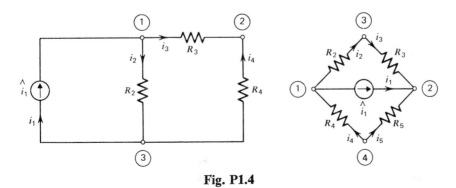

**Fig. P1.4**

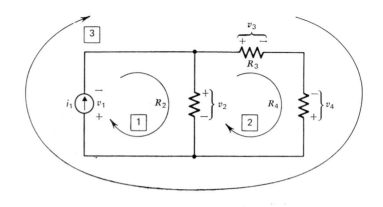

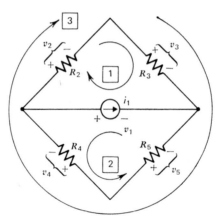

**Fig. P1.5**

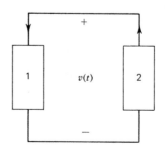

**Fig. P1.6**

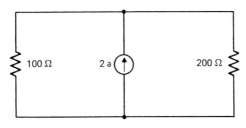

**Fig. P1.7**

**1.7**  What is the power *delivered* by the current source, in the network of Fig. P1.7?

**1.8**  Write the branch relationships for the network of Fig. P1.4*a*. Together with the independent KCL node equations and KVL loop equations obtained in Probs. 1.4 and 1.5 we have eight equations in terms of eight unknowns. Let $v_2$ and $v_4$ be basis voltages and reduce the set of eight equations to a set of two equations that are written solely in terms of these basis variables.

**1.9**  A voltage of $5 \sin 2\pi t$ volts is applied across the 5000 $\Omega$ resistor. Determine the current in the resistor and the power absorbed by the resistor. Sketch these quantities.

## CHAPTER 2

# Some Basic Considerations

In Chapter 1 we briefly presented a number of concepts so that we might analyze a simple network and indicate why these concepts were important. Now let us discuss in more depth some basic assumptions, constraints, and conventions that are important in the analysis of circuits. We begin with a more detailed investigation of Kirchhoff's laws.

## 2-1 THE CONSTRAINTS OF INTERCONNECTION

In Section 1-2 we stated that since we had assumed associated reference directions we could conveniently represent a two-terminal element, such as shown in Fig. 2-1a by an oriented branch, as shown in Fig. 2-1b, where the direction of the arrow indicates both the voltage drop and the current flow reference polarities. Since the constraints on element branch voltages and currents dictated by their interconnection in a circuit are independent of the nature of the elements involved, we can conveniently replace all elements of a network by directed branches. The configuration of directed branches that results is called a *directed graph* or simply a *graph*. For example, the network of two-terminal elements of Fig. 2-2a can be represented by the

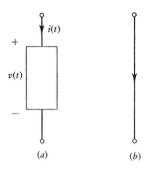

(a)    (b)

**Fig. 2-1    For the purpose of studying the constraints of interconnection it is convenient to represent the two-terminal lumped element of (a) by the directed branch of (b).**

graph of Fig. 2-2b. In such a graph we need not consider the various element types involved, but we must distinguish among the branches as well as assigning them an orientation. A sequential numbering scheme, where the $n_b$ branches are labeled $1, 2, 3, \ldots, n_b$, is usually employed, as in Fig. 2-2b. Observe that a given circuit may be associated with a number of superficially different *graphs*, which really differ only in the assigned branch orientations and labeling. All of the circuit graphs of Fig. 2-3 are equivalent and may represent the same circuit. The particular graph that would be employed in a specific situation is merely a matter of convenience.

The branch graph labeling is simply interpreted in terms of the associated reference directions. The $k$th branch ($k = 1, 2, \ldots, n_b$) is characterized by a voltage drop of $v_k(t)$ and a current flow of $i_k(t)$ in the direction of the arrow. The element type that the $k$th branch represents dictates a constraint between $v_k(t)$ and $i_k(t)$, but the interconnection itself dictates constraints among the voltages and currents. As stated in Chapter 1, *Kirchhoff's voltage law* and *Kirchhoff's current law* give the basic constraints of interconnection; these equations are independent of the two-terminal element types involved.

It is easiest to discuss Kirchhoff's laws in terms of a network's graph, rather than the lumped network itself. Before proceeding we make the convenient nonrestrictive assumption that we consider only *connected* graphs.

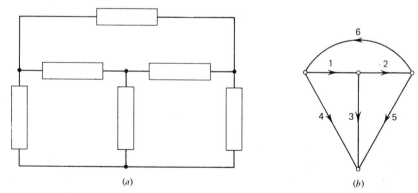

(a)    (b)

**Fig. 2-2    The graph (b) represents the circuit of two terminal elements (a).**

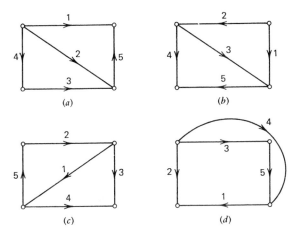

Fig. 2-3   A set of four equivalent circuit graphs, which differ only in branch numbering and orientation. The shape of these graphs may be distorted as shown here without altering their equivalence.

*A circuit graph is connected provided that there exists a path among the branches (regardless of their orientation) from any point in the graph to any other point in the graph.* The graphs of Figs. 2-2 and 2-3 are connected, those of Fig. 2-4 are not. This assumption is nonrestrictive because we may always consider each of the parts of an $n_p$ part disconnected graph separately.

Before continuing, let us review the definition of a node and a loop. Recall that a *node is a point of connection of the branches.* The graph of Fig. 2-5 has five nodes, which we distinguish by labeling them with encircled numbers. Usually we number the $n_n$ nodes ①, ②, ③, . . . , $(n_n)$ . A *loop is any closed path among the branches (regardless of their orientation) which begins and ends at the same node.* We will distinguish the loops of a network by labeling them with numbers enclosed in squares, that is, ①, ②, etc. Four possible oriented loops are indicated on the connected lumped circuit graph of Fig. 2-6.

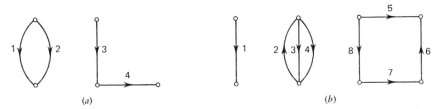

Fig. 2-4   (*a*) A disconnected lumped circuit graph with two separate parts. (*b*) A disconnected lumped circuit graph with three separate parts. It is not necessary, but it is convenient, to number the branches sequentially within each part.

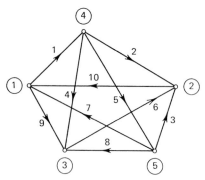

**Fig. 2-5   A connected lumped circuit graph with five nodes labeled by means of encircled numbers.**

We are now in a position to state a more general form of *Kirchhoff's current law (KCL); the algebraic sum of all of the instantaneous currents leaving (or entering) any closed surface is identically zero for all time.* We can look on each node of a graph as a special case of a closed surface. In other words, consider a sphere surrounding each node and let the radius of the sphere go to zero. Now we can consider the more specialized form of Kirchhoff's Current Law introduced in Chapter 1: *the algebraic sum of all of the instantaneous currents leaving (or entering) any node is identically zero for all time.*

**Example 2.1**

For the 4 node graph of Fig. 2-7 we have the following set of 4 Kirchhoff's current law node equations:[1]

(2.1a)  node ①:   $i_1(t) + i_2(t) \quad\quad + i_4(t) \quad\quad\quad\quad\quad\quad \equiv 0$

(2.1b)  node ②:   $\quad\quad -i_2(t) - i_3(t) \quad\quad\quad + i_5(t) \quad\quad\quad \equiv 0$

(2.1c)  node ③: $-i_1(t) \quad\quad + i_3(t) \quad\quad\quad\quad\quad\quad + i_6(t) \equiv 0$

(2.1d)  node ④:   $\quad\quad\quad\quad\quad\quad\quad\quad -i_4(t) - i_5(t) - i_6(t) \equiv 0$

Observe that the left hand side of this set of simultaneous linear equations sum identically to zero:

$$[i_1(t) + i_2(t) + i_4(t)] + [-i_2(t) - i_3(t) + i_5(t)]$$
$$+ [-i_1(t) + i_3(t) + i_6(t)] + [-i_4(t) - i_5(t) - i_6(t)] \equiv 0$$

We say that these equations are *linearly dependent* because any one of these four current constraint equations may be obtained from a linear combination

---

[1] We use the "identically equals" ($\equiv$) convention initially to indicate that these relations hold for all instances of time. This convention is soon dropped.

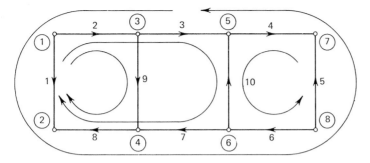

**Fig. 2-6   The illustration of four possible loops for a given connected lumped circuit graph.**

of the other three. In particular, the negative sum of any three of the above equations yields the fourth. Now consider any three of the equations, say (2.1a), (2.1b), and (2.1d):

$$(2.2a) \qquad i_1(t) + i_2(t) \qquad\qquad + i_4(t) \qquad\qquad\qquad \equiv 0$$

$$(2.2b) \qquad\qquad - i_2(t) - i_3(t) \qquad\qquad + i_5(t) \qquad\qquad \equiv 0$$

$$(2.2c) \qquad\qquad\qquad\qquad - i_4(t) - i_5(t) - i_6(t) \equiv 0$$

Observe that none of these equations can possibly be obtained as a linear combination of the others because each contains a term absent from the other equations. For instance, $i_1(t)$ appears only in equation (2.2a), $i_3(t)$ appears only in equation (2.2b), and $i_6(t)$ appears only in equation (2.2c). Such a set of equations is called *linearly independent.*   $\square$

In general the $n_n$ Kirchhoff's current law node equations for an $n_n$ node connected network are linearly dependent. We recognize that each branch current enters the set of all $n_n$ Kirchhoff's current law node equations only

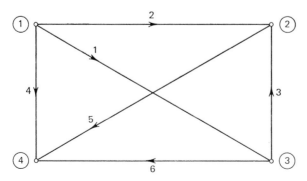

**Fig. 2-7   A 4 node connected lumped circuit graph used in Example 2.1 for the illustration of the Kirchhoff's current law node equations.**

twice, once with a positive sign (for the node it leaves) and once with a negative sign (for the node it enters); hence, the net sum of all of these equations must be zero. In the next section we will show that any $n_n - 1$ of the $n_n$ Kirchhoff's current law node equations of an $n_n$ node connected network constitutes a linearly independent set. We will postpone temporarily a discussion of the means of selecting the Kirchhoff's current law node equation that is most conveniently eliminated from consideration. Note that we may accept the redundancy of the extra current constraint equation as a simple check on the correct formulation of the overall set.

Another set of constraints of interconnection is given by the previously stated *Kirchhoff's voltage law (KVL)*: *the algebraic sum of all of the instantaneous voltage drops (or rises) encountered in traversing any lumped circuit loop is identically zero for all time.*

### Example 2.2

Consider again the 4 node graph of Fig. 2-7 that is redrawn in Fig. 2-8 along with four possible loops. The Kirchhoff's voltage law loop equations for the loops shown are:

(2.3a)   loop $\boxed{1}$:          $v_2(t) - v_3(t) - v_4(t) \qquad\quad + v_6(t) \equiv 0$

(2.3b)   loop $\boxed{2}$:          $v_2(t) \qquad\quad - v_4(t) + v_5(t) \qquad\quad \equiv 0$

(2.3c)   loop $\boxed{3}$:          $\qquad - v_3(t) \qquad\quad - v_5(t) + v_6(t) \equiv 0$

(2.3d)   loop $\boxed{4}$: $-v_1(t) + v_2(t) - v_3(t) \qquad\qquad\qquad \equiv 0$

Other loop equations are possible. For example, we could consider the loop formed by branches 1, 4, and 6. We see immediately that whereas there was an obvious maximum number of Kirchhoff's current law node equations,

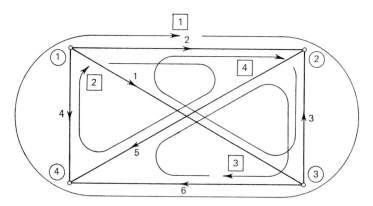

**Fig. 2-8   Four possible loops for the graph of Fig. 2-7 used in Example 2.2 for the illustration of the Kirchhoff's voltage law loop equations.**

the total number of possible loops for a given graph is not so obvious. More-over, we observe that not all of the Kirchhoff's voltage law loop equations are linearly independent. The loop equations for loops 2, 3, and 4 are linearly independent because each loop equation contains one branch voltage not found in either of the other two loop equations. That is, $v_4(t)$ appears only in (2.3b), $v_6(t)$ appears only in (2.3c) and $v_1(t)$ appears only in (2.3d). Loop equation (2.3a) is merely the sum of loop equations (2.3b) and (2.3c) so that it is not linearly independent, and, therefore, does not add any new informa-tion. This question arises: How many linearly independent Kirchhoff voltage law equations are there for an $n_n$ node $n_b$ branch graph? In the next section we show that there are $n_b - n_n + 1$ linearly independent Kirchhoff's voltage law equations.

Before proceeding, consider again the three independent Kirchhoff's voltage law loop equations (2.3b), (2.3c), and (2.3d):

$$(2.4a) \qquad v_2(t) \qquad - v_4(t) + v_5(t) \qquad\qquad \equiv 0$$

$$(2.4b) \qquad\qquad - v_3(t) \qquad\qquad - v_5(t) + v_6(t) \equiv 0$$

$$(2.4c) \qquad -v_1(t) + v_2(t) - v_3(t) \qquad\qquad\qquad \equiv 0$$

Each branch voltage can be defined by the difference between appropriate node voltages:

$$(2.5a) \qquad\qquad v_1(t) = v_①(t) - v_③(t)$$

$$(2.5b) \qquad\qquad v_2(t) = v_①(t) - v_②(t)$$

$$(2.5c) \qquad\qquad v_3(t) = v_③(t) - v_②(t)$$

$$(2.5d) \qquad\qquad v_4(t) = v_①(t) - v_④(t)$$

$$(2.5e) \qquad\qquad v_5(t) = v_②(t) - v_④(t)$$

and

$$(2.5f) \qquad\qquad v_6(t) = v_③(t) - v_④(t)$$

where $v_①(t)$, $v_②(t)$, $v_③(t)$, $v_④(t)$, $v_⑤(t)$, and $v_⑥(t)$ represent the node voltages or potentials of nodes 1–6 with respect to any arbitrary reference point. It can be seen by substitution of (2.5) into (2.4) that the node voltages satisfy the Kirchhoff voltage law equations:

$$[v_①(t) - v_②(t)] - [v_①(t) - v_④(t)] + [v_②(t) - v_④(t)] \equiv 0$$

$$-[v_③(t) - v_②(t)] - [v_②(t) - v_④(t)] + [v_③(t) - v_④(t)] \equiv 0$$

and

$$-[v_①(t) - v_③(t)] + [v_①(t) - v_②(t)] - [v_③(t) - v_②(t)] \equiv 0$$

Since the node voltages satisfy Kirchhoff's voltage law we can think of equa-tions (2.5) as an alternate form for expressing Kirchhoff's voltage law. We will find this form quite useful in the sequel.  □

## 2-2   LINEAR INDEPENDENCE OF KCL AND KVL EQUATIONS

Two important questions were raised in the previous section. How many linearly independent Kirchhoff's current law (KCL) equations and how many linearly independent Kirchhoff's voltage law (KVL) equations can be written for an $n_n$ node $n_b$ branch connected graph? How can these equations be written so as to guarantee their independence? To facilitate our discussion of these questions, we first introduce a few more elementary concepts concerned with graphs.

Often we will consider only part of a graph—or *subgraph of a graph. A subgraph of a graph is any set of branches and nodes belonging to a graph.* Several illustrations of subgraphs are presented in Fig. 2-9. *A tree of a graph is a connected subgraph that includes all the nodes of the graph but contains no loops.* Figure 2-10a shows several possible trees for the graph of Fig. 2-9a. Observe that the subgraph in Fig. 2-10b is not a tree for the graph of Fig. 2-9a because it contains a loop while the subgraph in Fig. 2-10c is not a tree for the graph of Fig. 2-9a because all the nodes are not included. Branches of a graph that are included in the tree are called *tree branches.* The branches of a graph that are not part of the tree are called *links* or *chords. (Note:* The

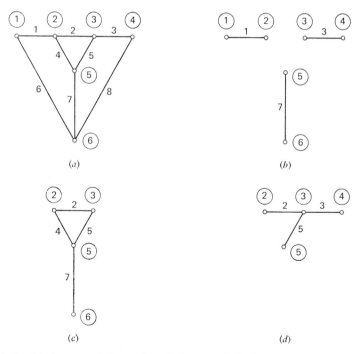

**Fig. 2-9**   (*a*) A connected graph and (*b, c,* and *d*) three possible subgraphs.

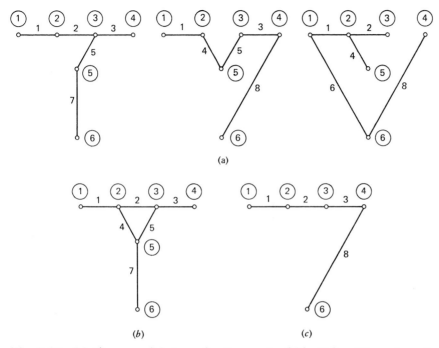

Fig. 2-10   (a) Three possible trees for the graph of Fig. 2-9a; (b) a subgraph that is not a tree because it contains a loop; (c) a subgraph that is not a tree because all the nodes of the graph are not included.

branches of a graph that are tree branches and the branches that are links depend on the choice of the tree.) We designate tree branches by solid lines and links by broken lines as shown in Fig. 2-11.

*Any tree associated with an* $n_n$ *node,* $n_b$ *branch connected graph contains exactly* $n_n - 1$ *tree branches.* This statement is easily proven by use of induction. Consider the simplest connected graph, one that has two nodes and one branch as shown in Fig. 2-12a. The only tree possible for this graph is the graph itself so the tree has one tree branch. It is easy to see that even if a two node graph has two or more branches, as shown in Fig. 2-12b, only one branch can be a tree branch since once the tree branch is chosen all remaining

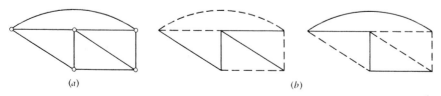

Fig. 2-11   (a) A graph and (b) two possible choices of trees. Solid lines denote tree branches and broken lines denote links.

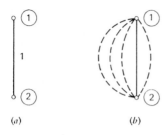

(a)          (b)

**Fig. 2-12  Two graphs used for the inductive proof of the statement that an $n_n$ node connected graph has $n_n - 1$ tree branches.**

branches form loops with it. Therefore, the hypothesis is true for any 2 node graph. We now assume that the hypothesis is true for any $2, 3, \ldots ,$ $n_n - 1$ node connected graph, and investigate its validity for an $n_n$ node connected graph. Consider any $n_n - 1$ node connected graph. This graph has, by assumption, $n_n - 2$ tree branches. If we add one node to this graph, at least one branch must also be added in order for the graph to remain connected. Since a tree must be a connected subgraph that contains all nodes but no loops, a single additional branch must be added to the original tree. Therefore, the $n_n$ node network has $(n_n - 2) + 1 = n_n - 1$ tree branches. Because of this, any $n_n$ node $n_b$ branch graph contains

$$n_b - (n_n - 1) = n_b - n_n + 1$$

links. We are now in a position to investigate the questions posed at the beginning of this section.

Consider the 8 branch 6 node graph of Fig. 2-13a. We know from the previous section that $n_n = 6$ Kirchhoff's current law node equations can be written for this network, but *at most* only 5 of these are linearly independent. It remains to be determined exactly how many linearly independent Kirchhoff's current law equations there are. We proceed by drawing a tree for the graph of Fig. 2-13a, as shown in Fig. 2-13b. Now consider the closed surfaces shown in Fig. 2-13c. Each of these closed surfaces is drawn so as to intersect, or "cut," one tree branch, the remaining branches cut are links. We can write a KCL equation for the set of branches that pass through, or are cut, by each of these closed surfaces. Such a set of branches is referred to as a *cutset*. Formally, a *cutset is any set of branches of a connected graph such that if they are removed from the graph the graph is separated into exactly two parts*. If the cutset is such that it contains only one tree branch with all other branches being links, it is called a *fundamental cutset*.

### Example 2.3

We now write the fundamental cutset equations for the graph of Fig. 2-13c. For convenience, let us assume that the positive direction of current for each cutset (closed surface) is given by the positive direction of current flow associated with the included tree branch. Then we have

$$i_1(t) + i_2(t) = 0$$

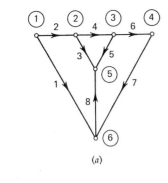

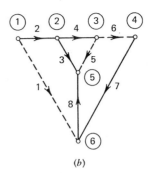

(a)                    (b)

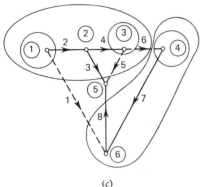

(c)

**Fig. 2-13** (a) **The 6 node, 8 branch connected graph of Example 2.3;** (b) **a tree for the graph in** (a); (c) **surfaces used to find the fundamental cutsets associated with the tree of** (b).

as the cutset equation associated with tree branch 2;

$$i_1(t) + i_3(t) + i_5(t) + i_6(t) = 0$$

as the cutset equation associated with tree branch 3;

$$i_4(t) - i_5(t) - i_6(t) = 0$$

as the cutset equation associated with tree branch 4;

$$-i_6(t) + i_7(t) = 0$$

as the cutset equation associated with tree branch 7; and

$$-i_1(t) - i_6(t) + i_8(t) = 0$$

as the cutset equation associated with tree branch 8. Now observe that these equations are linearly independent because each contains one branch current not found in any of the other equations: the current of the tree branch that defines the fundamental cutset.   □

It is easy to see the extension of this result to any arbitrary $n_n$ node $n_b$ branch connected graph. An $n_n$ node $n_b$ branch graph has an $n_n - 1$ branch tree that can be used to define $n_n - 1$ fundamental cutsets. The $n_n - 1$ Kirchhoff's current law equations that can be written for the fundamental cutsets are linearly independent because each equation contains a tree branch current not found in any of the other equations. Therefore, at least $n_n - 1$ linearly independent KCL equations exist. But, at most $n_n - 1$ linearly independent KCL node equations exist. Hence, we are led to conclude that *exactly $n_n - 1$ linearly independent KCL equations can be written for an $n_n$ node $n_b$ branch connected graph.* Two methods for choosing linearly independent KCL equations are: choose any of the $n_n - 1$ KCL node equations or pick a tree and write the $n_n - 1$ KCL fundamental cutset equations.

Let us now consider the question of the number of linearly independent Kirchhoff's voltage law equations. We will write KVL equations for the *fundamental loops* of the graph in Fig. 2-13. *A fundamental loop is a loop whose branches consist of only one link and as many tree branches as necessary to complete the loop.* Since there are $n_b - n_n + 1$ links in an $n_b$ branch $n_n$ node connected graph, there must be $n_b - n_n + 1$ fundamental loops. For convenience we take the positive direction of current flow in a fundamental loop to b  the same as the positive direction of current flow of the included link. The KVL fundamental loop equations for the graph of Fig. 2-13 are

$$v_1(t) - v_2(t) - v_3(t) + v_8(t) = 0$$

for the fundamental loop defined by link 1;

$$-v_3(t) + v_4(t) + v_5(t) = 0$$

for the fundamental loop defined by link 5; and

$$-v_3(t) + v_4(t) + v_6(t) + v_7(t) + v_8(t) = 0$$

for the fundamental loop defined by link 6. These equations are linearly independent because each contains a link voltage not found in any other equation.

The extension of the above to an $n_n$ node, $n_b$ branch network is straightforward. Such a network has $n_b - n_n + 1$ links and therefore $n_b - n_n + 1$ fundamental loops for which a Kirchhoff s voltage law equation can be written. The $n_b - n_n + 1$ KVL fundamental loop equations are linearly independent because each equation contains a link voltage not found in any of the other equations. *We conclude, therefore, that for an $n_n$ node $n_b$ branch connected graph there are at least $n_b - n_n + 1$ linearly independent Kirchhoff's voltage law equations.* The proof that there are at most $n_b - n_n + 1$ linearly independent KVL equations is beyond the scope of this text. The interested reader should consult the references. One method of generating linearly independent KVL equations is from the fundamental loops. Another method that works only for *planar graphs* is to use the *meshes*. This second method will be described in Section 2.4.

## 2-3 BASIS VARIABLES

Consider again the KCL fundamental cutset equations for the graph of Figure 2-13:

(2.6a)    $i_1(t) + i_2(t)$                                                    $= 0$

(2.6b)    $i_1(t)$         $+ i_3(t)$         $+ i_5(t) + i_6(t)$              $= 0$

(2.6c)                           $i_4(t) - i_5(t) - i_6(t)$                   $= 0$

(2.6d)                                    $- i_6(t) + i_7(t)$                $= 0$

and

(2.6e)  $-i_1(t)$                          $- i_6(t)$          $+ i_8(t) = 0$

Since there are only $n_n - 1 = 5$ equations relating the $n_b = 8$ unknown branch currents, there must be $n_b - (n_n - 1) = n_b - n_n + 1 = 3$ independent currents. These independent, or *basis* currents may be assigned arbitrary values. Once values for the basis currents have been assigned, all other currents in the network are fixed by the Kirchhoff's current law constraints. For example, we may arbitrarily let

$$i_1(t) = 2$$
$$i_5(t) = 5$$

and

$$i_6(t) = 1$$

The remaining currents $i_2(t)$. $i_3(t)$, $i_4(t)$, $i_7(t)$, and $i_8(t)$ are then fixed by the KCL fundamental cutset equations (2.6) and are found to be as follows. From (2.6a)

$$i_2(t) = -i_1(t) = -2$$

from (2.6b)

$$i_3(t) = -i_1(t) - i_5(t) - i_6(t) = -8$$

from (2.6c)

$$i_4(t) = i_5(t) + i_6(t) = 6$$

from (2.6d)

$$i_7(t) = i_6(t) = 1$$

and from (2.6e)

$$i_8(t) = i_6(t) + i_1(t) = 3$$

Care must be used when choosing which of the three currents are to be basis currents. The currents must be independent. For example, the currents

$i_3(t)$, $i_5(t)$, and $i_8(t)$ do not form a basis set of currents because they are not independent, as can be seen by writing the KCL node equation for node 5:

$$-i_3(t) - i_5(t) - i_8(t) = 0$$

To guarantee independence, branches whose currents form a basis set must not form a cutset, otherwise a KCL cutset equation can be written relating the branch currents. The set of $n_b - n_n + 1$ links do not form a cutset. Therefore, a good choice for a basis set of currents is the link currents. Other choices also exist.

From this discussion we see that a possible objective for a network analysis procedure would be to ascertain the values of the link currents, all other branch currents could then be found from the KCL fundamental cutset equations.

The KVL fundamental loop equations for the graph of Fig. 2-13 were found to be

(2.7a) $v_1(t) - v_2(t) - v_3(t)$ $\qquad\qquad\qquad\qquad\qquad + v_8(t) = 0$

(2.7b) $\qquad\qquad - v_3(t) + v_4(t) + v_5(t)$ $\qquad\qquad\qquad = 0$

and

(2.7c) $\qquad\qquad - v_3(t) + v_4(t) \qquad + v_6(t) + v_7(t) + v_8(t) = 0$

Here we have $n_b - n_n + 1 = 3$ equations that relate to the $n_b = 8$ unknown branch voltages. Therefore, $n_b - (n_b - n_n + 1) = n_n - 1 = 5$ branch voltages must be independent. These independent, or basis, voltages may be assigned arbitrary values; the KVL fundamental loop equations then fix all other branch voltages. As an example, arbitrarily choose

$$v_2(t) = 2$$
$$v_3(t) = 5$$
$$v_4(t) = -1$$
$$v_7(t) = -3$$

and
$$v_8(t) = 3$$

Then from (2.7a)
$$v_1(t) = v_2(t) + v_3(t) - v_8(t) = 4$$

from (2.7b)
$$v_5(t) = v_3(t) - v_4(t) = 6$$

and from (2.7c)
$$v_6(t) = v_3(t) - v_4(t) - v_7(t) - v_8(t) = 6$$

The voltages chosen to form the basis set must not be from a set of branches that form a loop, otherwise a KVL loop equation constraining some of the

**Fig. 2-14   Ground symbol used to denote the datum or reference node of a network. The voltage of the datum node is assumed to be zero.**

voltages can be written. Since tree branches do not form loops, the $n_n - 1$ tree branch voltages form a basis set.

In order for a set of $n_n - 1$ voltages to be a basis set they must be independent, and it must be possible to determine all branch voltages from them. Therefore, another suitable choice for a basis set of voltages is the set of any $n_n - 1$ node voltages. To see that $n_n - 1$ node voltages form a basis set of voltages for the graph of Fig. 2-13, consider equations (2.7) again. Each tree branch voltage can be expressed as a difference between two node voltages:

$$v_2(t) = v_①(t) - v_②(t)$$

$$v_3(t) = v_②(t) - v_⑤(t)$$

$$v_4(t) = v_②(t) - v_③(t)$$

$$v_7(t) = v_④(t) - v_⑥(t)$$

and

$$v_8(t) = v_⑥(t) - v_⑤(t)$$

Observe that all node voltages must appear in these equations because a tree must contain all nodes of the graph. Moreover, since all link voltages can be expressed in terms of tree branch voltages, link voltages can also be expressed in terms of node voltages. Finally, since there are $n_n - 1$ tree branch voltages expressed in terms of $n_n$ node voltages, one node voltage is arbitrary. Therefore, we can choose any node to be the *datum* or *reference* node and arbitrarily set its voltage to zero. We will indicate the datum node in a network by the ground symbol shown in Fig. 2-14.

We will find the node voltages a convenient set of variables to solve for, and a considerable emphasis will be placed on them in the sequel.

## 2-4   PLANAR NETWORKS AND MESHES*

As indicated in the previous section, an $n_n$ node $n_b$ branch network has $n_b - n_n + 1$ basis currents. One set of independent currents that can be used as a basis is the set of link currents. In this section we describe an alternative choice: the set of mesh currents. In order to proceed we must discuss the concept of a *planar* graph.

A *planar graph* is a graph that can be drawn on a plane surface in such a way that no two branches intersect at a point that is not a node. The graphs

---

* Starred sections may be skipped without loss of continuity.

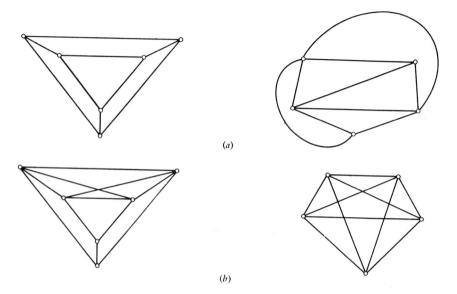

(a)

(b)

**Fig. 2-15** (*a*) **Some planar graphs and** (*b*) **some nonplanar graphs.**

of Fig. 2-15*a* are planar while the graphs of Fig. 2-15*b* are not. Observe that a planar graph can be drawn to look nonplanar as illustrated in Fig. 2-16. Therefore, we must caution against jumping to conclusions about planarity before careful examination of the graph.

Any loop of a planar graph for which there is no branch in its interior is called an *interior mesh* or simply a *mesh*.

The interior meshes for the graph of Fig. 2-17*a* are shown in Fig. 2-17*b*. An *exterior mesh* is a loop of planar graph for which there are no branches in its exterior. Figure 2-17*c* shows the exterior mesh associated with the planar graph of Fig. 2-17*a*.

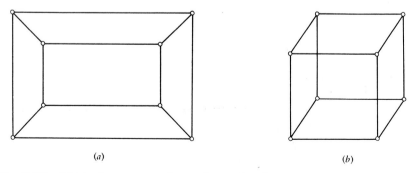

(a)                                        (b)

**Fig. 2-16** (*a*) **A planar network graph can be drawn to look nonplanar, as shown in** (*b*).

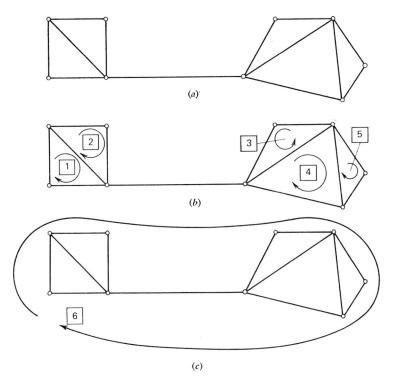

**Fig. 2-17** (*a*) **A planar graph;** (*b*) **the 5 interior meshes of the graph;** (*c*) **the exterior mesh of the graph.**

It can be shown[2] that an $n_n$ node $n_b$ branch graph has a total of $n_b - n_n + 1$ independent interior meshes. Since a KVL equation may be written for every loop, we can write $n_b - n_n + 1$ KVL *mesh equations*. In order to facilitate writing the KVL mesh equations, we assume there is a current flowing in each mesh. If the current in the branch belonging to a mesh flows in the same direction as the mesh current the voltage drop is positive; if the branch current is opposite to the mesh current, the voltage drop is negative.

**Example 2.4**

We wish to write the KVL mesh equations for the planar graph of Fig. 2-18*a*. The assumed mesh current directions are shown in Fig. 2-18*b*. The KVL mesh equations are

$$mesh\ 1\qquad -v_1 + v_2 + v_3 = 0$$

$$mesh\ 2\qquad -v_3 - v_4 + v_7 = 0$$

$$mesh\ 3\qquad -v_4 - v_5 + v_6 = 0$$

[2] See the references.

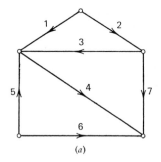 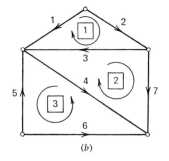

Fig. 2-18   (*a*) The planar network graph of Example 2.4 to illustrate writing KVL mesh equations; (*b*) the assumed mesh current directions.

Observe that these equations are linearly independent because each contains a voltage absent from the others. Moreover, these are the same equations that would have been obtained if we had chosen as a tree branches 1, 3, 4, and 5 and written the KVL fundamental loop equations.

**Example 2.5**

In the event the student believes that mesh equations are really fundamental loop equations, consider the KVL equations for the planar graph of Fig. 2-19 for the meshes shown.

$$\textit{mesh 1} \qquad v_1 + v_3 + v_8 + v_{10} \; = 0$$

$$\textit{mesh 2} \qquad v_2 - v_3 + v_4 + v_7 \; = 0$$

$$\textit{mesh 3} \qquad -v_4 - v_5 - v_6 - v_{11} \; = 0$$

$$\textit{mesh 4} \qquad v_9 - v_{10} + v_{11} + v_{12} = 0$$

$$\textit{mesh 5} \qquad v_6 - v_7 - v_8 - v_9 \; = 0$$

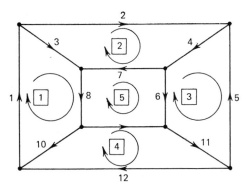

Fig. 2-19   A planar graph for which it is not possible to choose a tree such that the mesh equations correspond to fundamental loop equations.

There is no way to choose a tree for this network so that the fundamental loop equations are identical to the mesh equations. Moreover, even though each voltage that appears in mesh equation 5 also appears in one of the other mesh equations, the entire set of mesh equations is linearly independent. Finally, observe that whereas fundamental loop currents are physically measurable quantities, since they correspond to link currents that can be measured by an ammeter, mesh currents need not be physically measurable. In particular, mesh current 5 cannot be measured directly.

## 2-5  SUMMARY

In this chapter we have presented some basic concepts that form the background for a more detailed study of circuit analysis. The concept of a network *graph* was introduced so that we might discuss the *constraints of interconnection, Kirchhoff's voltage law,* and *Kirchhoff's current law* in the abstract, and emphasize the independence of these constraints from those imposed on the voltages and currents by the elements themselves. By choosing a *tree* and considering the *fundamental cutsets* and *fundamental loops* associated with the tree, we indicated that for an $n_n$ node $n_b$ branch network, there are $n_n - 1$ *linearly independent KCL equations* and $n_b - n_n + 1$ *linearly independent KVL equations.* Since the $n_n - 1$ linearly independent KCL equations are expressed in terms of the $n_b$ branch currents there are $n_b - n_n + 1$ *independent, or basis, currents.* Similarly, since the $n_b - n_n + 1$ linearly independent KVL equations are expressed in terms of the $n_b$ branch voltages there are $n_n - 1$ *independent, or basis, voltages.*

Any $n_n - 1$ *node voltages* or *tree branch voltages* form a basis set of voltages and any $n_b - n_n + 1$ *link currents* form a basis set of voltages. If the graph is *planar* the *mesh currents* form a basis set of currents.

## PROBLEMS

**2.1**   Draw a graph for each of the networks of Fig. P2.1 and determine the number of nodes, $n_n$, and the number of branches, $n_b$.

**2.2**   Determine which of the following sets of equations are linearly independent and which are linearly dependent.

(a) $i_1 - i_2 + i_3 \quad = 0$

$\quad\quad i_2 - i_3 \quad = 0$

(b) $v_1 + v_2 - v_3 + v_4 = 0$

$\quad\quad v_1 - v_2 \quad = 0$

$\quad\quad - v_3 + v_4 = 0$

(c) $2x_1 + 3x_2 \quad\quad = 0$

$\quad\quad - 2x_2 + 3x_3 = 0$

$\quad 4x_1 + 4x_2 + 3x_3 = 0$

(d) $2x_1 + 3x_2 \quad\quad = 0$

$\quad\quad - 2x_2 + 3x_3 = 0$

$\quad 4x_1 + 4x_2 - 3x_3 = 0$

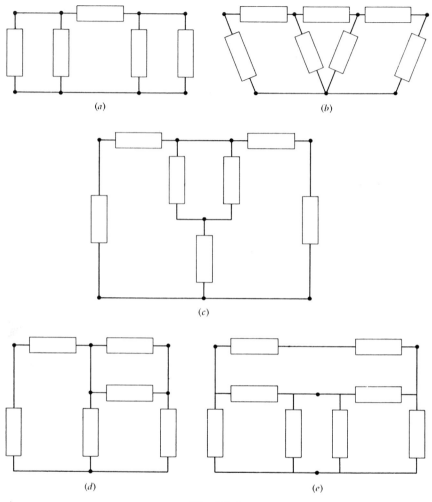

**Fig. P2.1**

**2.3**    Write the KCL equations for *each* node of the graphs shown in Fig. P2.3. Show that the set of equations for each graph is linearly dependent.

**2.4**    Write the KVL loop equations for each loop shown in the networks of Fig. P2.4. Determine whether or not the set of loop equations written for each network is linearly independent.

**2.5**    For each of the networks of Fig. P2.5 express the branch voltages in terms of the node voltages. For each loop drawn show that the node voltages satisfy KVL.

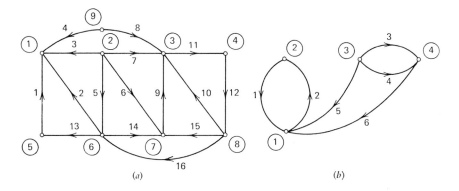

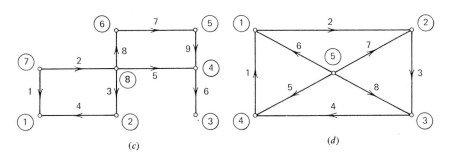

**Fig. P2.3**

**2.6** Write the KCL equations for the closed surfaces shown in the network of Fig. P2.6, also write the KCL equation for each node. Show that the KCL equation obtained for each closed surface can be found by appropriate combinations of the KCL node equations.

**2.7** For each of the graphs of Fig. P2.7 draw three different trees. Observe that there are $n_n - 1$ tree branches in each tree.

**2.8** In Fig. P2.8 a graph and three possible trees are shown. Write the fundamental cutset and fundamental loop equations for each tree. Why are the set of fundamental cutset equations and the set of fundamental loop equations independent? (*Note:* In some cases for nonplanar networks the closed surface that passes through a cutset may interest another tree branch twice. Observe that this tree branch current would enter the fundamental cutset equation twice, once with a positive sign and once with a negative sign, thereby canceling out.

**2.9** Consider again Prob. 2.8 and the fundamental cutset equations written for tree number 1. If $i_2 = 3$, $i_3 = 2$, $i_4 = 5$, $i_8 = 2$, $i_{11} = 5$,

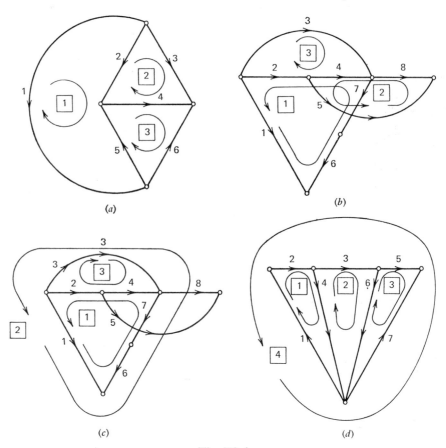

(a)

(b)

(c)

(d)

**Fig. P2.4**

and $i_{12} = 4$ amps determine the currents in branches 1, 5, 6, 7, 9, 10, and 13. Since all remaining currents can be obtained from knowledge of the link currents, the link currents do indeed form a basis set. Now try to find all the branch currents by using the fundamental cutset equations associated with tree 2 and the currents specified above for branches 2, 3, 4, 8, 11, and 12. Repeat for the fundamental cutset equations associated with tree 3. This exercise demonstrates that any set of basis currents can be used with any set of $n_n - 1$ independent KCL equations to determine the remaining currents.

**2.10**   Given the network of Prob. 2.8 and the six currents $i_2 = 4$, $i_3 = 5$, $i_5 = 2$, $i_7 = 3$, $i_9 = 1$, and $i_{12} = 3$ try to find the remaining currents using the KCL fundamental cutset equations for tree 1 obtained in Prob. 2.8. This exercise illustrates that although there are only 6

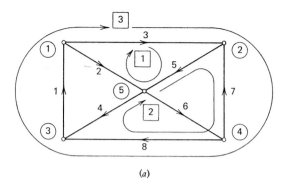

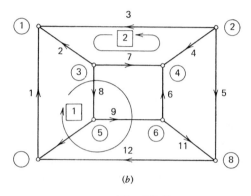

(a)

(b)

**Fig. P2.5**

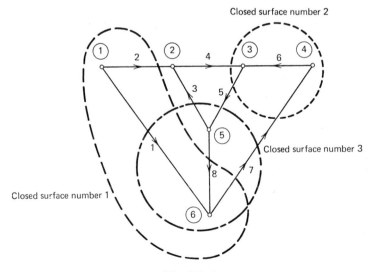

Closed surface number 2

Closed surface number 3

Closed surface number 1

**Fig. P2.6**

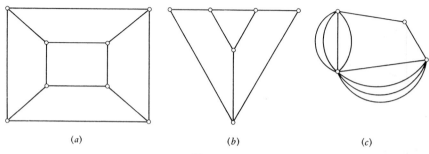

(a)                          (b)                          (c)

**Fig. P2.7**

basis currents, it is possible to choose the wrong 6 currents to specify. Specifically, an independent set of currents (one for which a KCL equation cannot be written) must be specified.

**2.11**    Consider again the graph and trees of Prob. 2.8. If $v_1 = 1$, $v_5 = 5$, $v_6 = 6$, $v_7 = 7$, $v_9 = 9$, $v_{10} = 10$, and $v_{13} = 13$ volts, determine the branch voltages of branches 2, 3, 4, 8, 11, and 12, by using the fundamental loop equations written for tree 1. Try to find these branch voltages by using the fundamental loop equations written for trees 2 and 3. Thus, any basis set of voltages can be used with any $n_b - n_n + 1$ independent KVL equations to determine the remaining voltages.

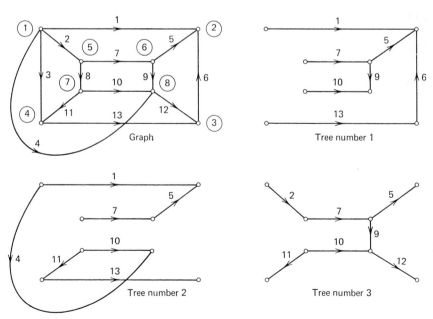

**Fig. P2.8**

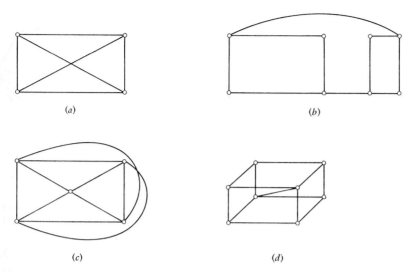

**Fig. P2.13**

**2.12**    Express all the tree branch voltages of tree 3 of Prob. 2.8 in terms of the node voltages. Show that all the link voltages are also expressible in terms of node voltages by substituting into the fundamental loop equations associated with tree 3.

**2.13**    Which of the graphs of Fig. P2.13 are planar and which are non-planar?

**2.14**    For each of the planar graphs of Fig. P2.14 write the KVL mesh equations for all interior meshes. In addition, write the KVL mesh equation for the outer mesh. Show that the KVL equation for the outer mesh could be obtained by a suitable combination of the interior mesh equations and, therefore, is linearly dependent on them.

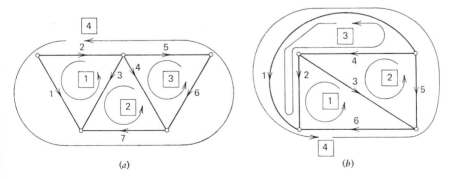

**Fig. P2.14**

# Analysis of Networks Containing Resistors and Current Sources

We are now in a position to analyze any network that contains resistors and current sources. Analysis of this class of networks is extremely important, since we will see later that it forms the basis for analysis of even the most complicated networks. The techniques introduced in Chapter 1 will be reviewed first by investigating the behavior of some simple resistive networks. Then the concepts of Chapter 2 will be employed to establish a general analysis procedure.

For notational convenience we will stop showing the explicit dependence of voltage and current on time. In other words, the branch relationship for a resistor will be written as

$$v = Ri$$

instead of

$$v(t) = Ri(t)$$

The student should bear in mind that every expression written, unless otherwise stated, holds for every instant of time. Henceforth, the time argument will only be used when required for clarity.

## 3-1  SIMPLE RESISTIVE NETWORKS

In this section we will study the behavior of some simple resistive networks. As a result of these studies several useful network properties will be shown. Moreover, the analysis of some simple networks will lay the groundwork for the systematic method of network analysis introduced in the next section.

We start with an investigation of some implications of Kirchhoff's laws on simple resistive networks. Consider the two resistive networks of Fig. 3-1. The behavior of the network of Fig. 3-1a is characterized by

$$v_a = Ri_a$$

For the network of Fig. 3-1b, Kirchhoff's voltage law, yields

$$v_b = v_1 + v_2$$

Upon substitution of the branch relationships

$$v_1 = R_1 i_b$$

and

$$v_2 = R_2 i_b$$

we have

$$v_b = R_1 i_b + R_2 i_b = (R_1 + R_2)i_b$$

Observe that if $R = R_1 + R_2$ we would be unable to distinguish between these two networks by making voltage-current measurements at the terminals $T_1 - T_2$.

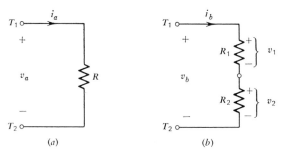

(a)    (b)

**Fig. 3-1**  These networks are equivalent when viewed from terminals $T_1$ and $T_2$ for $R = R_1 + R_2$.

For the case of $R = R_1 + R_2$, these networks are said to be *equivalent*. More precisely, *two networks are said to be equivalent at a pair of terminals if the voltage-current relationships for the two networks are identical at these terminals.* We conclude then that two resistors with values $R_1$ and $R_2$ in series behave as a single equivalent resistance $R_{eq}$ where

$$R_{eq} = R_1 + R_2$$

This result is easily generalized for $n$ resistors connected in series. The networks of Fig. 3-2a and Fig. 3-2b are equivalent if

$$R_{eq} = R_1 + R_2 + \cdots + R_n = \sum_{j=1}^{n} R_j$$

If the resistors were described as conductors $(G_j = 1/R_j \quad j = 1, 2, \ldots, n)$, then an expression for the behavior of conductors in series emerges:

$$\frac{1}{G_{eq}} = \frac{1}{G_1} + \frac{1}{G_2} + \cdots + \frac{1}{G_n} = \sum_{j=1}^{n} \frac{1}{G_j}$$

In particular, for the case of two conductors in series

$$\frac{1}{G_{eq}} = \frac{1}{G_1} + \frac{1}{G_2}$$

or

$$G_{eq} = \frac{G_1 G_2}{G_1 + G_2}$$

Now consider the two resistive networks of Fig. 3-3. For the network of Fig. 3-3b, Kirchhoff's current law yields

$$i_b = i_1 + i_2$$

On substitution of the branch relationships

$$i_1 = \frac{1}{R_1} v_b$$

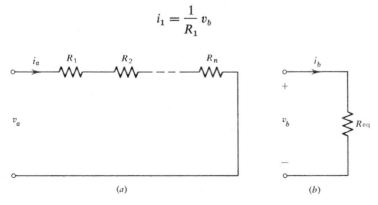

(a)                                    (b)

**Fig. 3-2** These networks are equivalent for $R_{eq} = R_1 + R_2 + \cdots + R_n$.

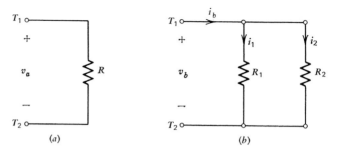

(a)        (b)

**Fig. 3-3** These networks are equivalent for $1/R_{\mathrm{eq}} = 1/R_1 + 1/R_2$.

and

$$i_2 = \frac{1}{R_2} v_b$$

we have

$$i_b = \frac{1}{R_1} v_b + \frac{1}{R_2} v_b$$

$$= \left( \frac{1}{R_1} + \frac{1}{R_2} \right) v_b$$

Clearly, if

$$\frac{1}{R} = \frac{1}{R_1} + \frac{1}{R_2}$$

or

$$R = \frac{R_1 R_2}{R_1 + R_2}$$

this network is equivalent to the network of Fig. 3-3a. The student should verify that the networks of Figs. 3-4a and 3-4b are equivalent if

$$\frac{1}{R_{\mathrm{eq}}} = \frac{1}{R_1} + \frac{1}{R_2} + \cdots + \frac{1}{R_n} = \sum_{j=1}^{n} \frac{1}{R_j}$$

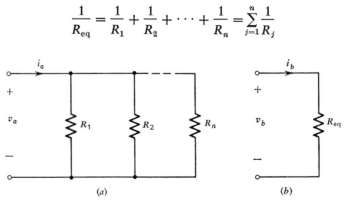

(a)        (b)

**Fig. 3-4** These networks are equivalent if $1/R_{\mathrm{eq}} = 1/R_1 + 1/R_2 + \cdots + 1/R_n$.

and that if the resistors were described as conductors, then

$$G_{eq} = G_1 + G_2 + \cdots + G_n = \sum_{j=1}^{n} G_j$$

The equivalent network concept can be used as the basis of a limited analysis scheme.

### Example 3.1

Suppose we are interested only in the current source voltage, $v_1$, for the network of Fig. 3-5a. Since the internal voltages and currents are not of interest, we can find $v_1$ by using an equivalent network as shown in Fig. 3-5b. The equivalent resistance $R_{eq}$ is found by reducing the series and parallel combinations of resistors of the original network. For instance, the series combination of resistors $R_4$ and $R_5$ is equivalent to a single resistance of value $R_{eq1} = R_4 + R_5 = 12\ \Omega$. This $12\ \Omega$ resistance, $R_{eq1}$, forms a parallel combination of resistors with $R_6$, the equivalent resistance of which is equal to

$$R_{eq2} = \frac{R_{eq1}R_6}{R_{eq1} + R_6} = 4\ \Omega$$

A series combination of $R_{eq2}$ and $R_3$ results and can be replaced by the equivalent resistor

$$R_{eq3} = R_{eq2} + R_3 = 12\ \Omega$$

Finally, $R_{eq3}$ forms a parallel combination with $R_2$ that is equivalent to the

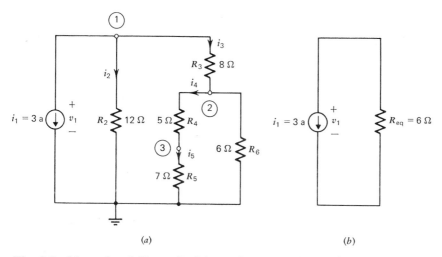

(a)                                    (b)

**Fig. 3-5** Network of Example 3.1 to show how the equivalent resistance concept can be employed to solve for $v_1$.

resistor

$$R_{eq} = \frac{R_{eq3}R_2}{R_{eq3} + R_2} = 6\,\Omega$$

The voltage $v_1$ is now determined as seen from Fig. 3-5b:

$$v_1 = -i_1 R_{eq} = -18\text{ v}$$

We must not lose sight of the fact that we are actually using Kirchhoff's laws to solve for $v_1$ even though the KVL and KCL equations are not formulated explicitly. This technique would not be as convenient if we also desired the node voltages $v_{②}$ and $v_{③}$. (*Note.* Node voltage $v_{①}$ is equal to branch voltage $v_1$.) Consider finding $v_{②}$ and $v_{③}$. Since $v_1 = -18$ volts, the current flowing through $R_2$ is

$$i_2 = \frac{v_2}{R_2} = \frac{v_{①}}{R_2} = -1.5\text{ a}$$

The current flowing through $R_3$ is found from the KCL node equation for node ①:

$$i_3 = -i_1 - i_2 = -1.5\text{ a}$$

$v_{②}$ can now be found:

$$v_{②} = v_{①} - v_3 = v_{①} - i_3 R_3 = -6\text{v}$$

The current flowing through $R_4$ and $R_5$ is

$$i_4 = i_5 = \frac{v_{②}}{R_4 + R_5} = -.5\text{ a}$$

Finally $v_{③}$ can be determined:

$$v_{③} = v_{②} - v_4 = v_{②} - R_4 i_4 = R_5 i_5 = -3.5\text{ v} \quad \square$$

In the next section a more systematic method of analysis is proposed.

## 3-2   NODAL EQUILIBRIUM EQUATIONS

From previous sections we know that for any $n_n$ node $n_b$ branch network comprised of resistors and current sources, we can write $n_n - 1$ linearly independent KCL node equations, $n_b - n_n + 1$ linearly independent KVL loop equations, and $n_b$ branch relationships. These $2\,n_b$ equations are in terms of the $2\,n_b$ unknown branch currents and voltages. We also know that there are $n_n - 1$ basis voltages and $n_b - n_n + 1$ basis currents. If the $n_n - 1$ basis voltages were known, all other voltages can be determined from the KVL equations, and all branch currents can be found from the branch relationships. Similarly, if any $n_b - n_n + 1$ basis currents are known, all other currents can be determined from the KCL equations, and all branch voltages can be found from the branch relationships. Therefore, from the $2\,n_b$ equations that we can write we wish to form a set of equations that can be used to solve for either a basis set of voltages or a basis set of currents. For the present we will choose the $n_n - 1$ linearly independent node voltages (excluding the datum node) as the basis set of variables. In a later section other choices of basis will be used for demonstration purposes, but our emphasis will always be on the node voltage basis. The reasoning behind this emphasis will become clear in the sequel.

In this section we will discuss an orderly method for reducing the set of $2\,n_b$ equations that can be written for a network to a set of $n_n - 1$ linearly independent equations that can be solved for the node voltages. As a motivational example, consider the network of Fig. 3-6a.

**Example 3.2**

The graph of the network of Fig. 3-6a is shown in Fig. 3-6b; we have arbitrarily chosen the positive directions of current flow as shown. Since $n_n = 4$ and $n_b = 6$ there are $n_n - 1 = 3$ linearly independent KCL equations and $n_b - n_n + 1 = 3$ linearly independent KVL equations. If we choose node ④ as the datum and assume that the positive current leaves a node, then, the 3 linearly independent KCL node equations are:

(3.1a)     *node* ①     $i_1 - i_2 - i_3 \qquad\qquad = 0$

(3.1b)     *node* ②     $\qquad\quad i_2 \quad + i_4 + i_5 \quad = 0$

(3.1c)     *node* ③     $\qquad\qquad\quad i_3 \quad\; - i_5 - i_6 = 0$

Instead of writing three linearly independent KVL loop equations in terms of the six branch voltages, we express the KVL loop equations indirectly by writing equations that express the six branch voltages in terms of the three

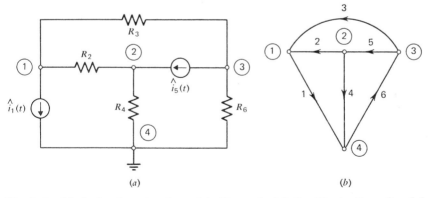

(a)                               (b)

**Fig. 3-6** (a) **A simple network used in Example 3.2 for illustration of nodal analysis techniques.** (b) **Its graph.**

independent node voltages $v_①$, $v_②$, and $v_③$:

| (3.2a) | *branch 1* | $v_1 = v_①$ |
|---|---|---|
| (3.2b) | *branch 2* | $v_2 = -v_① + v_②$ |
| (3.2c) | *branch 3* | $v_3 = -v_① \qquad + v_③$ |
| (3.2d) | *branch 4* | $v_4 = \qquad v_②$ |
| (3.2e) | *branch 5* | $v_5 = \qquad v_② - v_③$ |
| (3.2f) | *branch 6* | $v_6 = \qquad - v_③$ |

To see that the above relations are actually expressions for the three independent KVL loop equations, we observe that the basis set of node voltages $v_①$, $v_②$, and $v_③$ can be related directly to the basis set of branch voltages $v_1$, $v_4$, and $v_6$:

$$v_① = v_1$$
$$v_② = v_4$$

and

$$v_③ = -v_6$$

Three linearly independent KVL loop equations can be found in terms of this new basis from (3.2b), (3.2c), and (3.2e):

$$v_1 + v_2 - v_4 = 0$$
$$v_1 + v_3 + v_6 = 0$$

and

$$-v_4 + v_5 - v_6 = 0$$

These equations are linearly independent because each contains a branch voltage not found in the other equations. In other words, even though we have written $n_b$ equations to express KVL, only $n_b - n_n + 1$ are independent.

The only additional linearly independent equations that we can write for this network are the branch relationships:

| (3.3a) | *branch 1* | $i_1 = \hat{i}_1$ |
|---|---|---|
| (3.3b) | *branch 2* | $v_2 = R_2 i_2$ |
| (3.3c) | *branch 3* | $v_3 = R_3 i_3$ |
| (3.3d) | *branch 4* | $v_4 = R_4 i_4$ |
| (3.3e) | *branch 5* | $i_5 = -\hat{i}_5$ |
| (3.3f) | *branch 6* | $v_6 = R_6 i_6$ |

Observe that the minus sign that appears in the branch relationship for branch 5 arises because the current source current flows in a direction opposite to the assumed positive direction of current flow.

We have, in effect, written a total of $2\ n_b$ linearly independent equations ($n_n - 1$ KCL equations, $n_b - n_n + 1$ KVL equations, and $n_b$ branch relationships) in terms of $2\ n_b$ unknowns ($n_b$ branch currents, $n_b$ branch voltages). It is possible to solve this set of equations as is, and we will discuss this technique in a later chapter. Presently though, we wish to reduce this set of equations to a set of equations written solely in terms of the basis set of variables, that is, the $n_n - 1$ node voltages. The first step is to express the KCL node equations in terms of the branch voltages. Substitution of the branch relationships into the KCL equations yields the set of equations:

$$\hat{\imath}_1 - \frac{1}{R_2} v_2 - \frac{1}{R_3} v_3 \qquad\qquad\qquad = 0$$

$$\frac{1}{R_2} v_2 \qquad\quad + \frac{1}{R_4} v_4 - \hat{\imath}_5 \qquad = 0$$

$$\frac{1}{R_3} v_3 \qquad\quad + \hat{\imath}_5 - \frac{1}{R_6} v_6 = 0$$

The KVL equations are now used to express the branch voltages in the above equations in terms of the node voltages:

$$\hat{\imath}_1 - \frac{1}{R_2}(-v_① + v_②) - \frac{1}{R_3}(-v_① + v_③) = 0$$

$$\frac{1}{R_2}(-v_① + v_②) + \frac{1}{R_4}(v_②) - \hat{\imath}_5 \qquad\qquad = 0$$

$$\frac{1}{R_3}(-v_① + v_③) + \hat{\imath}_5 - \frac{1}{R_6}(-v_③) \qquad = 0$$

Combining and rearranging terms yields the following *nodal equilibrium equations*

$$\left(\frac{1}{R_2} + \frac{1}{R_3}\right) v_① - \frac{1}{R_2} v_② - \frac{1}{R_3} v_③ = -\hat{\imath}_1$$

$$-\frac{1}{R_2} v_① + \left(\frac{1}{R_2} + \frac{1}{R_4}\right) v_② \qquad = \hat{\imath}_5$$

$$-\frac{1}{R_3} v_① + \left(\frac{1}{R_3} + \frac{1}{R_6}\right) v_③ \qquad = -\hat{\imath}_5$$

These equations are referred to as the nodal equilibrium equations so as to distinguish them from the KCL node equations that are written in terms of the branch currents. Thus, we have three equations in terms of three unknowns. These simultaneous algebraic equations can be solved in several ways, for example, using *Cramer's rule*. We will delay a thorough discussion of how these equations are solved until Sections 3-3 and 3-4. Before proceeding it should be pointed out that although it may appear that an excessive amount of manipulation is required to arrive at the nodal equilibrium equations, it is quite easy to write these equations by inspection, as will be seen shortly.   □

The steps used to reduce the $n_n - 1$ linearly independent KCL equations, $n_b - n_n + 1$ linearly independent KVL equations, and $n_b$ branch relationships to a set of $n_n - 1$ linearly independent equations in terms of the $n_n - 1$ basis set of node voltages are easily formalized.

Given an $n_n$ node $n_b$ branch connected network comprised of lumped, linear time-invariant resistors, and current sources, the $n_n - 1$ linearly independent nodal equilibrium equations are found as follows.

1. Draw the network graph and arbitrarily assign directions of positive current flow.
2. Choose a datum node and write a KCL node equation for each of the remaining $n_n - 1$ nodes. (We assume currents that leave nodes are positive.)
3. Express all $n_b$ branch voltages in terms of the $n_n - 1$ independent node voltages. These are essentially the KVL equations.
4. Write the branch relationship for each element in the network, that is,

$$v = Ri \qquad \text{or} \qquad i = \frac{1}{R} v$$

   for each resistor;

$$i = Gv$$

   for each conductor; and

$$i = \pm \hat{\imath}$$

   for each current source.
5. Substitute the branch relations found in steps (4) into the KCL node equations found in step (2).
6. Substitute the KVL equations found in step (3) into the equations found in step (5). We now have $n_n - 1$ KCL equations in terms of the node voltages.

The equations that result are the Kirchhoff's current law node equations expressed in terms of the node voltages.


**Example 3.3**

As another illustration of the method outlined above, consider the 8 node, 12 branch network of Fig. 3-7a. An oriented graph for this network is shown in Fig. 3-7b. Let node ⑧ be the datum; the KCL node equations for

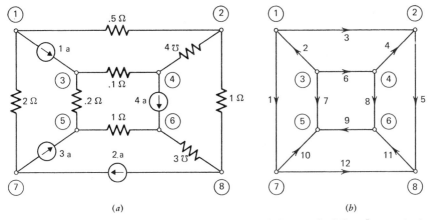

**Fig. 3-7** (a) **An 8 node, 12 branch network used in Example 3.3 to demonstrate the steps used for writing the nodal equilibrium equations.** (b) **Its graph.**

the remaining nodes are:

| | | | |
|---|---|---|---|
| (3.4a) | *node* ① | $i_1 - i_2 + i_3$ | $= 0$ |
| (3.4b) | *node* ② | $-i_3 - i_4 + i_5$ | $= 0$ |
| (3.4c) | *node* ③ | $i_2 + i_6 + i_7$ | $= 0$ |
| (3.4d) | *node* ④ | $i_4 + i_6 + i_8$ | $= 0$ |
| (3.4e) | *node* ⑤ | $-i_7 - i_9 - i_{10}$ | $= 0$ |
| (3.4f) | *node* ⑥ | $-i_8 + i_9 - i_{11}$ | $= 0$ |
| (3.4g) | *node* ⑦ | $-i_1 + i_{10} + i_{12} = 0$ | |

All branch voltages can be expressed in terms of the node voltages (since node ⑧ is the datum $v_⑧ \equiv 0$):

| | |
|---|---|
| (3.5a) | $v_1 = v_① - v_⑦$ |
| (3.5b) | $v_2 = v_③ - v_①$ |
| (3.5c) | $v_3 = v_① - v_②$ |
| (3.5d) | $v_4 = v_④ - v_②$ |
| (3.5e) | $v_5 = v_②$ |
| (3.5f) | $v_6 = v_③ - v_④$ |
| (3.5g) | $v_7 = v_③ - v_⑤$ |
| (3.5h) | $v_8 = v_④ - v_⑥$ |
| (3.5i) | $v_9 = v_⑥ - v_⑤$ |
| (3.5j) | $v_{10} = v_⑦ - v_⑤$ |
| (3.5k) | $v_{11} = -v_⑥$ |
| (3.5l) | $v_{12} = v_⑦$ |

It is interesting to digress briefly and show that a set of $n_b - n_n + 1 = 5$ linearly independent KVL loop equations are actually represented by these equations. We first choose a basis set of *branch* voltages. There are $n_n - 1 = 7$ basis or independent voltages. The voltages $v_5$, $v_{11}$, and $v_{12}$ are certainly independent, because they are each specified by the single independent node voltages $v_②$, $v_⑥$, and $v_⑦$. Therefore, $v_5$, $v_{11}$, and $v_{12}$ can be considered to be part of the basis set. Observe that any branch voltage that cannot be specified completely in terms of the node voltages $v_②$, $v_⑥$, and $v_⑦$ are also independent of the branch voltages $v_5$, $v_{11}$, and $v_{12}$ and can also be part of the basis set. From expression (3.5a) we see that branch voltage $v_1$ is not completely specifiable in terms of $v_②$, $v_⑥$, and $v_⑦$ and may be taken as a basis voltage. The next branch voltage picked as a basis voltage must be independent of $v_①$, $v_②$, $v_⑥$, and $v_⑦$. From (3.5c), branch voltage $v_3$ is specified in terms of $v_①$ and $v_②$ and cannot be part of the basis.

Proceeding in this manner, one of many possible basis sets of branch voltages emerges $\{v_1, v_2, v_4, v_5, v_7, v_{11}, v_{12}\}$ (*Note:* As might be expected, branches 1, 2, 4, 5, 7, 11, and 12 form a tree for the network.) Since expressions (3.5a, b, d, e, g, k, and l) were used to determine the independence of the basis variables, we turn to expressions (3.5c, f, h, i, and j) to yield the five linearly independent KVL loop equations. We proceed by first expressing the basis set of node voltages in terms of the basis set of branch voltages. From (3.5e, k, and l)

$$v_② = v_5$$

$$v_⑥ = -v_{11}$$

and

$$v_⑦ = v_{12}$$

from (3.5a)

$$v_① = v_1 + v_⑦$$

$$= v_1 + v_{12}$$

from (3.5b)

$$v_③ = v_2 + v_①$$

$$= v_1 + v_2 + v_{12}$$

from (3.5d)

$$v_④ = v_4 + v_②$$

$$= v_4 + v_5$$

and from (3.5g)

$$v_⑤ = -v_7 + v_③$$

$$= v_1 + v_2 - v_7 + v_{12}$$

The KVL loop equations now follow directly:

$$-v_1 + v_3 + v_5 - v_{12} = 0$$

from (3.5c);

$$-v_1 - v_2 + v_4 + v_5 + v_6 - v_{12} = 0$$

from (3.5f)

$$-v_4 - v_5 + v_8 - v_{11} = 0$$

from (3.5h);

$$v_1 + v_2 - v_7 + v_9 + v_{11} + v_{12} = 0$$

from (3.5i); and

$$v_1 + v_2 - v_7 + v_{10} = 0$$

from (3.5j). (*Note.* These equations are the KVL fundamental loop equations associated with the tree mentioned above.) It is easily verified that these equations are linearly independent.

The branch relationships are obtained directly from the original network. For current source branches we must check the orientation for positive current flow from the graph. The branch relationships are:

| | | | | |
|---|---|---|---|---|
| (3.6a) | *branch 1* | $i_1 = 2v_1$ | | |
| (3.6b) | *branch 2* | $i_2 = -1$ | | |
| (3.6c) | *branch 3* | $v_3 = .5i_3$ | or | $i_3 = 2v_3$ |
| (3.6d) | *branch 4* | $i_4 = 4v_4$ | | |
| (3.6e) | *branch 5* | $v_5 = 1i_5$ | or | $i_5 = v_5$ |
| (3.6f) | *branch 6* | $v_6 = .1i_6$ | or | $i_6 = 10v_6$ |
| (3.6g) | *branch 7* | $v_7 = .2i_7$ | or | $i_7 = 5v_7$ |
| (3.6h) | *branch 8* | $i_8 = 4$ | | |
| (3.6i) | *branch 9* | $v_9 = 1i_9$ | or | $i_9 = v_9$ |
| (3.6j) | *branch 10* | $i_{10} = 3$ | | |
| (3.6k) | *branch 11* | $i_{11} = 3v_{11}$ | | |
| (3.6l) | *branch 12* | $i_{12} = -2$ | | |

Substitution of the branch relationships (3.6) into the KCL node equations (3.4) yields:

$$2v_1 + 1 + 2v_3 = 0$$
$$-2v_3 - 4v_4 + v_5 = 0$$
$$-1 + 10v_6 + 5v_7 = 0$$
$$+4v_4 - 10v_6 + 4 = 0$$
$$-5v_7 - v_9 - 3 = 0$$
$$-4 + v_9 - 3v_{11} = 0$$
$$-2v_1 + 3 - 2 = 0$$

The final step is to substitute the KVL voltage equations (3.5) into these expressions:

$$2(v_① - v_⑦) + 1 + 2(v_① - v_②) = 0$$
$$-2(v_① - v_②) - 4(v_④ - v_②) + v_② = 0$$
$$-1 + 10(v_③ - v_④) + 5(v_③ - v_⑤) = 0$$
$$+4(v_④ - v_②) - 10(v_③ - v_④) + 4 = 0$$
$$-5(v_③ - v_⑤) - (v_⑥ - v_⑤) - 3 = 0$$
$$-4 + (v_⑥ - v_⑤) - 3(-v_⑥) = 0$$
$$-2(v_① - v_⑦) + 3 - 2 = 0$$

Combining and rearranging terms yields:

(3.7a) $\quad 4v_① - 2v_② \qquad\qquad\qquad\qquad - 2v_⑦ = -1$

(3.7b) $\quad -2v_① + 7v_② \qquad\quad - 4v_④ \qquad\qquad\qquad = 0$

(3.7c) $\qquad\qquad\qquad 15v_③ - 10v_④ - 5v_⑤ \qquad\qquad = 1$

(3.7d) $\qquad - 4v_② - 10v_③ + 14v_④ \qquad\qquad\qquad = -4$

(3.7e) $\qquad\qquad - 5v_③ \qquad\quad + 6v_⑤ - v_⑥ \qquad = 3$

(3.7f) $\qquad\qquad\qquad\qquad\quad - v_⑤ + 4v_⑥ \qquad = 4$

(3.7g) $\quad -2v_① \qquad\qquad\qquad\qquad\qquad + 2v_7 = -1$

The above 7 nodal equilibrium equations are in terms of the 7 unknown node voltages. Solutions of simultaneous algebraic equations will be discussed in the next Section. □

The nodal equilibrium equations for any resistive network can be written by inspection. We know from Kirchhoff's Current Law that the algebraic sum of the currents leaving a node through resistors and conductors must equal the algebraic sum of the currents entering a node through current sources. Therefore, KCL node equations can be written in terms of resistance or conductance values (known coefficients), current source values (known forcing functions), and node voltages (unknowns). In general nodal equilibrium equations can be written in the following form.

(3.8) $\quad y_{11}v_①(t) + y_{12}v_②(t) + y_{13}v_③(t) + \cdots + y_{1m}v_Ⓜ(t) = i_①(t)$

$\qquad y_{21}v_①(t) + y_{22}v_②(t) + y_{23}v_③(t) + \cdots + y_{2m}v_Ⓜ(t) = i_②(t)$

$\qquad y_{31}v_①(t) + y_{32}v_②(t) + y_{33}v_③(t) + \cdots + y_{3m}v_Ⓜ(t) = i_③(t)$

$\qquad \cdots\cdots\cdots\cdots\cdots\cdots\cdots\cdots\cdots\cdots\cdots\cdots\cdots\cdots\cdots$

$\qquad y_{m1}v_①(t) + y_{m2}v_②(t) + y_{m3}v_③(t) + \cdots + y_{mm}v_m(t) = i_m(t)$

where $m = n_n - 1$; the coefficients of the node voltages are such that

$y_{kk}$ = sum of all *conductances* or "admittances" that

are connected to node ⓚ

and

$y_{jk}$ = the *negative* sum of all conductances directly

connected between nodes ⓙ and ⓚ

and

$i_{ⓚ}(t)$ = sum of all current source currents *entering* node ⓚ

A quick explanation of the subscripting scheme is in order. The first subscript, $j$, of $y_{jk}$ refers to the node number for which the KCL node equation is written. The second subscript, $k$, of $y_{jk}$ refers to the coefficient of the node voltage $v_{ⓚ}(t)$ in nodal equilibrium equation $j$. For example, $y_{36}$ is the coefficient of node voltage $v_{ⓖ}(t)$ in the nodal equilibrium equation for node ③. Similarly, $y_{55}$ is the coefficient of node voltage $v_5(t)$ in the nodal equilibrium equation written for node ⑤. Since not all the node voltages appear in all the nodal equilibrium equations, many of the coefficients, $y_{jk}$, are zero.

Consider again the network of Fig. 3-7 and its nodal equilibrium equations (3.7). In the nodal equilibrium equation for node ①, (3.7a), $y_{11} = 4$, $y_{12} = -2$, $y_{13} = y_{14} = y_{15} = y_{16} = 0$, and $y_{17} = -2$. Observe that $y_{11}$ is indeed the sum of all the conductances connected to node ①: 2 mhos due to the conductor in branch 1 and $1/.5 = 2$ mhos due to the conductor in branch 3. Similarly, $y_{12}$ and $y_{17}$ represent the negative sum of all conductance directly connected between nodes ① and ② and nodes ① and ⑦, respectively. Since there are no conductances directly connected between node ① and nodes ③, ④, ⑤, or ⑥, the coefficients $y_{13}$, $y_{14}$, $y_{15}$, and $y_{16}$ must all be zero. Finally, observe that there is a current source current of 1 amp leaving node ①, hence $i_{①}(t) = -1$ as seen in equation (3.7a). The reader should verify for himself that the remaining coefficients of equations (3.7) are in agreement with the general form of nodal equilibrium equations.

The nodal equilibrium equations, (3.8), are easily written by inspection. As an example, consider the single node of Fig. 3-8 that is part of some large network. The KCL node equation for this node, written in terms of the node voltages is

$$\underbrace{\frac{1}{R_1}(v_{③} - v_{②}) + G_3(v_{③} - v_{⑥}) + G_5 v_{③}}_{\text{sum of currents leaving through resistors}} = \underbrace{-i_7 - i_9 + i_{15}}_{\substack{\text{sum of currents} \\ \text{entering through} \\ \text{current sources}}}$$

This equation can be rewritten as

$$-\frac{1}{R_1} v_{②} + \left(\frac{1}{R_1} + G_3 + G_5\right) v_{③} - G_3 v_{⑥} = -i_7 - i_9 + i_{15}$$

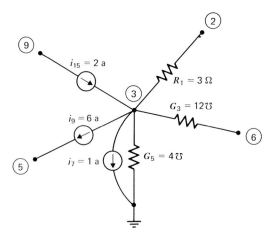

**Fig. 3-8 A single node 3, part of a larger network, for which we can write a KCL equation in terms of the node voltages.**

or, on substitution of element values,

$$-\frac{1}{3}v_{\text{②}} + 16\frac{1}{3}v_{\text{③}} - 12v_{\text{⑥}} = -5$$

The last two equations could have been written directly using expression (3.8).

**Example 3.4**

As another example, the nodal equilibrium equations for the network of Fig. 3-9 are:

$$node\ \text{①} \quad \left(\frac{1}{R_2} + G_3\right)v_{\text{①}} - G_3 v_{\text{②}} = \hat{i}_1 + \hat{i}_4$$

$$node\ \text{②} \quad -G_3 v_{\text{①}} + \left(G_3 + G_5 + \frac{1}{R_8}\right)v_{\text{②}} - \frac{1}{R_8}v_{\text{④}} = -\hat{i}_6$$

$$node\ \text{③} \quad \left(\frac{1}{R_7} + G_9\right)v_{\text{③}} - G_9 v_{\text{④}} = -\hat{i}_4 + \hat{i}_6$$

$$node\ \text{④} \quad -\frac{1}{R_8}v_{\text{②}} - G_9 v_{\text{③}} + \left(\frac{1}{R_8} + G_9 + \frac{1}{R_{10}} + \frac{1}{R_{11}}\right)v_{\text{④}} = 0 \quad \square$$

## 3-3 SOLUTION OF SIMULTANEOUS ALGEBRAIC EQUATIONS—CRAMER'S RULE

The $n_n - 1$ nodal equilibrium equations form a set of simultaneous algebraic equations in which the $n_n - 1$ node voltages are unknowns. In this

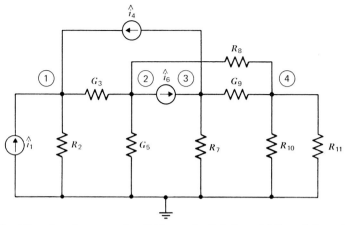

**Fig. 3-9  The 4 node network of Example 3.4 for which nodal equilibrium equations are written by inspection.**

section we consider the solution of simultaneous algebraic equations in general. The methods described in this section can, and will, be used to solve simultaneous algebraic equations that are more general than those obtained as nodal equilibrium equations for networks containing resistors and current sources. To emphasize this generality we use the following, somewhat standard, notation to denote the set of simultaneous algebraic equations under consideration:

$$
\begin{aligned}
a_{11}x_1 + a_{12}x_2 + \cdots + a_{1n}x_n &= b_1 \\
a_{21}x_1 + a_{22}x_2 + \cdots + a_{2n}x_n &= b_2 \\
&\cdots\cdots\cdots\cdots\cdots\cdots \\
a_{n1}x_1 + a_{n2}x_2 + \cdots + a_{nn}x_n &= b_n
\end{aligned}
$$

(3.9)

where the coefficients $a_{jk}$ and $b_k$, $j = 1, 2, \ldots, n$, $k = 1, 2, \ldots, n$, are known constants (for example, conductance values) and the $x_k$, $k = 1, 2, \ldots, n$, are the unknowns (for example, the node voltages). The above equations can also be written in terms of sums

$$
\sum_{k=1}^{n} a_{1k}x_k = b_1
$$

$$
\sum_{k=1}^{n} a_{2k}x_k = b_2
$$

$$
\cdots\cdots\cdots\cdots
$$

$$
\sum_{k=1}^{n} a_{nk}x_k = b_n
$$

or most succinctly as

$$
\sum_{k=1}^{n} a_{jk}x_k = b_j \qquad j = 1, 2, \ldots, n
$$

These alternative forms of expressing simultaneous algebraic equations will be useful when writing a computer program for their solution. There are several techniques available for the solution of simultaneous equations. Some of these are iterative, that is, where a solution is initially guessed and then continuously updated by a series of steps. Ultimately, the correct solution is determined. Other methods are direct, that is, the solution is found directly from manipulation of the original equations. We will be concerned only with direct methods, particularly two methods: *Cramer's rule* and *Gaussian elimination*. Students who are interested in other techniques for solving simultaneous algebraic equations should consult the references. In this section we discuss Cramer's rule. Gaussian elimination, which forms the basis of a computer program to be used throughout this text, is discussed in the next section. For convenience, before proceeding with the solution of (3.9) we introduce some nomenclature.

The coefficients of the unknowns in (3.9) form an array of numbers called a *matrix;*

$$\begin{pmatrix} a_{11} & a_{12} & \cdots & a_{1n} \\ a_{21} & a_{22} & \cdots & a_{2n} \\ \cdots & \cdots & \cdots & \cdots \\ a_{n1} & a_{n2} & \cdots & a_{nn} \end{pmatrix}$$

Since the number of rows equals the number of columns (there are $n$ of each) the matrix is said to be *square* or $(n \times n)$. An $(n \times n)$ matrix is called an $n$th *order* matrix. It is convenient to represent this array of numbers by a single boldface letter, for example **A**. In other words

$$\mathbf{A} = \begin{pmatrix} a_{11} & a_{12} & \cdots & a_{1n} \\ a_{21} & a_{22} & \cdots & a_{2n} \\ a_{n1} & a_{n2} & \cdots & a_{nn} \end{pmatrix}$$

Each of the quantities of **A** is called an element of **A** and the position of each element is characterized by a double subscript where the first subscript denotes the row number and the second subscript denotes the column number. Elements $a_{11}, a_{22}, a_{33} \cdots a_{nn}$ are said to be on the *main diagonal* of **A**. The *determinant* or $n$th *order determinant* of **A**, denoted by det (**A**) or $\Delta$, is a number that depends on the values of the elements of **A**. Straight line brackets are used to denote the determinant of a square matrix:

$$\Delta = \det (\mathbf{A}) = \begin{vmatrix} a_{11} & a_{12} & \cdots & a_{1n} \\ a_{21} & a_{22} & \cdots & a_{2n} \\ \cdots & \cdots & \cdots & \cdots \\ a_{n1} & a_{n2} & \cdots & a_{nn} \end{vmatrix}$$

Determinants are important in the solution of the set of simultaneous equations (3.9). The value of a determinant is found by a set of rules to be discussed below.

For motivation consider two simultaneous equations in two unknowns

(3.10a) $$a_{11}x_1 + a_{12}x_2 = b_1$$

and

(3.10b) $$a_{21}x_1 + a_{22}x_2 = b_2$$

The matrix of interest is

$$\mathbf{A} = \begin{pmatrix} a_{11} & a_{12} \\ a_{21} & a_{22} \end{pmatrix}$$

We are concerned with the second-order determinant

$$\Delta = \begin{vmatrix} a_{11} & a_{12} \\ a_{21} & a_{22} \end{vmatrix}$$

To solve for $x_1$ we can multiply (3.10a) equation by $a_{22}$:

$$a_{11}a_{22}x_1 + a_{12}a_{22}x_2 = a_{22}b_1$$

and the (3.10b) by $a_{12}$:

$$a_{12}a_{21}x_1 + a_{12}a_{22}x_2 = a_{12}b_2$$

and subtract the resulting equations:

$$(a_{11}a_{22} - a_{21}a_{12})x_1 = a_{22}b_1 - a_{12}b_2$$

so that

$$x_1 = \frac{a_{22}}{a_{11}a_{22} - a_{21}a_{12}} b_1 - \frac{a_{12}}{a_{11}a_{22} - a_{21}a_{12}} b_2$$

assuming the quantity $(a_{11}a_{22} - a_{21}a_{12})$ is nonzero. Similarly we could solve for $x_2$:

$$x_2 = \frac{-a_{21}}{a_{11}a_{22} - a_{21}a_{12}} b_1 + \frac{a_{11}}{a_{11}a_{22} - a_{21}a_{12}} b_2$$

The quantity $(a_{11}a_{22} - a_{21}a_{12})$ is known as the *determinant* of the matrix $\mathbf{A}$:

$$\det(\mathbf{A}) = (a_{11}a_{22} - a_{21}a_{12})$$

Thus we can say that the two simultaneous (3.10) can be solved for $x_1$ and $x_2$ if $\det(\mathbf{A})$ is nonzero, that is, they are "determinant."

The determinants of $n$th order square matrices, where $n$ is greater than two, are evaluated in terms of *minors* and *cofactors*.

The *minor* of any element of a determinant $a_{jk}$ is the determinant that remains when the row and column containing $a_{jk}$ ($j$th row and $k$th column) are deleted. For example, given the third-order determinant

$$\begin{vmatrix} a_{11} & a_{12} & a_{13} \\ a_{21} & a_{22} & a_{23} \\ a_{31} & a_{32} & a_{33} \end{vmatrix}$$

the minor of $a_{11}$ is

$$\begin{vmatrix} a_{22} & a_{23} \\ a_{32} & a_{33} \end{vmatrix}$$

the minor of $a_{22}$ is

$$\begin{vmatrix} a_{11} & a_{13} \\ a_{31} & a_{33} \end{vmatrix}$$

and the minor of $a_{23}$ is

$$\begin{vmatrix} a_{11} & a_{12} \\ a_{31} & a_{32} \end{vmatrix}$$

We designate the minor of $a_{jk}$ by $M_{jk}$. The *cofactor* of $a_{jk}$, denoted by $\Delta_{jk}$, is given by

$$\Delta_{jk} = (-1)^{j+k} M_{jk}$$

and is sometimes termed a *signed minor*. Observe that if $j + k$ is even the cofactor equals the minor and if $j + k$ is odd the cofactor is the negative of the minor.

The $n$th order determinant is equal to the sum of the products of the elements of *any* row or *any* column multiplied by their corresponding $(n-1)$st order cofactors. As an example we evaluate the third order determinant:

$$\Delta = \begin{vmatrix} a_{11} & a_{12} & a_{13} \\ a_{21} & a_{22} & a_{23} \\ a_{31} & a_{32} & a_{33} \end{vmatrix}$$

Expanding off of the first column yields

$$\Delta = a_{11}\,\Delta_{11} + a_{21}\,\Delta_{21} + a_{31}\,\Delta_{31}$$

$$= a_{11}M_{11} - a_{21}M_{21} + a_{31}M_{31}$$

$$= a_{11}\begin{vmatrix} a_{22} & a_{23} \\ a_{32} & a_{33} \end{vmatrix} - a_{21}\begin{vmatrix} a_{12} & a_{13} \\ a_{32} & a_{33} \end{vmatrix} + a_{31}\begin{vmatrix} a_{12} & a_{13} \\ a_{22} & a_{23} \end{vmatrix}$$

$$= a_{11}(a_{\cdot2}a_{33} - a_{23}a_{32}) - a_{21}(a_{12}a_{33} - a_{13}a_{32})$$

$$+ a_{31}(a_{12}a_{23} - a_{13}a_{22})$$

An equivalent value would be obtained if we had expanded off of any of the other rows or columns. A series of aids for the evaluation of determinants can easily be developed. For further discussion of these techniques consult any elementary math text.

The solution of the simultaneous equations

$$a_{11}x_1 + a_{12}x_2 + \cdots + a_{1n}x_n = b_1$$

$$a_{21}x_1 + a_{22}x_2 + \cdots + a_{2n}x_n = b_2$$

$$\cdots\cdots\cdots\cdots\cdots\cdots\cdots\cdots\cdots\cdots$$

$$a_{n1}x_1 + a_{n2}x_2 + \cdots + a_{nn}x_n = b_n$$

can now be undertaken using *Cramer's rule*. Cramer's rule gives the solution as

$$x_1 = \frac{D_1}{\Delta}, \; x_2 = \frac{D_2}{\Delta}, \ldots, \; x_n = \frac{D_n}{\Delta}$$

where

$$\Delta = \begin{vmatrix} a_{11} & a_{12} & \cdots & a_{1n} \\ a_{21} & a_{22} & \cdots & a_{2n} \\ \cdots & \cdots & \cdots & \cdots \\ a_{n1} & a_{n2} & \cdots & a_{nn} \end{vmatrix}$$

and $D_k$ $(k = 1, 2, \ldots, n)$ is the determinant obtained by replacing the $k$th column of $\Delta$ by $(b_1, b_2, \ldots, b_n)$. In other words,

$$D_1 = \begin{vmatrix} b_1 & a_{12} & \cdots & a_{1n} \\ b_2 & a_{22} & \cdots & a_{2n} \\ \cdots & \cdots & \cdots & \cdots \\ b_n & a_{n2} & \cdots & a_{nn} \end{vmatrix}$$

$$D_2 = \begin{vmatrix} a_{11} & b_1 & \cdots & a_{1n} \\ a_{21} & b_2 & \cdots & a_{2n} \\ \cdots & \cdots & \cdots & \cdots \\ a_{n1} & b_n & \cdots & a_{nn} \end{vmatrix}$$

etc. Observe that $\Delta$ must be nonzero for a unique solution to exist.

**Example 3.5**

Application of Cramer's rule is best illustrated by an example. The nodal equilibrium equations for the network of Fig. 3-10 are

$$5v_① - 2v_② \qquad\quad = 3$$

$$-2v_① + 7v_② - v_③ = -2$$

$$-v_② + 3v_③ = 2$$

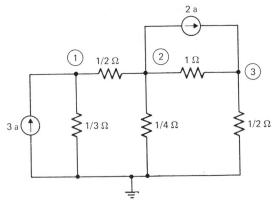

**Fig. 3-10    Network of Example 3.5.**

We first evaluate the determinant of coefficients:

$$\Delta = \begin{vmatrix} 5 & -2 & 0 \\ -2 & 7 & -1 \\ 0 & -1 & 3 \end{vmatrix}$$

Expanding off the first row yields

$$\Delta = 5 \begin{vmatrix} 7 & -1 \\ -1 & 3 \end{vmatrix} + 2 \begin{vmatrix} -2 & -1 \\ 0 & 3 \end{vmatrix}$$
$$= (5)(20) + (2)(-6)$$
$$= 88$$

From Cramer's rule, we have

$$v_① = \frac{\begin{vmatrix} 3 & -2 & 0 \\ -2 & 7 & -1 \\ 2 & -1 & 3 \end{vmatrix}}{88}$$

$$= \frac{3 \begin{vmatrix} 7 & -1 \\ -1 & 3 \end{vmatrix} + 2 \begin{vmatrix} -2 & -1 \\ 2 & 3 \end{vmatrix}}{88}$$

$$= \frac{52}{88}$$

$$v_② = \frac{\begin{vmatrix} 5 & 3 & 0 \\ -2 & -2 & -1 \\ 0 & 2 & 3 \end{vmatrix}}{88} = \frac{-2}{88}$$

and

$$v_{\text{③}} = \frac{\begin{vmatrix} 5 & -2 & 3 \\ -2 & 7 & -2 \\ 0 & -1 & 2 \end{vmatrix}}{88} = \frac{58}{88} \quad \square$$

## 3-4 SOLUTION OF SIMULTANEOUS ALGEBRAIC EQUATIONS BY GAUSSIAN ELIMINATION

It is not too difficult to see that as networks get larger, using Cramer's rule as a means of solution becomes increasingly more cumbersome and the chance of making an error becomes great. Therefore, we turn to the computer to solve these equations. However, it turns out that Cramer's rule is unnecessarily wasteful of computer time. A much more efficient scheme for solution of simultaneous equations is *Gaussian Elimination*.

We start with a simple example to illustrate the steps involved in the Gaussian elimination algorithm. (An *algorithm* is a set of instructions.)

### Example 3.6

Suppose we wish to find the solution to the following set of equations.

(3.11a)                $2x_1 + x_2 + 3x_3 = 2$

(3.11b)                $3x_1 - 2x_2 - x_3 = 1$

and

(3.11c)                $x_1 - x_2 + x_3 = -1$

We first express $x_1$ in terms of the other variables in (3.11a)

(3.12a)                $x_1 = 1 - \dfrac{1}{2} x_2 - \dfrac{3}{2} x_3$

and then substitute this result into (3.11b) and (3.11c)

(3.12b)                $-\dfrac{7}{2} x_2 - \dfrac{11}{2} x_3 = -2$

(3.12c)                $-\dfrac{3}{2} x_2 - \dfrac{1}{2} x_3 = -2$

Next we express $x_2$ in terms of the other variable, $x_3$, in (3.12b)

(3.13a)                $x_2 = \dfrac{4}{7} - \dfrac{11}{7} x_3$

and substitute this result into (3.12c)

(3.13b)
$$\frac{26}{14} x_3 = -\frac{16}{14}$$

or

(3.14)
$$x_3 = -\frac{8}{13}$$

We may now obtain the value of $x_2$ from (3.13a)

$$x_2 = \frac{20}{13}$$

and successively $x_1$ from (3.12a)

$$x_1 = \frac{15}{13}$$

Once $x_3$ has been found, the remainder of the process is called *back substitution*. □

We now proceed to discuss in general Gaussian elimination for the solution of the set of $n$ simultaneous linear algebraic equations

(3.15-1)
$$a_{11}x_1 + a_{12}x_2 + \cdots + a_{1n}x_n = a_{1,n+1}$$

(3.15-2)
$$a_{21}x_1 + a_{22}x_2 + \cdots + a_{2n}x_n = a_{2,n+1}$$

(3.15-3)
$$a_{31}x_1 + a_{32}x_2 + \cdots + a_{3n}x_n = a_{3,n+1}$$

$$\cdots \cdots \cdots \cdots \cdots \cdots \cdots \cdots \cdots \cdots$$

(3.15-n)
$$a_{n1}x_1 + a_{n2}x_2 + \cdots + a_{nn}x_n = a_{n,n+1}$$

Observe that we are now denoting the right hand side of the set of equations by $a_{1,n+1}, a_{2,n+1}, \ldots, a_{n,n+1}$ as opposed to using $b_1, b_2, \ldots, b_n$ as was done in (3.9).

Suppose that $a_{11}$ is nonzero (if it is not a simple reordering of equations would be required), then it is possible to divide equation (3.15-1) by $a_{11}$ to yield

(3.16-1)
$$x_1 + \frac{a_{12}}{a_{11}} x_2 + \cdots + \frac{a_{1n}}{a_{11}} x_n = \frac{a_{1,n+1}}{a_{11}}$$

The next step is to multiply (3.16-1) by $a_{21}$ and subtract the result from (3.15-2) to yield

(3.16-2)
$$\left( a_{22} - a_{21}\frac{a_{12}}{a_{11}} \right) x_2 + \cdots + \left( a_{2n} - a_{21}\frac{a_{1n}}{a_{11}} \right) x_n = a_{2,n+1} - a_{21}\frac{a_{1,n+1}}{a_{11}}$$

Similarly we can "eliminate" $x_1$ from equations (3.15-3)–(3.15-n) by multiplying (3.16-1) by $a_{k1}$ ($k = 3, 4, \ldots, n$) and subtracting the result from equation (3.15-k). The resulting set of equations can be written as

(3.17-1)
$$x_1 + a_{12}^{(1)}x_2 + \cdots + a_{1n}^{(1)}x_n = a_{1,n+1}^{(1)}$$

(3.17-2)
$$a_{22}^{(1)}x_2 + \cdots + a_{2n}^{(1)}x_n = a_{2,n+1}^{(1)}$$

$$\cdots$$

(3.17-n)
$$a_{n2}^{(1)}x_2 + \cdots + a_{nn}^{(1)}x_n = a_{n,n+1}^{(1)}$$

where the coefficients $a_{jk}^{(1)}$ are given explicitly by

$$a_{1k}^{(1)} = \frac{a_{1k}}{a_{11}} \qquad (k = 1, 2, \ldots, n, n + 1)$$

and

$$a_{jk}^{(1)} = a_{jk} - \frac{a_{j1}}{a_{11}} a_{1k} \qquad (j = 2, 3, \ldots, n; k = 1, 2, \ldots, n, n + 1)$$

Upon the assumption that the coefficient $a_{22}^{(1)}$, which resulted from the above operations, is nonzero, we divide it through the second equation and then subtract $a_{k2}^{(1)}$ times the result from each succeeding equation ($k = 3, 4, \ldots, n$) to obtain the equations

(3.18-1)
$$x_1 + a_{12}^{(1)}x_2 + a_{13}^{(1)}x_3 + \cdots + a_{1n}^{(1)}x_n = a_{1,n+1}^{(1)}$$

(3.18-2)
$$x_2 + a_{23}^{(2)}x_3 + \cdots + a_{2n}^{(2)}x_n = a_{2,n+1}^{(2)}$$

$$\cdots$$

(3.18-n)
$$a_{n3}^{(2)}x_3 + \cdots \quad a_{nn}^{(2)}x_n = a_{n,n+1}^{(2)}$$

in which $x_2$ has been eliminated from equations (3.18-3) to (3.18-n). Again, if $a_{22}^{(1)}$ were zero, we could effect a row interchange so as to place a nonzero element in the $(2, 2)$ position, provided that the equations were linearly independent. Column interchanges, as well as row interchanges, are possible, and sometimes desirable; however these entail a renumbering of the variables $x_j$.

The previous procedure is to be repeated for a total of $n$ steps. The final set of equations that results is

(3.19-1)
$$x_1 + a_{12}^{(1)}x_2 + a_{13}^{(1)}x_3 + \cdots \quad a_{1n}^{(1)}x_n = a_{1,n+1}^{(1)}$$

(3.19-2)
$$x_2 + a_{23}^{(2)}x_3 + \cdots + a_{2n}^{(2)}x_n = a_{2,n+1}^{(2)}$$

(3.19-3)
$$x_3 + \cdots + a_{3n}^{(3)}x_n = a_{3,n+1}^{(3)}$$

$$\cdots$$

(3.19-n)
$$x_n = a_{n,n+1}^{(n)}$$

where

(3.20)
$$a_{jk}^{(i)} + a_{jk}^{(i-1)} - \frac{a_{ji}^{(i-1)}}{a_{ii}^{(i-1)}} a_{ik}^{(i-1)}$$

and

$$i = 1, 2, \ldots, n; j = 1, 2, \ldots, n; \text{ and } k = 1, 2, \ldots, n + 1$$

We observe that the superscripts in parentheses on the elements $a_{jk}$ indicate the number of times the $k$th coefficient on the $j$th equation has been modified. It is easy to obtain the solution from (3.19) by *back substitution*,

$$(3.21) \qquad x_j = a_{j,n+1}^{(j)} - \sum_{k=j+1}^{n} a_{jk}^{(j)} x_k \qquad (j = n, n - 1, \ldots, 2, 1)$$

Gaussian elimination is easily programmed for a digital computer and is discussed in the next section. It is possible that numerical errors may arise if Gaussian elimination is used directly as indicated. The following example illustrates how roundoff errors may cause incorrect answers.

**Example 3.7**

Given the two simultaneous linear algebraic equations

$$.00034x_1 + 56.231x_2 = 0.0647$$

and

$$1.0112x_1 + 3.1334x_2 = -0.7259$$

Suppose that we are working with fixed-point arithmetic and that we may carry at most only five significant digits, as indicated by the coefficients above. The pair of equations that results from the first round of Gaussian elimination is

$$1.0000x_1 + 165390.x_2 = 190.29$$

and

$$-167240.x_2 = -193.15$$

from which we obtain the ostensible solution

$$x_2 = .00115 \qquad \text{and} \qquad x_1 = .09150$$

But the substitution of these values into the second of the original equations yields

$$(1.0112)(.09150) + (3.1334)(.00115) = .09613 \neq -0.7259$$

indicating gross error.   □

In general roundoff errors may occur if any of the divisors $a_{jj}^{(j-1)}$ is very small in magnitude relative to the elements $a_{kj}^{(j-1)}$, $j > k$. We may attempt to avoid such division by comparatively small numbers by rearranging the elements according to magnitude, a procedure called *pivoting*. In particular, at the $k$th round of Gaussian elimination, we must locate the element $a_{j-1,k-1}^{(j-1)}$, $k \geq j - 1$ with the greatest magnitude, and then interchange equations $j - 1$ and $k$. The resulting modified set of equations is then manipulated as before. Sets of equations for which such problems arise are termed *ill-conditioned* and in bad cases even this pivoting scheme may fail to alleviate such problems. However, it should be pointed out that the role that roundoff error plays in network related problems is not clearly understood and pivoting is usually employed for other purposes. Suffice it to say that roundoff errors can be minimized by using double precision arithmetic.

**Example 3.8**

Consider again the simultaneous algebraic equations of Example 3.7

$$.00034x_1 + 56.231x_2 = 0.0647$$

and

$$1.0112x_1 + 3.1334x_2 = -0.7259$$

If 16 significant digits are allowed, then the solution is

$$x_1 = -.53527 \quad \text{and} \quad x_2 = .001151066$$

which is "exact." □

There is, however, another instance when rearrangement of equations is required before Gaussian elimination may be undertaken; if the pivot location contains a zero. This situation is best illustrated by an example.

**Example 3.9**

Consider solution of the following set of simultaneous algebraic equations:

$$2x_1 + 4x_2 - x_3 = 1$$
$$x_1 + 2x_2 + x_3 = 2$$

and

$$x_1 - x_2 + x_3 = 0$$

After the first pass of Gaussian elimination we are left with the equations

$$x_1 + 2x_2 - \frac{1}{2}x_3 = \frac{1}{2}$$

$$\frac{3}{2}x_3 = \frac{3}{2}$$

and

$$-3x_2 + \frac{3}{2}x_3 = -\frac{1}{2}$$

The next step is to divide through by the coefficient of $x_2$ in the second equation, but this coefficient is zero. Thus, in order to proceed, we must rearrange the second two equations. Then we have

$$x_1 + 2x_2 - \frac{1}{2}x_3 = \frac{1}{2}$$

$$-3x_2 + \frac{3}{2}x_3 = -\frac{1}{2}$$

$$\frac{3}{2}x_3 = \frac{3}{2}$$

Gaussian elimination can then proceed. The second pass yields

$$x_1 + 2x_2 - \frac{1}{2}x_3 = \frac{1}{2}$$

$$x_2 - \frac{1}{2}x_3 = \frac{1}{6}$$

and

$$\frac{3}{2}x_3 = \frac{3}{2}$$

The third pass yields

$$x_1 + 2x_2 - \frac{1}{2}x_3 = \frac{1}{2}$$

$$x_2 - \frac{1}{2}x_3 = \frac{1}{6}$$

and

$$x_3 = 1$$

From back substitution we find that $x_3 = 1$, $x_2 = 2/3$, and $x_1 = -1/3$. □

The above example illustrates that during the course of elimination a zero pivot can arise. In this situation a reordering of equations is required. Note that if it were not possible to reorder the equations so as to find a nonzero pivot, the equations are not linearly independent.

### Example 3.10

Let us try to solve the following set of simultaneous algebraic equations using Gaussian elimination.

$$2x_1 + 4x_2 - x_3 = 1$$

$$x_1 + 2x_2 + x_3 = 2$$

and

$$x_1 + 2x_2 - 2x_3 = -1$$

After the first pass, these equations become

$$x_1 = 2x_2 - \frac{1}{2} x_3 = \frac{1}{2}$$

$$\frac{3}{2} x_3 = \frac{3}{2}$$

and

$$-\frac{3}{2} x_3 = -\frac{3}{2}$$

We can proceed no further. Observe that indeed the original equations are linearly dependent: the first minus the second equals the third.  □

## 3-5   THE GAUSSIAN ELIMINATION PROGRAM (GEP)

In this section we introduce a computer program that solves simultaneous algebraic equations by means of Gaussian elimination. The program is written in FORTRAN IV. If the reader is unfamiliar with FORTRAN he should skip the program description and just study the examples so that he will be able to use the program. Many of the problems in the sequel assume that this program, or one similar to it, is available for use. Throughout the discussion that follows, the reader is referred to the FORTRAN listing of the Gaussian elimination program, Fig. 3.11.

Discussion of the program is facilitated if we recall that we wish to solve the $N$ simultaneous algebraic equations

$$a_{11}x_1 + a_{12}x_2 + \cdots + a_{1N}x_N = b_1$$
$$a_{21}x_1 + a_{22}x_2 + \cdots + a_{2N}x_N = b_2$$
$$\cdots\cdots\cdots\cdots\cdots\cdots\cdots\cdots\cdots$$
$$a_{N1}x_1 + a_{N2}x_2 + \cdots + a_{NN}x_N = b_N$$

where the coefficients $a_{jk}$ $(j = 1, 2, \ldots, N;\ k = 1, 2, \ldots, N)$ and $b_j$ $(j = 1, 2, \ldots, N)$ are known and the $x_k$ $(k = 1, 2, \ldots, N)$ are unknown.

The program starts by reading in a title card. Whatever is on this card will be printed on the top of the output page.

Next, $N$, the number of equations to be solved is read. If $N$ is zero the program terminates; if $N$ is equal to one or greater than 30, an error message is printed and the program terminates; if $2 \leq N \leq 30$ the coefficients of the unknowns, $a_{jk}$, are read in and stored in the double subscripted array $A$ and the knowns on the right hand, $b_j$, are read in and stored in the single subscripted array $B$. All data read in is then printed out so that the user can check his input.

```
C                    GAUSSIAN ELIMINATION PROGRAM - 'GEP'
C
C
C
C
C       GEP IS A PROGRAM FOR SOLVING UP TO 30 SIMULTANEOUS LINEAR ALGEBRAIC
C       EQUATIONS WITH REAL COEFFICIENTS. THE PROGRAM USES GAUSSIAN ELIMINATION
C       WITH ROW PIVOTING.
C
C                                 INPUT FORMAT
C
C       THE FIRST CARD IS A TITLE CARD. THE CONTENTS OF THIS CARD WILL BE
C       PRINTED AT THE TOP OF THE OUTPUT.
C       THE NUMBER OF EQUATIONS SHOULD BE ENTERED IN THE FIRST 2 COLUMNS OF THE
C       SECOND DATA CARD (FOR THE PROGRAM TO RUN CORRECTLY THIS NUMBER SHOULD BE
C       BETWEEN 2 AND 30 AND RIGHT-ADJUSTED IN THE FIRST TWO COLUMNS).
C       THE REMAINING DATA CARDS CONTAIN THE COEFFICIENTS AND RIGHT-HAND-SIDE
C       OF THE EQUATIONS IN THE FOLLOWING FORMAT:
C                                 G10.3
C          EXPONENTS MUST BE RIGHT JUSTIFIED AND ENTRIES MAY BE CONTINUED ON
C          EXTRA CARDS UNTIL ALL THE COEFFICIENTS ARE ENTERED. IN THE NEXT 10
C          COLUMNS IMMEDIATELY FOLLOWING THE COEFFICIENTS ENTER THE RIGHT-
C          HAND-SIDE FOR THE EQUATION. THE REMAINING EQUATIONS ARE ENTERED IN
C          THE SAME FORMAT STARTING A NEW CARD FOR EACH EQUATION.
C          AFTER THE LAST SET OF DATA THERE SHOULD BE TWO BLANK CARDS INDICATING
C          TERMINATION OF INPUT DATA.
C
C
C       THIS PROGRAMS ASSUMES THAT:
C          READER=UNIT 5
C          PRINTER=UNIT 6
C
C
        IMPLICIT REAL*8 (A-H,O-Z)
        DIMENSION A(30,30),B(30),TITLE(10)
13      READ(5,30)TITLE
        WRITE(6,31)TITLE
        READ(5,18)N
        IF(N.EQ.0) GO TO 27
        IF(N.GT.30.OR.N.LT.2)GO TO 17
        WRITE(6,20)
        DO 12 J=1,N
        READ(5,19)(A(J,K),K=1,N),B(J)
12      WRITE(6,21)(A(J,K),K=1,N)
        WRITE(6,28)
        DO 29 J=1,N
29      WRITE(6,22)B(J)
        DO 5 I=1,N
        I1=I+1
C
C       AT THE ITH STEP, IF A(I,I) IS ZERO SEARCH FOR THE FIRST NONZERO ELEMENT
C       IN THE ITH COLUMN AND STORE ITS ROW NUMBER IN IPIV. IF ALL THE
C       ELEMENTS ARE ZERO,I.E. THE EQUATIONS ARE DEPENDENT,TERMINATE EXECUTION
C        AND PRINT OUT MESSAGE.
C
```

**Fig. 3-11   Listing of the program GEP.**

```
        IF(DABS(A(I,I)).LE.1.D-10) GO TO 1
        GO TO 15
   1    CONTINUE
        IF(I.EQ.N) GO TO 10
        DO 14 J=I1,N
        IF(DABS(A(J,I)).LE.1.D-10) GO TO 14
        IPIV=J
        GO TO 16
  14    CONTINUE
        GO TO 10
C
C    INTERCHANGE THE ROWS I AND IPIV
C
  16    DO 2 K=1,N
        PIV=A(IPIV,K)
        A(IPIV,K)=A(I,K)
   2    A(I,K)=PIV
        PIV=B(IPIV)
        B(IPIV)=B(I)
        B(I)=PIV
  15    IF(I.EQ.N) GO TO 3
C
C    FORWARD REDUCTION OF A AND B.
C
        DO 38 JI=I1,N
  38    A(I,JI)=A(I,JI)/A(I,I)
        B(I)=B(I)/A(I,I)
        DO 5 J=I1,N
        DO 4 K=I1,N
   4    A(J,K)=A(J,K)-A(J,I)*A(I,K)
   5    B(J)=B(J)-B(I)*A(J,I)
C
C    FIND THE SOLUTION BY BACK SUBSTITUTION. STORE THE SOLUTION
C    IN THE 'B' VECTOR.
C
   3    B(N)=B(N)/A(N,N)
        DO 6 K=2,N
        I=N-K+1
        L=I+1
        SUM=0.0
        DO 7 J=L,N
   7    SUM=SUM+A(I,J)*B(J)
   6    B(I)=B(I)-SUM
        GO TO 8
  10    WRITE(6,9)
   9    FORMAT(33HOEQUATIONS ARE LINEARLY DEPENDENT)
        GO TO 13
   8    WRITE(6,23)
        DO 11 I=1,N
        IF(I.LT.10)WRITE(6,24)I,B(I)
        IF(I.GE.10)WRITE(6,25)I,B(I)
  11    CONTINUE
```

**Fig. 3-11** (*Continued*)

```
           GO TO 13
    17     WRITE(6,26)
    27     STOP
    18     FORMAT(I2)
    19     FORMAT(8G10.3)
    20     FORMAT(11HOINPUT DATA,////,3X,12HCOEFFICIENTS,/)
    21     FORMAT(1H0,(T2,8(G12.5,3X)))
    22     FORMAT(1X,G12.5,/)
    23     FORMAT(/////,23HOVALUES OF THE UNKNOWNS,//)
    24     FORMAT(6HO     X,I1,4H  = ,G12.5)
    25     FORMAT(6HO     X,I2,3H = ,G12.5)
    26     FORMAT(77H1INITIAL DATA CARD DOES NOT HAVE A NUMBER BETWEEN 2 AND
           130 IN COLUMNS 1 AND 2)
    28     FORMAT(//,1H0,15HRIGHT-HAND-SIDE,//)
    30     FORMAT(10A8)
    31     FORMAT(1H1,10X,10A8,///)
           END
```

**Fig. 3-11** *(Continued)*

After the data has been read in, the actual Gaussian elimination procedure begins. The index $I$ is used to denote the pass number and corresponds to the superscript $i$ in equation (3.20). The first step in each pass is to determine if the pivot element is exceedingly small. If it is, equations $I + 1, I + 2, \ldots, N$ are searched to find one that has a nonzero coefficient in the $I$th column. If such an equation cannot be found, the equations are linearly dependent and execution is terminated. After an equation with a nonzero pivot is found and moved to the $I$th location, the $I$th unknown is eliminated from equations $I + 1, I + 2, \ldots, N$.

The above procedure is repeated $N$ times. The solution can then be found using back substitution. Finally, the solution is printed.

The program GEP expects the data to be inputed in the following manner. The first card is the title card. Whatever appears on this card (columns 1–80) will be printed on the top of output. The second data card contains the number of equations, $N$, to be solved (this is the same as the number of unknowns). This number, which must be greater than or equal to 2 and less than or equal to 30, should be in columns 1 and 2 and must be right justified. (In other words, if $2 \leq N \leq 9$ then this number is entered in column 2, and if $10 \leq N \leq 30$, then this number is entered in columns 1 and 2.)

The next set of cards to be read contain the coefficients $a_{jk}$ and $b_j$. These coefficients are entered in fields of 10 columns each. A decimal point must be used. If exponents are used they must be right justified. As an example, the number 1256.4 can be entered in any of the following ways:

| columns → 1 | 2 | 3 | 4 | 5 | 6 | 7 | 8 | 9 | 10 |
|---|---|---|---|---|---|---|---|---|---|
|   | 1 | 2 | 5 | 6 | . | 4 |   |   |    |
|   |   | 1 | 2 | 5 | . | 6 | 4 | E | 1  |
|   |   | 1 | 2 | . | 5 | 6 | 4 | E | + | 2  |
| 1 |   | 2 | 5 | 6 | 4 | . |   | E | − | 1  |

among other possibilities. Similarly the number $3.56 \times 10^{-12}$ can be entered in any of the following ways:

| columns → 1 | 2 | 3 | 4 | 5 | 6 | 7 | 8 | 9 | 10 |
|---|---|---|---|---|---|---|---|---|---|
| 3 | · | 5 | 6 | | | E | — | 1 | 2 |
| · | 0 | 0 | 3 | 5 | 6 | E | — | | 9 |
| 3 | 5 | 6 | · | | | E | — | 1 | 4 |

among other possibilities. The coefficients of the first equation, $a_{11}, a_{12}, \dots,$ $a_{1N}$ are entered in fields of 10 columns each, eight to a card. In the 10 column field immediately following the coefficient $a_{1N}$, the right hand side of the first equation, $b_1$, is entered. A new card is then started with the coefficients of the second equation. A new card is then started with the coefficients of the second equation.

More than one set of simultaneous equations can be submitted at a time. Additional data can be stacked behind the first set of data starting with a new title card. Two blank cards should be placed at the end of all data to indicate the end of the input.

**Example 3.11**

Consider again the network of Example 3.4 that is redrawn in Fig. 3-12a. The nodal equilibrium equations are

$$5v_① - 3v_② \qquad\qquad = \quad 3$$
$$-3v_① + 10v_② \qquad - 3v_④ = -6$$
$$6v_③ - v_④ = \quad 4$$
$$- 3v_② - v_③ + 9v_④ = \quad 0$$

The input to GEP is shown in Fig. 3-12b and the program output is shown in Fig. 3-12c. ☐

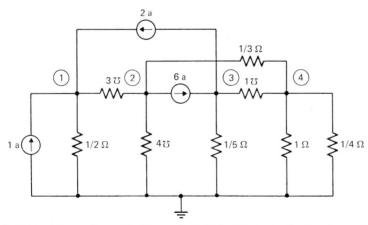

**Fig. 3-12(a)  Network used in Example 3.11 to illustrate the data input format expected by GEP.**

```
----- DATA FOR EXAMPLE 3.11 -----

CHAPTER 3 EXAMPLE 3.11
  4
 5.0        -3.0        0.0        0.0        3.0
-3.0        10.0        0.0       -3.0       -6.0
 0.0        0.0         6.0       -1.0        4.0
 0.0       -3.0        -1.0        9.0        0.0
```

**Fig. 3-12(*b*)  Input data for GEP for the network of (*a*).**

```
           CHAPTER 3 EXAMPLE 3.11



   INPUT DATA



      COEFFICIENTS


         5.0000         -3.0000         0.0            0.0

        -3.0000         10.000          0.0           -3.0000

         0.0            0.0             6.0000        -1.0000

         0.0            -3.0000        -1.0000         9.0000



   RIGHT-HAND-SIDE

       3.0000

      -6.0000

       4.0000

      0.0
```

```
   VALUES OF THE UNKNOWNS


      X1  =   0.26800

      X2  =  -0.55334

      X3  =   0.64792

      X4  =  -0.11245
```

**Fig. 3-12(*c*)  Output of GEP for input data shown in (*b*).**

## 3-6  SUMMARY

This chapter presents a detailed study of the analysis of networks that contain resistors and current sources.

A simple analysis procedure for some networks with a single source can be based upon the reduction of series and parallel combinations of resistors so that the original network is reduced to an equivalent network that contains one current source and one resistor. The equivalent network is easily analyzed. All the voltages and currents of the original network can then be found by going back through the steps that were used to obtain the equivalent network.

A more general analysis procedure arises from reduction of the linearly independent KCL and KVL equations, and branch relationships to a set of $n_n - 1$ equations that are expressed in terms of the node voltages. These equations, termed the *nodal equilibrium* equations, are easily written by inspection.

Two techniques that can be used for the solution of simultaneous algebraic equations are *Cramer's rule* and *Gaussian elimination*. Cramer's rule is useful for three or less equations; it becomes extremely cumbersome for more equations. *Gaussian elimination* is easily programmed for use on the digital computer. The program *GEP* implements the Gaussian elimination method and its use is described.

## PROBLEMS

**3.1**  Find equivalent resistances for the resistor combinations shown in Fig. P3.1.

**3.2**  Solve for the unknowns $v$ and $i$ in each of the networks of Fig. P3.2 by employing the relationships derived in the text for series and parallel combinations of resistors.

**3.3**  Try to solve for $i$ and $v$ in the network of Fig. P3.3 by the same method used in Prob. 3.2. This exercise demonstrates that solution by successively reducing series and parallel combinations of resistors may not always work. What method can be used to solve for $i$ and $v$? Use this technique to find $i$ and $v$.

**3.4**  Express the branch voltages of the network graph of Fig. P3.4 in terms of the node voltages. Demonstrate that this set of equations actually constitutes a linearly independent set of KVL loop equations.

**3.5**  For each of the networks shown in Fig. P3.5 write the nodal equilibrium equations by first writing all the independent KCL equations, all the independent KVL equations (i.e., express branch voltages in terms of node voltages) and branch relationships. Substitute the branch relationships into the KCL equations and then substitute in the KVL equations. The appropriate nodal equilibrium equations are then obtained by collecting terms.

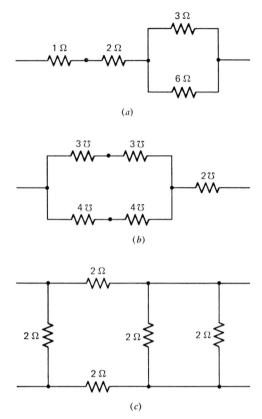

Fig. P3.1

**3.6**   Write the nodal equilibrium equations for the networks of Fig. P3.6 by inspection.

**3.7**   Use Cramer's rule to find solutions to the following sets of simultaneous algebraic equations.

(a)  $x_1 + 2x_2 + x_3 = 4$
$\quad$ $2x_1 + x_2 - x_3 = 5$
$\quad$ $2x_1 - 2x_2 + 4x_3 = -6$

(b) $2x_1 + x_2 + x_3 = 1$
$\quad$ $x_1 + 3x_2 + x_3 = 2$
$\quad$ $x_1 + x_2 + x_3 = 0$

(c) $7x_1 + 2x_2 + 4x_3 = 9$
$\quad$ $4x_1 + 6x_2 - x_3 = 23$
$\quad$ $2x_1 - 7x_2 + 7x_3 = 26$

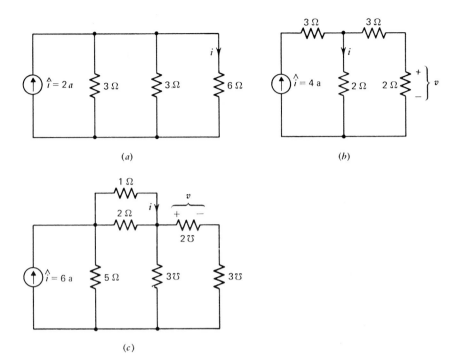

(a)

(b)

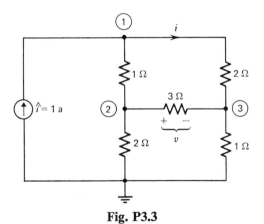

(c)

**Fig. P3.2**

**Fig. P3.3**

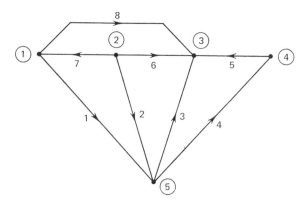

**Fig. P3.4**

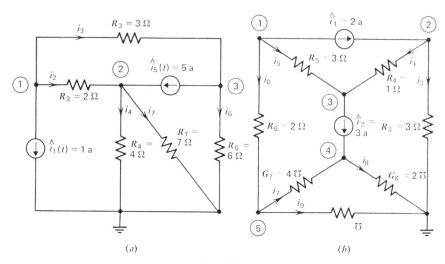

**Fig. P3.5**

**3.8**   Solve the sets of simultaneous equations of Prob. 3.7 using Gaussian
elimination. Compare the results obtained with those found in Prob.
3.7.

**3.9**   Determine which of the following sets of simultaneous equations are
linearly dependent by carrying out as many steps of Gaussian
elimination as possible. (Use pivoting if necessary.)

(a)   $x_1 \qquad + \; x_3 = 2$

$8x_1 + \; x_2 + 2x_3 = 0$

$-x_1 + \; x_2 \qquad = 1$

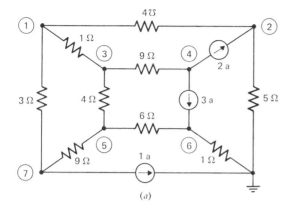

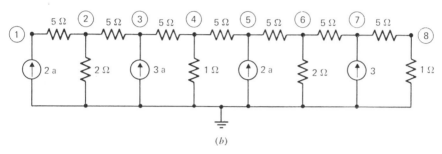

**Fig. P3.6**

(b)  $2x_1 - 3x_2 + x_3 = -4$

$x_1 + x_2 - 2x_3 = -7$

$-2x_1 + 2x_2 + 2x_3 = 18$

(c)  $8x_1 - 6x_2 + 2x_3 = 3$

$-6x_1 + 7x_2 - 4x_3 = 0$

$2x_1 - 4x_2 + 3x_3 = 4$

**3.10**  Determine the node voltages in the networks of Fig. P3.10 by writing nodal equilibrium equations and solving them using Cramer's rule.

**3.11**  Find the branch currents in resistance $R_3$ in each of the networks of Fig. P3.10.

**3.12**  What is the power delivered by each of the sources in Fig. P3.10.

**3.13**  Use the computer program GEP to find the solutions of the nodal equilibrium equations found in Probs. 3.5 and 3.6. Determine all branch voltages and currents from the node voltages and branch relationships.

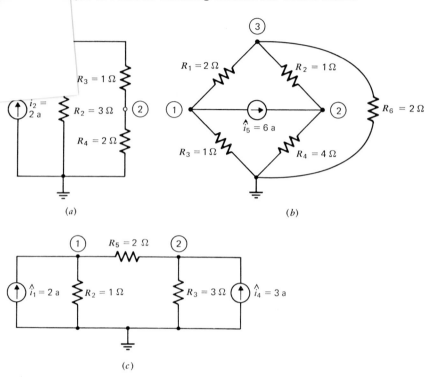

(a)  (b)

(c)

**Fig. P3.10**

**3.14** Determine the power absorbed by each branch of the networks of Fig. P3.5.

**3.15** Find all node voltages, branch voltages, and branch currents for networks of Fig. P3.15 by writing the nodal equilibrium equations by inspection and using GEP. Check to see if the KCL equations are satisfied for each node including the datum. Discuss your results.

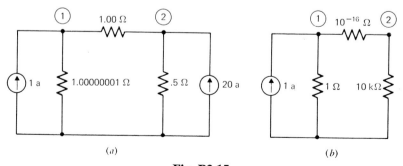

(a)  (b)

**Fig. P3.15**

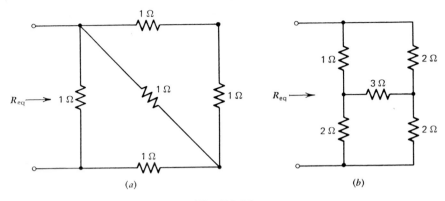

**Fig. P3.16**

**3.16**   Although it may not be possible to determine the equivalent resistance of a connection of resistors by reducing series and parallel
combinations, we can always find the equivalent resistance by exciting
the resistor combination by a unit valued current source and computing the voltage across the current source. The equivalent resistance is then equal to this voltage (why?). Using this technique
find the equivalent resistances for the resistor combinations shown
in Fig. P3.16.

**3.17**   Use GEP to determine the equivalent resistances of the resistor
combinations shown in Fig. P3.17.

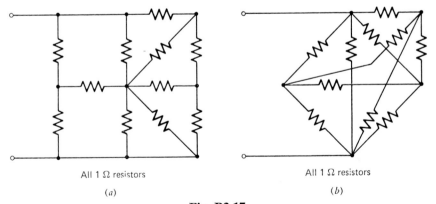

All 1 Ω resistors

(a)

All 1 Ω resistors

(b)

**Fig. P3.17**

## CHAPTER 4

# Voltage Sources

In this chapter we introduce another two-terminal element—the *independent voltage source*. We will find that writing nodal equilibrium equations for networks that contain voltage sources is no more difficult than writing nodal equilibrium equations for networks that contain only resistors and current sources.

## 4-1 DEFINITION OF THE VOLTAGE SOURCE ELEMENT

An *independent voltage source*, or simply *voltage source*, is a two-terminal element characterized by the branch relationship

$$v(t) = \hat{v}(t)$$

where $\hat{v}(t)$ is specified. The current through a voltage source is arbitrary. The symbol used to designate an independent voltage source is shown in Fig. 4-1.

## 4-2 NETWORKS CONTAINING GROUNDED VOLTAGE SOURCES

A voltage source that has one of its terminals connected to the datum node is said to be *grounded*. In this section we consider analysis of networks that

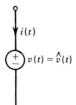

$i(t)$

$v(t) = \hat{v}(t)$

**Fig. 4-1  Symbol used to designate an independent voltage source.**

contain resistors, current sources, and grounded voltage sources. The analysis procedure is most easily illustrated with an example.

## Example 4.1

Consider the network shown in Fig. 4-2a. Recall that the first step in analysis of a network that is comprised of resistors and current sources, is to write a set of linearly independent KCL and KVL equations. Since these equations depend only on the network structure and not on the elements of the network, we can proceed in a similar manner when the network contains grounded voltage sources. The KCL node equations are obtained from the oriented graph of Fig. 4-2b:

$$i_1 - i_2 + i_3 + i_7 = 0$$
$$i_2 - i_4 - i_5 \qquad = 0$$

and

$$-i_3 + i_4 - i_6 - i_7 = 0$$

The KVL equations expressed in terms of the node voltages are:

$$v_1 = v_{①}$$
$$v_2 = v_{②} - v_{①}$$
$$v_3 = v_{①} - v_{③}$$
$$v_4 = v_{③} - v_{②}$$
$$v_5 = -v_{②}$$
$$v_6 = -v_{③}$$

and

$$v_7 = v_{①} - v_{③}$$

The only other linearly independent equations that can be written are the branch relationships

$$v_1 = R_1 i_1$$
$$v_2 = R_2 i_2$$
$$i_3 = \hat{\imath}_3$$
$$v_4 = R_4 i_4$$
$$v_5 = -\hat{v}_5$$
$$v_6 = R_6 i_6$$

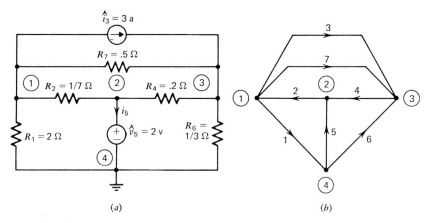

**Fig. 4-2** (*a*) **A network with a grounded voltage source used in Example 4.1; and** (*b*) **its graph.**

and

$$v_7 = R_7 i_7$$

Observe that the negative sign in the relationship for branch 5 results from the assumed orientation. Following the steps outlined in Chapter 3, we substitute the branch relationships into the KCL node equations:

$$\frac{1}{R_1} v_1 - \frac{1}{R_2} v_2 + \hat{i}_3 + \frac{1}{R_7} v_7 = 0$$

$$\frac{1}{R_2} v_2 - \frac{1}{R_4} v_4 - i_5 = 0$$

$$- \hat{i}_3 + \frac{1}{R_4} v_4 - \frac{1}{R_6} v_6 - \frac{1}{R_7} v_7 = 0$$

On substitution of the KVL equations, we have

$$\left( \frac{1}{R_1} + \frac{1}{R_2} + \frac{1}{R_7} \right) v_① - \frac{1}{R_2} v_② - \frac{1}{R_7} v_③ = -i_3$$

$$- \frac{1}{R_2} v_① + \left( \frac{1}{R_2} + \frac{1}{R_4} \right) v_② - \frac{1}{R_4} v_③ + i_5 = 0$$

$$- \frac{1}{R_7} v_① - \frac{1}{R_4} v_② + \left( \frac{1}{R_4} + \frac{1}{R_6} + \frac{1}{R_7} \right) v_③ = \hat{i}_3$$

There are *four* unknown quantities in the above three equations: the three independent node voltages $v_①$, $v_②$, and $v_③$ and the voltage source current $i_5$. The additional equation needed for the solution of the unknowns is the voltage source branch relationship that has not previously been used. The branch relationship is

$$v_5 = -\hat{v}_5$$

but from the KVL equations

$$v_5 = -v_\textcircled{2}$$

so that

$$v_\textcircled{2} = \hat{v}_5$$

In other words, the grounded voltage source constrains the voltage of the nondatum node to which it is connected to be equal to the value of the voltage source. Therefore, node voltage $v_\textcircled{2}$ is a known quantity, equal to $\hat{v}_5$. The nodal equilibrium equations become

$$\left(\frac{1}{R_1} + \frac{1}{R_2} + \frac{1}{R_7}\right)v_\textcircled{1} - \frac{1}{R_7}v_\textcircled{3} = -\hat{i}_3 + \frac{1}{R_2}\hat{v}_5$$

$$-\frac{1}{R_2}v_\textcircled{1} - \frac{1}{R_4}v_\textcircled{3} + i_5 = -\left(\frac{1}{R_2} + \frac{1}{R_4}\right)\hat{v}_5$$

$$-\frac{1}{R_7}v_\textcircled{1} + \left(\frac{1}{R_4} + \frac{1}{R_6} + \frac{1}{R_7}\right)v_\textcircled{3} = \hat{i}_3 + \frac{1}{R_4}\hat{v}_5$$

We now have three equations in three unknowns. The node voltages $v_\textcircled{1}$ and $v_\textcircled{3}$ and the voltage source current $i_5$. In essence, we have "traded" an unknown node voltage for an unknown voltage source current. Substitution of element values into the above equations yields

$$9.5v_\textcircled{1} - 2v_\textcircled{3} = 11$$

$$-7v_\textcircled{1} - 5v_\textcircled{3} + i_5 = -24$$

$$-2v_\textcircled{1} + 10v_\textcircled{3} = 13$$

the solution of which is $v_\textcircled{1} = 1.495$ v, $v_\textcircled{3} = 1.599$ v, and $i_5 = -5.544$ a.

If we are unconcerned about the voltage source current, the first and third equations,

$$\left(\frac{1}{R_1} + \frac{1}{R_2} + \frac{1}{R_7}\right)v_\textcircled{1} - \frac{1}{R_7}v_\textcircled{3} = -\hat{i}_3 + \frac{1}{R_2}\hat{v}_5$$

and

$$-\frac{1}{R_7}v_\textcircled{1} + \left(\frac{1}{R_4} + \frac{1}{R_6} + \frac{1}{R_7}\right)v_\textcircled{3} = \hat{i}_3 + \frac{1}{R_4}\hat{v}_5$$

can be solved simultaneously to yield solutions for $v_\textcircled{1}$ and $v_\textcircled{3}$. $\square$

From the above example, we see that the introduction of grounded voltage sources does not alter our method for obtaining nodal equilibrium equations. Again it is easy to determine a technique for writing nodal equilibrium equations by inspection. From Kirchhoff's current law, we know that the sum of the currents leaving a node through the resistive elements must equal the sum of the currents entering the node through current sources and voltage sources. Therefore, the KCL node equations can be written in terms of resistance or conductance values (known coefficients), the node voltages (unknowns), the current source values (known forcing functions), and the currents through voltage sources (unknowns). These equations can be written in the form

$$y_{11}v_① + y_{12}v_② + \cdots y_{1m}v_{ⓜ} = i_①$$

(4.1)
$$y_{21}v_① + y_{22}v_② + \cdots y_{2m}v_{ⓜ} = i_②$$

$$\cdots\cdots\cdots\cdots\cdots\cdots\cdots\cdots\cdots$$

$$y_{m1}v_① + y_{m2}v_② + \cdots y_{mm}v_{ⓜ} = i_{ⓜ}$$

where

$$m = n_n - 1$$

$y_{kk}$ = sum of all conductances connected between node ⓚ and all other nodes

$y_{jk}$ = negative sum of all conductances directly connected between nodes ⓙ and ⓚ

and

$i_{ⓚ}$ = sum of all current source currents entering node ⓚ plus the sum of all *unknown* voltage source currents entering node ⓚ

In addition to these equations, there is one equation for each grounded voltage source that equates a node voltage to a voltage source voltage. According to the method described above, the value of these constrained node voltages should be substituted into the nodal equilibrium equations. Finally, in order to prepare the simultaneous equations for computer solution, all unknowns should be moved to the left-hand side and all knowns moved to the right-hand side. An example will illustrate the steps involved.

**Example 4.2**

The nodal equilibrium equations for the network of Fig. 4-3 are

$$\frac{1}{R_2}v_① - \frac{1}{R_2}v_② \qquad\qquad\qquad\qquad = -i_1 + \hat{i}_3$$

$$-\frac{1}{R_2}v_① + \left(\frac{1}{R_2} + G_4 + \frac{1}{R_5} + G_7\right)v_② - \frac{1}{R_5}v_③ - G_7v_④ = 0$$

$$-\frac{1}{R_5}v_② + \left(\frac{1}{R_5} + G_6 + \frac{1}{R_8}\right)v_③ - \frac{1}{R_8}v_④ \qquad\qquad = -\hat{i}_3$$

$$-G_7v_② - \frac{1}{R_8}v_③ + \left(\frac{1}{R_8} + G_7\right)v_④ \qquad\qquad = i_9$$

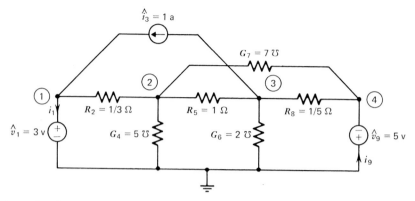

**Fig. 4-3** A network with grounded voltage sources used in Example 4.2 to demonstrate how nodal equilibrium equations can be written by inspection.

The node voltages constrained by grounded voltage sources are

$$v_① = \hat{v}_1$$

and

$$v_④ = -\hat{v}_9$$

Substituting these expressions into the nodal equilibrium equations and rearranging yields

$$-\frac{1}{R_2} v_② + i_1 \qquad\qquad = -\frac{1}{R_2}\hat{v}_1 + \hat{i}_3$$

$$\left(\frac{1}{R_2} + G_4 + \frac{1}{R_5} + G_7\right)v_② - \frac{1}{R_5} v_③ = \frac{1}{R_2}\hat{v}_1 - G_7\hat{v}_9$$

$$-\frac{1}{R_5} v_② + \left(\frac{1}{R_5} + \frac{1}{R_8} + G_6\right)v_③ \qquad = -\hat{i}_3 - \frac{1}{R_8}\hat{v}_9$$

$$-G_7 v_② - \frac{1}{R_8} v_③ - i_9 \qquad = \left(\frac{1}{R_8} + G_7\right)\hat{v}_9$$

The four equations can be solved simultaneously for the unknowns $v_②$, $v_③$, $i_1$, and $i_4$. Upon substitution of the element values, these equations become

$$-3v_② \qquad\qquad + i_1 \qquad = -8$$

$$16v_② - v_③ \qquad\qquad = -26$$

$$-v_② + 8v_③ \qquad\qquad = -46$$

$$-7v_② - 5v_③ \qquad - i_9 = 60$$

The solution of these equations is $v_② = -2.00$, $v_③ = -6.00$, $i_1 = -14.00$, and $i_9 = -16.00$.

Notice that the solution for the node voltages can be undertaken independently of the voltage source currents. In particular, if we were only interested in finding solutions for $v_②$ and $v_③$ only the equations

$$\left(\frac{1}{R_2} + G_4 + \frac{1}{R_5} + G_7\right)v_② - \frac{1}{R_5}v_③ = \frac{1}{R_2}\hat{v}_1 - G_7\hat{v}_9$$

and

$$-\frac{1}{R_5}v_② + \left(\frac{1}{R_5} + \frac{1}{R_8} + G_6\right)v_③ = -\hat{i}_3 - \frac{1}{R_8}\hat{v}_9$$

need be solved.   □

## 4-3  NETWORKS CONTAINING FLOATING VOLTAGE SOURCES

Although most voltage sources that appear in physical networks are grounded, it is useful to consider the analysis of networks in which voltage sources are connected between two nondatum nodes. Such sources will be referred to as *floating* voltage sources.

The analysis of networks that contain arbitrary connections of resistors, independent current sources, and independent voltage sources proceeds along in the same manner as described earlier. Again, the exact steps are probably best illustrated with an example.

### Example 4.3

Consider the network of Fig. 4-4. The nodal equilibrium equation for nodes ① and ⑤ are easily written by inspection

$$node\ ① \qquad \left(\frac{1}{R_1} + \frac{1}{R_2}\right)v_① - \frac{1}{R_2}v_② \qquad\qquad = -\hat{i}_3$$

$$node\ ⑤ \qquad -\frac{1}{R_4}v_② - \frac{1}{R_6}v_③ + \left(\frac{1}{R_4} + \frac{1}{R_6} + \frac{1}{R_9}\right)v_⑤ = 0$$

The grounded voltage source $\hat{v}_8$ establishes node voltage $v_④$ and its unknown current $i_8$ leaves node ④. Therefore, the nodal equilibrium equation for node ④ is

$$node\ ④ \qquad -\frac{1}{R_7}v_③ + \frac{1}{R_7}v_④ = -i_8$$

Temporarily we will allow $v_④$ to remain an unknown bearing in mind that

$$v_④ = \hat{v}_8$$

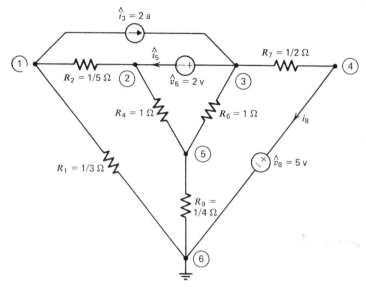

**Fig. 4-4   A network with a nongrounded voltage source used in Example 4.3.**

The nodal equilibrium equation for node ② must account for the current flowing through resistors $R_2$ and $R_4$ and through voltage source $\hat{v}_5$:

$$\frac{v_②-v_①}{R_2}+\frac{v_②-v_⑤}{R_4}-i_5=0$$

or

*node* ②     $-\dfrac{1}{R_2}v_① + \left(\dfrac{1}{R_2}+\dfrac{1}{R_4}\right)v_② - \dfrac{1}{R_4}v_⑤ = i_5$

Similarly, for node ③

*node* ③     $\left(\dfrac{1}{R_6}+\dfrac{1}{R_7}\right)v_③ - \dfrac{1}{R_7}v_④ - \dfrac{1}{R_6}v_⑤ = \hat{i}_3 - i_5$

These two expressions introduce an additional unknown, $i_5$, the current through voltage source $\hat{v}_5$. Another equation is needed. The additional equation is the voltage source branch relationship for $\hat{v}_5$ that can be written as

$$v_③ - v_② = \hat{v}_5$$

Collecting these equations, we have the nodal equilibrium equations

*node* ①     $\left(\dfrac{1}{R_1}+\dfrac{1}{R_2}\right)v_① - \dfrac{1}{R_2}v_② = -\hat{i}_3$

*node* ②     $-\dfrac{1}{R_2}v_① + \left(\dfrac{1}{R_2}+\dfrac{1}{R_4}\right)v_② - \dfrac{1}{R_4}v_⑤ = i_5$

(4.2) $\quad$ *node* ③ $\quad \left(\dfrac{1}{R_6} + \dfrac{1}{R_7}\right) v_③ - \dfrac{1}{R_7} v_④ - \dfrac{1}{R_6} v_⑤ = \hat{\imath}_3 - i_5$

$\quad$ *node* ④ $\quad - \dfrac{1}{R_7} v_③ + \dfrac{1}{R_7} v_④ = - i_8$

$\quad$ *node* ⑤ $\quad - \dfrac{1}{R_4} v_② - \dfrac{1}{R_6} v_③ + \left(\dfrac{1}{R_4} + \dfrac{1}{R_6} + \dfrac{1}{R_9}\right) v_⑤ = 0$

and the voltage source branch relationships

$$v_④ = \hat{v}_4$$

$$v_③ - v_② = \hat{v}_5$$

for a total of 7 equations in terms of the 7 unknowns: node voltages $v_①$, $v_②$, $v_③$, $v_④$, and $v_⑤$ and the voltages–source currents $i_8$ and $i_5$. Upon rearrangement of terms so that all unknowns appear on the left hand side of the equations and all knowns appear on the right hand side of the equations, we have the following set of simultaneous algebraic equations

$$\left(\frac{1}{R_1} + \frac{1}{R_2}\right) v_① - \frac{1}{R_2} v_② = - \hat{\imath}_3$$

$$- \frac{1}{R_2} v_① + \left(\frac{1}{R_2} + \frac{1}{R_4}\right) v_② - \frac{1}{R_4} v_⑤ - i_5 = 0$$

$$\left(\frac{1}{R_6} + \frac{1}{R_7}\right) v_③ - \frac{1}{R_7} v_④ - \frac{1}{R_6} v_⑤ + i_5 = \hat{\imath}_3$$

$$- \frac{1}{R_7} v_③ + \frac{1}{R_7} v_④ + i_8 = 0$$

$$- \frac{1}{R_4} v_② - \frac{1}{R_6} v_③ + \left(\frac{1}{R_4} + \frac{1}{R_6} + \frac{1}{R_9}\right) v_⑤ = 0$$

$$v_④ = \hat{v}_4$$

$$- v_② + v_③ = \hat{v}_5$$

We can submit these equations, as is, to our simultaneous equation solver computer program GEP for solution. The first five solution values printed would correspond to the node voltages $v_① - v_⑤$ and the last two solution values printed would correspond to the voltage source currents $i_5$ and $i_8$. Alternatively, we may reduce equations (4.2) to a set of 5 equations in terms of 5 unknowns, by elimination of $v_④$ which is fixed by the branch relationship

$$v_④ = \hat{v}_8$$

and $v_③$, which is dependent on $v_②$ through the relationship

$$v_③ = v_② + \hat{v}_5$$

Substitution of these constraints into the node equations yields

$$\left(\frac{1}{R_1}+\frac{1}{R_2}\right)v_{①}-\frac{1}{R_2}v_{②} \qquad\qquad =-\hat{\imath}_3$$

$$-\frac{1}{R_2}v_{①}+\left(\frac{1}{R_2}+\frac{1}{R_4}\right)v_{②}+\frac{1}{R_4}v_{⑤} \qquad = i_5$$

$$\left(\frac{1}{R_6}+\frac{1}{R_7}\right)(v_{②}+\hat{v}_5)-\frac{1}{R_7}\hat{v}_8-\frac{1}{R_6}v_{⑤}=\hat{\imath}_3-i_5$$

$$-\frac{1}{R_7}(v_{②}+\hat{v}_5)+\frac{1}{R_7}\hat{v}_8 \qquad\qquad =-i_8$$

and

$$-\frac{1}{R_4}v_{②}-\frac{1}{R_6}(v_{②}+\hat{v}_5)+\left(\frac{1}{R_4}+\frac{1}{R_6}+\frac{1}{R_9}\right)v_{⑤}=0$$

We can rewrite these equations so that all unknown quantities appear on the left-hand side and all known quantities appear on the right-hand side:

$$\left(\frac{1}{R_1}+\frac{1}{R_2}\right)v_{①}-\frac{1}{R_2}v_{②} \qquad\qquad =-\hat{\imath}_3$$

$$-\frac{1}{R_1}v_{①}\qquad +\left(\frac{1}{R_2}+\frac{1}{R_4}\right)v_{②}-\frac{1}{R_4}v_{⑤}-i_5=0$$

$$\left(\frac{1}{R_6}+\frac{1}{R_7}\right)v_{②}-\frac{1}{R_6}v_{⑤}+i_5=\hat{\imath}_3+\frac{1}{R_7}\hat{v}_8-\left(\frac{1}{R_6}+\frac{1}{R_7}\right)\hat{v}_5$$

$$-\frac{1}{R_7}v_{②}\qquad\qquad +i_8=-\frac{1}{R_7}\hat{v}_8+\frac{1}{R_7}\hat{v}_5$$

and

$$\left(-\frac{1}{R_4}-\frac{1}{R_6}\right)v_{②}+\left(\frac{1}{R_4}+\frac{1}{R_6}+\frac{1}{R_9}\right)v_{⑤}=\frac{1}{R_6}\hat{v}_5$$

Substitution of element values yields

$$8v_{①}-5v_{②} \qquad\qquad =-2$$

$$-5v_{①}+6v_{②}-\ v_{⑤}-i_5 \qquad =0$$

$$3v_{②}-\ v_{⑤}+i_5\qquad =6$$

$$-\ 2v_{②}\qquad\qquad +i_8=-6$$

$$-\ 2v_{②}+6v_{⑤}\qquad =2$$

that has the solution $v_1=.4$ v, $v_2=1.04$ v, $v_5=.68$ v, $i_5=3.56$ a, and $i_8=-3.92$ a. □

As might be expected, nodal equilibrium equations for networks comprised of resistors, current sources, and floating voltage sources, can be written by inspection. These equations take on the familiar form

$$y_{11}v_{①} + y_{12}v_{②} + \cdots + y_{1m}v_{⑩} = i_{①}$$

$$y_{21}v_{①} + y_{22}v_{②} + \cdots + y_{2m}v_{⑩} = i_{②}$$

(4.3)
$$\cdots\cdots\cdots\cdots\cdots\cdots\cdots\cdots$$

$$y_{m1}v_{①} + y_{m2}v_{②} + \cdots + y_{mm}v_{⑩} = i_{⑩}$$

where once again $m = n_n - 1$;

$y_{kk}$ = sum of all admittances connected between node $(k)$ and all other nodes

$y_{jk}$ = negative sum of all admittances directly connected between
$_{j \neq k}$     nodes $(j)$ and $(k)$

and

$i_{(k)}$ = sum of all known current source currents and unknown voltage source currents entering node $(k)$

In general, if there are $n_v$ voltage sources the $n_n - 1$ equations will be in terms of $n_n - 1 + n_v$ unknowns: the $n_n - 1$ node voltages and the $n_v$ voltage source currents. The $n_v$ additional equations needed for solution of the unknowns are obtained from the voltage source branch relationships written in terms of the node voltages.

The single node of Fig. 4-5, which is part of some larger network, will serve as an illustration of writing node equations by inspection. Summing all currents leaving node $(5)$, we have

$$\frac{1}{R_{19}} v_{⑤} - i_2 + \frac{1}{R_1}(v_{⑤} - v_{⑨}) + \hat{\imath}_{12} + \frac{1}{R_6}(v_{⑤} - v_{④})$$

$$+ i_4 - \hat{\imath}_7 = 0$$

or

$$-\frac{1}{R_6} v_{④} + \left(\frac{1}{R_{19}} + \frac{1}{R_1} + \frac{1}{R_6}\right) v_{⑤} - \frac{1}{R_1} v_{⑨} = -i_4 + i_2 + \overset{\circ}{\imath}_7 - \hat{\imath}_{12}$$

which is of the form suggested before. Note that the second equation could have been written directly.

When grounded voltage sources are encountered in a network, it will be our custom to substitute into the nodal equilibrium equations the voltage source value for the node voltage of the node connected to the ungrounded end of the voltage source. On the other hand, when floating voltage sources are encountered in a network we will append to the nodal equilibrium equations the constraints imposed on the node voltages by the voltage sources.

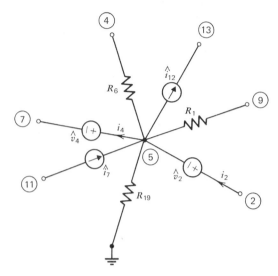

**Fig. 4-5** A node that is part of some larger network used to illustrate how node equations can be written by inspection.

## 4-4* SOLUTIONS OF NETWORKS USING A CURRENT BASIS

From what has preceded, we have seen that nodal analysis of a network that contains voltage sources is somewhat more complicated than the nodal analysis of a network that contains only current sources. In particular, if voltage sources are present, an unknown node voltage is "traded" for an unknown voltage source current, and additional manipulation is required in order to prepare the network equations so that they may be solved by GEP.

In this section we consider the analysis of networks using a current basis rather than a voltage basis. Recall from Chapter 2 that for an $n_n$ node, $n_b$ branch network there are $n_b - n_n + 1$ independent or basis currents. Two possible choices for this basis are the link currents and the mesh currents. Since meshes only exist for planar networks, we choose the link current basis. (Use of the mesh current basis is considered in the problems.)

The procedure for obtaining a set of equations solely in terms of the link currents is very similar to the procedure used to obtain the nodal equilibrium equations. Given a network and its oriented graph, the first step is to choose a tree. The tree should be chosen so that all current sources of the network appear as links and all voltage sources appear as tree branches. The reasons behind placing current sources in links and voltage sources in tree branches are considered in Prob. 4.11. Once a tree has been selected, the $n_n - 1$ linearly independent KCL fundamental cutset equations, and $n_b - n_n + 1$ linearly independent KVL fundamental equations are written. Also, the $n_b$

branch relationships are written. We now have $2n_b$ equations in terms of $2n_b$ unknowns: the $n_b$ branch voltages and the $n_b$ branch currents.

These $2n_b$ equations can be reduced to a set of $n_b - n_n + 1$ equations that are in terms of the $n_b - n_n + 1$ link currents. First, the branch relations, which express branch voltages in terms of branch currents are substituted into the KVL fundamental loop equations. Then the KCL fundamental cutset equations, which express all branch currents in terms of link currents, are substituted in. The desired set of equations result.

### Example 4.4

Consider the network of Fig. 4-6a and its oriented graph of Fig. 4-6b. The solid lines indicate a suitable tree. The fundamental cutset equations are

$$
\begin{array}{llll}
\textit{branch 1} & i_1 - i_2 & & = 0 \\
\textit{branch 4} & i_2 + i_3 + i_4 & & = 0 \\
\textit{branch 5} & -i_2 & +i_5 \quad -i_7 & = 0 \\
\textit{branch 6} & i_3 & +i_6 - i_7 & = 0 \\
\textit{branch 8} & -i_2 & -i_7 + i_8 = 0
\end{array}
$$

The fundamental loop equations are

$$
\begin{array}{llll}
\textit{link 2} & v_1 + v_2 & -v_4 + v_5 & +v_8 = 0 \\
\textit{link 3} & & v_3 - v_4 \quad -v_6 & = 0 \\
\textit{link 7} & & v_5 + v_6 + v_7 + v_8 = 0
\end{array}
$$

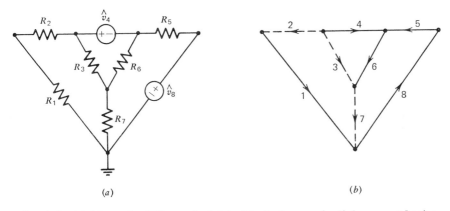

(a)                                             (b)

Fig. 4-6   (a) Network of Example 4.4 to illustrate use of a link current basis; (b) its graph.

The branch relationships are

$$branch\ 1 \quad v_1 = R_1 i_1$$
$$branch\ 2 \quad v_2 = R_2 i_2$$
$$branch\ 3 \quad v_3 = R_3 i_3$$
$$branch\ 4 \quad v_4 = \hat{v}_4$$
$$branch\ 5 \quad v_5 = R_5 i_5$$
$$branch\ 6 \quad v_6 = R_6 i_6$$
$$branch\ 7 \quad v_7 = R_7 i_7$$
$$branch\ 8 \quad v_8 = -\hat{v}_8$$

Observe that the minus sign that appears in the branch relation for branch 8 occurs because the assumed reference direction in the graph yields a voltage that is opposite to the actual polarity of the voltage source $\hat{v}_8$.

Substitution of the branch relations into the fundamental loop equations yields

$$R_1 i_1 + R_2 i_2 \qquad - \hat{v}_4 + R_5 i_5 \qquad\qquad - \hat{v}_8 = 0$$
(4.4)
$$R_3 i_3 - \hat{v}_4 \qquad - R_6 i_6 \qquad\qquad = 0$$
$$R_5 i_5 + R_6 i_6 + R_7 i_7 - \hat{v}_8 = 0$$

The link currents are $i_2$, $i_3$, and $i_7$. All other branch currents must be expressible in terms of these currents. The expressions we desire are obtained by rewriting the fundamental cutset equations:

$$i_1 = i_2$$
$$i_4 = -i_2 - i_3$$
$$i_5 = i_2 + i_7$$
$$i_6 = -i_3 + i_7$$

and

$$i_8 = i_2 + i_7$$

Substituting these equations into (4.4) and rearranging terms we have the following relations

$$(R_1 + R_2 + R_5)i_2 \qquad\qquad + R_5 i_7 = \hat{v}_4 + \hat{v}_8$$
$$(R_3 + R_6)i_3 \qquad\qquad - R_6 i_7 = \hat{v}_4$$
$$R_5 i_2 - \qquad R_6 i_3 + (R_5 + R_6 + R_7)i_7 = \hat{v}_8$$

These three equations can be solved for the three link currents. □

If fundamental loop analysis is used for a network that contains current sources, some of the unknown link currents become known current source currents while an equivalent number of current source voltages appear as unknowns in the equations.

### Example 4. 5

The network of Fig. 4-7a, which contains both a voltage source and a current source, was analyzed using a node voltage basis in Example 4.1. We wish to reanalyze it using a link current basis. The oriented graph is shown in Fig. 4-7b, where the solid lines indicate a suitable tree. The fundamental cutset equations are

$$\begin{array}{llll}
\text{branch 4} & i_1 - i_2 & + i_4 & - i_6 & = 0 \\
\text{branch 5} & -i_1 & & + i_5 + i_6 & = 0 \\
\text{branch 7} & i_1 - i_2 + i_3 & & + i_7 = 0
\end{array}$$

The fundamental loop equations are

$$\begin{array}{lllll}
\text{link 1} & v_1 & & - v_4 + v_5 & - v_7 = 0 \\
\text{link 2} & & v_2 & + v_4 & + v_7 = 0 \\
\text{link 3} & & v_3 & & - v_7 = 0 \\
\text{link 6} & & & v_4 - v_5 + v_6 & = 0
\end{array}$$

The branch relationships are

$$v_1 = R_1 i_1$$
$$v_2 = R_2 i_2$$
$$i_3 = \hat{i}_3$$
$$v_4 = R_4 i_4$$
$$v_5 = -\hat{v}_5$$
$$v_6 = R_6 i_6$$
$$v_7 = R_7 i_7$$

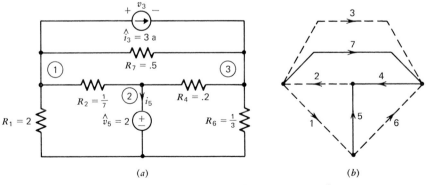

(a)                                         (b)

**Fig. 4-7   A network that contains both a voltage source and a current source used in Example 4.5.**

Substitution of the branch relationships into the fundamental loop equations yields

$$R_1 i_1 \qquad\qquad - R_4 i_4 - \hat{v}_5 \qquad\qquad - R_7 i_7 = 0$$

(4.5)
$$\qquad\qquad R_2 i_2 \qquad + R_4 i_4 \qquad\qquad + R_7 i_7 = 0$$

$$\qquad\qquad\qquad v_3 \qquad\qquad\qquad - R_7 i_7 = 0$$

and

$$R_4 i_4 + \hat{v}_5 + R_6 i_6 \qquad\qquad = 0$$

Observe that the current source voltage $v_3$ has not been removed from these equations, nor has the branch relation for the current source been used. Rewriting the fundamental cutset equations so as to express all tree branch currents in terms of the link currents $i_1$, $i_2$, $i_3$, and $i_6$ yields

$$i_4 = -i_1 + i_2 + i_6$$

$$i_5 = \quad i_1 - i_6$$

and

$$i_7 = -i_1 + i_2 - i_3$$

Substituting these expressions into (4.5) and rearranging terms yields

$$(R_1 + R_4 + R_7)i_1 \qquad - (R_4 + R_7)i_2 + R_7 i_3 \qquad - R_4 i_6 \qquad\qquad = \hat{v}_5$$

$$- (R_4 + R_7)i_1 + (R_2 + R_4 + R_7)i_2 - R_7 i_3 \qquad + R_4 i_6 \qquad\qquad = 0$$

$$R_7 i_1 \qquad\qquad - R_7 i_2 + R_7 i_3 \qquad + v_3 \qquad\qquad = 0$$

$$- R_4 i_1 \qquad\qquad + R_4 i_2 \qquad\qquad + (R_4 + R_6)i_6 \qquad = -\hat{v}_5$$

We have four equations in five unknowns: $i_1$, $i_2$, $i_3$, $i_6$, and $v_3$. But we have not used the current source branch relationship

$$i_3 = \hat{i}_3$$

On substitution of this equation, we have four equations in four unknowns:

$$(R_1 + R_4 + R_7)i_1 \qquad - (R_4 + R_7)i_2 \qquad - R_4 i_6 \qquad = \hat{v}_5 - R_7 \hat{i}_3$$

$$- (R_4 + R_7)i_1 + (R_2 + R_4 + R_7)i_2 \qquad + R_4 i_6 \qquad = R_7 \hat{i}_3$$

$$R_7 i_1 \qquad\qquad - R_7 i_2 \qquad\qquad + v_3 = -R_7 \hat{i}_3$$

$$- R_4 i_1 \qquad\qquad R_4 i_2 + (R_4 + R_6)i_6 \qquad = -\hat{v}_5$$

or

$$2.7 i_1 - .7 i_2 - .2 i_6 \qquad = .5$$

$$-.7 i_1 + .843 i_2 + .2 i_6 \qquad = 1.5$$

$$.5 i_1 - .5 i_2 \qquad\qquad + v_3 = -1.5$$

$$-.2 i_1 + .2 i_2 + .533 i_6 \qquad = -2$$

The solution of these equations is easily obtained using GEP: $i_1 = .747$, $i_2 = 3.54$, $i_6 = -4.8$, and $v_3 = -.104$. The reader should compute the branch voltages and currents and compare the results with those obtained in Example 4.1.   □

Throughout the remainder of the text we will generally confine ourselves to the node based analysis. Analysis using other basis will be considered in the problems only. The main reason for this decision is that fundamental loop analysis does not really buy us anything. The extra manipulation we ran into when using nodal analysis on networks that contain voltage sources also shows up when loop analysis is used on networks that contain current sources. In fact, fundamental loop analysis is more time consuming for two reasons: first a tree must be chosen, and on a computer this may require considerable time; second, the equations that result from fundamental loop analysis are not easily written by inspection, so that additional manipulation must be performed.

## 4-5  SUMMARY

This chapter introduces the *independent voltage source*. The voltage source is characterized by the branch relationship

$$v(t) = \hat{v}(t)$$

where $\hat{v}(t)$ is specified; the current through a voltage source is arbitrary.

The *grounded voltage* source is a voltage source with one terminal connected to the datum node. Although most voltage sources are grounded, a network may contain *floating voltage sources*.

Nodal analysis of networks that contain voltage sources proceeds along the same lines as the analysis of networks that contain only current sources and resistors. However, an additional unknown is introduced for each voltage source in the network. The additional equations needed for solution are the voltage source branch relationships expressed in terms of the node voltages. As before, the nodal equilibrium equations are easily written by inspection.

A current based analysis scheme was presented. This scheme was shown to require an increased amount of manipulation. Moreover, the equations that result are not easily written by inspection.

## PROBLEMS

**4.1**     Each of the networks of Fig. P4.1 contains grounded voltage sources. For each network, write all the independent KCL node equations,

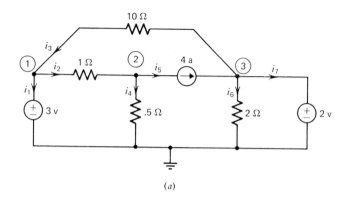

(a)

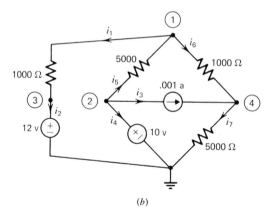

(b)

**Fig. P4.1**

all the independent KVL equations (i.e., express all branch voltages in terms of node voltages) and the branch relationships. Manipulate these equations to obtain a set of nodal equilibrium equations. The unknowns should be all the node voltages, excluding the voltages at nodes connected to the voltage sources, and the currents through the voltage sources.

**4.2**    Write the nodal equilibrium equations for the networks of Fig. P4.1 by inspection. First write a set of equations similar to (4.1) for these networks. These $n_n - 1$ equations are in terms of the $n_n - 1$ node voltages and $n_v$ unknown voltage source currents ($n_v$ is the number of voltage sources). Then write the $n_v$ voltage source imposed constraints on the node voltages. Substitute these constraints into the set of equations previously written and rearrange terms so that all the unknowns appear on the left and all the knowns appear on the right. Compare the results with those obtained in Prob. 4.1.

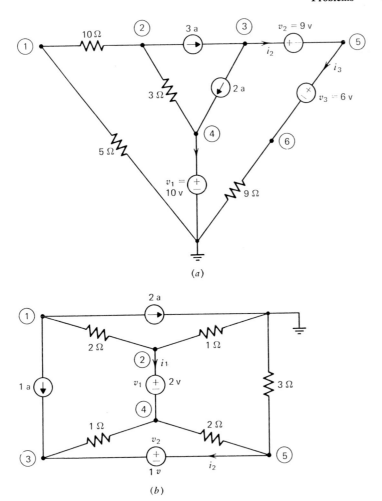

Fig. P4.3

**4.3** Write the set of nodal equilibrium equations by inspection for the networks of Fig. P4.3. Append to this set the constraints imposed on the node voltages by the voltage sources. Use GEP to solve for the unknowns.

**4.4** Consider again Prob. 4.3. By substituting the voltage source constraints into the nodal equilibrium equations, derive a set of $n_n - 1$ equations in terms of $n_n - 1$ unknowns. Use GEP to solve these equations and compare the results with those obtained in Prob. 4.3.

**4.5** Find the voltages or currents indicated in the networks of Fig. P4.5 by writing node equations and solving them using Cramers rule.

**4.6** What is the power delivered by the voltage source in Fig. P4.5a?

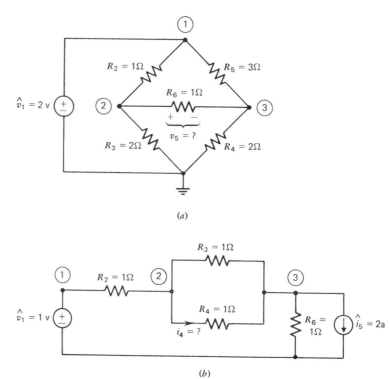

(a)

(b)

**Fig. P4.5**

**4.7** Find a relationship between $v_5$ and $\hat{v}_1$ in the network of Fig. P4.7. (Hint—reduce the series and parallel resistor combinations.)

**4.8** Find the value of $R$ in the network of Fig. P4.8a so that the current in the voltage source is the same as the current in the voltage source in the network of Fig. 4.8b.

**4.9** When nodal equilibrium equations are written for net works that contain voltage sources some node voltages are no longer considered to be unknowns while the currents through voltage sources are considered to be unknowns. If instead of reducing the KCL equations, KVL equations, and branch relationships to a set of equations in terms of node voltages we reduced them to a set of equations in terms of link currents, then the voltages across current sources replace link currents as unknowns. Consider the networks of Fig. P4.9. For each choose a tree such that all voltage sources appear in tree branches and all current sources appear in links (this step reduces equation manipulation but is not necessary for solution). Write the fundamental cutset and loop equations associated with the tree chosen. Also write the branch relationships. Reduce these sets of equations

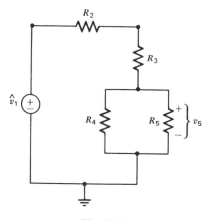

**Fig. P4.7**

by first substituting the branch relationships into the KVL fundamental loop equations and then substituting in the KCL fundamental cutset equations. The result should be a set of equations in
terms of link currents (as a basis set) and current source voltages.
Use the current source branch relationship to eliminate one of the
unknown link currents so that there are $n_b - n_n + 1$ equations in
terms of $n_b - n_n + 1$ unknowns.

**4.10**    The mesh currents of a network can also be used as a basis. For each
of the networks of Fig. P4.9 write the KVL mesh equations, KCL
equations by expressing each branch current in terms of mesh currents,
and the branch relationships. Reduce these equations to a set of
$n_b - n_n + 1$ equations that are written in terms of $n_b - n_n + 1$ mesh
currents and $n_I$ current source voltages. Use the current source
branch relationship to eliminate an unknown mesh current from these
equations.

**4.11**    We have indicated that less manipulation is required for a current
based analysis if all current sources were placed in links. To see this,

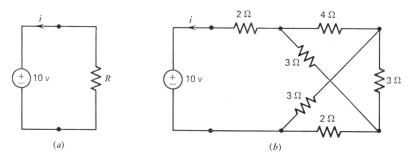

(a)                                (b)

**Fig. P4.8**

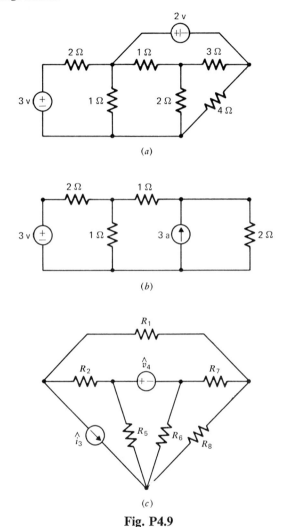

(a)

(b)

(c)

**Fig. P4.9**

choose a tree for the network of Fig. P4.9b so that the current source is a tree branch. Then proceed as indicated in Prob. 4.9 to derive a set of fundamental loop equilibrium equations.

**4.12**    Consider the network and two trees of Fig. P4.12. Write the fundamental cutset equations associated with the first tree. Write the fundamental loop equations associated with the second tree. Write the branch relationships. Reduce these equations to a set involving the link currents of the second tree. This exercise indicates that any linearly independent set of KCL equations can be used with any linearly independent set of KVL equations to derive a set of fundamental loop equilibrium equations.

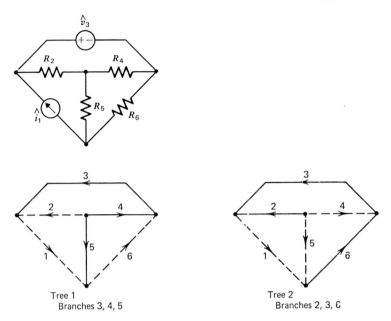

**Fig. P4.12**

**4.13**    An alternative to using the node voltages as a basis set of voltages is to use the tree branch voltages. Consider again the networks of Fig. P4.9 and the fundamental loop equations, fundamental cutset equations, and branch relations written in Prob. 4.9. For each network, substitute the branch relationships into the fundamental cutset equations. Then rewrite the fundamental loop equations so as to express all link voltages in terms of tree branch voltages and substitute these equations into the equations written above. The result is a set of $n_n - 1$ equations expressed in terms of the $n_n - 1$ tree branch voltages and the $n_v$ voltage source currents. Use the voltage source branch relationships to reduce the number of unknowns to $n_n - 1$.

**4.14**    Draw a network that has the following nodal equilibrium equations.

$$5v_① - 4v_② \qquad\qquad = -3$$
$$-4v_① + 17v_② - 8v_④ - i_6 = 3$$
$$17v_③ - 10v_④ + i_6 \qquad = 0$$
$$-8v_② - 10v_③ + 27v_④ \qquad = -12$$
$$v_② - v_④ \qquad\qquad = 6$$

## CHAPTER 5

# Some Useful Network Principles and Theorems

There are several easily proved network principles and theorems that can be used profitably to simplify network analysis. These theorems are *not* limited to the small class of resistive networks that we have been studying. It is worthwhile though to study some of these theorems and their use for resistive network analysis. Since much of the mathematical manipulation and clutter associated with more complex networks is avoided, the implications of what follows should be more transparent. In later chapters we will revisit this material and generalize it to much wider classes of networks.

## 5-1 VOLTAGE AND CURRENT SOURCE CONVERSIONS

From our previous work it is apparent that if voltage sources are present in a network somewhat more manipulation is needed for a solution by nodal analysis than if only current sources are present. (Similarly, if analysis is undertaken using a current basis, i.e., mesh or fundamental loop analysis, more manipulation is required for solution if current sources are present than if only voltage sources are present.) Some reduction in analysis effort is achieved by effecting a current source conversion for those voltage sources that form a series combination with a single resistor. This conversion is most easily envisioned in terms of the network of Fig. 5-1*a* in which a series voltage source-resistor combination is connected to an arbitrary network of resistors

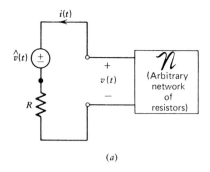

(a)

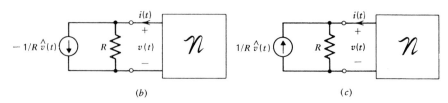

(b)                                    (c)

**Fig. 5-1 A current source conversion for network (a) yields network (b) or (c).**

denoted by $\mathcal{N}$. The current flowing through the series combination is $i(t)$ and the voltage across it is $v(t)$. It is easily verified that $i(t)$ and $v(t)$ can be related by

$$v(t) = \hat{v}(t) + i(t)R$$

This relationship can be rewritten as

$$i(t) = -\frac{1}{R}\hat{v}(t) + \frac{1}{R}v(t)$$

which can be interpreted as the relationship describing the parallel combination of a current source of value $-(1/R)\hat{v}(t)$ and resistance $R$ shown in Fig. 5-1b. Alternatively, this expression can be interpreted as the parallel combination of a current source of value $(1/R)\hat{v}(t)$ and resistance $R$ as shown in Fig. 5-1c. Observe that as far as the network $N$ is concerned, both the series voltage source-resistor combination or the parallel current source-resistor combination supply identical constraints on the voltage $v(t)$ and the current $i(t)$. Therefore, the branch voltages and currents of network $N$ are the same for both source-resistor combinations.

Although the behavior of the network $N$ is unchanged after a current-source conversion, there are two distinctions between the entire network of Fig. 5-1a and the entire network of Figs. 5-1b and 5-1c. First, the current flowing through $R$ in Fig. 5-1a has a different value than the current flowing through $R$ in Figs. 5-1b and 5-1c. Second, the networks of Figs. 5-1b and c have one less node than the network of Fig. 5-1a. In other words there is one less KCL nodal equilibrium equation and one less unknown node voltage for the networks of Figs. 5-1$b$ and $c$. On the other hand, the network of Fig. 5-1$a$ has one less independent loop than the networks of Figs. 5-1$b$ and 5-1$c$. However, in order to obtain the "solution" to the network of Fig. 5-1$a$ once the "solution" of the networks of Figs. 5-1$b$ and 5-1$c$ have been found, one additional expression is needed to solve for the voltage source current:

$$i(t) = \frac{v(t) - \hat{v}(t)}{R}$$

The following example illustrates the usefulness of the current source conversion.

## Example 5.1

Consider the network of Fig. 5-2$a$. The nodal equilibrium equations are easily written by inspection:

$$2v_① - 2v_② \qquad = -i_1$$
$$-2v_① + 8v_② - 5v_③ = 0$$
$$- 5v_② + 8v_③ = i_2$$

The additional equation needed for solution is the voltage source branch relationship

$$v_① = \hat{v}_1$$

Substitution of this branch relationship into the nodal equilibrium equations yields

$$-2v_② \qquad + i_1 = -2\hat{v}_1$$
$$8v_② - 5v_③ \qquad = 2\hat{v}_1$$
$$-5v_② + 8v_③ \qquad = i_2$$

which can be solved for the node voltages $v_②$ and $v_③$ and the voltage source current $i_1$. Observe that only the last two equations need be considered if only the node voltages are of interest. An alternate method of solution is to perform a current source conversion. The voltage source $\hat{v}_1$ and the .5 $\Omega$ resistor form a series combination. This series combination can be replaced by a parallel combination of .5 $\Omega$ resistor and a $2\hat{v}_1$ amp current source as shown in Fig. 5-2$b$. Node ① has been temporarily eliminated from consideration. The nodal equilibrium equations are

$$8v_② - 5v_③ = 2\hat{v}_1$$
$$-5v_② + 8v_③ = i_2$$

Observe that these equations are identical to those obtained for the unconverted network after substitution of the voltage source branch relationship. Thus, we see that the effect of a current source conversion is to eliminate from consideration the unknown voltage source current. ☐

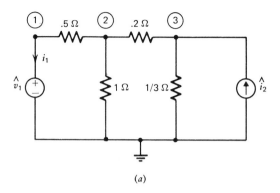

(a)

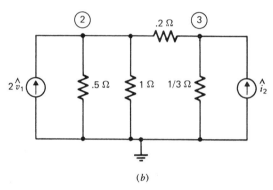

(b)

**Fig. 5-2**   (*a*) **A network with a series voltage source-resistor combination.** (*b*) **The network after a current source conversion. Observe that node 1 has been "eliminated" from this network.**

**Example 5.2**

As another example of the use of the current source conversion, consider the four node network shown in Fig. 5-3*a*. Using what was learned about the behavior of series and parallel combinations of resistors, $R_2$, $R_3$, and $R_4$ can be combined to yield the single resistor $R_{234}$ with resistance of $3\ \Omega$ as shown in Fig. 5-3*b*. Two current-source conversions are now easily effected: one for voltage source $\hat{v}_1$ and resistor $R_{234}$ and the other for voltage source $\hat{v}_2$ and resistor $R_6$. The network that results from these source conversions is shown in Fig. 5-3*c*. One nodal equilibrium equation is easily written to solve for $v_{\circled{3}}$:

$$\left(\frac{1}{3} + 1 + \frac{1}{2}\right) v_{\circled{3}} = 15 \text{ a}$$

so that

$$v_{\circled{3}} = \frac{45}{19} \text{ V} \quad \square$$

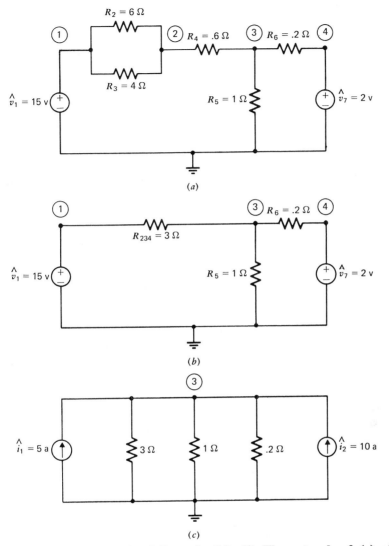

**Fig. 5-3** (*a*) **The network of Example 5.2.** (*b*) **The network of** (*a*) **after combining resistors** $R_2$, $R_3$, **and** $R_4$. (*c*) **after two current source conversions.**

The procedure of conversion from a voltage source in series with a resistor to a current source in parallel with a resistor can be reversed. We term the procedure of converting a current source in parallel with a resistor to a voltage source in series with a resistor a *voltage source conversion*. The steps involved are easily derived with the aid of Fig. 5-4. In the network of Fig. 5-4*a* we have

$$i(t) = \frac{1}{R} v(t) - \hat{i}(t)$$

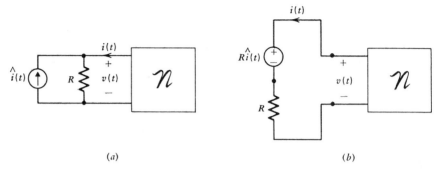

Fig. 5-4   **A voltage source conversion for network (*a*) yields network (*b*).**

This equation can be rewritten as

$$v(t) = R\hat{\imath}(t) + Ri(t)$$

which can be modeled as a voltage source in series with a resistor as shown in Fig. 5-4*b*.

**Example 5.3**

Let us consider again the network discussed in Example 5.1 that is redrawn in Fig. 5-5*a*. The network of Fig. 5-5*b* results after a voltage source conversion. Observe that this network has only two independent loops, whereas the original network has three independent loops. Thus, if a loop based analysis method is used, a set of equations is to be written in terms of a current basis and voltage source conversions could be used to reduce the number of equations to be solved.   □

**5-2   V-SHIFTS AND I-SHIFTS**

In the previous section we saw that the current source conversion could be used to reduce the number of nodal equilibrium equations to be solved for a network that contained voltage sources in series with resistors. However, not all voltage sources of a network appear in series with a resistor. For example, see the networks of Figs. 5-6*a* and 5-6*b*. In these networks a current source conversion cannot be effected directly. If we desire to perform a current source conversion on such a network there are two alternatives. The first is to assume that all voltage sources have a "small" valued series resistor (Certainly this assumption is valid for physical networks since all voltage sources have internal resistances.) Hopefully, the introduction of this small resistance will not drastically alter the results obtained. But, recall that the current source conversion requires division of the voltage source voltage by

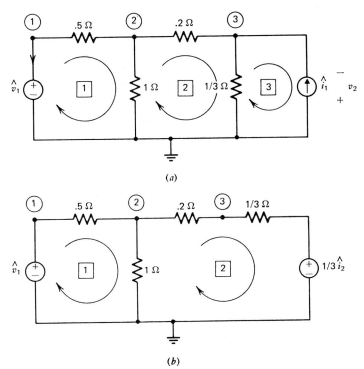

*(a)*

*(b)*

**Fig. 5-5** *(a)* **A network with a parallel current source-resistor combination.** *(b)* **The network after a voltage source conversion. Observe that the converted network has only two independent loops while the original network has three independent loops.**

the value of the series resistor. Therefore, if the series resistance is too small, a division by a small number must be performed. This operation is not desirable from a numerical viewpoint because it is prone to loss of accuracy.

**Example 5.4**

The network of Fig. 5-7a contains a voltage source that does not have a series resistor. We can insert a "small" resistor in series with the voltage source, as shown in Fig. 5-7b, and assume that any error introduced is negligible. ☐

An alternative method that can be used when voltage sources appear in a network without a series resistor is known as the *V-shift*.[1] It is possible to shift the voltage source to all the branches connected to either end of the voltage source without altering the behavior of the network. The V-shift principle is illustrated in Fig. 5-8. First, observe that the node voltages remain

[1] Some authors refer to this operation as the *E*-shift since the letter "*e*" is also used to designate voltage.

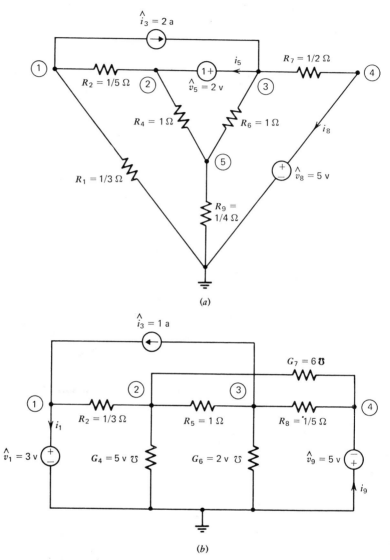

(a)

(b)

**Fig. 5-6   Two networks that contain voltage sources that do not have series resistors.**

the same except at a node through which the voltage source has been shifted. Fig. 5-8*b* shows the network of Fig. 5-8*a* after the voltage source is shifted through node ⑪ ; the node voltage at node ⑦ is the same as before. Observe that node voltage $v$ ⑪ appears at two points in the shifted network. Fig. 5-8*c* shows the network of Fig. 5-8*a* after the voltage source is shifted through node ⑦; the node voltage at node ⑪ is the same as before. Node voltage $v_⑦$

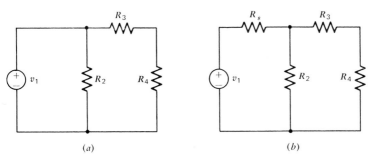

(a)                              (b)

**Fig. 5-7 Illustration of insertion of a small resistor in series with a voltage.**

appears at three points as indicated. In addition, the currents through the resistance branches in the networks of Figs. 5-8b and 5-8c are the same as the currents in the branches of the network of Fig. 5-8a. This is easily proven by writing KVL loop equations. The original current through the voltage source of the network of Fig. 5-8a is equal to either the sum of the currents through the two sources of the network of Fig. 5-8b or the sum of the currents through the three voltage sources of the network of Fig. 5-8c.

Once the V-shift has been performed, all voltage sources appear in series with resistors so that a current source conversion may be effected and the node equations easily written. An example serves to illustrate the steps involved.

### Example 5.5

Consider again the network of Example 4.3, redrawn in Fig. 5-9a. The only voltage source not forming a series combination with a resistor is $\hat{v}_5$. We may shift $\hat{v}_5$ through either node ② or node ③, as shown in Fig. 5-9b and 5-9d, respectively. Consider the network of Fig. 5-9b first. The network of Fig. 5-9c results after a current source conversion. The nodal equilibrium equations are

node ①   $$\left(\frac{1}{R_1} + \frac{1}{R_2}\right)v_① - \frac{1}{R_2}v_③ = -\frac{1}{R_2}\hat{v}_5 - \hat{i}_3$$

node ③   $$-\frac{1}{R_2}v_① + \left(\frac{1}{R_2} + \frac{1}{R_4} + \frac{1}{R_6} + \frac{1}{R_7}\right)v_③ - \left(\frac{1}{R_4} + \frac{1}{R_6}\right)v_⑤$$
$$= \left(\frac{1}{R_2} + \frac{1}{R_4}\right)\hat{v}_5 + \hat{i}_3 + \frac{1}{R_7}\hat{v}_8$$

node ⑤   $$-\left(\frac{1}{R_4} + \frac{1}{R_6}\right)v_③ + \left(\frac{1}{R_4} + \frac{1}{R_9} + \frac{1}{R_6}\right)v_⑤ = -\frac{1}{R_4}\hat{v}_5$$

Upon substitution of the element values, these equations become

$$8v_① - 5v_③ \qquad\quad = -12$$
$$-5v_① + 9v_③ - 2v_⑤ = 24$$
$$-2v_③ + 6v_⑤ = -2$$

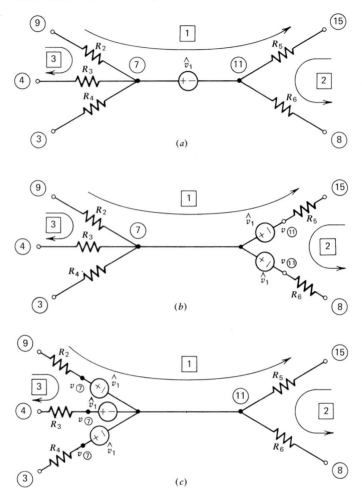

**Fig. 5-8   Illustration of the V-shift principle.**

that can be solved to yield $v_{\text{①}} = .4$ v, $v_{\text{③}} = 3.04$ v, and $v_{\text{⑤}} = .68$ v. Now let us find the node voltages for the nodes we have suppressed. Node voltage $v_{\text{④}}$ is equal to the voltage source voltage $\hat{v}_8$.

$$v_{\text{④}} = \hat{v}_8 = 5 \text{ v}$$

Node voltage $v_{\text{②}}$ is determined by node voltage $v_{\text{③}}$ and voltage source $\hat{v}_5$:

$$v_{\text{②}} = v_{\text{③}} - \hat{v}_5 = 1.04 \text{ v}$$

The only unknowns remaining are the voltage source currents $i_5$ and $i_8$. The current $i_5$ equals the sum of the branch currents $i_2$ and $i_4$ since these are the

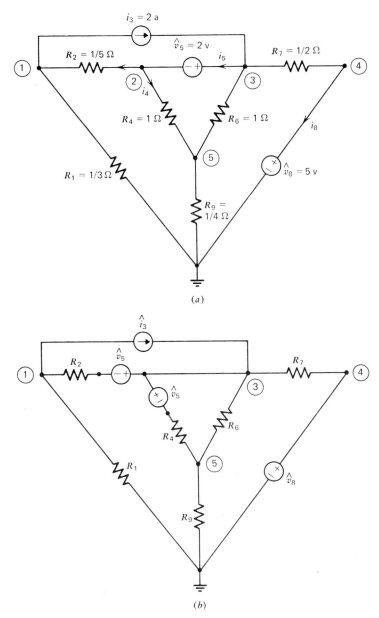

**Fig. 5-9** Networks used in Example 5.5 to demonstrate the use of the V-shift principle and current source conversions; (*a*) the original network; (*b*) the network after a V-shift through node 2; (*c*) the network after a current source conversion; (*d*) the network of (*a*) after a V-shift through node 3; (*e*) the network of (*d*) after a current source conversion.

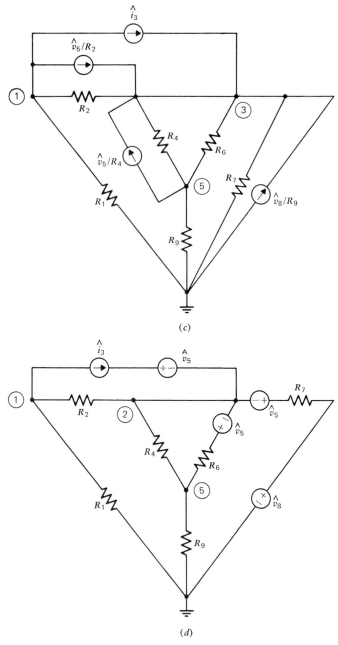

Fig. 5-9  (*Continued*)

branches from which $\hat{v}_5$ was shifted:

$$i_5 = i_2 + i_4 = \left(\frac{1}{R_2} + \frac{1}{R_4}\right)v_{\text{②}} - \frac{1}{R_2}v_{\text{①}} - \frac{1}{R_4}v_{\text{⑤}} = 3.56 \text{ a}$$

The current $i_8$ is simply the current flowing through $R_7$:

$$i_8 = \frac{1}{R_7}(v_{\text{③}} - v_{\text{④}}) = -3.92 \text{ a}$$

These results agree with those obtained previously.

Let us now consider the network of Fig. 5-9d. Observe that the voltage source in series with $\hat{i}_3$ does not affect the operation of the circuit and may as well be set to zero. This fact will become clearer when we discuss the principle of superposition. Furthermore, the two voltage sources in series with resistor $R_7$ may be combined into a single voltage source of value $\hat{v}_8 - \hat{v}_5$. After a current source conversion, the network of Fig. 5-9e results. The nodal equilibrium equations for this network are

$$node \; \text{①} \qquad \left(\frac{1}{R_1} + \frac{1}{R_2}\right)v_{\text{①}} - \frac{1}{R_2}v_{\text{②}} = -\hat{i}_3$$

$$node \; \text{②} \qquad -\frac{1}{R_2}v_{\text{①}} + \left(\frac{1}{R_2} + \frac{1}{R_4} + \frac{1}{R_6} + \frac{1}{R_7}\right)v_{\text{②}} - \left(\frac{1}{R_4} + \frac{1}{R_6}\right)v_{\text{⑤}}$$

$$= \hat{i}_3 - \frac{\hat{v}_5}{R_6} + \frac{1}{R_7}(\hat{v}_8 - \hat{v}_5)$$

$$node \; \text{⑤} \qquad -\left(\frac{1}{R_4} + \frac{1}{R_6}\right)v_{\text{②}} + \left(\frac{1}{R_4} + \frac{1}{R_6} + \frac{1}{R_9}\right)v_{\text{⑤}} = \frac{\hat{v}_5}{R_6}$$

or after substitution of element values

$$\begin{aligned} 8v_{\text{①}} - 5v_{\text{②}} \qquad\quad &= -2 \\ -5v_{\text{①}} + 9v_{\text{②}} - 2v_{\text{⑤}} &= 6 \\ -2v_{\text{②}} + 6v_{\text{⑤}} &= 2 \end{aligned}$$

which requires $v_{\text{①}} = .4$ v, $v_{\text{②}} = 1.04$ v, and $v_{\text{⑤}} = .68$ v. These results are seen to agree with our previous results. □

Just as the V-shift principle is useful when writing node equations, the *I-shift principle*, along with *voltage source conversions*, is useful when writing equilibrium equations in terms of a basis set of currents. In order to perform a voltage source conversion a current source must appear in parallel with a resistor. Current sources need not have parallel resistors as illustrated in Fig. 5-10. *The I-shift principle* places a current source across each resistor that forms a closed loop with the original current source. This operation does not in any way affect the behavior of the circuit. The I-shift technique is illustrated in Fig. 5-11. We leave further discussion of the I-shift to the problems.

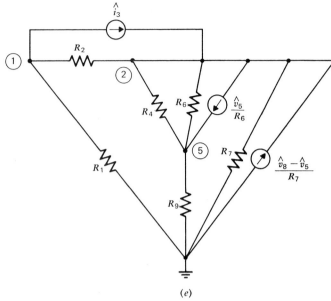

*(e)*

**Fig. 5-9**    *(Continued)*

## 5-3  THE PRINCIPLE OF SUPERPOSITION

The principle of *superposition* is a fundamental consequence of *linearity* and plays an important role in analysis. Before proceeding we must discuss the concept of linearity.

Consider the arbitrary lumped, two-terminal resistance element of Fig. 5-12. Let $v(t)$ denote the voltage across the element and $i(t)$ the current

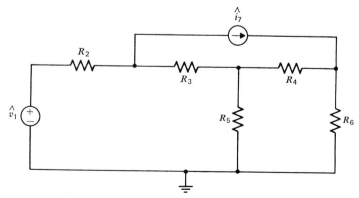

**Fig. 5-10**  **An example of a network containing a current source that does not form a parallel combination with a resistor.**

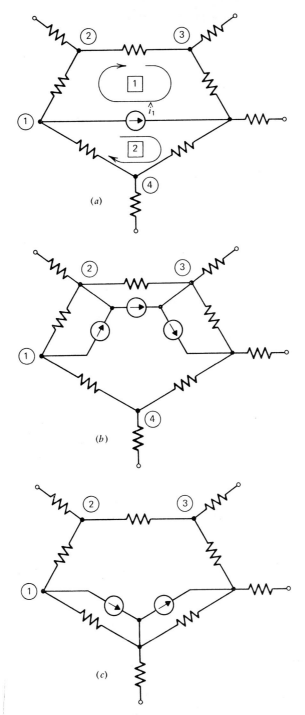

**Fig. 5-11   Illustration of the I-shift principle.**

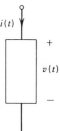

**Fig. 5-12    An arbitrary lumped, two-terminal element.**

through the element. Since the branch relationship relates $v(t)$ to $i(t)$ we say that the voltage $v(t)$ corresponds to the current $i(t)$.

The branch relationship of a two-terminal resistive element is *linear* if it is *homogeneous* and *additive*. *The branch relationship is homogeneous if the following condition exists; if the voltage $v(t)$ corresponds to the current $i(t)$ then the voltage $\alpha v(t)$ corresponds to the current $\alpha i(t)$ where $\alpha$ is any finite constant. The branch relationship is additive if the following condition exists; if the voltage $v_1(t)$ corresponds to the current $i_1(t)$ and the voltage $v_2(t)$ corresponds to the current $i_2(t)$ then the voltage $v_1(t) + v_2(t)$ corresponds to the current $i_1(t) + i_2(t)$.*

Observe that an element described by the branch relationship

$$v(t) = i^2(t)$$

is *nonlinear* because it is not homogeneous or additive. To check for homogeneity consider the voltage corresponding to the current $\alpha i(t)$:

$$\hat{v}(t) = (\alpha i(t))^2$$
$$= \alpha^2 i^2(t)$$
$$\neq \alpha i^2(t)$$

The check for additivity is also easily facilitated. If

$$v_1(t) = i_1{}^2(t)$$

and

$$v_2(t) = i_2{}^2(t)$$

consider the voltage associated with the current $i_1(t) + i_2(t)$:

$$v(t) = (i_1(t) + i_2(t))^2$$
$$= i_2{}^2(t) + i_2{}^2(t) + 2i_1(t)i_2(t)$$
$$\neq v_1(t) + v_2(t)$$

It is easy to verify that the resistive branch relationships

$$v(t) = Ri(t) \qquad R \text{ is a constant}$$
$$i(t) = Gv(t) \qquad G \text{ is a constant}$$

are linear. Finally, a network that is comprised solely of linear elements is said to be linear.

We now return to our discussion of superposition. Simply stated, the *principle of superposition* says that the branch voltage or current of any element in a network containing several independent sources is equal to the sum of the branch voltages or currents due to each source acting alone with all other sources set to zero. (*Note*. A zero-valued voltage source is a short circuit and a zero-valued current source is an open circuit. An example will illustrate this statement.

## Example 5.6

Consider the network of Fig. 5-13*a*. The equations required for solution of this network are

$$\frac{1}{R_2} v_① - \frac{1}{R_2} v_② = -i_1$$

$$-\frac{1}{R_2} v_① + \left(\frac{1}{R_2} + \frac{1}{R_3}\right) v_② = -\hat{i}_4$$

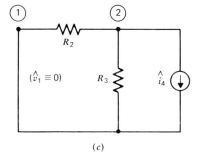

**Fig. 5-13  A simple network used to illustrate the principle of superposition. The node voltages, branch voltages, and branch currents of (*a*) are equal to the sum of the node voltages, branch voltages, and branch currents of (*b*) and (*c*).**

and

$$v_{①} = \hat{v}_1$$

or

$$-\frac{1}{R_2} v_{②} + i_1 = -\frac{1}{R_2} \hat{v}_1$$

and

$$\left(\frac{1}{R_2} + \frac{1}{R_3}\right) v_{②} = \frac{1}{R_2} \hat{v}_1 - \hat{i}_4$$

The solution is

$$v_{②} = \frac{R_2 R_3}{R_2 + R_3} \left(\frac{1}{R_2} \hat{v}_1 - \hat{i}_4\right)$$

and

$$i_1 = -\frac{1}{R_2 + R_3} (\hat{v}_1 + R_3 \hat{i}_4)$$

All branch voltages and currents can now be determined. The principle of superposition requires that these results be obtained from analysis of the network when each source acts independently. Figure 5-13$b$ shows the network when $\hat{v}_1$ acts alone, $\hat{i}_4$ being set to zero. The equations needed for solution are

$$\frac{1}{R_2} v'_{①} - \frac{1}{R_2} v'_{②} = -i'_1$$

$$-\frac{1}{R_2} v'_{①} + \left(\frac{1}{R_2} + \frac{1}{R_3}\right) v'_{②} = 0$$

and

$$v'_{①} = \hat{v}_1$$

or

$$-\frac{1}{R_2} v'_{②} + i'_1 = -\frac{1}{R_2} \hat{v}_1$$

and

$$\left(\frac{1}{R_2} + \frac{1}{R_3}\right) v'_{②} = \frac{1}{R_2} \hat{v}_1$$

The solution of these equations is

$$v'_{②} = \frac{R_3}{R_2 + R_3} \hat{v}_1$$

and

$$i'_1 = -\frac{1}{R_2 + R_3} \hat{v}_1$$

Figure 5-13c shows the network when $\hat{i}_4$ acts alone, $\hat{v}_1$ being set to zero. The equations needed for solution are

$$\left(\frac{1}{R_2} + \frac{1}{R_3}\right) v''_{\circled{2}} = -\hat{i}_4$$

and

$$i''_1 = \frac{1}{R_2} v''_{\circled{2}}$$

The solution is

$$v''_{\circled{2}} = -\frac{R_2 R_3}{R_2 + R_3} \hat{i}_4$$

and

$$i''_1 = -\frac{R_3}{R_2 + R_3} \hat{i}_4$$

Clearly

$$v_{\circled{2}} = v'_{\circled{2}} + v''_{\circled{2}}$$

and

$$i_1 = i'_1 + i''_1 \quad \square$$

Although it would appear that the use of superposition in network analysis entails excessive manipulation, we will find it indispensable when we consider the ac steady-state analysis of networks that contain sinusoidal sources of different frequencies.

## 5-4   THE SUBSTITUTION THEOREM

Given an arbitrary network (not necessarily linear) for which all the voltages and currents are known and unique, if the $k$th branch, no matter what it is, has associated with it a voltage $v_k(t)$ and current $i_k(t)$, it can be replaced at any instant of time by any one of the following *without* altering the behavior of the network:

1. An independent voltage source of value $v_k$.
2. An independent current source of value $i_k$.
3. A resistor of value $R = v_k/i_k$ or a conductor of value $G = i_k/v_k$.

This statement is known as the *substitution theorem* and is easily proven.

Let $v_1, v_2, \ldots, v_b$, and $i_1, i_2, \ldots, i_b$ be the branch voltages and currents, respectively, of the given network. By assumption, these voltages and currents are unique and satisfy the KVL equations, KCL equations, and the branch relationships. First, consider case $(i)$ where the $k$th branch is replaced by a voltage source of value $v_k$. Since the structure, or topology, of this modified network is exactly the same as the topology of the original network, the KVL and KCL equations are the same. All branch relationships, except the one for the $k$th branch, are the same for the modified network as for the given network. But the constraint imposed by the $k$th branch is that its voltage be equal to $v_k$, its current can be anything. Therefore, the unique voltages and currents of the original network satisfy all the constraints of the modified network and hence is the unique solution of the modified network. The proof of the substitution theorem for cases 2 and 3 follows similar arguments and is left as an exercise for the reader, (Prob. 5.13).

The substitution theorem is of utility for the proof of Thévenin's and Norton's theorems, which are to be discussed in the next section.

### Example 5.7

To illustrate the substitution theorem, consider again the network of Example 5.5, which is redrawn in Fig. 5-14$a$. The node voltages were found to be $v_① = .4$ v, $v_② = 1.04$ v, $v_③ = 3.04$ v, $v_④ = 5$ v, and $v_⑤ = .68$ v. Then

$$v_2 = v_① - v_② = -.64 \text{ v}$$

and

$$i_2 = \frac{v_2}{R_2} = -3.2 \text{ a}$$

According to the substitution theorem the voltages and currents of the network will not be altered if we replace resistor $R_2$ by a current source of value $-3.2a$. The modified network is shown in Fig. 5-14$b$. The nodal equilibrium equations for the modified network are

$$\begin{aligned}
3v_① &= -2 + 3.2 \\
v_② - v_⑤ - i_5 &= -3.2 \\
3v_③ - 2v_④ - v_⑤ + i_5 &= 2 \\
-2v_③ + 2v_④ + i_8 &= 0
\end{aligned}$$

and

$$-v_② - v_③ + 6v_⑤ = 0$$

and the two voltage source relationships are

$$-v_② + v_③ = 2$$

and

$$v_④ = 5$$

The solution of these equations yield the same results as those obtained for the original network.    □

## 5-5  THÉVENIN'S THEOREM

At times, networks are so complex that a complete analysis to determine circuit behavior could be very time consuming. For example, if we are

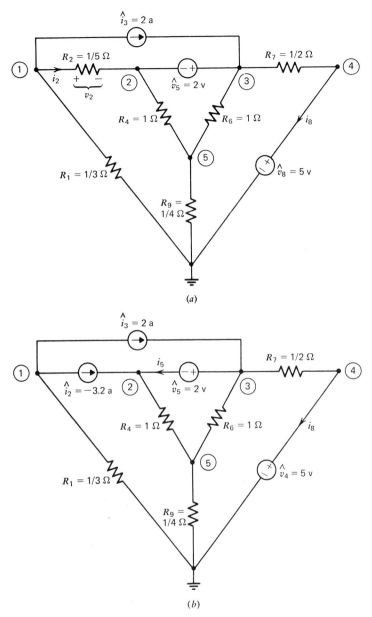

(a)

(b)

Fig. 5-14 (a) Network used in Example 5.7 to illustrate the substitution theorem. Both networks (a) and (b) have the same unique solution.

131

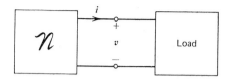

**Fig. 5-15   A linear network $N$ connected to an arbitrary load.**

interested in the voltage across the output of a network $\mathscr{N}$ under loading situations, as illustrated in Fig. 5-15, it would certainly be convenient if a complete analysis of $\mathscr{N}$ were not required each time a new load was tried especially if $\mathscr{N}$ represented a complex network. If $\mathscr{N}$ is a linear network then *Thévenin's theorem* could be invoked and a considerable savings in analysis achieved. *Thévenin's theorem* is stated as follows.

> Let $\mathscr{N}$ be a linear network connected by any two of its terminals to an arbitrary load. Assume that $\mathscr{N}$ has a unique solution when it is terminated by this load. Then as far as the load is concerned, the network $\mathscr{N}$ can be replaced by an exactly equivalent two-terminal network consisting of a single resistance connected in series with a voltage source. The value of the voltage source is equal to the open-circuit voltage of $\mathscr{N}$, $v_{oc}$. (The open circuit voltage is the voltage across the two terminals when the load is disconnected.) The value of the resistance is equal to the equivalent resistance of $\mathscr{N}$ seen across the two terminals when all independent sources in $\mathscr{N}$ are set to zero.

Before proving this theorem, an example should convince the reader of its utility.

**Example 5.8**

We wish to determine the voltage across $R_L$ in the network of Fig. 5-16*a* for $R_L = 1\,\Omega$, $2\,\Omega$, $3\,\Omega$, $4\,\Omega$, and $5\,\Omega$. Certainly one way to ascertain the desired quantities is to write the nodal equilibrium equations. Since we are unconcerned with the current through $\hat{v}_5$, we might as well perform a current source conversion. The resulting network is shown in Fig. 5-16*b*. The equations necessary for solution of this network are

$$\left(\frac{1}{8} + \frac{1}{4} + \frac{1}{R_L}\right)v_{②} - \frac{1}{4}v_{③} - \frac{1}{R_L}v_{④} = 2$$

$$-\frac{1}{4}v_{②} + \left(\frac{1}{4} + \frac{1}{20}\right)v_{③} \qquad\qquad = -1$$

and

$$-\frac{1}{R_L}v_{②} + \left(\frac{1}{R_L} + \frac{1}{3}\right)v_{④} \qquad\qquad = 1$$

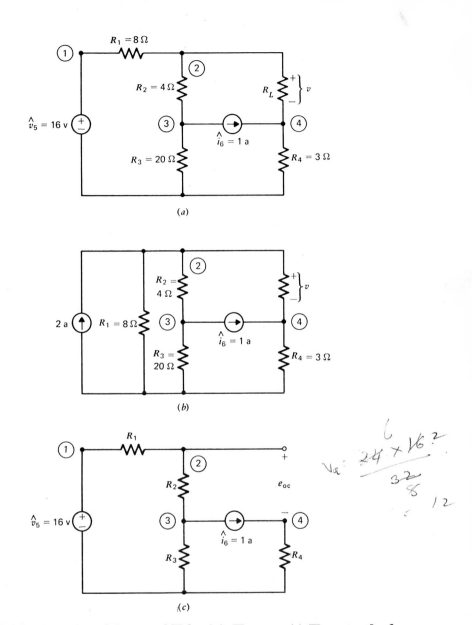

$$V_a = \frac{24 \times 16}{32} \cdot 2 = \frac{32}{8} = 12$$

Fig. 5-16 Illustration of the use of Thévenin's Theorem. (a) The network of Example 5.8 to be analyzed for various values of $R_L$. (b) Network of (a) after a current source conversion. (c) Network used to find open circuit voltage, $e_{oc}$. (d) Network used to find Thévenin resistance, $R_{th}$. (e) The Thévenin equivalent of (a) as seen by $R_L$.

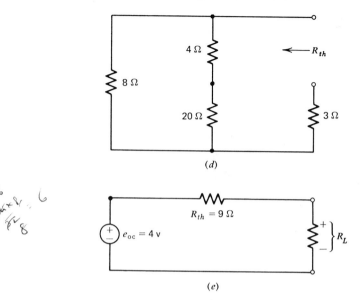

(d)

(e)

**Fig. 5-16**   (*Continued*)

This set of equations must now be solved once for each value of $R_L$. Use of Thévenin's theorem drastically reduces the computational effort. Let us find the Thévenin equivalent network "seen by" $R_L$. The open circuit voltage $v_{oc}$, is found by removing $R_L$, as shown in Fig. 5-16c. The reader should verify that $v_{oc} = 4$ v. The Thévenin resistance is obtained by calculating the equivalent resistance at the terminals of $R_L$ with all sources set to zero, as shown in Fig. 5-16d. It is easily verified that $R_{th} = 9\ \Omega$. Once the Thévenin equivalent, as seen by $R_L$, is found, the network of Fig. 5-16e can be used to evaluate the behavior of $R_L$. Certainly it is much easier to determine $v$ for various values of $R_L$ in the network of Fig. 5-16e than it is for the network of Fig. 5-16a. The reader should be aware that in effect two complete analyses were carried out to find the Thévenin equivalent, one on the network of Fig. 5-16c and one on the network of Fig. 5-16d.   □

Let us now prove Thévenin's theorem. Consider a linear network $\mathcal{N}$ connected to an arbitrary load as shown in Fig. 5-17a. Let $v$ represent the voltage across the load and $i$ the current through the load. By the substitution theorem, the voltages and currents in $\mathcal{N}$ are unchanged if the load is replaced with an independent current source of value $\hat{i}$, as shown in Fig. 5-17b. The voltage $v$ results from the current source $\hat{i}$ and any internal sources of $\mathcal{N}$. Since the entire network of Fig. 5-17b is linear, the superposition principle may be invoked. Therefore, the voltage $v$ is equal to the sum of two voltages $v_{oc}$ and $v_1$: $v_{oc}$ is the open-circuit voltage of $\mathcal{N}$, that is, the voltage resulting from the internal sources of $\mathcal{N}$ alone with $\hat{i} = 0$ (as illustrated in Fig. 5-17c),

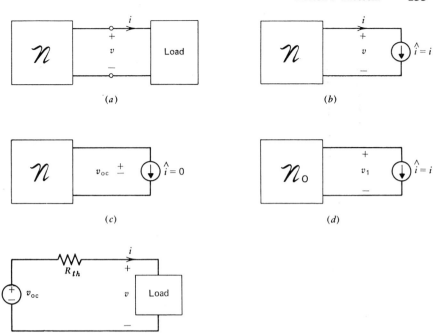

Fig. 5-17 (a) A linear network connected to an arbitrary load. (b) By the substitution theorem the load can be replaced by the current source $i$. The voltage $v$ is the sum of two voltages (c) $e_{oc}$, the voltage produced by the internal sources of $N$ with $i$ set to zero and (d) $v_1$, the voltage across $i$ when $i = i$ and all sources in $N$ are set to zero. (e) The Thévenin equivalent of (a).

and $v_1$ is the voltage resulting from $i$ acting alone with all internal sources of $\mathscr{N}$ set to zero (as illustrated in Fig. 5-17d)

$$v = v_{oc} + v_1$$

Since $N_0$ is purely resistive, it is equivalent at its terminals to a single resistance $R_{eq} = R_{th}$. Hence

$$v_1 = -iR_{th}$$

so that

$$v = v_{oc} - iR_{th}$$

which characterizes the situation of Fig. 5-17d and proves Thévenin's theorem.

## 5-6 NORTON'S THEOREM

Norton's theorem is the "dual" of Thévenin's theorem:

Any two-terminal linear network, $\mathscr{N}$, may be replaced by an exactly equivalent two-terminal network consisting of a constant

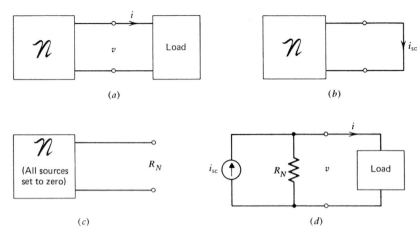

**Fig. 5-18**    (*a*) **To find the Norton equivalent of a linear network** *N* **connected to an arbitrary load;** (*b*) **we find the short circuit current** $i_{sc}$ **by shorting the output terminals;** (*c*) **the equivalent resistance of** *N* **with all internal sources set to zero** (*d*) **The Norton equivalent network is an exact replacement for** *N*.

current source connected in parallel with a single resistance element. The value of the current source is equal to the short circuit current of $\mathscr{N}$, $i_{sc}$. (The short circuit current is equal to the current flowing out of $\mathscr{N}$ when the two terminals are shorted.) The value of the resistance, $R_N$, is equal to the equivalent resistance of $\mathscr{N}$ seen across the two terminals when all independent sources in $\mathscr{N}$ are set to zero.

Norton's theorem is illustrated in Fig. 5-18. The proof of this theorem follows the proof of Thévenin's theorem and is left as an exercise for the reader, (Prob. 5.15).

**Example 5.9**

Consider again the network of Fig. 5-16*b* (Example 5.8) which is redrawn in Fig. 5-19*a*. To find the Norton equivalent "seen" by $R_L$ we must determine the short circuit current $i_{sc}$ and the Norton resistance $R_N$. $i_{sc}$ is found by replacing $R_L$ by a short circuit as shown in Fig. 5-19*b*: $i_{sc} = 4/9$. $R_N$ is found by setting all sources to zero and computing the equivalent resistance seen by $R_L$ as shown in Fig. 5-19*c*: $R_N = 9 \, \Omega$. The Norton equivalent is shown in Fig. 5-19*d*.    □

Observe that any linear network $\mathscr{N}$ has both a Thévenin equivalent and a Norton equivalent as shown in Fig. 5-20. Moreover

$$R_N = R_{th}$$

and since

$$v = v_{oc} - iR_{th}$$

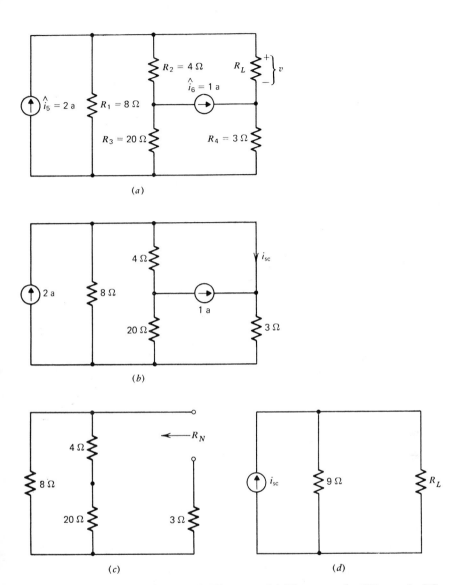

**Fig. 5-19** **Illustration of Norton's Theorem.** (*a*) **The network of Example 5.9.**
(*b*) **The network used to find the short circuit current,** $i_{sc}$; (*c*) **The network used**
**to find Norton resistance** $R_N$; (*d*) **the Norton equivalent.**

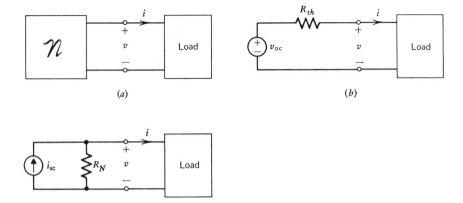

(a)                                                (b)

(c)

**Fig. 5-20** (a) A linear network $N$ can be replaced by both a (b) Thévenin equivalent and a (c) Norton equivalent. Note $R_N = R_{\text{th}}$ and $e_{\text{oc}} = R_{\text{th}}i_{\text{sc}}$.

and

$$v = (i_{\text{sc}} - i)R_N$$

we must have

$$v_{\text{oc}} = i_{\text{sc}}R_N$$

Therefore, if the Thévenin equivalent is known so is the Norton equivalent. In other words, a current source conversion of the Thévenin equivalent yields the Norton equivalent.

It should be emphasized that we have performed *two* separate circuit analyses to determine either the Thévenin or Norton equivalents. For large networks considerable hand computation may be involved. The following example shows how GEP can be employed to find a Thévenin or Norton equivalent of network.

**Example 5.10**

We wish to find the Norton equivalent of the network of Fig. 5-21a as seen across terminals $T$–$T$. (This network has been considered in Examples 5.5 and 5.6.) To find the short circuit current $i_{\text{sc}}$ we can analyze the network of Fig. 5-20b. The nodal equilibrium equations for this network are

$$\left(\frac{1}{R_1} + \frac{1}{R_2}\right)v_{\textcircled{1}} - \frac{1}{R_2}v_{\textcircled{2}} + i_{\text{sc}} = -\hat{i}_3$$

$$-\frac{1}{R_2}v_{\textcircled{1}} + \left(\frac{1}{R_2} + \frac{1}{R_4}\right)v_{\textcircled{2}} - \frac{1}{R_4}v_{\textcircled{5}} - i_5 = 0$$

$$\left(\frac{1}{R_6} + \frac{1}{R_7}\right)v_{\textcircled{3}} - \frac{1}{R_7}v_{\textcircled{4}} - \frac{1}{R_6}v_{\textcircled{5}} + i_5 = \hat{i}_3$$

$$-\frac{1}{R_7}v_{\textcircled{3}} + \frac{1}{R_7}v_{\textcircled{4}} + i_8 = 0$$

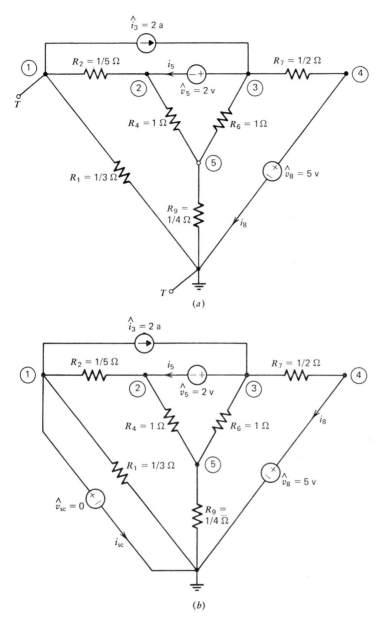

Fig. 5-21 (a) Network used in Example 5.10 to illustrate how a Norton equivalent for a complex network can be found using GEP; (b) the network used to find $i_{sc}$; (c) the network used to find $R_N$; (d) the Norton equivalent.

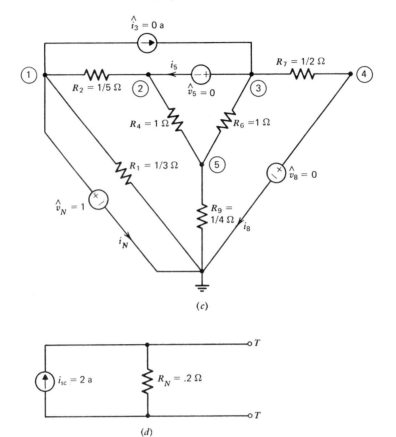

(c)

(d)

**Fig. 5-21**   (*Continued*)

and

$$-\frac{1}{R_4}v_{②} - \frac{1}{R_6}v_{③} + \left(\frac{1}{R_4} + \frac{1}{R_6} + \frac{1}{R_9}\right)v_{⑤} = 0$$

and the three voltage source equations are

$$v_{①} = 0$$

$$v_{③} - v_{②} = \hat{v}_5$$

and

$$v_{④} = \hat{v}_8$$

On substitution of the element values and elimination of $v_①$ and $v_④$ ($v_① = 0$ and $v_④ = 5$) these equations become

$$-5v_②\qquad\qquad\qquad + i_{sc} = -2$$
$$6v_②\qquad - v_⑤ - i_5 \qquad = 0$$
$$3v_③ - v_⑤ + i_5 \qquad = 12$$
$$- 2v_③ \qquad\qquad + i_8 = -10$$
$$- v_② - v_③ + 6v_⑤ \qquad = 0$$

and

$$- v_② + v_③ \qquad\qquad = 2$$

The solution obtained from GEP yields $i_{sc} = 1.03$.

To find the Norton resistance $R_N$ all sources must be set to zero. Then if a voltage source of value $v_N$ is placed across terminals $T$–$T$ and the current through it equals $i_N$ then

$$R_N = \frac{-v_N}{i_N}$$

where the negative sign emerges due to our assumed reference directions. (An alternate method would be to place a unity-valued current source across $T$–$T$, then $R_N$ equals the voltage across this source.) The network that results is shown in Fig. 5-21c. The nodal equilibrium equations for this network are

$$\left(\frac{1}{R_1} + \frac{1}{R_2}\right)v_① - \frac{1}{R_2}v_② \qquad\qquad + i_N = 0$$

$$-\frac{1}{R_2}v_① + \left(\frac{1}{R_2} + \frac{1}{R_4}\right)v_② \qquad\qquad - i_5 = 0$$

$$\left(\frac{1}{R_6} + \frac{1}{R_7}\right)v_③ - \frac{1}{R_7}v_④ - \frac{1}{R_6}v_⑤ + i_5 = 0$$

$$-\frac{1}{R_7}v_③ + \frac{1}{R_7}v_④ \qquad\qquad + i_8 = 0$$

and

$$-\frac{1}{R_4}v_② - \frac{1}{R_6}v_③ + \left(\frac{1}{R_4} + \frac{1}{R_6} + \frac{1}{R_9}\right)v_⑤ = 0$$

and the voltage source relationships are

$$v_① \qquad\qquad = \hat{v}_N$$
$$-v_② + v_③ = \hat{v}_5$$

and

$$v_④ \qquad\qquad = \hat{v}_8$$

On substitution of the element values and elimination of $v_①$ and $v_④$ ($v_① = 1$ and $v_④ = 0$) these equations become

$$-5v_② \qquad\qquad\qquad + i_N = -8$$
$$6v_② \qquad\qquad - i_5 \qquad = 5$$
$$3v_③ - v_⑤ + i_5 \qquad = 0$$
$$- 2v_③ \qquad\qquad + i_8 = 0$$
$$- v_② - v_③ + 6v_⑤ \qquad = 0$$

and

$$-v_② + v_③ \qquad\qquad\qquad = 0$$

Solution using GEP yields $i_N = -5.12$. Therefore

$$R_N = \frac{-v_N}{i_N} = \frac{1}{5.12} = .196 \ \Omega$$

The Norton equivalent is shown in Fig. 5-21c.    □

## 5-7 TELLEGEN'S THEOREM

*Tellegen's theorem* is a very interesting result that will form the basis of a very powerful method for calculating sensitivities discussed in Chapter 6. In its simplest form, Tellegen's theorem states that *if $v_1(t)$, $v_2(t)$, . . . , $v_{n_b}(t)$ are the $n_b$ branch voltages and $i_1(t)$, $i_2(t)$, . . . , $i_{n_b}(t)$ are the $n_b$ branch currents of a given $n_b$ branch network comprised of arbitrary two terminal lumped elements then*

(5.1) 
$$\sum_{k=1}^{n_b} v_k(t) i_k(t) = 0$$

for all time. This statement may be looked upon as a statement of the conservation of energy or power. In other words, the total power generated by the network must equal the total power consumed by the network.

### Example 5.11

To see that (5.1) indeed holds, and is a statement of conservation of power, consider the network shown in Fig. 5-22. The nodal equilibrium equations are

$$\left(\frac{1}{2} + \frac{1}{16}\right)v_① - \frac{1}{16}v_② = 1$$

and

$$-\frac{1}{16}v_① + \left(\frac{1}{16} + \frac{1}{6}\right)v_② = 0$$

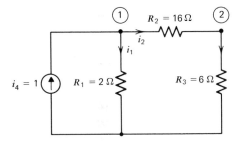

**Fig. 5-22 The network used in Example 5.11 to illustrate that Tellegen's theorem is a statement of the conservation of power.**

Thus $v_① = 1.83$ and $v_② = .5$. The branch voltages are:

$$v_1 = v_① = 1.83$$

$$v_2 = v_① - v_② = 1.33$$

$$v_3 = v_② = .5$$

and

$$v_4 = -v_① = -1.83$$

and the branch currents are

$$i_1 = \frac{v_1}{R_1} = .915$$

$$i_2 = \frac{v_2}{R_2} = .083$$

$$i_3 = \frac{v_3}{R_3} = .083$$

and

$$i_4 = 1$$

Substituting into the Tellegen's theorem relationship yields

$$v_1 i_1 + v_2 i_2 + v_3 i_3 + v_4 i_4$$

$$= (1.83)(.915) + (1.33)(.083) + (.5)(.083) + (-1.83)(1),$$

$$= 0$$

Observe that we can interpret this result as stating that the power supplied by the source

$$v_4 i_4 = -1.83 \text{ watts}$$

equals the power absorbed by the resistors

$$v_1 i_1 + v_2 i_2 + v_3 i_3 = 1.83 \text{ watts} \quad \square$$

The fact that

$$\sum_{k=1}^{n_b} v_k(t) i_k(t) = 0$$

may not be very impressive to the reader but its implications are profound.

The proof of Tellegen's theorem is based entirely on Kirchhoff's voltage and current laws. Like KVL and KCL, Tellegen's theorem is valid for all networks, regardless of which elements appear in the branches. For convenience, we write the KVL equations as

$$v_k(t) = \sum_{j=1}^{n_{n-1}} a_{jk} v_{\textcircled{j}}(t), \qquad k = 1, 2, \ldots, n_b$$

where $v_{\textcircled{j}}, j = 1, 2, \ldots, n_n - 1$, are the node voltages and

$$a_{jk} = \begin{cases} +1 \text{ if branch number } k \text{ leaves node number } \textcircled{j} \\ -1 \text{ if branch number } k \text{ enters node number } \textcircled{j} \\ 0 \text{ if branch number } k \text{ doesn't touch node number } \textcircled{j} \end{cases}$$

Observe that since each branch leaves one node and enters another node only two of the $a_{jk}$ are nonzero for each $k$. Moreover, the coefficients $a_{jk}$ completely characterize the network graph since they contain all information regarding the interconnection of branches and nodes. The set of coefficients $a_{1k}, a_{2k}, \ldots, a_{n_n-1,k}$ describe the connection of each branch $k$. For example, suppose in a 6 node network $a_{24} = +1, a_{54} = -1$, and $a_{14} = a_{34} = a_{44} = 0$, then branch 4 leaves node $\textcircled{2}$ and enters node $\textcircled{5}$. Likewise, the set of coefficients $a_{j1}, a_{j2}, \ldots, a_{j,n_b}$ describe the connection of all branches to node $\textcircled{j}$. For example, suppose in a 5 branch network $a_{31} = +1, a_{33} = -1, a_{34} = -1$, and $a_{32} = a_{35} = 0$, then branch 1 leaves node $\textcircled{3}$, branches 3 and 4 enter node $\textcircled{3}$ and branches 2 and 5 are not connected to node $\textcircled{3}$. With the above interpretation of the coefficients $a_{jk}$, the KCL node equations can be expressed most succinctly as

$$\sum_{k=1}^{n_b} a_{jk} i_k(t) = 0 \qquad \text{for} \qquad j = 1, 2, \ldots, n_n - 1$$

It is now possible to present a simple proof of Tellegen's theorem. We wish to show

$$\sum_{k=1}^{n_b} v_k(t) i_k(t) = 0$$

Consider the sum

$$\sum_{k=1}^{n_b} v_k(t) i_k(t)$$

From above, KVL can be written as

$$v_k(t) = \sum_{j=1}^{n_n-1} a_{jk} v_{\textcircled{j}}(t) \qquad k = 1, 2, \ldots, n_b$$

so that

$$\sum_{k=1}^{n_b} v_k(t) i_k(t) = \sum_{k=1}^{n_b} i_k(t) \sum_{j=1}^{n_n-1} a_{jk} v_{(j)}(t)$$

$$= \sum_{j=1}^{n_n-1} v_{(j)}(t) \sum_{k=1}^{n_b} a_{jk} i_k(t)$$

But from KCL

$$\sum_{k=1}^{n_b} a_{jk} i_k(t) = 0$$

which yields the desired result:

$$\sum_{k=1}^{n_b} v_k(t) i_k(t) = 0$$

Tellegen's theorem can be generalized as follows. *If* $v_1(t), v_2(t), \ldots, v_{n_b}(t)$ *are the $n_b$ branch voltages of an $n_b$ branch network and $\hat{\imath}_1(\tau), \hat{\imath}_2(\tau), \ldots, \hat{\imath}_{n_b}(\tau)$ are the $n_b$ branch currents of another $n_b$ branch network that has the same topology (structure) as the first, then*

(5.2)
$$\sum_{k=1}^{n_b} v_k(t) \hat{\imath}_k(\tau) = 0$$

*for all times $t$ and $\tau$.*

In other words, Tellegen's theorem remains valid even if applied to *two different networks*, assuming they have the same topology, even if the voltages and currents are measured at different times. A special case, of course, is that Tellegen's theorem can be applied to the voltages of a network measured at time $t_1$ and the currents of the same network measured at time $t_2$:

$$\sum_{k=1}^{n_b} v_k(t_1) i_k(t_2) = 0$$

The proof of the more general form of Tellegen's theorem follows the proof given above. We wish to prove

$$\sum_{k=1}^{n_b} v_k(t) \hat{\imath}_k(\tau) = 0$$

KVL applied to the first network is

$$v_k(t) = \sum_{j=1}^{n_n-1} a_{jk} v_{(j)}(t) \qquad k = 1, 2, \ldots, n_b$$

so

$$\sum_{k=1}^{n_b} v_k(t) \hat{\imath}_k(\tau) = \sum_{k=1}^{n_b} \hat{\imath}_k(\tau) \sum_{j=1}^{n_n-1} a_{jk} v_{(j)}(t)$$

$$= \sum_{j=1}^{n_n-1} v_{(j)}(t) \sum_{k=1}^{n_b} a_{jk} \hat{\imath}_k(\tau)$$

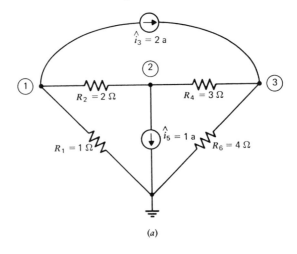

(a)

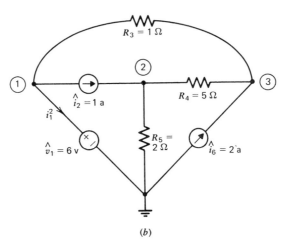

(b)

**Fig. 5-23   Networks used in Example 5.12 to illustrate the validity of Tellegen's theorem.**

We recognize that the KCL equations for the second network can be written as

$$\sum_{k=1}^{n_b} \hat{a}_{jk} \hat{i}_k(\tau) = 0 \qquad j = 1, 2, \dots, n_n - 1$$

for all $\tau$, but since the first and second network have the same topology

$$\hat{a}_{jk} = a_{jk}$$

for $j = 1, 2, \dots, n_n - 1$ and $k = 1, 2, \dots, n_b$. Therefore

$$\sum_{k=1}^{n_b} a_{jk} \hat{i}_k(\tau) = 0$$

and, finally,

$$\sum_{k=1}^{n_b} v_k(t)\hat{i}_k(\tau) = \sum_{j=1}^{n_n-1} v_{\ j}\ (t)\sum_{k=1}^{n_b} a_{jk}\hat{i}_k(\tau)$$
$$= 0$$

**Example 5.12**

As an illustration of the fact that Tellegen's theorem remains valid for two different networks that have the same topology, consider the networks of Figs. 5-23a and 5-23b. The nodal equilibrium equations for the network in Fig. 5-23a are

$$\frac{3}{2}v^1_{①} - \frac{1}{2}v^1_{②} = -2$$

$$-\frac{1}{2}v^1_{①} + \frac{5}{6}v^1_{②} - \frac{1}{3}v^1_{③} = -1$$

and

$$-\frac{1}{3}v^1_{②} + \frac{7}{12}v^1_{③} = 2$$

where the superscript 1 is used to denote the first network. The solution of these equations by use of Cramer's rule yields

$$v^1_{①} = \frac{\begin{vmatrix} -2 & -\dfrac{1}{2} & 0 \\[2mm] -1 & \dfrac{5}{6} & -\dfrac{1}{3} \\[2mm] 2 & -\dfrac{1}{3} & \dfrac{7}{12} \end{vmatrix}}{\begin{vmatrix} \dfrac{3}{2} & -\dfrac{1}{2} & 0 \\[2mm] -\dfrac{1}{2} & \dfrac{5}{6} & -\dfrac{1}{3} \\[2mm] 0 & -\dfrac{1}{3} & \dfrac{7}{12} \end{vmatrix}} = \frac{\dfrac{-17}{24}}{\dfrac{10}{24}} = \frac{-17}{10}$$

$$v^1_{②} = \frac{\begin{vmatrix} \dfrac{3}{2} & -2 & 0 \\[2mm] -\dfrac{1}{2} & -1 & -\dfrac{1}{3} \\[2mm] 0 & 2 & \dfrac{7}{12} \end{vmatrix}}{\dfrac{10}{24}} = \frac{-11}{10}$$

and

$$v_{\textcircled{3}}^1 = \frac{\begin{vmatrix} \dfrac{3}{2} & -\dfrac{1}{2} & 2 \\[2mm] -\dfrac{1}{2} & \dfrac{5}{6} & -1 \\[2mm] 0 & -\dfrac{1}{3} & 2 \end{vmatrix}}{\dfrac{10}{24}} = \frac{28}{10}$$

The branch voltages are

$$v_1^1 = v_{\textcircled{1}}^1 = -\frac{17}{10}$$

$$v_2^1 = v_{\textcircled{1}}^1 - v_{\textcircled{2}}^1 = \frac{-6}{10}$$

$$v_3^1 = v_{\textcircled{1}}^1 - v_{\textcircled{3}}^1 = -\frac{45}{10}$$

$$v_4^1 = v_{\textcircled{2}}^1 - v_{\textcircled{3}}^1 = -\frac{39}{10}$$

$$v_5^1 = v_{\textcircled{2}}^1 = -\frac{11}{10}$$

and

$$v_6^1 = -v_{\textcircled{3}}^1 = -\frac{28}{10}$$

We now find the branch currents in the network of Fig. 5.23*b*. The nodal equilibrium equations are

$$-\quad v_{\textcircled{3}}^2 + i_1^2 = -7$$

$$\frac{7}{10} v_{\textcircled{2}}^2 - \frac{1}{5} v_{\textcircled{3}}^2 \quad = \quad 1$$

$$-\frac{1}{5} v_{\textcircled{2}}^2 + \frac{6}{5} v_{\textcircled{3}}^2 \quad = \quad 8$$

where the superscript 2 denotes the second network. Solution by Cramer's rule yields

$$v^2_{\textcircled{2}} = \frac{\begin{vmatrix} -7 & -1 & 1 \\ 1 & -\dfrac{1}{5} & 0 \\ 8 & \dfrac{6}{5} & 0 \end{vmatrix}}{\begin{vmatrix} 0 & -1 & 1 \\ \dfrac{7}{10} & -\dfrac{1}{5} & 0 \\ -\dfrac{1}{5} & \dfrac{6}{5} & 0 \end{vmatrix}} = \frac{\dfrac{14}{5}}{\dfrac{40}{50}} = \frac{140}{40}$$

$$v^2_{\textcircled{3}} = \frac{\begin{vmatrix} 0 & -7 & 1 \\ \dfrac{7}{10} & 1 & 0 \\ -\dfrac{1}{5} & 8 & 0 \end{vmatrix}}{\dfrac{40}{50}} = \frac{\dfrac{29}{5}}{\dfrac{40}{50}} = \frac{290}{40}$$

and

$$i^2_1 = \frac{\begin{vmatrix} 0 & -1 & -7 \\ \dfrac{7}{10} & -\dfrac{1}{5} & 1 \\ -\dfrac{1}{5} & \dfrac{6}{5} & 8 \end{vmatrix}}{\dfrac{40}{50}} = \frac{\dfrac{10}{50}}{\dfrac{40}{50}} = \frac{10}{40}$$

Note that $v^2_{\textcircled{1}} = \hat{v}_1 = 60$. The remaining branch currents can now be found:

$$i^2_2 = 1$$

$$i^2_3 = \frac{v^2_{\textcircled{1}} - v^2_{\textcircled{3}}}{R_3} = -\frac{50}{40}$$

$$i^2_4 = \frac{v^2_{\textcircled{2}} - v^2_{\textcircled{3}}}{R_4} = -\frac{30}{40}$$

$$i^2_5 = \frac{v^2_{\textcircled{2}}}{R_5} = \frac{70}{40}$$

and

$$i_6^2 = 2$$

Substituting the voltages of the first network and the currents of the second network into the expression

$$v_1^1 i_1^2 + v_2^1 i_2^2 + v_3^1 i_3^2 + v_4^1 i_4^2 + v_5^1 i_5^5 + v_6^1 i_6^2$$

should produce a sum of zero:

$$-\left(\frac{17}{10}\right)\left(\frac{10}{40}\right) - \left(\frac{6}{10}\right)\left(\frac{40}{40}\right) + \left(\frac{45}{10}\right)\left(\frac{50}{40}\right) + \left(\frac{39}{10}\right)\left(\frac{30}{40}\right)$$

$$-\left(\frac{11}{10}\right)\left(\frac{70}{40}\right) - \left(\frac{28}{10}\right)\left(\frac{80}{40}\right) = 0 \quad \square$$

## 5-8 SUMMARY

Several useful techniques that can be used to aid in the analysis of networks have been introduced. Whenever a voltage source of value $\hat{v}$ appears in series with a resistor of value $R$ it can be converted to a current source of value $\hat{i} = \hat{v}/R$ in parallel with a resistor of value $R$. This operation is called a *current source conversion*. Similarly, a *voltage source conversion* is the operation of replacing a parallel combination of a current source of value $\hat{i}$ and a resistor of value $R$ by a series combination of a voltage source of value $\hat{v} = \hat{i}R$ and a resistor of value $R$.

The *V-shift principle* is a method for transforming a network that contains voltage sources that do not have series resistors to an equivalent network in which all voltage sources have series resistors. The *I-shift principle* is a method for transforming a network that contains current sources that does not have parallel resistors to an equivalent network in which all current sources have parallel resistors.

The *substitution theorem* states that if any branch of a network with voltage $v$ and current $i$ is replaced by either a voltage source of voltage $v$, a current source of current $i$, or a resistor of resistance $v/i$, all the voltages and currents in the network remain the same.

*Thévenin's and Norton's theorems* are techniques whereby whole sections of a linear network may be replaced by either a series voltage source-resistor combination or a parallel current source-resistor combination without affecting the remaining parts of the network.

The student should be able to see that by using the methods of this chapter all networks containing resistors, current sources, and voltage sources can be transformed into an equivalent network that contains only resistors and current sources.

*Tellegen's theorem* is a statement of conservation of power in that it shows that the sum of the products of branch voltages and currents of a network must equal zero.

It must be emphasized that the techniques presented here are useful for a much wider class of networks than those discussed. We presented them here so that the student may start using them early and we will consider these ideas again in later sections.

## PROBLEMS

**5.1** Compare the number of independent KCL equations and independent KVL equations that can be written for each of the networks of Fig. P5.1 with the number of independent equations that can be written after a current source conversion has been made for each series voltage source-resistor combination. Observe that for each current source conversion one new independent loop has been added and one node eliminated.

**5.2** Write the nodal equilibrium equations for the network in Fig. P5.2 along with the constraints imposed by the voltage sources. Reduce this set of equations to one in which the voltage source currents do not appear, that is, a set that involves only the node voltages. Then perform a current source conversion on each voltage source and write the nodal equilibrium equations for the converted network. Compare this set to the one derived earlier, they should be identical.

**5.3** Perform current source conversions on each of the networks in Fig. P5.3 and determine the currents or voltages indicated.

**5.4** Determine the power absorbed in $R$ in the network of Fig. P5.4 by performing a current source conversion.

**5.5** Perform a voltage source conversion on each parallel current source-resistor combination in each of the networks of Fig. P5.5. Compare the number of independent KCL equations and independent KVL equations that can be written for the original networks with the number that can be written for the converted networks.

**5.6\*** Write the set of equilibrium equations in terms of mesh currents for the network of Fig. P5.6 along with the constraints imposed by the current sources. Reduce this set of equations to one in which the current source voltages do not appear. Then perform a voltage source conversion for each parallel current source-resistor combination. Write the mesh equations for this converted network and compare with those obtained above.

**5.7** Use the V-shift principle and current source conversion to write nodal equilibrium equations for the networks of Fig. P5.7. What is the current in $R_3$?

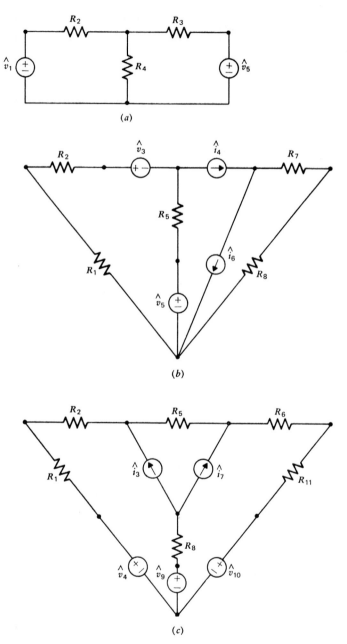

Fig. P5.1

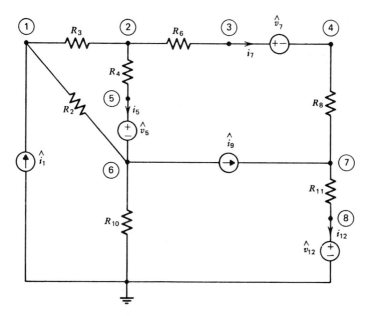

**Fig. P5.2**

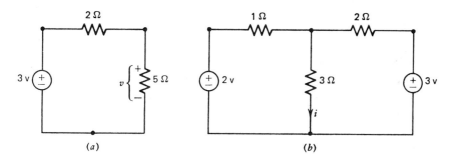

(a)                                    (b)

**Fig. P5.3**

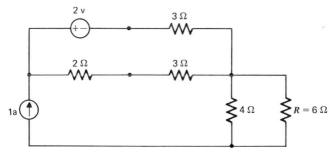

**Fig. P5.4**

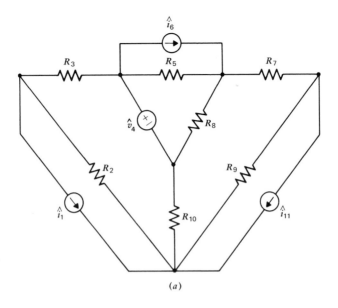

(a)

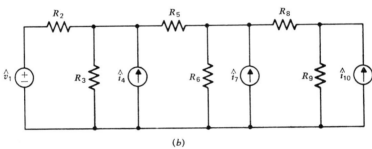

(b)

**Fig. P5.5**

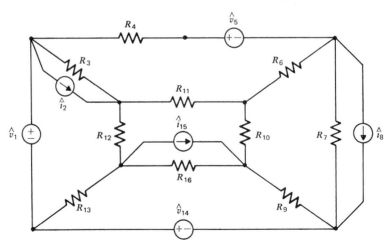

**Fig. P5.6**

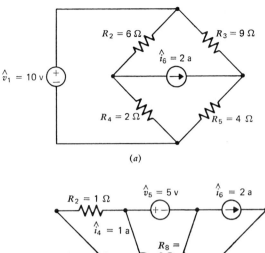

(a)

(b)

Fig. P5.7

**5.8**    Write the nodal equilibrium equations and the constraints imposed by the voltage sources for the network of Fig. P5.8. Use GEP to solve these equations. Then, using the V-shift principle and current source conversions, transform this network to an equivalent one that contains only current sources and resistors. Write the nodal equilibrium equations for the transformed network and use GEP to solve them. Compare the two sets of results.

**5.9**    Repeat Prob. 5.8 for the network of Fig. P5.9.

**5.10**    The I-shift is similar to the V-shift and in terms of a link current basis for networks that contain current sources. Recall from Chapter 4 that voltages across current sources replace link currents as unknowns in this case. The I-shift is used for those current sources that are not in parallel with a single resistor. Note if a current source appears in parallel with a resistor a voltage source conversion can be performed. In an I-shift a current source, equal in value to the current source being shifted, is placed in parallel with each of the branches that form a closed loop with the current source being shifted. The original current source is then set to zero, that is, open circuited. The I-shift is illustrated in Fig. P5.10a. For the networks Figs. P5.10b and P5.10c perform an I-shift and a voltage source

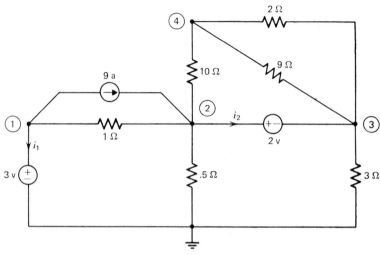

**Fig. P5.8**

conversion. Then choose a tree such that all voltage sources appear in tree branches. Write the fundamental loop and cutset equations and branch relationships. Reduce this set of equations to a set involving only link currents.

**5.11**    Determine expressions for the voltage $v$ and the current $i$ in terms of the sources for each of the networks of Fig. P5.11. (*Hint*. Use superposition.)

**5.12**    Consider the network shown in Fig. P5.12. Use the superposition principle to solve for the node voltages by first considering the behavior of the network with all current sources set to zero and then the behavior of the network with all voltage sources set to zero. In other words, set all current sources to zero and write the nodal

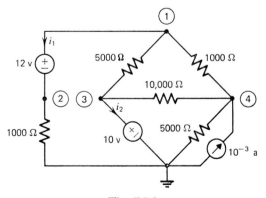

**Fig. P5.9**

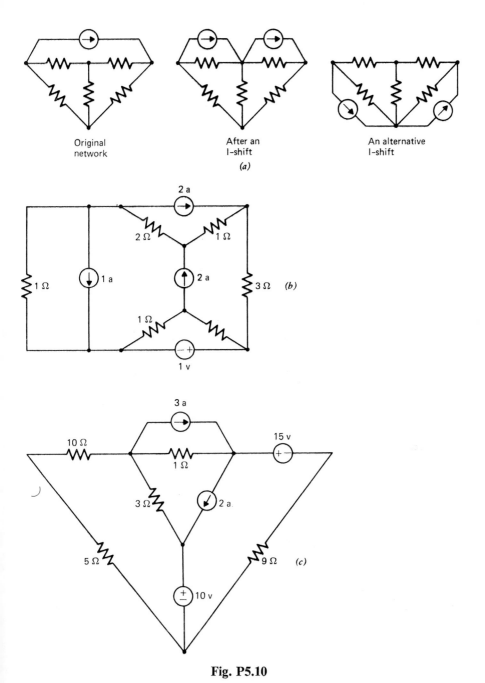

Original
network

After an
I-shift

An alternative
I-shift

(a)

2 a

2 Ω

1 Ω

1 Ω

1 a

2 a

3 Ω    (b)

1 Ω

1 v

3 a

10 Ω

15 v

1 Ω

3 Ω

2 a.

5 Ω

9 Ω    (c)

10 v

**Fig. P5.10**

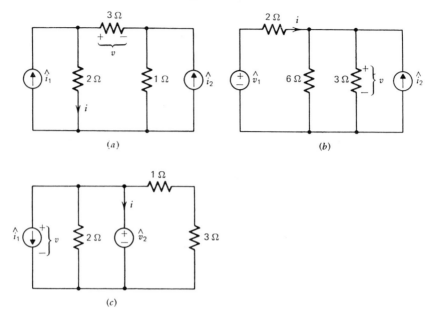

**Fig. P5.11**

equilibrium equations. Use GEP to solve for the node voltages. Then set the current sources to their original values and set all voltage sources to zero. Again write the nodal equilibrium equations and use the GEP to solve for the node voltages. Add the two sets of node voltages to obtain the node voltages that would result if all sources acted together.

**5.13**    Prove that the behavior of an arbitrary network that has a unique set of known branch voltages and currents and whose $k$th branch

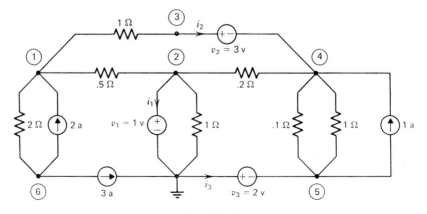

**Fig. P5.12**

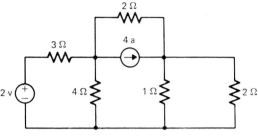

**Fig. P5.14**

has associated with its voltage $v_k$ and current $i_k$, is unaltered if the $k$th branch is replaced by either an independent current source of value $i_k$ or a resistor of value $R = v_k/i_k$.

**5.14**   Find the current in the 1 ohm and 3 ohm resistors in the network of Fig. P5.14. Then replace the 1 ohm resistor with a current source equal to this value and show that the current in the three ohm resistor is unchanged.

**5.15**   Prove Norton's theorem. (*Hint.* Use the substitution theorem to replace the load with a voltage source of value $\hat{v}$ and follow the reasoning given for the proof of Thévenin's theorem.)

**5.16**   Find the Thévenin and Norton equivalents as seen by the load resistor $R_L$ in the network of Fig. P5.16.

**5.17**   Use GEP to help find the Thévenin and Norton equivalent networks as seen by $R_L$ in Fig. P5.17.

**5.18**   Find all branch voltages and currents in the network of Fig. P5.18 and verify Tellegen's theorem.

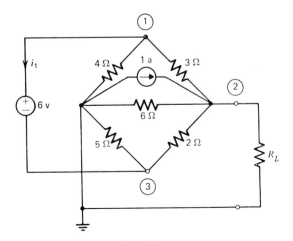

**Fig. P5.16**

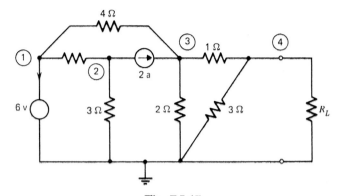

**Fig. P5.17**

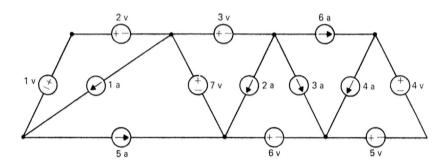

**Fig. P5.18**

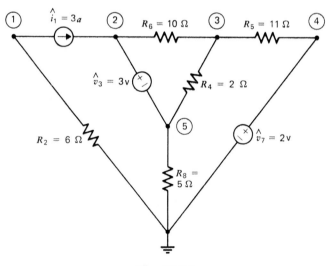

**Fig. P5.19**

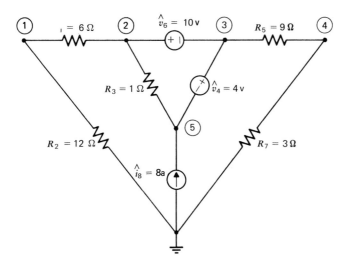

**Fig. P5.20**

**5.19**  Calculate all the branch voltages and currents in the network of Fig. P5.19 and verify Tellegen's theorem.

**5.20**  The network of Fig. P5.20 has the same topology as the network of Fig. P5.19. Calculate the branch voltages and currents of the network in Fig. P5.20. Verify that Tellegen's theorem is valid for the branch currents of the network in Fig. P5.19 and the branch voltages of the network in Fig. P5.20. Similarly, verify that Tellegen's theorem is valid for the branch voltages of the network in Fig. P5.19 and the branch currents of the network in Fig. P5.20.

## CHAPTER 6

# Worst Case Analysis and Network Sensitivities*

Until now we have been concerned solely with the analysis of resistive networks. One may rightfully ask "for what purpose was the network built, and how can a network be designed to perform a given task." Although a thorough investigation of network design is beyond the scope of this text, we consider two useful design tools in this chapter: *worst case analysis* and *network sensitivities*.

The reader is probably aware that the value given for a physical element is only an estimate of the exact value. Usually in addition to specifying the element value, the manufacturer will also specify a *tolerance*. The tolerance defines a range of possible values the element can have. For instance, a 1000 Ω resistor with a tolerance of 20 percent can have actually any value between 800 Ω (20 percent less than 1000 Ω) and 1200 Ω (20 percent greater than 1000 Ω). Similarly, a 500 Ω 10 percent tolerance resistor can have any value between 450 Ω and 550 Ω. The value specified by the manufacturer is sometimes referred to as the *nominal value*. Since a network designer must use physical elements to build his network, he must ascertain if the network will perform satisfactorily for all allowable values of each element. If the performance is unacceptable the designer will have to place tighter tolerance limits on some of the elements. The price paid for tighter tolerances is more

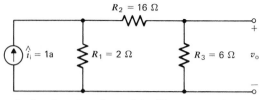

**Fig. 6-1    A simple network used to discuss worst case analysis.**

expensive elements. If costs are to be kept down, the designer would like to use elements with the largest tolerances possible. Thus we are motivated to determine the worst possible condition, that is, what set of possible element values will degrade the network's performance the most.

## 6-1  WORST CASE ANALYSIS

Suppose we wished to construct the network of Fig. 6-1. The nominal values of the resistors that produce the desired output voltage of $v_o = .5$ v are shown. The resistors used to build the circuit are usually sorted so that their values fall within various tolerances such as 20 10, .5, and 1 percent. The tighter the tolerance, the more expensive the resistor. In order to keep costs to a minimum, we should determine the largest tolerance each resistor can be allowed to have and still have the network perform within acceptable limits. This problem is a difficult one to solve. A simpler problem that has a similar flavor is to determine the range of possible values of the output for the range of possible values of the resistors, that is, find the *worst case*.

Assume that the network of Fig. 6-1 is to be constructed using resistors having 10 percent tolerance. Let $\hat{\imath}_i$ denote the "input" and $v_o$ denote the "output." In other words, $R_1$ might have any value between 1.8 $\Omega$, which is 10 percent less than the nominal 2 $\Omega$, and 2.2 $\Omega$, which is 10 percent greater than the nominal 2 $\Omega$. We can express the possible range of values of $R_1$ as

$$1.8 \leq R_1 \leq 2.2$$

where the symbol $\leq$ means "less than or equal to." Similarly, the possible ranges of values of $R_2$ and $R_3$ are

$$14.4 \leq R_2 \leq 17.6$$

and

$$5.4 \leq R_3 \leq 6.6$$

respectively. It is easy to verify that the relationship between the output $v_o$ and the input $\hat{\imath}_i$ is

$$v_o = \frac{R_1 R_3}{R_1 + R_2 + R_3} \hat{\imath}_1$$

and the *nominal* value of $v_o$ is .5 volts.

We now wish to ascertain the range of values of $v_o$ for the range of values of $R_2$, $R_3$, and $R_4$. This information is important. If the possible variation of $v_o$ from nominal is too great, then the tolerances of $R_1$, $R_2$, and $R_3$ should be tightened and hence, the cost of manufacturing the network will increase. On the other hand, if $v_o$ is relatively insensitive to the possible values of the resistors, then maybe less accurate, and more inexpensive, resistors could be used. One method that might be employed to determine the range of values of $v_o$ is to randomly choose sets of possible values of $R_1$, $R_2$, and $R_3$ and analyze the network for each set. Such a scheme forms the basis of a technique known as a *Monte Carlo analysis* and if enough sets of element values are chosen an accurate estimate of the range of values of $v_o$ can be determined. Clearly though, this method is time consuming because one analysis is needed for each change of element values. Another method is to try and estimate how the output varies with respect to small variations of each element value. Then based on this information, we predict the worst case. The second method is much faster than the first because fewer analyses are required although, as will be seen later, it may not yield the "true" worst case. We will investigate this method in some detail because the techniques involved are extremely useful in network design.

## 6-2  AN INTRODUCTION TO NETWORK SENSITIVITIES

The first step towards determining the worst case is to ascertain the relationship between variation of the output and variation, or *perturbation*, of each element. This relation is called the *sensitivity* of the output with respect to an element. For the present we will study sensitivity in terms of the arbitrary resistive network of Fig. 6-2, and return to the discussion of the network of Fig. 6-1 later.

Let $v_o$ denote the output voltage of a network $\mathcal{N}$, which is composed of resistors $R_1$, $R_2$, ..., $R_n$. Suppose that the input current is $\hat{i}_i$. (The subscripts o and i will be used throughout this chapter to denote output and input, respectively.) We note that $v_o$ is a function of the input current and the resistor values:

$$v_o = v_o(R_1, R_2, \ldots, R_n; \hat{i}_i)$$

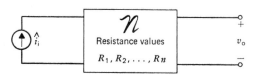

Fig. 6-2   A resistance network whose output, $V_{\text{out}}$, is a function of the input, $\hat{i}_i$, and the resistor values, $R_1$, $R_2$, ..., $R_n$.

Let $R_1^\circ, R_2^\circ, \ldots, R_n^\circ$, represent the nominal element values and $v_o^\circ$ the corresponding nominal output:

$$v_o^\circ = v_o(R_1^\circ, R_2^\circ, \ldots, R_n^\circ; \hat{\imath}_i)$$

Formally, the *sensitivity* of $\mathcal{N}$ with respect to any resistance value $R_k$ is the partial derivative of $v_o^\circ$ $(R_1, R_2, \ldots, R_n; \hat{\imath}_i)$ with respect to $R_k$ evaluated at the nominal resistor values $R_1^\circ, R_2^\circ, \ldots, R_n^\circ$, multiplied by $(R_k/V_o)$ when $i_1 = 1$:

$$(6.1) \qquad S_{R_k}^{v_o} \equiv \left(\frac{R_k}{V_o}\right) \frac{\partial v_o(R_1 R_2, \ldots, R_n; \hat{\imath}_i)}{\partial R_k}\Bigg|_{\substack{R_1=R_1^\circ \\ R_2=R_2^\circ \\ \vdots \\ R_n=R_n^\circ \\ \hat{\imath}_1=1}}$$

The sensitivity as defined by (6.1) is also called the *normalized sensitivity*. The *unnormalized sensitivity*

$$(6.2) \qquad \hat{S}_{R_k}^{v_o} = \frac{v_o(R_1, R_2, \ldots, R_n; \hat{\imath}_i)}{\partial R_k}\Bigg|_{\substack{R_1=R_1^\circ \\ R_2=R_2^\circ \\ \vdots \\ R_n=R_n^\circ \\ \hat{\imath}_1=1}}$$

is also of value.

**Example 6.1**

To illustrate that the sensitivity as defined in (6.1) or (6.2) does indeed predict how the output varies when an element value is varied, consider again the network of Fig. 6-1, redrawn in Fig. 6-3. We indicated previously that

$$v_o = \frac{R_1 R_3}{R_1 + R_2 + R_3} \hat{\imath}_i$$

Since $R_1 = 2\,\Omega$, $R_2 = 16\,\Omega$, $R_3 = 6\,\Omega$ then for $\hat{\imath}_i = 1a$, $v_o = .5$ v. The unnormalized sensitivity of $v_o$ with respect to $R_1$ is

$$\hat{S}_{R_1}^{v_o} = \frac{\partial v_o}{\partial R_1}\Bigg|_{\substack{R_1=2 \\ R_2=16 \\ R_3=6 \\ \hat{\imath}_1=1}} = \frac{R_3(R_2 + R_3)\hat{\imath}_i}{(R_1 + R_2 + R_3)^2}\Bigg|_{\substack{R_1=2 \\ R_2=16 \\ R_3=6 \\ \hat{\imath}_1=1}} = \frac{11}{48} = .229$$

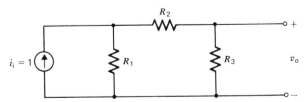

**Fig. 6-3   Network used in Example 6.2.**

The normalized sensitivity is

$$S_{R_1}^{v_o} = \frac{R_1}{v_o} \hat{S}_{R_1}^{v_o} = \left(\frac{2}{.5}\right)(.229) = .916$$

Similarly,

$$\hat{S}_{R_2}^{v_o} = \frac{-(R_1 R_3)\hat{\imath}_i}{(R_1 + R_2 + R_3)^2}\Biggr|_{\substack{R_1=2 \\ R_2=16 \\ R_3=6 \\ i_i=1}} = -\frac{1}{48} = -.021$$

and

$$S_{R_2}^{v_o} = \frac{R_2}{v_o} \hat{S}_{R_2}^{v_o} = \left(\frac{16}{.5}\right)(-.021) = -.672$$

and

$$\hat{S}_{R_3}^{v_o} = \frac{R_1(R_1 + R_2)\hat{\imath}_i}{(R_1 + R_2 + R_3)^2}\Biggr|_{\substack{R_1=2 \\ R_2=16 \\ R_3=6 \\ i_i=1}} = \frac{3}{48} = .063$$

and

$$S_{R_3}^{v_o} = \frac{R_3}{v_o} \hat{S}_{R_3}^{v_o} = \left(\frac{6}{.5}\right)(.063) = .756$$

We now show that the sensitivities do indeed indicate how the output changes with respect to small changes in the parameters. Consider a 10 percent change in the value of $R_1$, that is, let $R_1 = 2.2$ then

$$v_o(R_1, R_2^\circ, R_3^\circ, \hat{\imath}_i) = \frac{13.2}{24.2} = .545$$

Thus an increase of resistance value from nominal causes an increase in the output. This result is indicated by the fact that the sensitivity of the output with respect to $R_1$ is positive. Observe that a positive sensitivity also indicates that a small decrease in the parameter value from nominal will cause a decrease in the output. For a 10 percent decrease in $R_1$, that is, $R_1 = 1.8$, we have

$$v_o(R_1, R_2^\circ, R_3^\circ, \hat{\imath}_i) = \frac{10.8}{23.8} = .453$$

which is a decrease from nominal. Now consider a 10 percent change increase in the value of $R_2$, that is, let $R_2 = 17.6$, then

$$v_o(R_1^\circ, R_2, R_3^\circ, \hat{\imath}_i) = \frac{12}{25.6} = .469$$

which represents a decrease from nominal. This result, that a small increase in $R_2$ causes a decrease in the output is predicted by the fact that the sensitivity is negative. The reader should verify that a negative sensitivity also indicates that a small decrease in a parameter from nominal causes an increase in the output. Finally, consider a 10 percent change in the value of $R_3$. That is, let $R_3 = 6.6$, so that

$$v_o(R_1^\circ, R_2^\circ, R_3, \hat{\imath}_i) = \frac{13.2}{24.6} = .537$$

The output increases as indicated by the positive sensitivity. Finally, observe that the magnitudes of the sensitivities indicate the relative effect on the output due to changes in the parameters. The effect of a 10 percent change in $R_1$ on $v_o$ is greater than the effect of a 10 percent change in $R_3$ as predicted by the fact that $S_{R_1}^{v_o}$ is larger than $S_{R_3}^{v_o}$.   □

In the previous discussion we have been careful to indicate that the sensitivity may only predict how *small* changes in a parameter affect the output. Sensitivity is derivative information and as such only predicts *local* behavior of a function. It will be seen later that the estimates of local behavior may not correctly predict "large change" behavior.

For a large network it might be difficult to write an explicit function for the output voltage in terms of the network elements. Therefore, we could not explicitly find the sensitivities by taking partial derivatives; an alternate method must be found. One technique that might be employed is to perturb each element value and reanalyze the network. Comparison of the outputs for the perturbed and unperturbed cases could possibly indicate the unnormalized sensitivity.

**Example 6.2**

Consider again the network of Example 6.1 with nominal element values $R_1^\circ = 2\,\Omega$, $R_2^\circ = 16\,\Omega$, and $R_3^\circ = 6\,\Omega$. Then

$$v_o^\circ = \frac{R_1^\circ R_3^\circ}{R_1^\circ + R_2^\circ + R_3^\circ}\,\hat{\imath}_i$$

and $v_o^\circ = .5$. To find the sensitivity of $v_o$ with respect to $R_1$, perturb the value of $R_1$ by say 10 percent, so that the new value of $R_1$ is $R_1^1 = R_1^\circ + .1R_1^\circ = 2.2$, then

$$v_o^1 = \frac{13.2}{24.2} = .545$$

(*Note:* The superscript 1 is used to denote that $R_1$ has been altered.)
The approximate sensitivity then is

$$\frac{(v_o^1 - v_o^\circ)}{(R_1^1 - R_1^\circ)} = \frac{.045}{.2} = .225$$

as compared with the actual unnormalized sensitivity of $\hat{S}_{R_1}^{v_o} = .229$. Similarly, to find the sensitivity of $v_0$ with respect to $R_2$, we perturb $R_2$ by

10 percent so that $R_2^2 = 17.6$ and analyze the network:

$$v_o{}^2 = \frac{12}{25.6} = .469$$

The approximate sensitivity is

$$\frac{v_o^2 - v_o^\circ}{R_2^2 - R_2^\circ} = \frac{-.031}{1.6} = -.0195$$

as compared with the actual unnormalized sensitivity of $\mathcal{S}_{R_3}^{v_o} = -.021$. Finally, perturb $R_3$ by 10 percent, $R_3^3 = 6.6$, so that

$$v_o^3 = \frac{13.2}{24.6} = .537$$

and the approximate sensitivity is

$$\frac{v_o^3 - v_o^\circ}{R_3^3 - R_3^\circ} = \frac{.037}{.6} = .062$$

as compared with the actual sensitivity of $\mathcal{S}_{R_3}^{v_o} = .063$.   □

These results are close to being accurate but several drawbacks to the perturbation method exist. First, there is some question as to the amount each element value should be perturbed to obtain good results. Observe that if we perturbed $R_2$ by 20 percent, instead of 10 percent so that $R_2 = 19.2$ then

$$v_o^2 = \frac{12}{27.2} = .441$$

and

$$\frac{v_0^2 - v_0^\circ}{R_2^2 - R_2^0} = \frac{-.059}{3.2} = -.0184$$

which is a poorer approximation to $\mathcal{S}_{R_2}^{v_o}$. It should be observed that for Example 6.2 a perturbation of 10 percent works fairly well, for other networks 10 percent variations may prove useless.

In addition to the problem posed above, the perturbation method requires one network analysis for each element, since elements are varied one at a time. For large networks the perturbation method would require a considerable amount of time. A more efficient method is needed. Such a method is discussed in Section 6-4. In the following section we will show how the sensitivities are used to predict the worst case.

## 6-3  THE USE OF SENSITIVITIES TO PREDICT THE WORST CASE

In the last section we saw that if the sensitivity of the output with respect to a resistor was positive, the output increased (decreased) if the resistor's value increased (decreased) by a small amount. Similarly, if sensitivity of the output with respect to a resistor was negative, the output decreased (increased) if the resistor's value increased (decreased) by a small amount. Assuming that the sensitivities accurately predict large change behavior and there are no coupling effects (i.e., the sensitivity with respect to one element does not change sign due to a change in value of another element) the greatest decrease in the output would occur if the value of each resistor with a negative sensitivity was increased to the upper limit of its tolerance range, while the value of each resistor with a positive sensitivity was decreased to the lower limit of its tolerance range. Similarly, the greatest increase in the output would occur if the value of each resistor with a negative sensitivity was decreased to the lower limit of its tolerance range, while the value of each resistor with a positive sensitivity was increased to the upper limit of its tolerance range. Thus the worst case, that is, lower and upper limits of the range of possible output values, can be ascertained by analyzing the network twice: once with the element values adjusted so as to give the greatest decrease in its output and once with the element values adjusted so as to give the greatest increase in its output.

## Example 6.3

The network of Fig. 6-4a serves as an example to illustrate the procedures involved in worst case analysis. We found in Example 6.1 that the normalized sensitivities of $v_o$ with respect to the resistors $R_1$, $R_2$, and $R_3$ are

$$S_{R_1}^{v_o} = .916$$

$$S_{R_2}^{v_o} = -.672$$

and

$$S_{R_3}^{v_o} = .756$$

Since $S_{R_1}^{v_o}$ and $S_{R_3}^{v_o}$ are positive, if $R_1$ and $R_3$ increase in value, $v_o$ should also increase. Similarly, since $S_{R_2}^{v_o}$ is negative, if $R_2$ increases in value, $v_o$ should decrease. Therefore, the "worst cases" result from increasing, $R_1$ and $R_3$ and decreasing $R_2$, or decreasing $R_1$ and $R_3$ and increasing $R_2$. The ranges of $R_1$, $R_2$, and $R_3$ are (10 percent tolerance)

$$1.8 \leq R_1 \leq 2.2$$

$$14.4 \leq R_2 \leq 17.6$$

and

$$5.4 \leq R_3 \leq 6.6$$

The smallest $v_o$ results when $R_1 = 1.8$, $R_2 = 17.6$, and $R_3 = 5.4$ as shown in Fig. 6-4b, then

$$v_o = \frac{R_1 R_3}{R_1 + R_2 + R_3} \hat{i}_i = \frac{9.7}{24.8} = .392 \text{ v}$$

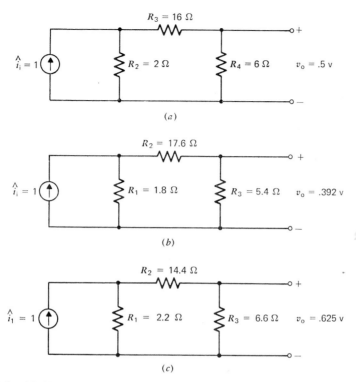

**Fig. 6-4** (*a*) **Network used in Example 6.3 to illustrate worst case analysis. The elements of this network have 10 percent tolerances. The lower bound of** $v_0$ **is found from network** (*b*) **and upper bound of** $v_0$ **is found from network** (*c*)**.**

The largest $v_0$ results when $R_1 = 2.2$, $R_2 = 14.4$, and $R_3 = 6.6$ as shown in Fig. 6-4*c*, then

$$v_0 = \frac{R_1 R_3}{R_1 + R_2 + R_3} \hat{\imath}_i = \frac{14.5}{23.2} = .625 \text{ v}$$

In a design situation, if the range of possible values of $v_0$ is too large, then tighter tolerances are needed. Observe that since the $S_{R_1}^{v_0}$ is larger than either $S_{R_2}^{v_0}$ or $S_{R_3}^{v_0}$ it might be sufficient to tighten the tolerance of just $R_1$. □

### Example 6.4

The network of Fig. 6-5 can be used to illustrate that worst case analysis-based upon sensitivity analysis may in fact lead to erroneous conclusions. Each resistor has a 10 percent tolerance. An expression for $v_0$ in terms of the

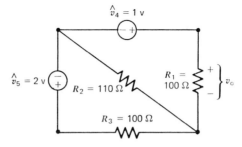

**Fig. 6-5** **Network used in Example 6.4 to illustrate that sensitivity information may not correctly predict the worst case.**

inputs $\hat{v}_4$ and $\hat{v}_5$ can easily be determined by use of superposition:

$$v_o = \frac{R_1}{R_1 + (R_2 R_3/(R_2 + R_3))}\, \hat{v}_4 - \frac{R_1 R_2/(R_1 + R_2)}{R_3 + (R_1 R_2/(R_1 + R_2))}\, \hat{v}_5$$

$$= \frac{R_1(R_2 + R_3)\hat{v}_4 - R_1 R_2 \hat{v}_5}{R_1 R_2 + R_1 R_3 + R_2 R_3}$$

On substitution of the voltage source values, the above expression becomes

$$v_o = \frac{R_1(R_3 - R_2)}{R_1 R_2 + R_1 R_3 + R_2 R_3}$$

The sensitivities of $v_o$ with respect to $R_1$, $R_2$, and $R_3$ are found by taking the partial derivatives:

$$S_{R_1}^{v_o} = \frac{\partial v_o}{\partial R_1} = \frac{R_2 R_3 (R_3 - R_2)}{(R_1 R_2 + R_1 R_3 + R_2 R_3)^2}$$

$$S_{R_1}^{v_o} = \frac{\partial v_o}{\partial R_2} = \frac{-R_1(2R_1 R_3 + R_3^2)}{(R_1 R_2 + R_1 R_3 + R_2 R_3)^2}$$

and

$$S_{R_1}^{v_o} = \frac{\partial v_o}{\partial R_3} = \frac{R_1(2R_1 R_2 + R_2^2)}{(R_1 R_2 + R_1 R_3 + R_2 R_3)^2}$$

On substitution of the nominal element values, ($R_1^\circ = 100$, $R_2^\circ = 100$, and $R_3^\circ = 110$) the sensitivities become

$$S_{R_1}^{v_o} = 1.07 \times 10^{-4}$$

$$S_{R_2}^{v_o} = -3.33 \times 10^{-3}$$

and

$$S_{R_3}^{v_o} = 2.93 \times 10^{-3}$$

and the nominal output voltage is

$$v_o = -3.13 \times 10^{-2}$$

Based on these sensitivities we would expect the minimum output to occur if $R_1$ and $R_3$ were decreased to their lower limits and $R_2$ increased to its upper limit, that is, since each element has a 10 percent tolerance the minimum $v_o$ should occur for $R_1 = 90$, $R_2 = 110$, and $R_3 = 100$. The output for this set of resistor values is

$$v_o = -3.01 \times 10^{-2}$$

But observe that if we choose the element values $R_1 = 100$, $R_2 = 110$, and $R_3 = 100$, then

$$v_o = -3.13 \times 10^{-2}$$

which is less than what was indicated as the minimum. The trouble lies with the fact that the sensitivity of $v_o$ with respect to $R_1$ has changed signs. Since $R_2$ is now greater than $R_3$, $S_{R_1}^{v_o} < 0$. Thus the minimum $v_o$ should occur when $R_1$ and $R_2$ are at their upper limits and $R_3$ at its lower limit, viz, $R_1 = 110$, $R_2 = 110$, and $R_3 = 100$. Then

$$v_o = -3.23 \times 10^{-2} \quad \square$$

This example illustrates that a sensitivity predicts local behavior due to changes in one parameter and may not predict the effects due to large changes in elements or, the coupling effects due to changes in several elements. It also illustrates that we can improve our confidence in the sensitivity information by recalculating the sensitivities at the indicated extreme (maximum or minimum). If the sensitivity of an element has changed signs, the opposite limit of the element's value should be chosen. If the signs of the sensitivities have not changed, then the results are accepted.

Although "worst case" analysis based upon sensitivity information may not yield the true worst case, because the sensitivities may not correctly predict the large change behavior, this method is widely used since it is relatively inexpensive information to obtain. We see now the importance of the sensitivities. Recall the two problems with calculating sensitivities using perturbations. First, the proper choice of size for the perturbation is unknown and could be critical for accurate results. Second, one complete network analysis is required for each element that is perturbed. A much more efficient scheme is presented next.

## 6-4 THE ADJOINT NETWORK APPROACH TO SENSITIVITIES

The perturbation method is too inefficient and inaccurate to be used for sensitivity calculations. The technique to be described in this section is highly efficient and extremely accurate. Since we seek the sensitivity of the output to changes of element values, we desire some expression that relates the output to the network elements. We will find that use of the Tellegen's theorem

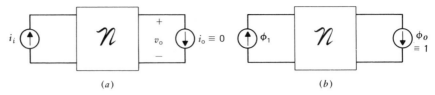

(a)                                    (b)

**Fig. 6-6**  (*a*) **The sensitivity of** $V_{\text{out}}$ **with respect to all resistors in** $N$ **can be ascertained after two network analyses, one on G and one on the adjoint network** $\hat{N}$ (*b*).

equation introduced in Chapter 5, Equation (5.1), along with the branch relationships yield the desired results.

Let $v_0$ denote the output voltage and $\hat{i}_i$ the input current of a given network $\mathcal{N}$ comprised of resistors as shown in Fig. 6-6*a*. Observe that $v_0$ is measured across a zero-valued current source $\hat{i}_0 = 0$. The introduction of this source does not affect the operation of the network but merely provides a convenient branch across which to measure the output. Assume $\mathcal{N}$ contains $n_b$ branches, $n$ of which are resistance branches numbered $1, 2, \ldots, n$; and two of which are source branches numbered $n + 1$ and $n + 2 = n_b$. We will also be considering a second network, termed the adjoint network, denoted by $\hat{\mathcal{N}}$ that has exactly the same topology (structure) as $\mathcal{N}$.

If $v_k$ and $i_k$ are used to denote the $k$th branch voltage and current, respectively, in $\mathcal{N}$ and $v_k$ and $\phi_k$ are used to denote the $k$th branch voltage and current respectively, in $\hat{\mathcal{N}}$, then upon application of Tellegen's theorems

$$\sum_{k=1}^{n_b} v_k \phi_k = 0$$

and

$$\sum_{k=1}^{n_b} i_k v_k = 0$$

We are interested in how variations of the element values of $\mathcal{N}$ affect the output, $v_0$. The branch relationship of the $k$th ($k = 1, 2, \ldots, n$) branch of $\mathcal{N}$ is

$$v_k = R_k i_k$$

If each resistor is perturbed slightly, then the branch voltages and currents will be changed. If $\Delta R_k$ denotes the change in resistance, $\Delta v_k$ the change in voltage, and $\Delta i_k$ the change in current of the $k$th resistance branch, then

$$(v_k + \Delta v_k) = (R_k + \Delta R_k)(i_k + \Delta i_k)$$

for $k = 1, 2, \ldots, n_b$. Expanding this expression, we have

$$v_k + \Delta v_k = R_k i_k + \Delta R_k i_k + R_k \Delta i_k + \Delta i_k \Delta R_k$$

but

$$v_k = R_k i_k$$

and $\Delta i_k \Delta R_k$ is a second order term that we assume is negligible so that

$$\Delta v_k \simeq \Delta R_k i_k + R_k \Delta i_k$$

For the input branch, $i_i$ remains fixed so that $\Delta i_i = 0$; similarly, $i_o$ is fixed at zero so that $\Delta i_o = 0$. Since the variations of the element values of $\mathcal{N}$ do not change its topology, Tellegen's theorem can still be applied between the varied original network and $\hat{\mathcal{N}}$:

$$\sum_{k=1}^{n_b}(v_k + \Delta v_k)\hat{\phi}_k = 0$$

and

$$\sum_{k=1}^{n_b}(i_k + \Delta i_k)\hat{v}_k = 0$$

But

$$\sum_{k=1}^{n_b}v_k\phi_k = 0$$

and

$$\sum_{k=1}^{n_b}i_k v_k = 0$$

so that

$$\sum_{k=1}^{n_b}\Delta v_k\phi_k = 0$$

and

$$\sum_{k=1}^{n_b}\Delta i_k v_k = 0$$

which can be combined to yield

$$\sum_{k=1}^{n_b}(\Delta v_k\phi_k - \Delta i_k v_k) = 0$$

or

$$\Delta v_i\phi_i - \Delta i_i v_i + \Delta v_o\phi_o - \Delta i_o v_o + \sum_{k=1}^{n}(\Delta v_k\phi_k - \Delta i_k v_k) = 0$$

Recall that

$$\Delta i_i = 0 \qquad \text{and} \qquad \Delta i_o = 0$$

and

$$\Delta v_k = \Delta R_k i_k + R_k \Delta i_k$$

for $k = 1, 2, \ldots, n$; hence

$$\sum_{k=1}^{n}(\Delta v_k\phi_k - \Delta i_k v_k) = \sum_{k=1}^{n}[(\Delta R_k i_k + R_k \Delta i_k)\phi_k - \Delta i_k v_k]$$

$$= \sum_{k=1}^{n}[(R_k\phi_k - v_k)\Delta i_k + (i_k\phi_k)\Delta R_k]$$

Therefore, we have the relationship

(6.3)     $$\Delta v_i \phi_i + \Delta v_o \phi_o + \sum_{k=1}^{n} [(R_k \phi_k - v_k) \Delta i_k + i_k \phi_k \Delta R_k] = 0$$

What we desire is to find an expression that relates changes in the output, $\Delta v_o$, to changes in the elements $\Delta R_k$. The previous expression does contain $\Delta v_o$ and $\Delta R_k$ as well as some additional terms. These additional terms can be eliminated by appropriately choosing elements in $\hat{\mathcal{N}}$. Recall that to this juncture the only restriction placed on $\hat{\mathcal{N}}$ is that it have the same network graph as $\mathcal{N}$. Observe that if we choose the element of the branch in $\hat{\mathcal{N}}$ that corresponds to the $k$th $(k = 1, 2, \ldots, n)$ resistance branch of $\mathcal{N}$ to be a resistor of value $R_k$ than the branch relationship for this branch is

$$v_k = R_k \phi_k \qquad k = 1, 2, \ldots, n$$

Therefore,

$$R_k \phi_k - v_k = 0 \qquad k = 1, 2, \ldots, n$$

so that expression (6.3) becomes

$$\Delta v_i \phi_i + \Delta v_o \phi_o + \sum_{k=1}^{n} i_k \phi_k \Delta R_k = 0$$

$\Delta v_i$ can be eliminated from this expression by choosing the branch of $\hat{\mathcal{N}}$ that corresponds to the current source $\hat{i}_i$ in $\mathcal{N}$ to be a zero-valued current source, then $\phi_i = 0$. Moreover, we can let the branch in $\hat{\mathcal{N}}$ that corresponds to the current source $i_o$ in $\mathcal{N}$ be a unity valued current source so that $\phi_o = 1$. Finally, this expression reduces to

$$\Delta v_o = \sum_{k=1}^{n} - i_k \phi_k \Delta R_k$$

If only the $l$th resistor is perturbed then

$$\Delta R_k = 0 \text{ for all } k \neq l$$

and

$$\Delta v_o = -i_l \phi_l \Delta R_l$$

Therefore, the unnormalized sensitivity for $R_l$ is

(6.6)     $$\hat{S}_{R_l}^{v_o} = \frac{\Delta v_o}{\Delta R_l} = -i_l \phi_l$$

and the normalized sensitivity for $R_l$ is

(6.7)     $$S_{R_l}^{v_o} = \frac{R_l}{v_o} \frac{\Delta v_o}{\Delta R_l} = -\frac{R_l}{v_o} i_l \phi_l$$

$l = 1, 2, \ldots, n$.

However, this result indicates that all sensitivities can be evaluated after *only two* network analyses: the analysis of the original network $\mathcal{N}$ yields the

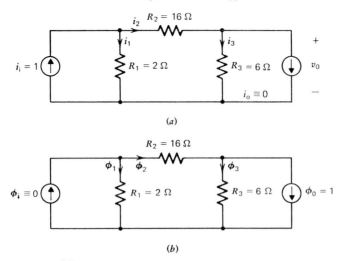

(a)

(b)

**Fig. 6-7   Illustration of the adjoint network method of calculating sensitivities (a) a given network and (b) its adjoint.**

currents $i_l$, $l = 1, 2, \ldots, n$, and analysis of the adjoint network yields the currents $\phi_l$, $l = 1, 2, \ldots, n$. The sensitivities result from the product of these currents.

**Example 6.5**

We illustrate the adjoint network method for calculating sensitivities by considering the network shown in Fig. 6-7a. This network is the same one considered in Example 6.3 except for the addition of a zero-valued current source across $R_3$. The reader should verify that $i_1 = .917$, $i_2 = .083$, and $i_3 = .083$. The associated adjoint network is shown in Fig. 6-7b. The student should verify that $\phi_1 = -.25$, $\phi_2 = .25$, and $\phi_3 = -.75$. Therefore, the unnormalized sensitivities are

$$\hat{S}_{R_1}^{v_0} = -i_1\phi_1 = -(.917)(-.25) = .229$$
$$\hat{S}_{R_2}^{v_0} = -i_2\phi_2 = -(.083)(.25) = -.021$$

and

$$\hat{S}_{R_3}^{v_0} = -i_3\phi_3 = -(.083)(-.75) = .063$$

which is in excellent agreement with previous results.   □

In a later section we will discuss further the significance of the adjoint network and some of its properties. It should be emphasized that the adjoint network approach is accurate since it will always yield the exact partial derivatives and, it is efficient since *only two* network analyses are needed no matter how many elements there are.

## 6-5  SUMMARY

In this chapter we have discussed some basic design aids: *worst case* analysis and *network sensitivities*. Generally, a *worst case analysis* is undertaken to determine what the widest possible range of output values is due to the tolerance ranges of the elements. The sensitivities are used to indicate how the output varies with respect to the element values. If the sensitivity is positive, the output increases (decreases) if the value of the element increases (decreases) and, if the sensitivity is negative the output decreases (increases) if the value of the element increases (decreases). To find the upper limit of the output, the elements with positive sensitivities should be increased and the elements with negative sensitivities should be decreased. To find the lower limit of the output, the elements with the positive sensitivities should be decreased and the elements with the negative sensitivities should be increased.

Two methods were described for calculation of the network sensitivities. The first, called the *perturbation method*, required that a small change be made in each parameter and the network reanalyzed. This method requires one analysis for each element in addition to the original analysis and is prone to inaccuracies. The second method, called the *adjoint network method*, requires only one analysis in addition to the original analysis and yields accurate results.

## PROBLEMS

**6.1**    Write an expression that relates the output $v_0$ to the input current $\hat{\imath}_i$ in the network of Fig. P6.1 (Do not substitute the element values.) By differentiating this expression, find the unnormalized sensitivity of $v_0$ to each of the resistors. Assume that $R_1 = 2\ \Omega$, $R_2 = 3\ \Omega$, and

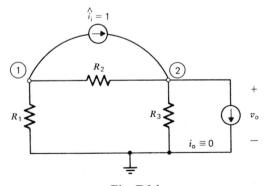

**Fig. P6.1**

$R_3 = 1\,\Omega$. Based on the sensitivities, predict how changes in the value of each resistor affects $v_0$. Verify your prediction by actually perturbing the element values one at a time by 10 percent and recalculating $v_0$.

**6.2**  Repeat Prob. 6.1 for the network of Fig. P6.2.

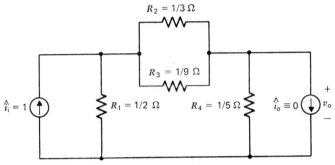

R_2 = 1/3 Ω

R_3 = 1/9 Ω

$\hat{i}_i = 1$         $R_1 = 1/2\,\Omega$         $R_4 = 1/5\,\Omega$         $\hat{i}_0 \equiv 0$   $v_0$

**Fig. P6.2**

**6.3**  Determine the sensitivity of $v_0$ with respect to each element in the networks of Figs. P6.1 and P6.2 using the perturbation method. Use 1 percent, 10 percent, and 20 percent perturbations and compare the results with those obtained in Probs. 6.1 and 6.2. What conclusions can be drawn with regard to the accuracy of the perturbation method.

**6.4**  Calculate the sensitivity of $v_0$ with respect to each element value in the network of Fig. P6.4 by using 10 percent perturbations. (Use of GEP should help with the computations.)

**6.5**  Assume that each of the elements used to build the networks of Figs. P6.1, P6.2, and P6.4 have 20 percent tolerances. Using the sensitivities calculated in Probs. 6.1, 6.2, and 6.4 determine the range of possible values of $v_0$.

**6.6**  Compute the sensitivities of $v_0$ to each element in the network of Fig. P6.6 by first writing an expression for $v_0$ in terms of the elements and then taking partial derivatives. Calculate the worst case for 20 percent element tolerances. With all the elements adjusted so that the "maximum" $v_0$ is obtained, move the value of $R_1$ to its opposite boundary. What is $v_0$ now? Similarly, with all elements adjusted so that the "minimum" $v_0$ is obtained, move the value of $R_1$ to its opposite value and recompute $v_0$. What conclusions can be drawn?

**6.7**  Use the adjoint network approach to compute the sensitivities of $v_0$ with respect to each circuit element in the networks of Figs. P6.1, P6.2, and P6.4. Compare these results with those obtained previously by direct differentiation and perturbation.

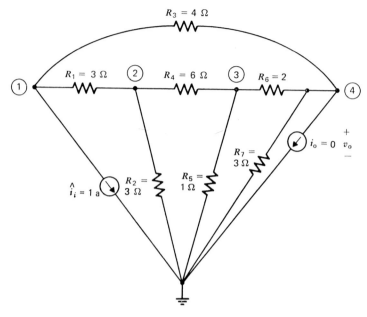

**Fig. P6.4**

**6.8**   When calculating sensitivities we have always set the input current $\hat{i}_i$ equal to unity. What would happen if another value of the input current was used, for example $\hat{i}_i = 2$?

**6.9**   The output of the network in Fig. P6.9 is a current $i_o$ that is measured as the current flowing through a zero-valued voltage source. Show how the adjoint network method may be used to determine the sensitivity of $i_o$ to each element.

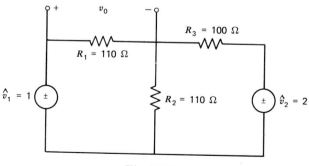

**Fig. P6.6**

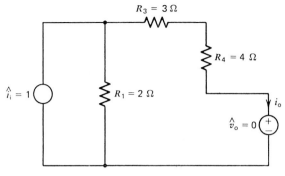

**Fig. P6.9**

**6.10**   By following the derivation presented in the text for resistor elements show that the adjoint element corresponding to a conductor
described by the branch relationship

$$i_G = Gv_G$$

is a conductor described by

$$\phi_G = Gv_G$$

and that the sensitivity of an output voltage $v_o$ to a conductor is

$$\frac{\partial v_o}{\partial G} = v_G v_G$$

**6.11**   Verify the results of Prob. 6.10 by finding the sensitivity of $v_o$ to $G_1$
and $G_2$ in the network of Fig. P6.11 by (a) expressing $v_o$ in terms of

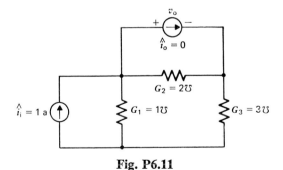

**Fig. P6.11**

$G_1$ and $G_2$ taking derivatives explicitly, and substituting in element
values; and (b) using the adjoint network approach. Compare your
results.

**6.12**   Determine the worst case values of $v_o$ for the network of Fig. P6.12.
Assume all elements have 5 percent tolerances.

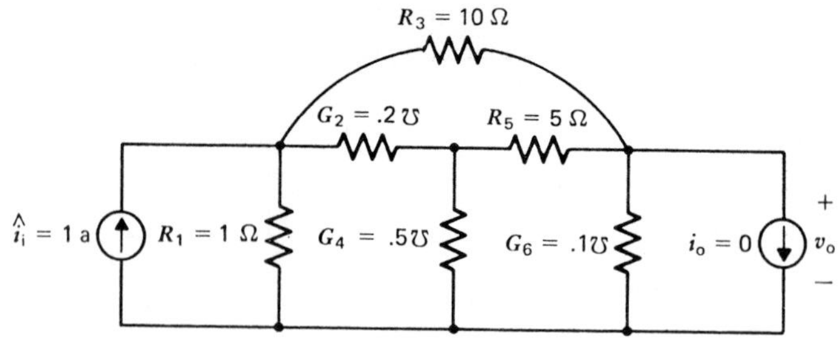

**Fig. P6.12**

CHAPTER **7**

# Simple Nonlinear Resistive Networks

A nonlinear network is one that contains at least one nonlinear element. Nonlinear elements are important in circuit design and are used for a variety of tasks. Some of these functions will be discussed later. In this chapter we will introduce the two terminal nonlinear resistive element. The most common nonlinear resistor is the ideal diode. The analysis techniques discussed so far are incapable of analyzing nonlinear networks and so a new method will be introduced. We will find though that the analysis methods of the earlier chapters form the basis for nonlinear network analysis.

## 7-1 THE DIODE

In general, a nonlinear resistor is characterized by a branch relationship of the form

$$v = f(i)$$

or

$$i = g(v)$$

where $f$ and $g$ are nonlinear functions of $i$ and $v$, respectively. In other words, if voltage $v_1$ causes a current $i_1$ to flow in the element and voltage $v_2$ causes

183

current $i_2$ to flow in the element, then a voltage $\alpha v_1 + \beta v_2$ (where $\alpha$ and $\beta$ are finite constants) *will not* cause a current $\alpha i_1 + \beta i_2$ to flow. Symbolically,

$$\alpha v_1 + \beta v_2 \neq f(\alpha i_1 + \beta i_2)$$

or,

$$\alpha i_1 + \beta i_2 \neq g(\alpha v_1 + \beta v_2)$$

For example, the branch relationship

$$v = f(i) = 6i^2 + 2i$$

describes a nonlinear resistor, since

$$v_1 = f(i_1) = 6i_1{}^2 + 2i_1$$

and

$$v_2 = f(i_2) = 6i_2{}^2 + 2i_2$$

but

$$f(\alpha i_1 + \beta i_2) = 6(\alpha i_1 + \beta i_2)^2 + 2(\alpha i_1 + \beta i_2)$$
$$= 6\alpha^2 i_1{}^2 + 2\alpha i_1 + 6\beta^2 i_2{}^2 + 2\beta i_2$$
$$+ 12\alpha\beta i_1 i_2$$

which is not equal to

$$f(\alpha i_1) + f(\beta i_2) = 6\alpha^2 i_1{}^2 + 2\alpha i_1$$
$$+ 6\beta^2 i_2{}^2 + 2\beta i_2$$

Although we can mathematically describe many types of nonlinear resistors, only a few types actually occur physically. One of the most common physical nonlinear resistors is the semiconductor diode that is characterized by the equation

(7.1) $$i = I_S(e^{qv/kT} - 1)$$

where $I_S$ is called the reverse saturation current, $q$ is electronic charge ($1.6 \times 10^{-19}$ coulombs), $k$ is Boltzmann's constant ($1.38 \times 10^{-23}$ joules/°K), and $T$ is temperature in degrees Kelvin. Usually $T = 300°K$ (room temperature) and

$$\frac{q}{kT} = 40 \, (\text{joules/coulomb})^{-1} = 40 \, (\text{volts})^{-1}$$

The diode symbol and a graphical $i - v$ characterization are shown in Figs. 7-1a and 7-1b. To see that the diode is really a nonlinear element, assume voltage $v_1$ corresponds to current $i_1$ and voltage $v_2$ corresponds to current $i_2$, that is

$$i_1 = I_S(e^{qv_1/kT} - 1)$$

and

$$i_2 = I_S(e^{qv_2/kT} - 1)$$

One of the consequences of linearity is that voltage $v_1 + v_2$ correspond to current $i_1 + i_2$. But if voltage $v_1 + v_2$ appears across the diode, the current

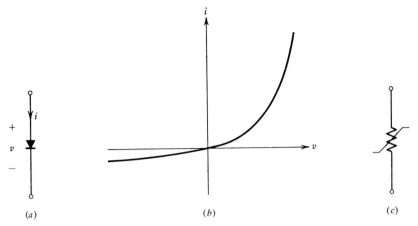

**Fig. 7-1** (*a*) **The diode symbol and** (*b*) **a graphical** *i–v* **characterization.** (*c*)
**Symbol used to indicate a general nonlinear resistor.**

through it is

$$i = I_S(e^{q(v_1+v_2)kT} - 1) = I_S(e^{qv_1/kT}e^{qv_2/kT} - 1)$$

which does not correspond to the current $i_1 + i_2$:

$$i_1 + i_2 = I_S(e^{qv_1/kT} + e^{qv_2/kT} - 2)$$

Therefore, the diode is nonlinear. As will be seen shortly, the inclusion of a
diode in a network causes the analysis procedure to be considerably more
complex. We observe that the diode relationship (7.1) can also be expressed
as

(7.2) $$v = \frac{kT}{q}\ln\frac{i}{I_S + 1}$$

At times it will be convenient to consider other nonlinear resistors, not all
of which are physically available. The symbol of Fig. 7-1*c* is used to denote
a general nonlinear resistor.

## 7-2  SINGLE DIODE NETWORKS AND GRAPHICAL SOLUTION

Consider the simple diode network of Fig. 7-2*a*. KVL and KCL remain
valid for nonlinear networks. There is one independent KCL node equation:

$$-i_1 + i_2 + i_3 = 0$$

and two independent KVL equations that we express in terms of the three
equations relating the branch voltages to the node voltages:

$$v_1 = -v_①$$

$$v_2 = v_①$$

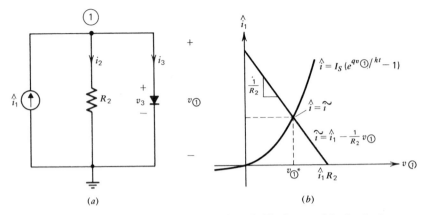

**Fig. 7-2**  (*a*) **A simple diode network and** (*b*) **the graphical solution.**

and

$$v_3 = v_①$$

The three branch relationships are

$$i_1 = \hat{\imath}_1$$

$$v_2 = R_2 i_2$$

and

$$i_3 = I_S(e^{qv_3/kT} - 1)$$

Substitution of the branch relationships into the KCL node equation yields

$$-\hat{\imath}_1 + \frac{1}{R_2} v_2 + I_S(e^{qv_3/kT} - 1) = 0$$

Finally, on substitution of the KVL equations, we have

$$-\hat{\imath}_1 + \frac{1}{R_2} v_① + I_S(e^{qv_①/kT} - 1) = 0$$

or

(7.3)    $$\frac{1}{R_2} v_① + I_S(e^{qv_①/kT} - 1) = \hat{\imath}_1$$

It is impossible to proceed any further. An analytic solution does not exist for this nonlinear equation. The student should be impressed with the fact that even the simple nonlinear network of Fig. 7-2 is not amenable to the analysis techniques we presently have at our disposal.

In the next section we will describe an approximation scheme that can be used to "solve" nonlinear equations such as (7.3). For the remainder of this section we will describe a graphical method of solution.

Equation (7.3) can be rewritten as

$$(7.4) \qquad I_S(e^{qv_①/kT} - 1) = \hat{i}_1 - \frac{1}{R_2} v_①$$

where the left-hand side is a current due to the diode and, the right-hand side is a current due to the current source and resistor. We may view the problem of finding $v_①$ as follows. Let $\hat{i}$ denote the current flowing in the diode,

$$(7.5) \qquad \hat{i} = I_S(e^{qv_①/kT} - 1)$$

and $\bar{i}$ the current flowing through the current source-resistor combination,

$$(7.6) \qquad \bar{i} = \hat{i}_1 - \frac{1}{R_2} v_1$$

The network of Fig. 7-2 requires that

$$(7.7) \qquad \hat{i} = \bar{i}$$

To find the value of $v_1$ such that (7.7) is valid we plot on a two-dimensional graph $\hat{i}$ as a function of $v_①$ and of $\bar{i}$ as a function of $v_1$. At the intersection of these curves $\hat{i} = \bar{i}$, and the value of $v_①$ at this point is the solution of (7.4) and hence (7.3).

Figure 7-2b illustrates the graphical solution. The diode curve $\hat{i}$ is easily plotted as a function of $v_①$. The curve $\bar{i}$ is easily constructed by recognizing that it is a straight line with an intercept of $\hat{i}_1$ on the vertical axis (when $v_① = 0$) and a slope of $-1/R_2$.

By using Norton's theorem, the graphical method of analysis can be used to solve rather complex resistive networks that contain a single diode.

### Example 7.1

We wish to solve for the node voltages of the network shown in Fig. 7-3a. The nodal equilibrium equations for nodes ① and ③ are easily written:

$$node\ ① \qquad -\frac{1}{3} v_② \qquad\qquad + i_1 = \frac{-5}{3}$$

and

$$node\ ③ \qquad 19v_③ - 9v_④ \qquad = 2$$

The nodal equilibrium equations for nodes ② and ④ involve the diode current $i_8$:

$$i_8 = .1(e^{40v_8} - 1)$$

This current can be expressed in terms of the node voltages since

$$v_8 = v_④ - v_②$$

Therefore,

$$i_8 = .1(e^{40(v_④ - v_②)} - 1)$$

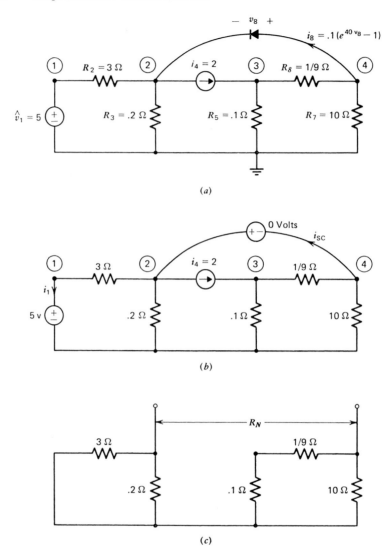

(a)

(b)

(c)

**Fig. 7-3** (*a*) Nonlinear network used in Example 7.1 to further illustrate the graphical method of solution; (*b*) network used to determine the short circuit current, $i_{sc}$, for the Norton equivalent; (*c*) network used to find $R_N$; (*d*) the Norton equivalent; (*e*) graphical solution.

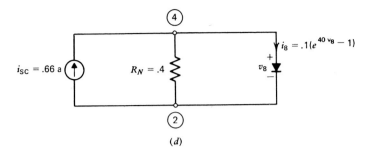

(d)

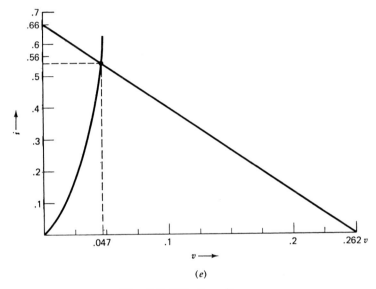

(e)

**Fig. 7-3 (Continued)**

The nodal equilibrium equations for nodes ② and ④ are now easy to write:

$$node \ ② \qquad \frac{16}{3} v_{②} - .1[e^{40(v_{④} - v_{②})} - 1] \qquad = -\frac{1}{3}$$

and

$$node \ ④ \qquad -9v_{③} + 9.1v_{④} + .1[e^{40(v_{④} - v_{②})} - 1] = 0$$

Clearly, we are unable to use GEP to solve these nonlinear algebraic equations. What we propose is to determine the Norton equivalent network seen by the diode and then use graphical analysis to find node voltages $v_{②}$ and $v_{④}$. Then node equation ① can be used to solve for $i_1$ and node equation ④ can be used to solve for $v_{③}$.

The short circuit current is found from the network of Fig. 7-3$b$. The nodal equilibrium equations of this network are:

$$\text{node } \textcircled{1} \qquad -\frac{1}{3}v_{\textcircled{2}} \qquad\qquad\qquad + i_1 = -\frac{5}{3}$$

$$\text{node } \textcircled{2} \qquad \frac{16}{3}v_{\textcircled{2}} \qquad\qquad\qquad - i_{sc} = -\frac{1}{3}$$

$$\text{node } \textcircled{3} \qquad 19v_{\textcircled{3}} - 9\ v_{\textcircled{4}} \qquad\quad = \quad 2$$

$$\text{node } \textcircled{4} \qquad - 9v_{\textcircled{3}} + 9.1v_{\textcircled{4}} + i_{sc} = \quad 0$$

In addition, we have the voltage source relationship

$$-v_{\textcircled{2}} + v_{\textcircled{4}} = 0$$

These equations are easily solved by use of GEP: $v_{\textcircled{2}} = .06$, $v_{\textcircled{3}} = .13$, $v_{\textcircled{4}} = .06$, $i_1 = -1.6$, and $i_{sc} = .66$. Only $i_{sc}$ is of interest.

The Norton resistance $R_N$ is obtained from the network of Fig. 7-3$c$:

$$R_N = \frac{(3)(.2)}{3 + .2} + \frac{\left(\dfrac{1}{9} + .1\right)(10)}{\left(\dfrac{1}{9} + .1\right) + 10}$$

$$= .4$$

As far as the diode is concerned, the behavior of the network of Fig. 7-3$d$, is identical to the behavior of the network of Fig. 7-3$a$. We have labeled the nodes of the network in Fig. 7-3$d$ so as to indicate their correspondence with the nodes of the network in Fig. 7-3$a$.

The network of Fig. 7-3$d$ can be analyzed graphically as shown in Fig. 7-3$e$. We find that $v_8 = .047$ and $i_8 = .56$. Observe that the original nodal equilibrium equations may be expressed as

$$\text{node } \textcircled{1} \qquad -\frac{1}{3}v_{\textcircled{2}} \qquad\qquad\qquad + i_1 = -\frac{5}{3}$$

$$\text{node } \textcircled{2} \qquad \frac{16}{3}v_{\textcircled{2}} \qquad\qquad\qquad = -\frac{1}{3} + i_8$$

$$\text{node } \textcircled{3} \qquad 19v_{\textcircled{3}} - 9\ v_{\textcircled{4}} \qquad\quad = \quad 2$$

$$\text{node } \textcircled{4} \qquad -9v_{\textcircled{3}} + 9.1\ v_{\textcircled{4}} \qquad = -\ i_8$$

or, on substitution of the calculated value of $i_8 = .56$

$$\text{node } ① \quad -\frac{1}{3}v_{②} \qquad\qquad +i_1 = -\frac{5}{3}$$

$$\text{node } ② \quad \frac{16}{3}v_{②} \qquad\qquad = \quad .23$$

$$\text{node } ③ \quad 19v_{③} - 9\ v_{④} \qquad = \quad 2$$

$$\text{node } ④ \quad -9v_{③} + 9.1v_{④} \qquad = \ -.56$$

Solution of these equations yields the unknown node voltages and voltage source currents of the original network: $v_{②} = .04$, $v_{③} = .14$, $v_{④} = .08$, and $i_1 = -1.7$. Observe that we have in effect used the substitution theorem to replace the diode by a current source of an appropriate value. ☐

## 7-3  THE NEWTON-RAPHSON ALGORITHM

The solution of nonlinear networks graphically is cumbersome and inaccurate when a single nonlinear element is present and, next to impossible if many nonlinear elements are present. Moreover, graphical solutions are not easily programmed and, therefore, not amenable for computer usage. In this section we describe a technique, called the *Newton-Raphson* method, that is often used to solve nonlinear equations. The method is *iterative*, in other words, an estimate of the solution is made and then by following a prescribed set of rules, or algorithm, the estimate is updated. The process is then repeated and the estimate updated again and again until either the "exact" solution has been found or the estimate starts to *diverge*, that is, successive estimates become further and further apart.

Before proceeding, it is expedient to review the *Taylor series expansion* of a function $f(x)$ about a point $x^*$. (We will use the word *point* interchangably with *value*.) By definition, if $f(x)$ possesses derivatives of all orders at $x = x^*$, the *Taylor series expansion* of $f(x)$ about the point $x = x^*$ is

$$f(x) = f(x^*) + \frac{df}{dx}\bigg|_{x^*}(x - x^*) + \frac{1}{2!}\frac{d^2f}{dx^2}\bigg|_{x^*}(x - x^*)^2$$

$$+ \frac{1}{3!}\frac{d^3f}{dx^3}\bigg|_{x^*}(x - x^*)^3 + \cdots$$

$$= f(x^*) + \sum_{k=1}^{\infty}\frac{1}{k!}\frac{d^{(k)}f}{dx^{(k)}}\bigg|_{x^*}(x - x^*)^k$$

where

$$\frac{d^{(k)}f}{dx^{(k)}}\bigg|_{x^*}$$

is used to denote the $k$th derivative of $f(x)$ evaluated at the point $x = x^*$. (*Note.* The $k$ derivatives of $f(x)$ are taken and then $x^*$ is substituted in.)

### Example 7.2

Suppose we wish to expand

$$f(x) = 3x^2 + e^x$$

in a Taylor series about the point $x = 1$. We have

$$f(x) = f(1) + \frac{df}{dx}\Big|_1 (x - 1) + \frac{1}{2}\frac{d^2f}{dx^2}\Big|_1 (x - 1)^2 + \cdots$$

$$= 5.72 + 8.72(x - 1) + \frac{1}{2}(2.72)(x - 1)^2 + \cdots$$

If third and higher order terms of the Taylor series are neglected, then $f(x)$ is approximated by $\hat{f}(x)$:

$$f(x) \simeq \hat{f}(x) = 1.36x^2 + 6x - 1.64$$

Table 7-1 compares the actual value of $f(x)$ and the approximate value of $f(x)$, that is $\hat{f}(x)$, for various values of $x$. Observe that as $x$ moves further away from 1, the approximation to $f(x)$ by the first three terms of the Taylor series gets worse. $\square$

An alternate form for expressing the Taylor series expansion is to define

$$\Delta x = x - x^*$$

so that

$$x = x^* + \Delta x$$

Then we can write

$$f(x^* + \Delta x) = f(x^*) + \sum_{k=1}^{\infty} \frac{1}{k!}\frac{d^{(k)}f}{dx^{(k)}}(\Delta x)^k$$

**Table 7-1**

| $x$ | $f(x)$ | $\hat{f}(x)$ |
|------|--------|--------|
| 1    | 5.72   | 5.72   |
| 1.05 | 6.17   | 6.16   |
| 1.1  | 6.63   | 6.61   |
| .95  | 5.29   | 5.29   |
| .90  | 4.89   | 4.86   |
| 1.5  | 11.2   | 10.4   |
| 2    | 19.4   | 15.8   |
| 0    | 1.00   | −1.64  |

The Newton-Raphson algorithm is most easily motivated by trying to solve some nonlinear equation. Suppose we desire the solution of

$$x^3 + 3x - 4 = 0$$

Denote the left-hand side of this equation by $f(x)$:

$$f(x) = x^3 + 3x - 4$$

We seek to find the solution, that is, a value of $x$, denoted by $x^*$, such that

$$f(x^*) = 0$$

For any other value of $x$,

$$f(x) \neq 0$$

Let $x^0$ denote the initial estimate of $x^*$. Unless we are extremely lucky, $x^0 \neq x^*$ in general, so that

$$f(x^0) \neq 0$$

Now let $x^1$ be the first updated estimate of $x^*$. We can think of obtaining $x^1$ by adding a small change, $\Delta x^0$, to the initial estimate $x^0$, that is,

$$x^1 = x^0 + \Delta x^0$$

and

$$f(x^1) = f(x^0 + \Delta x^0)$$

Now $f(x^0 + \Delta x^0)$ can be expanded in a Taylor series about the initial estimate $x^0$:

$$f(x^0 + \Delta x^0) = f(x^0) + \frac{df}{dx}\bigg|_{x^0} \Delta x^0 + \frac{1}{2} \frac{d^2 f}{dx^2}\bigg|_{x^0} (\Delta x^0)^2 + \cdots$$

Next we make the assumption that $\Delta x^0$ is small so that $(\Delta x^0)^2$ and higher powers of $(\Delta x^0)$ are negligible. It is important to bear in mind that this assumption may not always be valid. Under the assumption that $\Delta x^0$ is small

$$f(x^0 + \Delta x^0) \simeq f(x^0) + \frac{df}{dx}\bigg|_{x^0} \Delta x^0$$

We now ask what value of $\Delta x^0$ will cause $x^1 = x^0 + \Delta x^0$ to be the correct solution? If $x^1$ is the correct solution, then

$$f(x^1) = 0$$

so that

$$f(x^0 + \Delta x^0) = f(x^0) + \frac{df}{dx}\bigg|_{x^0} \Delta x^0 = 0$$

The second equation can be solved for $\Delta x^0$:

$$\Delta x^0 = \frac{-f(x^0)}{\dfrac{df}{dx}\bigg|_{x^0}}$$

or, since $\Delta x^0 = x^1 - x^0$,

$$x^1 = x^0 - \frac{f(x^0)}{\left.\frac{df}{dx}\right|_{x^0}}$$

Let us return to our original problem of trying to find the solution of

$$f(x) = x^3 + 3x - 4 = 0$$

Suppose our initial estimate is $x^0 = 2$. Observe that $f(2) = 10$. Our updated estimate is obtained from

$$x^1 = x^0 - \frac{f(x^0)}{\left.\frac{df}{dx}\right|_{x^0}}$$

But

$$\frac{df}{dx} = 3x^2 + 3$$

so that

$$\left.\frac{df}{dx}\right|_{x=2} = 15$$

and

$$x^1 = 2 - \frac{10}{15} = 1.33$$

Observe that $f(1.33) = 2.35$. Clearly $x^1$ is a better estimate of the true solution than $x^0$.

We can repeat the procedure, assuming that 1.33 was our original estimate and calculating a new $x^1$. Rather than considering $x^1$ to be $x^0$ and starting over again we can obtain a "cleaner result" with a slight modification. Let $x^l$ denote the $l$th estimate of the solution and $x^{l+1}$ denote the $(l + 1)$st estimate of the solution. Then

$$x^{l+1} = x^l + \Delta x^l$$

Now expanding $f(x^{l+1})$ in a Taylor series about $x^l$, we have

$$f(x^{l+1}) = f(x^l + \Delta x^l) = f(x^l) + \left.\frac{df}{dx}\right|_{x^l} \Delta x^l$$

$$+ \frac{1}{2} \left.\frac{d^2 f}{dx^2}\right|_{x^l} (\Delta x^l)^2 + \cdots$$

Under the assumption that $\Delta x^l$ is small and that $x^{l+1}$ is the solution we seek, that is,

$$f(x^{l+1}) = 0$$

an expression for $\Delta x^l$ is obtained

$$\Delta x^l = \frac{-f(x^l)}{\dfrac{df}{dx}\bigg|_{x=x^l}}$$

so that

(7.8)
$$x^{l+1} = x^l - \frac{f(x^l)}{\dfrac{df}{dx}\bigg|_{x=x^l}}$$

Expression (7.8) is an iterative expression known as the *Newton-Raphson* algorithm.

Again consider the problem of solving

$$f(x) = x^3 + 3x - 4 = 0$$

We started with $x^0 = 2$ and found that $x^1 = 1.33$. We now find the second updated estimate of the solution from (7.8) with $l = 1$:

$$x^2 = x^1 - \frac{f(x^1)}{\dfrac{df}{dx}\bigg|_{x=x^1}}$$

$$= 1.33 - \frac{2.35}{8.31}$$

$$= 1.05$$

Observe that $f(1.05) = .25$. Thus $x^2 = 1.05$ is an even better estimate of the true solution which is $x^* = 1$. A few examples will illustrate further the use of the Newton-Raphson algorithm.

### Example 7.3

The Newton-Raphson algorithm can be used to determine the value of $\sqrt{2}$. Consider the equation

$$f(x) = x^2 - 2 = 0$$

Let the initial guess be

$$x^0 = 1$$

The estimate is obtained from

$$x^1 = x^0 - \frac{f(x^0)}{\dfrac{df(x)}{dx}\bigg|_{x=x^0}}$$

where

$$f(x^0) = (1)^2 - 2 = -1$$

**Table 7-2**

| $l$ | $x^l$ | $f(x^l)$ | $\left.\dfrac{df}{dx}\right|_{x=x^l}$ | $\Delta x^l$ |
|-----|-------|----------|---------------------------------------|--------------|
| 0 | 1 | −1 | 2 | .5 |
| 1 | 1.5 | .25 | 3 | −.083 |
| 2 | 1.43 | .04 | 2.86 | −.014 |
| 3 | 1.416 | .005 | 2.84 | −.00176 |

and

$$\frac{df(x)}{dx} = 2x$$

so that

$$\left.\frac{df(x)}{dx}\right|_{x=x^0} = (2)(1) = 2$$

Therefore

$$x^1 = 1 - \frac{(-1)}{(2)} = 1.5$$

Table 7-2 illustrates a few more steps of the process.

Observe that $x^l$ is converging to the value $1.414\ldots$. If the initial guess was $x^0 = -1$ then the $x^l$ would converge to $-1.414\ldots$ as seen in Table 7-3. Before leaving this example we will show that the Newton-Raphson algorithm is not guaranteed to converge. If the initial guess of $x^0 = 0$ were used

$$\left.\frac{df}{dx}\right|_{x=x^0} = 0$$

and the algorithm fails.    □

**Table 7-3**

| $l$ | $x^l$ | $f(x^l)$ | $\left.\dfrac{df}{dx}\right|_{x=x^l}$ | $\Delta x^l$ |
|-----|-------|----------|---------------------------------------|--------------|
| 0 | −1 | −1 | −2 | −.5 |
| 1 | −1.5 | .25 | −3 | .083 |
| 2 | −1.43 | .04 | −2.86 | .014 |
| 3 | −1.416 | .005 | −2.84 | .00176 |
| 4 | −1.414 | | | |

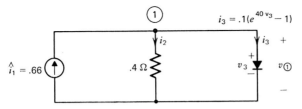

**Fig. 7-4   Nonlinear network used in Example 7.4 to illustrate the use of the Newton-Raphson algorithm.**

**Example 7.4**

The nodal equilibrium equation of the network in Fig. 7-4 is

$$f(v_①) = \frac{1}{.4} v_① + .1(e^{40v_①} - 1) - .66 = 0$$

This nonlinear equation can be solved for $v_1$ using the Newton-Raphson algorithm:

$$v_①^{l+1} = v_①{}^l - \frac{f(v_①{}^l)}{\dfrac{df}{dv_①}\Big|_{v_①{}^l}}$$

The derivative of $f(v_①)$ evaluated at $v_①{}^l$ is

$$\frac{df}{dv_①}\Big|_{v_①{}^l} = \frac{1}{.4} + 4e^{40v_①{}^l}$$

so that

$$v_①^{l+1} = v_①{}^l - \frac{\left(\dfrac{1}{.4} v_①{}^l + .1(e^{40v_①{}^l} - 1) - .66\right)}{\dfrac{1}{.4} + 4e^{40v_①{}^l}}$$

Table 7-4 lists $v_①{}^l$, $f(v_①{}^l)$ and $df/dv_①|_{v_①{}^l}$ for an initial estimate of $v_①{}^0 = 0$. Observe that the solution, $v_① = .047$, is the same as was obtained graphically in Example 7.1.

We have neglected to say anything with regard to the number of iterations required to achieve convergence of the Newton-Raphson algorithm. Clearly, the further away from the solution that we start the more iterations are required. As seen in Table 7-4, for the initial estimate of $v_1{}^0 = 0$, five iterations were required. If the initial estimate was $v_1{}^0 = .264$ (which is the voltage across the .4 $\Omega$ resistor if the diode was disconnected) then 11 iterations are needed. This case is shown in Table 7-5. $\square$

**Table 7-4**  $v_①{}^0 = 0$

| $l$ | $v_①{}^l$ | $f(v_①{}^l)$ | $df/dv_①\|_{v_①{}^l}$ |
|---|---|---|---|
| 0 | 0.0 | −.66 | 6.5 |
| 1 | 0.102 | 5.3 | 235 |
| 2 | 0.079 | 1.79 | 96.6 |
| 3 | 0.060 | .512 | 47.4 |
| 4 | 0.050 | .091 | 31.6 |
| 5 | 0.047 | .005 | |

The reader should bear in mind that the Newton-Raphson algorithm is not guaranteed to converge. Even though we are not guaranteed of convergence, the Newton-Raphson algorithm is very powerful and is widely used.

A graphical interpretation of the Newton-Raphson algorithm can be made that may yield additional insight. Consider the function $f(x)$ shown in Fig. 7-5. Let $x^*$ denote the value of $x$ for which $f(x) = 0$. $x^k$ denotes the estimate of $x^*$ after $k$ Newton-Raphson iterations. The $(k + 1)$st estimate is

$$x^{k+1} = x^k - \frac{f(x^k)}{\dfrac{df(x)}{dx}\bigg|_{x=x^k}}$$

The term $df(x)/dx|_{x=x^k}$ is the slope of tangent of $f(x)$ about the point $x^k$. Recall that the Newton-Raphson scheme was derived by assuming that

$$f(x^{k+1}) = 0$$

**Table 7-5**  $v_①{}^0 = .264$

| $l$ | $v_①{}^l$ | $f(v_①{}^l)$ | $df/dv_①\|_{v_①{}^l}$ |
|---|---|---|---|
| 0 | .264 | 3856. | $1.54 \times 10^5$ |
| 1 | .239 | 1419. | $5.67 \times 10^4$ |
| 2 | .214 | 522 | $2.09 \times 10^4$ |
| 3 | .189 | 192 | 7688 |
| 4 | .164 | 70.5 | 2835 |
| 5 | .139 | 25.8 | 1051 |
| 6 | .115 | 9.34 | 395 |
| 7 | .091 | 3.28 | 155 |
| 8 | .070 | 1.05 | 67.9 |
| 9 | .054 | .257 | 37.7 |
| 10 | .048 | .030 | 29.3 |
| 11 | .047 | .0006 | |

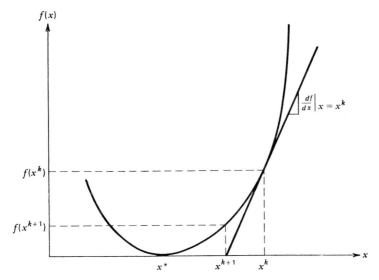

**Fig. 7-5   Graphical interpretation of the Newton-Raphson algorithm.**

therefore, $x^{k+1}$ is the value of $x$ for which the tangent line of $f(x)$ at $x = x^k$ intercepts the $x$-axis.

## 7-4   A NETWORK INTERPRETATION OF THE NEWTON-RAPHSON ALGORITHM

The fundamental principle behind the Newton-Raphson algorithm is the Taylor series expansion that is used to "linearize" the nonlinear expression about an estimate of the solution. By expanding the diode nonlinearity in a Taylor series expansion we can obtain a network interpretation of the Newton-Raphson algorithm that will aid our investigation of multiple diode networks.

Assume for the present that a diode is characterized by the branch relationship

$$i = f(v)$$

Let $v^l$ and $v^{l+1}$ denote the $l$th and $(l + 1)$st estimate of the diode voltage respectively. Furthermore, let $i^l$ and $i^{l+1}$ denote the diode currents that correspond to the diode voltages $v^l$ and $v^{l+1}$, respectively,

$$i^{l+1} = f(v^{l+1})$$

and

$$i^l = f(v^l)$$

Expansion of $f(v^{l+1})$ in a Taylor series about $v^l$ yields

(7.9)
$$i^{l+1} = f(v^{l+1}) = f(v^l) + \frac{df}{dv}\bigg|_{v^l} (v^{l+1} - v^l)$$

$$+ \frac{1}{2} \frac{d^2f}{dv^2}\bigg|_{v^l} (v^{l+1} - v^l)^2 + \cdots$$

Under the assumption that $|v^{l+1} - v^l|$ is small, so that second and higher order powers of $(v^{l+1} - v^l)$ are negligible, then the expression becomes

(7.10)
$$i^{l+1} = f(v^l) + \frac{df}{dv}\bigg|_{v^l} (v^{l+1} - v^l)$$

or

(7.11)
$$i^{l+1} = f(v^l) - \frac{df}{dv}\bigg|_{v^l} v^l + \frac{df}{dv}\bigg|_{v^l} v^{l+1}$$

Remember, at the $(l + 1)$st iteration all terms with a superscript $l$ are known. In particular $v^l$ is known and, therefore, $f(v^l)$ and $df/dv|_{v^l}$ are also known. Moreover, since $df/dv$ has units of conductance,

$$\frac{df}{dv}\bigg|_{v^l} v^l$$

has units of current. Thus the first two terms on the right hand side of expression (7.11) are known and have units of current ($f(v^l)$ is the diode current at the $l$th iteration). With this observation, Equation (7.11) may be interpreted at the $(l + 1)$st iteration as the branch relationship of a current source of value

$$f(v^l) - \frac{df}{dv}\bigg|_{v^l} v^l$$

in parallel with a conductance of value $df/dv_{v^l}$. This interpretation is shown in Fig. 7-6. For reasons that will become clear, we will refer to this interpretation as the *large signal model* or the *linearized model*. The following example illustrates the use of the network interpretation of the Newton-Raphson method for analysis.

**Example 7.5**

The nonlinear element in the network of Fig. 7-7a is described by

$$i_3 = f(v_3) = v_3^3 + \frac{1}{2} v_3^2$$

(We are not using the ordinary diode relationship for two reasons: first, manipulations of an exponential becomes cumbersome and may obscure the

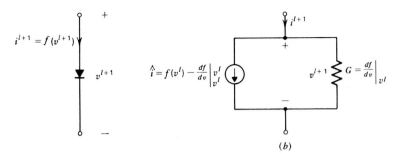

(b)

**Fig. 7-6** (a) Upon a Taylor series expansion of $f(v^{l+1})$ about $v^l$, the network interpretation (b) can be made.

method; second, we wish to show that the technique described here can be used for a wide variety of nonlinear resistors.) At the $(l + 1)$st iteration the nonlinear element is described by

$$i_3^{l+1} = f(v_3^{l+1}) = (v_3^{l+1})^3 + \frac{1}{2}(v_3^{l+1})^2$$

Upon a Taylor series expansion of this expression about $v_3^l$ and under the assumption that $|v_3^{l+1} - v_3^l|$ is small, the nonlinear element is described by

$$i_3^{l+1} = f(v_3^l) + \left.\frac{df}{dv}\right|_{v_3^l} (v_3^{l+1} - v_3^l)$$

$$= (v_3^l)^3 + \frac{1}{2}(v_3^l)^2 + [3(v_3^l)^2 + v_3^l](v_3^{l+1} - v_3^l)$$

$$= \left[-2(v_3^l)^3 - \frac{1}{2}(v_3^l)^2\right] + (3(v_3^l)^2 + v_3^l)v_3^{l+1}$$

The network containing the large signal model is shown in Fig. 7-7b. We have also used the superscript $l + 1$ on voltages and currents associated with other branches of the network, except the current source current, since they may also change from one iteration to the next. The current source current $\hat{i}_1$ is fixed and remains the same for each iteration. We are now in a position to proceed with the iterative analysis process. First, an initial estimate of the voltage of the nonlinear resistor is made. Suppose $v_3^0 = 1$, then Fig. 7-7c shows the linearized model and the network to be analyzed for the first iteration. The nodal equilibrium equation for this network is

$$9v_{\text{①}}^1 = 3$$

so that

$$v_{\text{①}}^1 = \frac{1}{3}$$

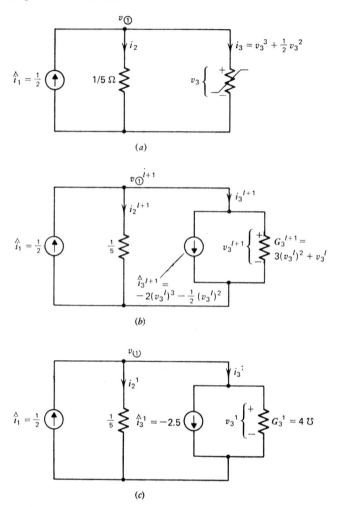

**Fig. 7-7** (*a*) Network used in Example 7.5 to illustrate solution by Newton-Raphson iteration; (*b*) the network interpretation of the Newton-Raphson algorithm; (*c*) to (*e*) the network in (*b*) for the first through third iterations.

and

$$v_3{}^1 = v_①{}^1 = \frac{1}{3}$$

The current source and conductance of the linearized model are now "updated":

$$\hat{i}_3{}^2 = -.13 \qquad \text{and} \qquad G_3{}^2 = \frac{2}{3}$$

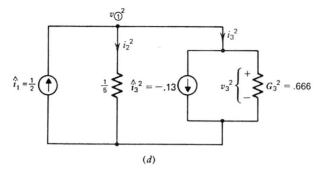

(d)

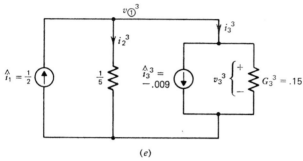

(e)

**Fig. 7-7    (Continued)**

The network to be analyzed for the second iteration is shown in Fig. 7-7d. The corresponding nodal equilibrium equation is

$$5.667v_{①}^2 = .63$$

so that

$$v_3^2 = v_{①}^2 = .111$$

and

$$\hat{i}_3^3 = -.009 \quad \text{and} \quad G_3^3 = .15$$

The next network to be analyzed, shown in Fig. 7-7e, has the nodal equilibrium equation

$$5.15v_{①}^3 = .509$$

so that

$$v_3^3 = v_{①}^3 = .099$$

An additional iteration will show that this value is correct. ☐

The astute student will have observed that we have indicated two techniques of using the Newton-Raphson algorithm to solve a network that contains a nonlinear resistor. First, we may write nodal equilibrium equations, some of which are nonlinear. The Newton-Raphson algorithm is then used to linearize the nonlinear equations about estimates of the node voltages.

Alternatively, the Newton-Raphson algorithm can be used to linearize the nonlinear resistance branch relationship about an estimate of the branch voltage. Then linear nodal equilibrium equatiŏns can be written. The following example illustrates that both techniques are equivalent.

**Example 7.6**

The nodal equilibrium equation for the nonlinear network of Fig. 7-8a is

$$\frac{1}{R_2} v_{①} + f(v_{①}) - \hat{\imath}_1 = 0$$

Application of the Newton-Raphson algorithm yields the following iterative solution for $v_{①}$:

$$v_{①}^{l+1} = v_{①}^{l} - \frac{\dfrac{1}{R_2} v_{①}^{l} + f(v_{①}^{l}) - \hat{\imath}_1}{\dfrac{1}{R_2} + \dfrac{df}{dv}\bigg|_{v_{①}^{l}}}$$

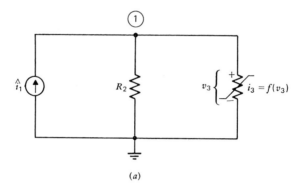

(a)

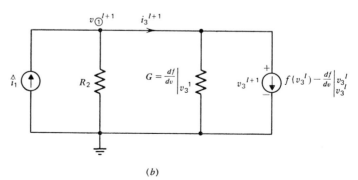

(b)

**Fig. 7-8    Networks used in Example 7.6 to show that the two different schemes of using the Newton-Raphson algorithm yield the same results.**

or

$$v_{①}^{l+1} = \frac{\hat{i}_1 - f(v_{①}^l) + \left.\dfrac{df}{dv}\right|_{v_{①}}^l}{\dfrac{1}{R_2} + \left.\dfrac{df}{dv}\right|_{v_{①}}^l}$$

If the Newton-Raphson algorithm is applied to the nonlinear resistor branch relationship

$$i_3 = f(v_3)$$

we have

$$i_3^{l+1} = \left[ f(v_3^l) - \left.\frac{df}{dv}\right|_{v_3^l} v_3^l \right] + \left.\frac{df}{dv}\right|_{v_3^l} v_3^{l+1}$$

which can be interpreted as a current source of value $f(v_3^l) - df/dv|_{v_3^l} v_3^l$ in parallel with a conductance of value of $df/dv|_{v_3^l}$ as shown in Fig. 7-8b. The nodal equilibrium equation for the linearized network is

$$\left( \frac{1}{R_2} + \left.\frac{df}{dv}\right|_{v_3^l} \right) v^{l+1} = \hat{i}_1 - \left[ f(v_3^l) - \left.\frac{df}{dv}\right|_{v_3^l} v_3^l \right]$$

which can be written as

$$v_{①}^{l+1} = \frac{\hat{i}_1 - f(v_3^l) + \left.\dfrac{df}{dv}\right|_{v_3^l} v_3^l}{\dfrac{1}{R_2} + \left.\dfrac{df}{dv}\right|_{v_3^l}}$$

and is identical to the previous result since

$$v_3^l = v_{①}^l \qquad \qquad \square$$

Before ending this section, we will show how the network interpretation of the Newton-Raphson algorithm can be viewed graphically. We have seen how the diode network of Fig. 7-9a can be solved graphically as the intersection of the diode characteristic curve, and the load line of the current source resistor combination. Let the point "$l$" on the diode curve denote the estimate of the diode voltage and current after $l$ Newton-Raphson iterations. We have seen that to evaluate the $(l + 1)$st estimate of the diode voltage and current we essentially replace the diode with a parallel conductor-current source combination and analyze the resulting linear network. The value of the conductor is

$$\left.\frac{df}{dv_3}\right|_{v_3^l}$$

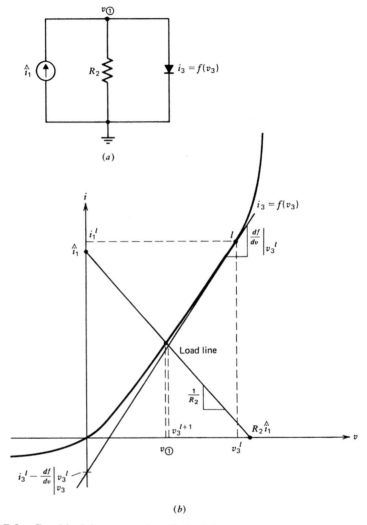

**Fig. 7-9    Graphical interpretation if the Newton-Raphson algorithm applied to the solution of a nonlinear circuit.**

which is the slope of the tangent line to the diode characteristic about the point $v_3{}^l$. The value of the current source,

$$f(v_3{}^l) - \frac{df}{dv_3}\Big|_{v_3{}^l} v_3{}^l = i_3{}^l - \frac{df}{dv_3}\Big|_{v_3{}^l} v_3{}^l$$

is the intercept of this tangent line with the current axis $i$, as illustrated in Fig. 7-9$b$. The intercept of this tangent line with the load line due to $\hat{\imath}_1$ and $R_2$ is the $(l+1)$st estimate $v_3^{l+1}$.

## 7-5  ANALYSIS OF NETWORKS THAT CONTAIN SEVERAL NONLINEAR RESISTORS

The network interpretation given for the Newton-Raphson algorithm in the previous section can be used as the basis of a general analysis procedure for networks that contain many two-terminal nonlinear resistors. Basically this procedure is as follows. Make an estimate of the branch voltage of each nonlinear resistor. Expand the branch relationship of each nonlinear element in a Taylor series about this estimate. Neglecting second and higher order terms, interpret this expansion as a conductance in parallel with a current source. Analyze the linear network that results and update the linearized model associated with each nonlinear element. Repeat the analysis. Continue this procedure until the difference between the estimated voltages on successive iterations becomes small. The following example illustrates the steps involved.

**Example 7.7**

The network of Fig. 7-10 has two nonlinear resistors characterized by

$$i_3 = f_3(v_3)$$

and

$$i_6 = f_6(v_6)$$

Let $v_3{}^l$ and $v_6{}^l$ be the $l$th estimates and $v_3^{l+1}$ and $v_6^{l+1}$ the $(l + 1)$st estimates of the voltages of these elements. Then

$$i_3^{l+1} = f_3(v_3^{l+1})$$

and

$$i_6^{l+1} = f_6(v_6^{l+1})$$

A Taylor series expansion of the nonlinear branch relations about the $l$th estimates of the element voltages yields

$$i_3^{l+1} = f_3(v_3{}^l) + \frac{df_3}{dv_3}\bigg|_{v_3{}^l} (v_3^{l+1} - v_3{}^l)$$

$$= f_3(v_3{}^l) - \frac{df_3}{dv_3}\bigg|_{v_3{}^l} v_3{}^l + \frac{df_3}{dv_3}\bigg|_{v_3{}^l} v_3^{l+1}$$

and

$$i_6^{l+1} = f_6(v_6{}^l) + \frac{df_6}{dv_3}\bigg|_{v_6{}^l} (v_6^{l+1} - v_6{}^l)$$

$$= f_6(v_6{}^l) - \frac{df_6}{dv_3}\bigg|_{v_6{}^l} v_6{}^l + \frac{df_6}{dv_3}\bigg|_{v_6} v_6^{l+1}$$

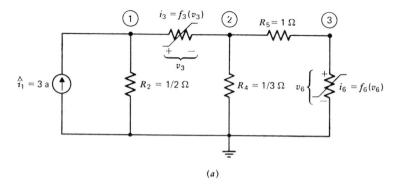

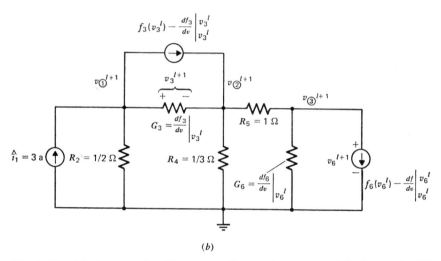

**Fig. 7-10** **(a) A network with two nonlinear elements used in Example 7.7 and (b) the linearized equivalent network.**

where all second and higher order terms have been neglected. This expression may be interpreted as the branch relationships of parallel conductance—current source combinations as shown in Fig. 7-10$b$. Solution of the nodal equilibrium equations of the linearized network yields values for the $(l + 1)$st estimates of the nonlinear element voltages

$$v_3^{l+1} = v_{\textcircled{1}}^{l+1} - v_{\textcircled{2}}^{l+1}$$

and

$$v_6^{l+1} = v_{\textcircled{3}}^{l+1}$$

Suppose the nonlinear resistors were described by

$$i_3 = f_3(v_3) = v_3^3 + 2v_3$$

and

$$i_6 = f_6(v_6) = \frac{1}{2}v_6^2 + 3v_6$$

then

$$\frac{df_3}{dv_3}\bigg|_{v_3{}^l} = 3(v_3{}^l)^2 + 2$$

and

$$\frac{df_6}{dv_6}\bigg|_{v_6{}^l} = v_6{}^l + 3$$

The nodal equilibrium equations of the linearized network become after some manipulation:

$$\left[\frac{5}{2} + 3(v_3{}^l)^2\right]v_{\textcircled{1}}^{l+1} - [3(v_3{}^l)^2 + 2]v_{\textcircled{2}}^{l+1} \qquad = 3 + 2(v_3{}^l)^3$$

$$-[3(v_3{}^l)^2 + 2]v_{\textcircled{1}}^{l+1} + [6 + 3(v_3{}^l)^2]v_{\textcircled{2}}^{l+1} - v_{\textcircled{3}}^{l+1} = -2(v_3{}^l)^3$$

and

$$-v_{\textcircled{2}}^{l+1} + [4 + v_6{}^l]v_{\textcircled{3}}^{l+1} = \frac{1}{2}(v_6{}^l)^2$$

A solution to these equations can be obtained as follows. For $l = 0$, and the initial estimates for $v_3$ and $v_6$ of $v_3{}^0 = 0$ and $v_6{}^0 = 0$, the equations become

$$\frac{5}{2}v_{\textcircled{1}}^{1} - 2v_{\textcircled{2}}^{1} \qquad = 3$$

$$-2v_{\textcircled{1}}^{1} + 6v_{\textcircled{2}}^{1} - v_{\textcircled{3}}^{1} = 0$$

$$- v_{\textcircled{2}}^{1} + 4v_{\textcircled{3}}^{1} = 0$$

Either GEP or Cramer's rule can be used to obtain the solution of $v_{\textcircled{1}}^1 = 1.66$, $v_{\textcircled{2}}^1 = .58$, and $v_{\textcircled{3}}^1 = .14$. The first updated estimate of the nonlinear element voltages can be made:

$$v_3{}^1 = v_{\textcircled{1}}^1 - v_{\textcircled{2}}^1 = 1.08$$

and

$$v_6{}^1 = v_{\textcircled{3}}^1 = .14$$

We can now write the linearized nodal equilibrium equations for $l = 1$:

$$6v_{\textcircled{1}}^2 - 5.5v_{\textcircled{2}}^2 \qquad = 5.52$$

$$-5.5v_{\textcircled{1}}^2 + 9.5v_{\textcircled{2}}^2 - v_{\textcircled{3}}^2 = -2.52$$

$$- v_{\textcircled{2}}^2 + 4.14v_{\textcircled{3}}^2 = .01$$

which has the solution $v_{①}{}^2 = 1.47$, $v_{②}{}^2 = .60$, and $v_{③}{}^2 = .15$. Thus

$$v_3{}^2 = v_{①}{}^2 - v_{②}{}^2 = .87$$
$$v_6{}^2 = v_{③}{}^2 = .15$$

and the linearized nodal equilibrium equations for $l = 2$ become

$$4.77v_{①}{}^3 - 4.27v_{②}{}^3 \qquad\qquad = 4.32$$
$$-4.27v_{①}{}^3 + 8.27v_{②}{}^3 - \qquad v_{③}{}^3 = -1.32$$
$$- \qquad v_{②}{}^3 + 4.15v_{③}{}^3 = .01$$

from which we find that $v_{①}{}^3 = 1.45$, $v_{②}{}^3 = .61$, and $v_{③}{}^3 = .15$. Thus

$$v_3{}^3 = v_{①}{}^3 - v_{②}{}^3 = .84$$

and

$$v_6{}^3 = v_{③}{}^3 = .15$$

which are just about the correct values of $v_{③}$ and $v_{⑥}$.    □

Example 7.7 demonstrates that although the method for solving nonlinear resistive networks is fairly simple, using GEP for each iteration is very time consuming and tedious. In the next section we introduce the FORTRAN subroutine GAUSS that, like GEP, solves simultaneous algebraic equations by Gaussian elimination. However, unlike GEP, it is a subroutine, to be called from a main program. We will show how GAUSS can easily be incorporated into a program written by the user, to solve nonlinear networks.

## 7-6    THE SUBROUTINE GAUSS

The computer program GAUSS is a FORTRAN subroutine that can be called from another program to solve a set of simultaneous algebraic equations by Gaussian elimination. Figure 7-11 is a listing of GAUSS. Subroutine GAUSS is essentially identical to GEP, as can be seen by comparing the listing of GAUSS with the listing of GEP (Fig. 3-12). The main difference is that GEP has input and output statements.

GAUSS is called from another program by using the statement

CALL GAUSS (A, B, N, M)

A is a double-subscripted double precision array, dimensioned M × M in the calling program, B is a single-subscripted double precision array, dimensioned M in the calling program, and N is the number of equations to be solved. A contains the coefficients of the unknowns of the algebraic equations; B contains the coefficients on the right hand side of the algebraic equations. On return from GAUSS B contains the values of the unknowns and A is destroyed. Use of the two integers M and N allows the dimensions of A and

```
C
C
C                        SUBROUTINE 'GAUSS'
C
C
C
C     GAUSS IS A SUBROUTINE FOR SOLVING SIMULTANEOUS LINEAR
C     ALGEBRAIC EQUATIONS WITH REAL COEFFICIENTS. THE PROGRAM
C     USES GAUSSIAN ELIMINATION WITH ROW PIVOTING.
C
C
C     GAUSS IS CALLED FROM ANOTHER PROGRAM BY USING THE STATEMENT
C
C               CALL GAUSS(A,B,N,M)
C
C     A IS A DOUBLE-SUBSCRIPTED REAL*8 ARRAY
C     DIMENSIONED MXM IN THE CALLING PROGRAM, B IS A
C     SINGLE-SUBSCRIPTED REAL*8 ARRAY, DIMENSIONED
C     M IN THE CALLING PROGRAM, AND N IS THE NUMBER OF
C     EQUATIONS TO BE SOLVED. A CONTAINS THE COEFFICIENTS OF
C     THE UNKNOWNS OF THE ALGEBRAIC EQUATIONS; B CONTAINS
C     THE COEFFICIENTS ON THE RIGHT HAND SIDE OF THE ALGEBRAIC
C     EQUATIONS; UPON RETURN FROM GAUSS, B CONTAINS THE
C     SOLUTION AND A CONTAINS THE ELEMENTS OF THE LOWER AND
C     UPPER TRIANGULAR MATRICES OF A. USE OF THE TWO INTEGERS
C     M AND N ALLOWS THE DIMENSIONS OF A AND B TO BE SET
C     ONCE IN THE MAIN PROGRAM AND THE ACTUAL NUMBER OF
C     EQUATIONS TO BE SOLVED TO VARY (NOTE M≥N),DEPENDING
C     ON THE PROBLEM.
C
C
      SUBROUTINE GAUSS(A,B,N,M)
      IMPLICIT REAL*8(A-H,O-Z)
      DIMENSION A(M,M),B(M)
      DO 5 I=1,N
      I1=I+1
      IF(DABS(A(I,I)).LE.1.D-10) GO TO 1
      GO TO 15
   1  CONTINUE
      IF(I.EQ.N) GO TO 10
      DO 14 J=I1,N
      IF(DABS(A(J,I)).LE.1.D-10) GO TO 14
      IPIV=J
      GO TO 16
  14  CONTINUE
      GO TO 10
  16  DO 2 K=1,N
      PIV=A(IPIV,K)
      A(IPIV,K)=A(I,K)
   2  A(I,K)=PIV
      PIV=B(IPIV)
      B(IPIV)=B(I)
      B(I)=PIV
  15  IF(I.EQ.N) GO TO 3
      DO 8 JI=I1,N
   8  A(I,JI)=A(I,JI)/A(I,I)
      B(I)=B(I)/A(I,I)
```

**Fig. 7-11  Listing of the subroutine GAUSS.**

```
        DO 5 J=I1,N
        DO 4 K=I1,N
    4   A(J,K)=A(J,K)-(A(J,I)*A(I,K))
    5   B(J)=B(J)-(B(I)*A(J,I))
    3   B(N)=B(N)/A(N,N)
        DO 6 K=2,N
        I=N-K+1
        L=I+1
        SUM=0.D0
        DO 7 J=L,N
    7   SUM=SUM+A(I,J)*B(J)
    6   B(I)=B(I)-SUM
        GO TO 11
   10   WRITE(6,9)
    9   FORMAT(' EQUATIONS ARE LINEARLY DEPENDENT')
        STOP
   11   RETURN
        END
```

**Fig. 7-11** (*Continued*)

B to be set once in the main program, and the actual number of equations to be solved to vary (*note* N ≤ M), depending on the problem.

**Example 7.8**

As a first example of the use of GAUSS, suppose we wished to solve the linear network of Fig. 7-12. The nodal equilibrium equations and the voltage source branch relationship are

$$\left(\frac{1}{R_1}+\frac{1}{R_3}\right)v_① - \frac{1}{R_3}v_② - \frac{1}{R_1}v_④ = \hat{i}_2$$

$$-\frac{1}{R_3}v_① + \left(\frac{1}{R_3}+\frac{1}{R_4}\right)v_② + i_5 = 0$$

$$\left(\frac{1}{R_6}+\frac{1}{R_7}\right)v_③ - \frac{1}{R_7}v_④ - i_5 = 0$$

$$-\frac{1}{R_1}v_① - \frac{1}{R_7}v_③ + \left(\frac{1}{R_1}+\frac{1}{R_7}+\frac{1}{R_8}\right)v_④ = 0$$

$$v_② - v_③ = \hat{v}_5$$

The listing of a simple program, which calls GAUSS, capable of solving these equations is shown in Fig. 7-13a. The first part of the program initializes the element values. Then the arrays that are used to store the left and right hand side coefficients are initially set to zero. Explicit expressions that define these coefficients follow. Finally, GAUSS is called to solve the equations and the results are printed. Observe that although the array CIN was initially used to store the right hand side of the equations to be solved, upon return from GAUSS it contains the solution, namely the node voltages $v_①, v_②, v_③$, and $v_④$ and the voltage source current $i_5$.

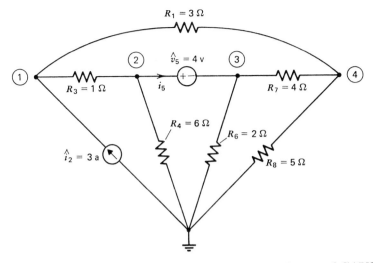

**Fig. 7-12   Network used in Example 7.8 to illustrate the use of GAUSS.**

Since one of the goals of this text is to familiarize the student with the structure of general purpose circuit simulation programs, we show an alternate program listing in Fig. 7-13$b$. This second program solves the same equations but is broken up into subroutines. The main program is more similar in structure to that of a general purpose simulation program. At this point some of the subroutines are rather trivial but it is important to observe the general format. This format will be used repeatedly throughout the remaining chapters of the text. The results of the program are shown in Fig. 7-13$c$.   □

### Example 7.9

Suppose we wish to analyze the network of Fig. 7-12 for various values of $R_3$; in particular for $R_3$ equal to 1 $\Omega$, 3 $\Omega$, 5 $\Omega$, and 10 $\Omega$. This task is easily accomplished by modifying the program listing of Fig. 7-13$b$ as shown in Fig. 7-14$a$. The output is shown in Fig. 7-14$c$.   □

We have seen that the analysis of nonlinear networks is really nothing more than repeated analysis of a linear network whose elements change from one iteration to the next. The subroutine GAUSS can be profitably employed in the analysis of nonlinear networks.

### Example 7.10

Consider again the nonlinear network of Example 7.7 (Fig. 7-10) that is redrawn in Fig. 7-15$a$. The two nonlinear resistors are described by

$$i_3 = f_3(v_3) = v_3{}^3 + 2v_3$$

```
C       CHAPTER 7 EXAMPLE 7.8
C
C       DIMENSIONS
C
        IMPLICIT REAL*8(A-H,O-Z)
        DIMENSION Y(5,5),CIN(5),VN(5)
C
C       READ IN ELEMENT VALUES
C
        READ(5,3) R1,R3,R4,R6,R7,R8,CI2,V5
C
C       SET UP AND SOLVE NODE EQUATIONS
C
C
C       INITIALIZE COEFFICIENT ARRAYS
C
        DO 1 J=1,5
        CIN(J)=0.
        DO 1 K=1,5
      1 Y(J,K)=0.
C
C       SET UP NONZERO COEFFICIENTS OF NODAL EQUILIBRIUM EQUATIONS
C       EQUATION ONE-LEFT HAND SIDE
C
        Y(1,1)=(1./R1)+(1./R3)
        Y(1,2)=-1./R3
        Y(1,4)=-1./R1
C       RIGHT HAND SIDE
        CIN(1)=CI2
C
C       SIMILARLY FOR THE REMAINING EQUATIONS
C
        Y(2,1)=-1./R3
        Y(2,2)=(1./R3)+(1./R4)
        Y(2,5)=1.
C
        Y(3,3)=(1./R6)+(1./R7)
        Y(3,4)=-1./R7
        Y(3,5)=-1.
C
        Y(4,1)=-1./R1
        Y(4,3)=-1./R7
        Y(4,4)=(1./R1)+(1./R7)+(1./R8)
C
        Y(5,2)=1.
        Y(5,3)=-1.
        CIN(5)=V5
C
C       SOLVE NODE EQUATIONS
C
        CALL GAUSS(Y,CIN,5,5)
C
C       PRINT SOLUTION
C
        DO 2 I=1,4
      2 VN(I)=CIN(I)
```

**Fig. 7-13** (*a*) Listing of a program that uses GAUSS to analyze the network of Fig. 7-12; (*b*) the program shown in (*a*) broken into subroutines to more nearly represent the structure of a general purpose analysis program; (*c*) the output of the program.

214

```
      C5=CIN(5)
      PRINT 4
      PRINT 5, (VN(I),I=1,4),C5
    3 FORMAT(8G10.0)
    4 FORMAT(' RESULTS FOR EXAMPLE 7.8'//)
    5 FORMAT(' NODE VOLTAGES(1-4)'/4(5X,G12.5)//' VOLTAGE SOURCE CURRENT
     +'/5X,G12.5)
      END
```

----- DATA FOR PROGRAM OF EXAMPLE 7.8 -----

    3.        1.        6.        2.        4.        5.        3.        4.

**Fig. 7-13a**

```
C     CHAPTER 7 EXAMPLE 7.8
C
C     DIMENSIONS
C
      IMPLICIT REAL*8(A-H,O-Z)
      DIMENSION Y(5,5),CIN(5),VN(5)
C
C     READ IN ELEMENT VALUES
C
      CALL READ(R1,R3,R4,R6,R7,R8,CI2,V5)
C
C     SET UP AND SOLVE NODE EQUATIONS
C
      CALL NODEQN(R1,R3,R4,R6,R7,R8,CI2,V5,Y,CIN)
      CALL GAUSS(Y,CIN,5,5)
C
C     PRINT SOLUTION
C
      DO 2 I=1,4
    2 VN(I)=CIN(I)
      C5=CIN(5)
      CALL PRINT(VN,C5)
      END



      SUBROUTINE READ(R1,R3,R4,R6,R7,R8,CI2,V5)
      IMPLICIT REAL*8(A-H,O-Z)
      READ(5,3) R1,R3,R4,R6,R7,R8,CI2,V5
    3 FORMAT(8G10.0)
      RETURN
      END



      SUBROUTINE PRINT(VN,C5)
      IMPLICIT REAL*8(A-H,O-Z)
      DIMENSION VN(5)
      PRINT 1
      PRINT 2, (VN(I),I=1,4),C5
    1 FORMAT(' RESULTS FOR EXAMPLE 7.8'//)
    2 FORMAT(' NODE VOLTAGES(1-4)'/4(5X,G12.5)//' VOLTAGE SOURCE CURRENT
     +'/5X,G12.5)
      RETURN
      END
```

**Fig. 7-13b**

```
      SUBROUTINE NODEQN(R1,R3,R4,R6,R7,R8,CI2,V5,Y,CIN)
      IMPLICIT REAL*8(A-H,O-Z)
      DIMENSION Y(5,5),CIN(5)
C
C     INITIALIZE COEFFICIENT ARRAYS
C
      DO 1 J=1,5
      CIN(J)=0.
      DO 1 K=1,5
    1 Y(J,K)=0.
C
C     SET UP NONZERO COEFFICIENTS OF NODAL EQUILIBRIUM EQUATIONS
C     EQUATION ONE-LEFT HAND SIDE
C
      Y(1,1)=(1./R1)+(1./R3)
      Y(1,2)=-1./R3
      Y(1,4)=-1./R1
C     RIGHT HAND SIDE
      CIN(1)=CI2
C
C     SIMILARLY FOR THE REMAINING EQUATIONS
C
      Y(2,1)=-1./R3
      Y(2,2)=(1./R3)+(1./R4)
      Y(2,5)=1.
C
      Y(3,3)=(1./R6)+(1./R7)
      Y(3,4)=-1./R7
      Y(3,5)=-1.
C
      Y(4,1)=-1./R1
      Y(4,3)=-1./R7
      Y(4,4)=(1./R1)+(1./R7)+(1./R8)
C
      Y(5,2)=1.
      Y(5,3)=-1.
      CIN(5)=V5
      RETURN
      END
```

**Fig. 7-13b** *(Continued)*

```
RESULTS FOR EXAMPLE 7.8


NODE VOLTAGES(1-4)
     7.9779        6.2647        2.2647        4.1176

VOLTAGE SOURCE CURRENT
     0.66912
```

**Fig. 7-13c**

```
C      CHAPTER 7 EXAMPLE 7.9
C
C      DIMENSIONS
C
       IMPLICIT REAL*8(A-H,O-Z)
       DIMENSION Y(5,5),CIN(5),VN(5),R3(4)
C
C      READ IN ELEMENT VALUES
C      R3 HAS BEEN DIMENSIONED R3(4) TO ALLOW
C      STORAGE OF FOUR VALUES
C
       CALL READ(R1,R3,R4,R6,R7,R8,CI2,V5)
C
C      SET UP AND SOLVE NODE EQUATIONS
C
       DO 2 JJ=1,4
       CALL NODEQN(R1,R3(JJ),R4,R6,R7,R8,CI2,V5,Y,CIN)
       CALL GAUSS(Y,CIN,5,5)
       DO 1 I=1,4
     1 VN(I)=CIN(I)
       C5=CIN(5)
       CALL PRINT(VN,C5,R3(JJ),JJ)
     2 CONTINUE
       END



       SUBROUTINE READ(R1,R3,R4,R6,R7,R8,CI2,V5)
       IMPLICIT REAL*8(A-H,O-Z)
       DIMENSION R3(4)
       READ(5,1) R1,(R3(JJ),JJ=1,4),R4,R6,R7,R8,CI2,V5
     1 FORMAT(8G10.0/3G10.0)
       RETURN
       END



       SUBROUTINE PRINT(VN,C5,R3,JJ)
       IMPLICIT REAL*8(A-H,O-Z)
       DIMENSION VN(5)
       PRINT 1
       PRINT 2,JJ,R3
       PRINT 3,(VN(I),I=1,4),C5
     1 FORMAT(' RESULTS FOR EXAMPLE 7.9'//)
     2 FORMAT(' RUN NO. ',I4,3X,' R3=',G12.5)
     3 FORMAT(' NODE VOLTAGES(1-4)'/4(5X,G12.5)//' VOLTAGE SOURCE CURR
      +'/5X,G12.5///)
       RETURN
       END
```

**Fig. 7-14** *(a)* **Listing of a program that uses GAUSS to analyze the network of Fig. 7-12 for four values of** $R_3$**; and** *(b)* **the output.**

```
      SUBROUTINE NODEQN(R1,R3,R4,R6,R7,R8,CI2,V5,Y,CIN)
      IMPLICIT REAL*8(A-H,0-Z)
      DIMENSION Y(5,5),CIN(5)
C
C     INITIALIZE COEFFICIENT ARRAYS
C
      DO 1 J=1,5
      CIN(J)=0.
      DO 1 K=1,5
1     Y(J,K)=0.
C
C     SET UP NONZERO COEFFICIENTS OF NODAL EQUILIBRIUM EQUATIONS
C     EQUATION ONE-LEFT HAND SIDE
C
      Y(1,1)=(1./R1)+(1./R3)
      Y(1,2)=-1./R3
      Y(1,4)=-1./R1
C     RIGHT HAND SIDE
      CIN(1)=CI2
C
C     SIMILARLY FOR THE REMAINING EQUATIONS
C
      Y(2,1)=-1./R3
      Y(2,2)=(1./R3)+(1./R4)
      Y(2,5)=1.
C
      Y(3,3)=(1./R6)+(1./R7)
      Y(3,4)=-1./R7
      Y(3,5)=-1.
C
      Y(4,1)=-1./R1
      Y(4,3)=-1./R7
      Y(4,4)=(1./R1)+(1./R7)+(1./R8)
C
      Y(5,2)=1..
      Y(5,3)=-1.
      CIN(5)=V5
      RETURN
      END
```

----- DATA FOR PROGRAM OF EXAMPLE 7.9 -----

| | | | | | | | |
|---|---|---|---|---|---|---|---|
| 3. | 1. | 3. | 5. | 10. | 6. | 2. | 4. |
| 5. | 3. | 4. | | | | | |

**Fig. 7-14a**

and

$$i_6 = f_6(v_6) = \frac{1}{2} v_6^2 + 3v_6$$

The linearized equivalent network, shown in Fig. 7-15b, has the nodal equilibrium equations

$$\left[\frac{1}{R_2} + G_3^l\right] v_{\circled{1}}^{l+1} - G_3^{l+1} v_{\circled{2}}^{l+1} = 3 - i_3^l$$

$$-G_3^l v_{\circled{1}}^{l+1} + \left[G_3^l + \frac{1}{R_4} + \frac{1}{R_5}\right] v_{\circled{2}}^{l+1} - \frac{1}{R_5} v_{\circled{3}}^{l+1} = i_3^l$$

$$-\frac{1}{R_5} v_{\circled{2}}^{l+1} + \left[\frac{1}{R_5} + G_6^l\right] = i_6^l$$

RESULTS FOR EXAMPLE 7.9

RUN NO.    1    R3=   1.0000
NODE VOLTAGES(1-4)
         7.9779              6.2647              2.2647              4.1176

VOLTAGE SOURCE CURRENT
      0.66912


RESULTS FOR EXAMPLE 7.9

RUN NO.    2    R3=   3.0000
NODE VOLTAGES(1-4)
         9.9607              6.0337              2.0337              4.8876

VOLTAGE SOURCE CURRENT
      0.30337


RESULTS FOR EXAMPLE 7.9

RUN NO.    3    R3=   5.0000
NODE VOLTAGES(1-4)
        11.186               5.8909              1.8909              5.3636

VOLTAGE SOURCE CURRENT
      0.77273D-01


RESULTS FOR EXAMPLE 7.9

RUN NO.    4    R3=  10.000
NODE VOLTAGES(1-4)
        12.865               5.6954              1.6954              6.0154

VOLTAGE SOURCE CURRENT
     -0.23231

**Fig. 7-14$b$**

where $G_3{}^l = 3(v_3{}^l)^2 + 2$, $i_3{}^l = -2(v_3{}^l)^3$, $G_6{}^l = v_6{}^l + 3$, and $i_6{}^l = -\frac{1}{2}(v_6{}^l)^3$. At the $(l + 1)$st iteration, $v_3{}^l$ and $v_3{}^l$ and $v_6{}^l$ are known numbers and $v_1^{l+1}$, $v_2^{l+2}$, and $v_3^{l+1}$ are unknown. A program that calls GAUSS can be written to solve these equations iteratively. Such a program is shown in Fig. 7-16$a$. The program iterates until the difference between successive estimates of $v_3$ and $v_6$ are less than .01. Observe that we have chosen the initial estimates of the nonlinear resistor voltages to be $v_3{}^0 = v_6{}^0 = 0$. The results of the program are shown in Fig 7-16$b$.  □

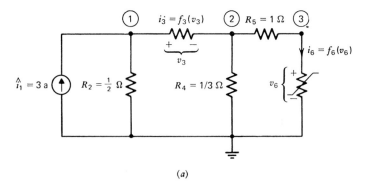

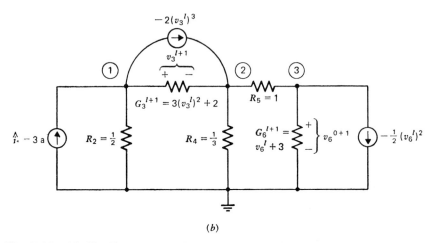

(b)

**Fig. 7-15** (*a*) **Nonlinear network used in Example 7.10 to illustrate how Gauss is used, and** (*b*) **its linearized equivalent.**

## 7-7 SMALL SIGNAL ANALYSIS

The techniques introduced thus far can be used to establish a special type of analysis procedure, known as *small signal analysis*. A small signal analysis cannot only save considerable time over a conventional, or "large signal" analysis, but, as will be seen later, is the fundamental procedure to be used for studying the behavior of a large class of networks.

To introduce small signal analysis, consider the network of Fig. 7-17, which is similar to the network of Fig. 7-2a except for the addition of the current source $\Delta \hat{\imath}_1$. The current source $\Delta \hat{\imath}_1$ can be thought of as representing a change in the current source $\hat{\imath}_1$ from some nominal value. For the case when $\Delta \hat{\imath} = 0$, the node voltage $v_1$ is the same as that found in Section 7-2; it is the

```
C      CHAPTER 7 EXAMPLE 7.10
C
C      DIMENSIONS
C
       IMPLICIT REAL*8(A-H,O-Z)
       DIMENSION Y(3,3),CIN(3)
C
C      INITIALIZE ITERATION COUNTER AND
C        CONVERGENCE DETECTOR
C
       I=1
       KNVRG=0
C
C      READ IN ELEMENT VALUES AND INITIAL
C        GUESS FOR V3 AND V6
C
       CALL READ(CI1,R2,R4,R5,V3,V6)
C
C      SET UP AND SOLVE NODE EQUATIONS
C
     1 CALL NODEQN(CI1,R2,R4,R5,V3,V6,Y,CIN)
       CALL GAUSS(Y,CIN,3,3)
C
C      CALCULATE NEW ESTIMATE OF V3 AND V6
C
       V3T=CIN(1)-CIN(2)
       V6T=CIN(3)
     2 CALL PRINT(V3T,V6T,I,KNVRG)
C
C      CHECK FOR CONVERGENCE
C
       ERR1=V3T-V3
       ERR2=V6T-V6
       IF(DABS(ERR1).LT..01.AND.DABS(ERR2).LT..01) GO TO 3
C
C      CONVERGENCE HAS NOT OCCURRED. INCREASE ITERATION
C        COUNTER,UPDATE VOLTAGES AND ITERATE
C
       I=I+1
       V3=V3T
       V6=V6T
       GO TO 1
C
C      CONVERGENCE HAS OCCURRED
C
     3 KNVRG=1
       GO TO 2
       END
```

**Fig. 7-16** (*a*) **A program for the analysis of the nonlinear network of Fig. 7-15,** (*b*) **the output.**

```
      SUBROUTINE READ(CI1,R2,R4,R5,V3,V6)
      IMPLICIT REAL*8(A-H,O-Z)
      READ(5,1) CI1,R2,R4,R5,V3,V6
      PRINT 2
      PRINT 3, V3,V6
      PRINT 4
    1 FORMAT(7G10.0)
    2 FORMAT(' RESULTS FOR EXAMPLE 7.10'//)
    3 FORMAT(' INITIAL GUESS',4X,' V3=',G12.5,4X,' V6=',G12.5/)
    4 FORMAT(//' ITERATION NO.',8X,' V3',13X,' V6'/)
      RETURN
      END




      SUBROUTINE NODEQN(CI1,R2,R4,R5,V3,V6,Y,CIN)
      IMPLICIT REAL*8(A-H,O-Z)
      DIMENSION Y(3,3),CIN(3)
C
C     INITIALIZE COEFFICIENT ARRAYS
C
      DO 1 J=1,3
      CIN(J)=0.
      DO 1 K=1,3
    1 Y(J,K)=0.
C
C     SET UP NODAL EQUILIBRIUM EQUATION COEFFICIENTS
C
      G3=3.*(V3**2)+2.
      G6=V6+3.
      CI3=-2.*(V3**3)
      CI6=(-1./2.)*(V6**2)
      Y(1,1)=(1./R2)+G3
      Y(1,2)=-G3
      CIN(1)=CI1-CI3
      Y(2,1)=-G3
      Y(2,2)=G3+(1./R4)+(1./R5)
      Y(2,3)=-1./R5
      CIN(2)=CI3
      Y(3,2)=-1./R5
      Y(3,3)=(1./R5)+G6
      CIN(3)=-CI6
      RETURN
      END




      SUBROUTINE PRINT(V3T,V6T,I,KNVRG)
      IMPLICIT REAL*8(A-H,O-Z)
      IF(KNVRG.EQ.0) GO TO 3
      PRINT 1,V3T,V6T
      STOP
    3 PRINT 2,I,V3T,V6T
    1 FORMAT(/' FINAL SOLUTION'/' V3=',G12.5,5X,' V6=',G12.5)
    2 FORMAT(5X,I4,10X,G12.5,5X,G12.5,5X,G12.5)
      RETURN
      END

----- DATA FOR PROGRAM OF EXAMPLE 7.10 -----

       3.        .5      .3333      1.        0.        0.
```

**Fig. 7-16a**

RESULTS FOR EXAMPLE 7.10

INITIAL GUESS     V3= 0.0                    V6= 0.0


ITERATION NO.              V3                V6

    1            0.59212           0.78942D-01
    2            0.54453           0.82200D-01
    3            0.54359           0.82278D-01

FINAL SOLUTION
V3= 0.54359              V6= 0.82278D-01


**Fig. 7-16b**


solution of the nonlinear equation (7-4):

$$\hat{\imath}_1 - \frac{v_①}{R_2} - I_S(e^{qv_①/kT} - 1) = 0$$

Again let $v_①{}^*$ denote the solution of this equation. Now suppose $\Delta \hat{\imath}_1 \neq 0$. Then the node voltage $v_①$ is the solution of the nonlinear equation

$$(\hat{\imath}_1 + \Delta \hat{\imath}_1) - \frac{1}{R_2} - I_S(e^{qv_①/kT} - 1) = 0$$

Certainly this equation can be solved exactly for $v_①$ using the Newton-Raphson algorithm. But if we were interested in finding $v_①$ for various values of $\Delta \hat{\imath}_1$, then using the Newton-Raphson algorithm each time would be quite time consuming. Alternatively we can approximate what $v_①$ would be for various values of $\Delta \hat{\imath}_1$, if $\Delta \hat{\imath}_1$ is small compared with $i_1$, by performing a *small signal analysis*.

For notational convenience let

$$f(v) \equiv I_S(e^{qv/kT} - 1)$$

then, the voltage $v_①$ we seek is the solution of the equation

$$(\hat{\imath}_1 + \Delta \hat{\imath}_1) - \frac{v_①}{R_2} - f(v_①) = 0$$

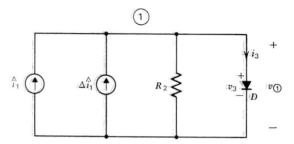

**Fig. 7-17   A network used to represent the network of Fig. 7-2a for changes in the current source from $\hat{i}_1$.**

We know that $v_{①}{}^*$ is the solution when $\Delta \hat{i} = 0$:

$$\hat{i}_1 - \frac{v_{①}{}^*}{R_2} - f(v_{①}{}^*) = 0$$

A change of the input from $\hat{i}_1$ to $\hat{i}_1 + \Delta \hat{i}_1$ causes the voltage $v_{①}{}^*$ to change from $v_{①}{}^*$ to $v_{①} = v_{①}{}^* + \Delta v_{①}$. Hence

$$\hat{i}_1 + \Delta \hat{i}_1 - \frac{1}{R_2}(v_{①}{}^* + \Delta v_{①}) - f(v_{①}{}^* + \Delta v_{①}) = 0$$

The function $f(v_{①}{}^* + \Delta v_{①})$ can be expanded in a Taylor series about $v_{①}{}^*$:

$$f(v_{①}{}^* + \Delta v_{①}) = f(v_{①}{}^*) + \frac{df}{dv}\bigg|_{v=v_{①}{}^*} \Delta v_{①} + \frac{1}{2}\frac{d^2 f}{dv^2}\bigg|_{v=v_{①}{}^*} (\Delta v_{①})^2$$

$$+ \frac{1}{3!}\frac{d^3 f}{dv^3}\bigg|_{v=v_{①}{}^*} (\Delta v_{①})^3 + \cdots$$

We now make the assumption that small changes in $\hat{i}_1$ cause small changes in $v_{①}{}^*$, that is, if $\Delta \hat{i}_1$ is small compared to $\hat{i}_1$ then $\Delta v_{①}$ is small compared to $v_{①}{}^*$. A detailed discussion of the validity of this assumption is beyond the scope of this text. Suffice it to say that for the network under consideration this assumption is valid. Therefore, $(\Delta v_{①})^2$, $(\Delta v_{①})^3$, etc. can be assumed to be negligibly small and

$$f(v_{①}{}^* + \Delta v_{①}) \simeq f(v_{①}{}^*) + \frac{df}{dv}\bigg|_{v=v_{①}{}^*} \Delta v_{①}$$

The nodal equilibrium equation

$$(\hat{i}_1 + \Delta \hat{i}_1) - \frac{1}{R_2}(v_{①}{}^* + \Delta v_{①}) - f(v_{①}{}^* + \Delta v_{①}) = 0$$

can be written as

$$(\hat{\imath}_1 + \Delta\hat{\imath}_1) - \frac{1}{R_2}(v_①^* + \Delta v_①) - f(v_①^*) - \frac{df}{dv}\bigg|_{v=v_①^*}\Delta v_① = 0$$

After rearranging terms, we have

$$\hat{\imath}_1 - \frac{1}{R_2}v_①^* - f(v_①^*) + \Delta\hat{\imath}_1 - \frac{1}{R_2}\Delta v_① - \frac{df}{dv}\bigg|_{v=v_①^*}\Delta v_① = 0$$

But

$$\hat{\imath}_1 - \frac{1}{R_2}v_①^* - f(v_①^*) = 0$$

because $v_①^*$ is the node voltage corresponding to $\hat{\imath}_1$, therefore, we are left with

$$\Delta\hat{\imath} - \frac{1}{R_2}\Delta v_① - \frac{df}{dv}\bigg|_{v=v_①^*}\Delta v_① = 0$$

We may interpret this expression as the nodal equilibrium equation of the network of Fig. 7-18a. This *linear* network is obtained from the original nonlinear network by setting the "large-signal" current source $\hat{\imath}_1$ to zero and replacing the diode with a conductor whose value is equal to the slope of the diode characteristic curve about the *operating point* $v_①^*$ as shown in Fig. 7-18b. Since the voltage $\Delta v_①$, and current $\Delta\hat{\imath}_1$ are small compared to $v_①^*$ and $\hat{\imath}_1$ this network is called the *small signal equivalent*. Analysis of this network will indicate how small changes in $\hat{\imath}_1$ (denoted by $\Delta\hat{\imath}_1$) in the network of Fig. 7-17

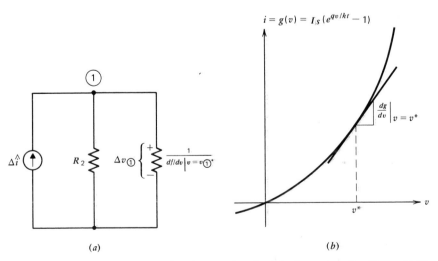

$$i = g(v) = I_s(e^{qv/kt} - 1)$$

(a)          (b)

**Fig. 7-18** (a) Small-signal circuit associated with the network of Fig. 7-17. (b) The small-signal resistance of the diode is the inverse of the slope of the diode characteristic about the operating point $v^*$.

affect the voltage $v_①$*. Instead of performing a Newton-Raphson solution for each different value of $\hat{i}_1 + \Delta\hat{i}_1$, a single Newton-Raphson solution of the network of Fig. 7-17 is performed with $\Delta\hat{i} = 0$. Once $v_①$* is found, $\hat{i}_1$ is set to zero and the diode is replaced with its small-signal resistance. The linear network that results (Fig. 7-18a) is analyzed to determine $\Delta v_①$ for each value of the change $\Delta\hat{i}_1$. The results are added to $v_①$*.

## Example 7.11

Consider the network of Fig. 7-19a; the nonlinear resistor is characterized in the $i$-$v$ plane as shown in Fig. 7-19b. The nominal value of the current $i_1$ is 10 amp. Since $\Delta\hat{i}_1 = \sin t$, it varies between $+1$ and $-1$. Since the total current $\hat{i}_1 + \Delta\hat{i}$ varies 10 percent from its nominal value of 10 amp, we can try to employ small-signal analysis. First we analyze the network of Fig. 7-19c. The nodal equilibrium equation of this network is

$$\hat{i}_1 - \frac{1}{R_3} v_① - v_①{}^2 = 0 \qquad \text{for} \qquad v > 0$$

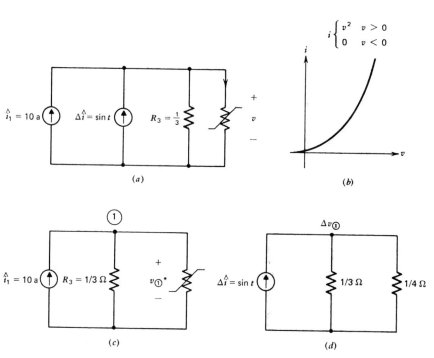

(a)

(b)

(c)

(d)

**Fig. 7-19** Networks used in Example 7.11 to demonstrate small-signal analysis. (a) The original network; (b) the nonlinear resistor characteristic; (c) the large-signal network; (d) the small-signal network.

The student should verify that the solution is

$$v_①^* = 2$$

(*Note.* $v_① = -5$ is not a solution because we must have $v > 0$.) The small-signal circuit is shown in Fig. 7-19d. Observe that the small-signal resistance associated with the nonlinear resistor is obtained from

$$\left.\frac{df(v)}{dv}\right|_{v=2} = \left.\frac{d}{dv}(v^2)\right|_{v=2} = \left.2v\right|_{v=2} = 4$$

Analysis of this network yields

$$\Delta v_① = \frac{1}{7}\sin t$$

Finally, the node voltage of the original network is

$$v_① = v_①^* + \Delta v_①$$

$$= 2 + \frac{1}{7}\sin t$$

We can check the validity of the small-signal assumption by analyzing the original network exactly for the case when $i_1 + \Delta_i$ has it largest deviation from the nominal value of 10 amps, that is, when $i_1 + \Delta i = 11$ amp. The equation to be solved is

$$11 - 3v_① - v_①^2 = 0$$

The actual solution is

$$v_① = 2.14$$

as compared to approximate $v_① = 2 + \frac{1}{7} = 2.1333$ predicted by the small-signal analysis. □

## 7-8  SOME PRACTICAL CONSIDERATIONS FOR THE ANALYSIS OF DIODE NETWORKS

In Section 7-4 we discussed how the Newton-Raphson algorithm could be employed for the solution of diode networks and provided a network interpretation. Specifically at the beginning of each iteration, the diode is replaced by a parallel combination of a conductance and a current source as illustrated in Fig. 7-20a. The value of the conductance is equal to the slope of the diode characteristic about previous estimate of the diode's voltage. The value of the current source is equal to the value of intercept of the tangent line to the diode characteristic at the previous estimate with the ordinate (current axis)

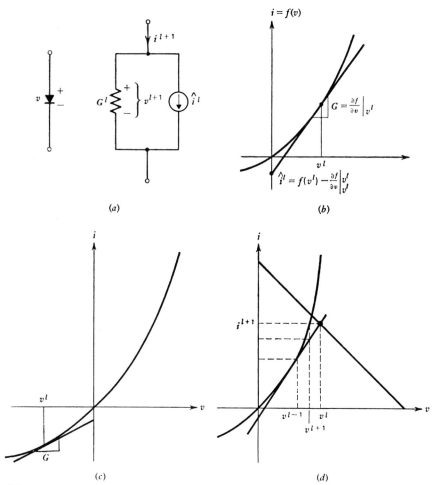

**Fig. 7-20** (*a*) The diode and its linearized equivalent; (*b*) computation of model parameters from diode characteristic; (*c*) some estimates of the diode voltage result is very small values of *G*; (*d*) some estimates of the diode voltage result in extremely high diode currents.

as shown in Fig. 7-20*b*. We now consider two limitations of the digital computer that necessitates a reformulation of the diode's model. The first limitation is due to *roundoff errors*. The accuracy of a computer is limited by roundoff errors. The IBM 360 retains only a maximum of 16 digits for every number. When two numbers are added or substracted that are more than 16 orders of magnitude different, the smaller number is lost in roundoff. Therefore, we must impose reasonable boundaries on the resistor in the large signal diode model since this value will be added to or substracted from other conductance values when we set up the nodal equilibrium equations. Consider the situation of Fig. 7-20c. The slope of the diode curve about $v^l$ is very small and as $v^l$ moves further to the left this number decreases more. The possibility exists for unrealistically small numbers to crop up for values of $1/R$. To avoid this situation we can put a lower boundary on $1/R$ of, for example, $10^{-12}$ mhos.

Now consider the situation of Fig 7-20d. Here the $l$th estimate of the diode voltage, $v^l$, is such that $i^l = f(v^l)$ is much too large. This situation can arise as illustrated. Practically speaking, the diode voltage can never be greater than about .7 v (for $Si$). To avoid this situation, we can take as our next estimate of $v$:

$$v^{l+1} = \frac{kT}{q} \ln \left( \frac{i^l}{I_S + 1} \right)$$

As indicated on the diagram this procedure corresponds to moving horizontally from the estimated operating point to the diode characteristic rather than vertically. These problems are discussed again in Chapter 10 when we study transistor networks.

## 7-9  SUMMARY

In this chapter we introduced the *nonlinear resistor*. The most common physical nonlinear resistor is the *semiconductor diode* characterized by

$$i = I_S(e^{qv/kT} - 1)$$

Inclusion of a nonlinear element in a network greatly increases the complexity of solution.

For simple single nonlinear element networks a *graphical method* of solution can be used. However, the graphical method is not very accurate and is not amenable to use on the computer.

Since an analytic solution is usually impossible to obtain for a nonlinear network, some *iterative technique* must be employed. In particular, the *Newton-Raphson algorithm* is used. The Newton-Raphson algorithm is essentially the repeated *linearization* (via a *Taylor series expansion*) of the nonlinear branch relationship about an estimate of the operating point. It is shown that the linearized branch relationship can be interpreted as the branch relationship of a linear resistor in parallel with a current source. Once all nonlinear elements have been replaced by their linearized equivalent models, a linear network analysis is performed, a new estimate of the operating point of each nonlinear element is made, and if convergence has not been achieved, the linearized models are updated. The whole procedure is repeated until convergence has been achieved.

The subroutine GAUSS, which solves simultaneous algebraic equations using Gaussian elimination, was introduced. Several illustrations of how to incorporate GAUSS into a main program were presented.

Finally, the concept of *small-signal analysis* and its usefulness for approximating the behavior of nonlinear networks to small variations in the inputs was discussed.

## PROBLEMS

**7.1**    Show that the following branch relationships are nonlinear

(a) $i = v^2$

(b) $i = v + \sin v$

(c) $v = \ln i$

**7.2**    Determine $v_1$ in the network of Fig. P7.2 by graphical means.

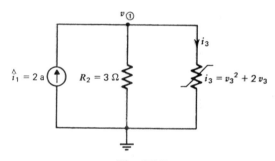

**Fig. P7.2**

**7.3**    Find all the branch voltages and currents in the network of Fig. P7.3 using a graphical solution (*Hint.* Use a Norton equivalent.)

**7.4**    For each of the following, compute the Taylor series approximation

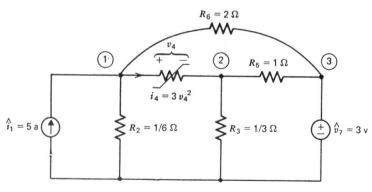

**Fig. P7.3**

about the point specified, and using the number of terms indicated:

(a) $f(x) = (1 + x)^{1/3}$, about $x = 0$, using 4 terms

(b) $f(x) = 2x^2 + 3x + 5$, about $x = -2$, using 2 terms

(c) $f(x) = e^x + \cos x$, about $x = 0$, using 5 terms

**7.5**   Use the Newton-Raphson algorithm to solve for $x$ in each of the following nonlinear equations. (Find the positive solution only.)

(a) $x^4 + 6x^2 - 1 = 0$

(b) $3x^5 - 2x^3 + x - 2 = 0$

(c) $x^{12} - 11x^{11} + 8x^7 - 2 = 0$

**7.6**   Write the nonlinear nodal equilibrium equation for the network of Fig. P7.2 and solve for $v_①$ using the Newton-Raphson algorithm. Compare the result with those obtained in Prob. 7.2.

**7.7**   Write a set of nonlinear nodal equilibrium equations for the network of Fig. P7.3. Using the first two terms of a Taylor series expansion, linearize these equations about an estimate of the values of the node voltages and demonstrate how these equations can be used iteratively to solve for the actual node voltages. Carry out a few iterations and compare the results with those obtained in Prob. 7.3.

**7.8**   Determine a linearized model for the nonlinear elements described by

(a) $i = 3v^2 + 2v$

(b) $v = 3i^2 + 2i$

(c) $i = e^v$

(d) $v = \ln i$

**7.9**   Determine the linearized models associated with the nonlinear networks of Fig. P7.2 and P7.3. Write the nodal equilibrium equations for these networks when the nonlinear element is replaced by its linearized model. Show that these equations are identical to those obtained in Probs. 7.6 and 7.7 when the nonlinear equations were solved using the Newton-Raphson algorithm.

**7.10**   Determine $v_1$ iteratively in the network of Fig. P7.2 by replacing the nonlinear resistor by a linearized equivalent model. Use GEP and compare the results with those obtained graphically in Prob. 7.2.

**7.11**   Consider again the network of Fig. 7-12 and the program used to determine the node voltages of this network shown in Fig. 7-13$b$. Use the program to determine the node voltages for the network if the element values were changed to $R_1 = 15\,\Omega$, $i_2 = .1$a, $R_3 = 100\,\Omega$, $R_4 = 50\,\Omega$, $\hat{v}_1 = 10$ v, $R_6 = 1000\,\Omega$, $R_7 = 500\,\Omega$, and $R_8 = 150\,\Omega$. What is the power absorbed by $R_1$ and $R_2$ and delivered by $v_5$?

**7.12**   Suppose the network of Fig. 7-12 was modified by moving resistor

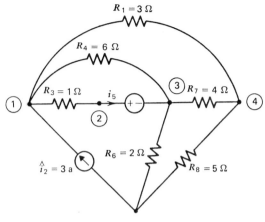

**Fig. P7.12**

$R_4$ as shown in Fig. P7.12. Modify the subroutine NODEQN in Fig. 7-13$b$ to reflect this change and determine the current that flows in $R_1$.

**7.13**    The program in Fig. 7-16$a$ was written to solve the nonlinear network shown in Fig. 7-15$a$. The nonlinear elements are described by

$$i_3 = v_3{}^3 + 2v_3$$

and

$$i_6 = \frac{1}{2} v_6{}^2 + 3v_6$$

Modify this program so that it solves the network with nonlinear elements described by

$$i_3 = v_3{}^4 + 6v_3{}^2 - 1$$

and

$$i_6 = 3v_6{}^5 - 2v_6{}^3 + v_6 - 2$$

Use the program to solve the network.

**7.14**    Write a program that uses GAUSS to solve for the node voltages of the network in Fig. P7.3.

**7.15**    The nonlinear element in the network of Fig. P7.15 is characterized by

$$v = -2i + \frac{1}{3} i^3$$

It is known that if $\Delta v_{in} = 0$ then the loop current is $i = 5$ a. Determine the loop current if $\Delta v_{in} = \cos t$. (*Hint*. Use small signal analysis.)

**7.16**    Determine the branch voltage of the nonlinear resistor in the network

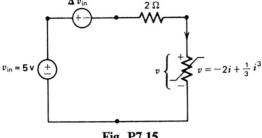

**Fig. P7.15**

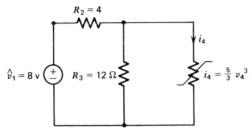

**Fig. P7.16**

of Fig. P7.16 by writing a nonlinear nodal equilibrium equation and solving it by the Newton-Raphson algorithm. An estimate of the correct value can be made after only a few iterations. (*Hint.* Start with an initial estimate of $v_4 = .5$ volts.) What is the voltage across the nonlinear resistor if the source voltage $\hat{v}_1$ is changed to 7.8 volts, 8.1 volts, and 8.2 volts?

**7.17**   In Example 7.10 the program of Fig. 7-16a was used to determine the voltages across the nonlinear elements in the network of Fig. 7-15a. Using this operating point (see Fig. 7.15b) find the small signal equivalent circuit and determine the approximate voltages that would appear across the nonlinear elements if $\hat{i}_1 = 3.1$ a. Verify your results by running the program of Fig. 7-16a with $\hat{i}_1 = 3.1$ a.

**7.18**   Write a program that calls GAUSS to solve the diode network of Fig. P7.18. Make sure that the linearized model is modified according to the discussion in Section 7-8. Assume that the diode branch relationship is $i_D = 10^{-14} \, (e^{40v_D} - 1)$.

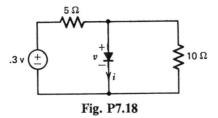

**Fig. P7.18**

CHAPTER **8**

# Linear Multiterminal Elements and Port Descriptions

Until now we have only considered two-terminal resistive circuit elements. In this chapter we introduce elements that have more than two terminals. Similar to resistors, the elements to be introduced are *instantaneous elements*, in other words, an instantaneous change in the voltage causes a corresponding instantaneous change in the current. Instantaneous elements are also referred to as *memoryless elements* and are characterized by algebraic equations. By way of contrast, in Chapter 11 we will introduce two elements, the capacitor and inductor, which are described by differential equations. Such elements have "memory."

We initially restrict our attention to linear multiterminal elements. In Chapter 10, nonlinear multiterminal elements will be investigated. We start by considering the description of general linear multiterminal devices.

## 8-1 TERMINAL PAIRS AND PORTS

A multiterminal element is most easily envisioned as a black box with several accessible terminals available for connection as illustrated in Fig. 8-1. The simplest multiterminal element is the two-terminal element of Fig. 8-2 where $i_a = i_b$.

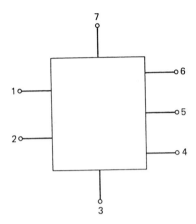

**Fig. 8-1   A multiterminal element. Connection of this element is made through the accessible terminals.**

Often terminals of a multiterminal element will be grouped by pairs, as illustrated in Fig. 8-3. A pair of terminals of such an element are said to constitute a *port*, if the current entering the top terminal of the pair equals the current leaving the bottom terminal of the pair. Usually a port will have a voltage and current associated with it.

## 8-2  TWO-PORTS AND THEIR DESCRIPTIONS

Most multiterminal elements encountered will have three or four terminals. A three-terminal element will usually be connected so that it could be considered to be a four-terminal element as shown in Fig. 8-4. The usual method of characterizing such four-terminal elements is by using a two-port description.

Consider the four-terminal element of Fig. 8-5. Terminals 1 and 1′ constitute port 1 and terminals 2 and 2′ constitute port 2. Let $v_1$ and $v_2$ denote the voltages across ports 1 and 2, respectively. We will refer to $v_1$ and $v_2$ as *port voltages*. Similarly, $i_1$ and $i_2$ are called *port currents*. The current entering

**Fig. 8-2   The two-terminal element is a special case of the multiterminal element when $i_a = i_b$.**

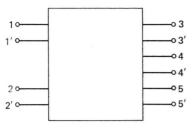

**Fig. 8-3   A multiterminal element whose terminals are grouped in pairs.**

terminal 1 (port 1) is $i_1$ and $i_2$ is the current entering terminal 2 (port 2). The assumed directions of positive voltage and current flow shown in Fig. 8-5 are standard and follow our previous conventions.

Recall that for the case of a two-terminal element, a single relationship involving the two variables, branch voltage and branch current, could be written. In particular, branch relationships for resistors, could be written as

$$v = Ri$$

or

$$i = \frac{1}{R}v$$

if $R \neq 0$. The current $i$ is the independent variable in the first expression and the voltage $v$ the dependent variable. In the second expression the voltage is the independent variable while the current is the dependent variable.

For a two-port (or a four-terminal element) two independent relationships can be written involving the four variables: the port voltages $v_1$ and $v_2$ and the port currents $i_1$ and $i_2$. If $v_1$ and $v_2$ can be chosen as dependent variables and $i_1$ and $i_2$ as independent variables, the two port is said to have a *z-parameter description* (or open circuit impedance description) and is characterized by the equations

$$v_1 = z_{11}i_1 + z_{12}i_2$$
$$v_2 = z_{21}i_1 + z_{22}i_2$$

(We will see later that not all two ports have $z$-parameter descriptions.) In other words, the voltage of port 1 is dependent on the current of port 2 as

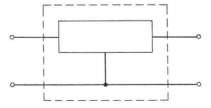

**Fig. 8-4   A three-terminal element that can be considered as a four-terminal element.**

**Fig. 8-5  A two-port network showing positive voltage and current conventions.**

well as the current of port 1 and the voltage of port 2 is dependent on the current of port 1 and the current of port 2.

Another possible method of describing a two-port is to choose $i_1$ and $i_2$ as dependent variables and $v_1$ and $v_2$ as independent variables; the *y-parameter description* (or short circuit admittance description) results:

$$i_1 = y_{11}v_1 + y_{12}v_2$$

$$i_2 = y_{21}v_1 + y_{22}v_2$$

(Again, not all two ports have a *y*-parameter description). Other possible methods of characterizing two-ports are summarized in Table 8-1.

At times it will be convenient to refer to port 1 as the *input port* and port 2 as the *output port*. Two-ports are also referred to as coupling elements because variables at one port depend on or are "coupled" to variables at the other port.

**Table 8-1**  Two-port Parameters

| Name | Dependent Variables | Independent Variables | Describing Equations |
|---|---|---|---|
| Open circuit impedance | $v_1, v_2$ | $i_1, i_2$ | $v_1 = z_{11}i_1 + z_{12}i_2$ <br> $v_2 = z_{21}i_1 + z_{22}i_2$ |
| Short-circuit admittance | $i_1, i_2$ | $v_1, v_2$ | $i_1 = y_{11}v_1 + y_{12}v_2$ <br> $i_2 = y_{21}v_1 + y_{22}v_2$ |
| Transmission | $v_1, i_1$ | $v_2, i_2$ | $v_1 = Av_2 - Bi_2$ <br> $i_1 = Cv_2 - Di_2$ |
| Inverse transmission | $v_2, i_2$ | $v_1, i_1$ | $v_2 = \hat{A}v_1 - \hat{B}i_1$ <br> $i_2 = \hat{C}v_1 - \hat{D}i_1$ |
| Hybrid | $v_1, i_2$ | $i_1, v_2$ | $v_1 = h_{11}i_1 + h_{12}v_2$ <br> $i_2 = h_{21}i_1 + h_{22}v_2$ |
| Inverse hybrid | $i_1, v_2$ | $v_1, i_2$ | $i_1 = g_{11}v_1 + g_{12}i_2$ <br> $v_2 = g_{21}v_1 + g_{22}i_2$ |

## 8-3  IDEAL TRANSFORMERS, DEPENDENT SOURCES, AND GYRATORS

We now introduce some specific two-port elements. None of these two-ports are physical elements; they cannot be bought in any store. However, the behavior of some physical elements, such as transistors, can be closely approximated, or *modeled*, by a suitable connection of two-terminal and two-port elements. The ability to model the behavior of actual elements by ideal elements provides motivation for the study of ideal elements.

An *ideal transformer* is a two-port element described by the branch relationships

$$v_1 = \left(\frac{n_1}{n_2}\right) v_2$$

and

$$i_1 = -\left(\frac{n_2}{n_1}\right) i_2$$

where $n_1$ and $n_2$ are constants and $n_1/n_2$ is known as the *turns ratio*. The ideal transformer circuit symbol is shown in Fig. 8-6. The dots shown are used to indicate the terminal at which current is assumed to flow into the transformer. Port 1 is known as the *primary* and port 2 the *secondary*. The ideal transformer is not a physical element, but can be closely approximated in practice for time-varying signals. In theory though, the ideal transformer functions for even constant signals.

Controlled sources are like independent sources in that they fix the voltage across a branch or the current in a branch. The difference is that the value of the controlled source depends on the voltage or current in some other branch. There are four types of controlled sources. The *voltage controlled voltage source* is characterized by the two-port equations

$$v_2 = \mu v_1$$

and

$$i_1 = 0$$

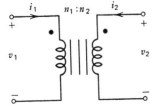

**Fig. 8-6  Symbol for an ideal transformer.**

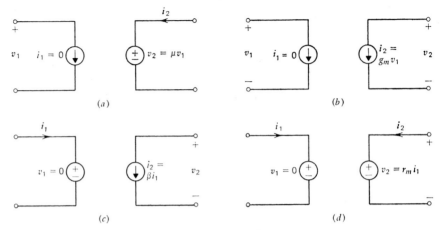

(a)                              (b)

(c)                              (d)

**Fig. 8-7**   Circuit models of the four types of controlled sources. (a) the voltage-controlled voltage source; (b) the voltage-controlled current source; (c) the current-controlled current source; (d) the current-controlled voltage source.

A circuit symbol for the voltage-controlled source is shown in Fig. 8-7a.[1] The controlling voltage $v_1$ is taken as the voltage across a zero-valued current source (open circuit). $\mu$ is a dimensionless number and is called the *voltage amplification factor*.

The *voltage controlled current source*, shown symbolically in Fig. 8-7b is characterized by the equations

$$i_2 = g_m v_1$$

and

$$i_1 = 0$$

where $g_m$ is called the *transductance* and has units of mhos. The *current controlled current source* shown symbolically in Fig. 8-7c is characterized by the equations

$$i_2 = \beta i_1$$

and

$$v_1 = 0$$

where $\beta$ is called the *current amplification factor* and is dimensionless. The last type of controlled source is the *current controlled voltage source* shown symbolically in Fig. 8-7d, and characterized by the equations

$$v_2 = r_m i_1$$

[1] We use here a two-branch model for all controlled sources. Some authors prefer not to show explicitly the controlling branch but rather take the controlling voltage or current to be associated with a two-terminal element. A zero-valued source can always be added without altering circuit operation. In the sequel we will sometimes forgo explicitly showing the zero-valued source in the controlling branch.

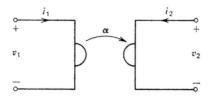

**Fig. 8-8    Symbol for a gyrator.**

and

$$v_1 = 0$$

where $r_m$ is called the *transresistance* and has units of ohms.

The last two-port element to be introduced here is the *gyrator*. The gyrator is described by

$$i_1 = \alpha v_2$$

and

$$i_2 = -\alpha v_1$$

where $\alpha$ is called the *gyration ratio* and has units of mhos. The circuit symbol of a gyrator is shown in Fig. 8-8. As with transformers, port 1 is sometimes referred to as the *primary* and port 2 as the *secondary*.

## 8-4  SOME SIMPLE NETWORKS CONTAINING TWO-PORTS

As motivation for the discussion of the general method of analysis of networks containing two ports we investigate some simple networks first.

### Example 8.1

Consider the network of Fig. 8-9. Noting that the currents in the primary and secondary of the transformer are $i_3$ and $i_4$, respectively, we can write the following nodal equilibrium equations:

*node* ①
$$\frac{1}{R_2} v_① - \frac{1}{R_2} v_② + i_1 = 0$$

*node* ②
$$-\frac{1}{R_2} v_① + \frac{1}{R_2} v_② + i_3 = 0$$

*node* ③
$$\frac{1}{R_5} v_③ + i_4 = 0$$

These three equations are in terms of six unknowns: $v_①$, $v_②$, $v_③$, $i_1$, $i_3$, and $i_4$. But three branch relationships still remain unused: the voltage source branch relationship, which can be written as

$$v_① = \hat{v}_1$$

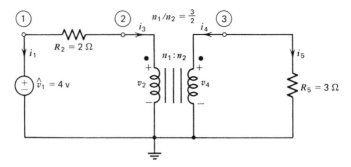

**Fig. 8-9   A simple network, considered in Example 8.1, which contains a transformer.**

and the two transformer branch relationships, which can be written as

$$i_3 = -\frac{n_2}{n_1} i_4$$

and

$$v_3 = \frac{n_1}{n_2} v_4$$

or, in terms of the node voltages,

$$v_{②} = \frac{n_1}{n_2} v_{③}$$

Substitution of these expressions into the nodal equilibrium equations yields

$$-\left(\frac{1}{R_2}\right)\left(\frac{n_1}{n_2}\right) v_{③} + i_1 = -\frac{1}{R_2} \hat{v}_1$$

$$\left(\frac{1}{R_2}\right)\left(\frac{n_1}{n_2}\right) v_{③} - \left(\frac{n_2}{n_1}\right) i_4 = \frac{1}{R_2} \hat{v}_1$$

and

$$\left(\frac{1}{R_5}\right) v_{③} + i_4 = 0$$

that is, three equations in terms of three unknowns. On substitution of element values, these equations become

$$-\frac{3}{4} v_{③} + i_1 \qquad = -2$$

$$\frac{3}{4} v_{③} \qquad -\frac{2}{3} i_4 = \quad 2$$

and

$$\frac{1}{3} v_{③} \qquad + \frac{2}{3} i_4 = 0$$

The last two equations can be used to solve for $v_{③}$ and $i_4$: $v_{③} = \frac{7 \cdot 2}{3 \cdot 5}$, $i_4 = -\frac{2 \cdot 4}{3 \cdot 5}$; the first equation is used to determine $i_1$ from knowledge of $v_{③}$: $i_1 = \frac{1 \cdot 6}{3 \cdot 5}$. Node voltage $v_{②}$ is obtained from the transformer branch relationship

$$v_{②} = \frac{n_1}{n_2} v_{③}$$

$$= \left(\frac{3}{2}\right)\left(\frac{72}{35}\right) = \frac{108}{35} \quad \square$$

The concept of an equivalent resistance has been useful in our previous studies. In Chapter 2, we described a technique for circuit analysis based on reduction of series and parallel combinations of resistors to a single equivalent resistance. In order to replace a portion of a network by a Thévenin or Norton equivalent, an equivalent Thévenin or Norton resistance for that portion of the network must be found. If, as is usually the case, this equivalent resistance cannot be found by reducing series and parallel combinations of resistors, a set of nodal equilibrium equations can be written and used to determine the equivalent resistance (see Example 5.10). Since the transformer is a memoryless element, it is possible to determine an equivalent resistance for a network that contains resistors and transformers.

**Example 8.2**

We wish to determine the equivalent resistance seen by the voltage source in the network of Fig. 8-10, which was considered previously in Example 8.1.

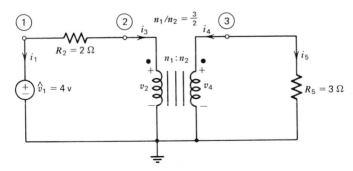

**Fig. 8-10   The network used in Example 8.2 to demonstrate an equivalent resistance computation.**

Therefore, the current $i_1$ must be determined. The nodal equilibrium equations were found to be

$$-\frac{1}{R_2}\left(\frac{n_1}{n_2}\right)v_{\circled{3}} + i_1 = -\frac{1}{R_2}\hat{v}_1$$

$$\left(\frac{1}{R_2}\right)\left(\frac{n_1}{n_2}\right)v_{\circled{3}} - \left(\frac{n_2}{n_1}\right)i_4 = \frac{1}{R_2}\hat{v}_1$$

and

$$\left(\frac{1}{R_5}\right)v_{\circled{3}} + i_4 = 0$$

The last two equations can be solved for $v_{\circled{3}}$:

$$\left[\left(\frac{1}{R_2}\right)\left(\frac{n_1}{n_2}\right) + \left(\frac{n_2}{n_1}\right)\left(\frac{1}{R_5}\right)\right]v_{\circled{3}} = \frac{1}{R_2}\hat{v}_1$$

Substitution of this result into the first equation yields, after suitable manipulation,

$$i_1 = \frac{-n_2{}^2}{n_2{}^2 R_2 + n_1{}^2 R_5}\hat{v}_1$$

Thus the equivalent resistance is

$$R_{eq} = -\frac{\hat{v}_1}{i_1} = R_2 + \frac{n_1{}^2}{n_2{}^2}R_5$$

(the negative sign arises because the current flowing through the equivalent resistance is opposite to $i_1$.) Observe that the transformer has the property of "transforming" the effect of a resistor. $R_5$ is the resistor across the secondary. The transformer reflects $n_1{}^2/n_2{}^2 R_5$ back to the primary!   □

Analysis of networks that contain controlled sources and gyrators is conceptually no more difficult than the analysis of networks that contain transformers.

### Example 8.3

The network of Fig. 8-11 contains a voltage controlled voltage source. We write the nodal equilibrium equations by inspection as if the two sources of the controlled source model were independent:

*node* ①          $(G_2 + G_4)v_{\circled{1}} - G_4 v_{\circled{2}} = \hat{\imath}_1$

*node* ②          $-G_4 v_{\circled{1}} + (G_4 + G_5)v_{\circled{2}} = 0$

*node* ③          $G_7 v_{\circled{3}} + i_6 = 0$

*node* ④          $G_8 v_{\circled{4}} - i_6 = 0$

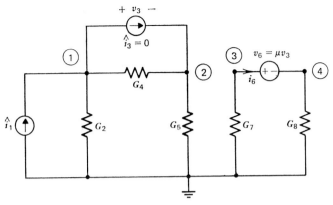

**Fig. 8-11   A network with a voltage-controlled voltage source analyzed in Example 8.3.**

(Since $\hat{i}_3 = 0$ it does not appear in these equations.) We have four equations and five unknowns. The additional equation needed is the voltage controlled voltage source branch relationship

$$v_6 = \mu v_3$$

or in terms of node voltages

$$v_③ - v_④ = \mu(v_① - v_②)$$

On substitution, the nodal equilibrium equations become:

$$(G_2 + G_4)v_① - G_4 v_② = \hat{i}_1$$

$$-G_4 v_① + (G_4 + G_5)v_② = 0$$

$$\mu G_7 v_① - \mu G_7 v_② + G_7 v_④ + i_6 = 0$$

$$G_8 v_④ - i_6 = 0$$

These four equations are in terms of four unknowns: $v_①$, $v_②$, $v_④$, and $i_6$.   □

### Example 8.4

The network of Fig. 8-12 contains a voltage controlled current source. Proceeding as we have previously, the nodal equilibrium equations are written as if all sources were independent:

$$\left(\frac{1}{R_2} + \frac{1}{R_3}\right)v_① - \frac{1}{R_3}v_② = \hat{i}_1$$

$$-\frac{1}{R_3}v_① + \left(\frac{1}{R_3} + \frac{1}{R_6}\right)v_② - i_5 = 0$$

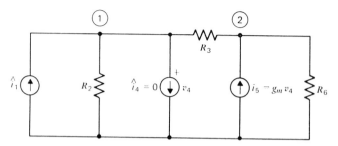

**Fig. 8-12   A network with a voltage-controlled current source analyzed in Example 8.4.**

(The current $i_4 = 0$ and, therefore, is omitted from these equations.) There are three unknowns in these equations: $v_①$, $v_②$, and $i_5$. The additional equation required for solution is the controlled source relationship

$$i_5 = g_m v_4$$

or in terms of the node voltages

$$i_5 = g_m v_①$$

On substitution of this expression, the nodal equilibrium equations become

$$\left(\frac{1}{R_2} + \frac{1}{R_3}\right)v_① - \frac{1}{R_3}v_② = \hat{\imath}_1$$

$$-\left(\frac{1}{R_3} + g_m\right)v_① + \left(\frac{1}{R_3} + \frac{1}{R_6}\right)v_② = 0$$

which can be solved for $v_①$ and $v_②$.   □

**Example 8.5**

The network of Fig. 8-13 contains a current controlled current source. The nodal equilibrium equations under the assumption that all sources are independent are

$$i_1 = -\frac{1}{R_2}\hat{v}_1$$

$$i_3 + i_4 = \frac{1}{R_2}\hat{v}_1$$

$$\frac{1}{R_5}v_③ - i_4 = 0$$

where we have recognized that

$$v_② = \hat{v}_3 = 0$$

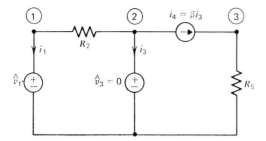

**Fig. 8-13 A network with a current-controlled current source analyzed in Example 8.5.**

The additional equation needed for solution is the controlled source branch relationship

$$i_4 - \beta i_3 = 0$$

The three nodal equilibrium equations and the controlled source branch relationship can be solved simultaneously to yield solutions for the unknowns $v_{\circled３}$, $i_1$, $i_3$, and $i_4$.   □

## Example 8.6

The network of Fig. 8-14 contains a current controlled voltage source. The following set of nodal equilibrium equations can be written by inspection:

$$\frac{1}{R_2} v_{\circled１} + i_3 = \hat{\imath}_1$$

$$\frac{1}{R_4} v_{\circled２} - i_3 = 0$$

$$\frac{1}{R_6} v_{\circled３} + i_5 = 0$$

and

$$\frac{1}{R_7} v_{\circled４} - i_5 = 0$$

There are six unknowns in the four nodal equilibrium equations. The two additional equations needed are the two branch relationships associated with the current controlled voltage source:

$$\hat{v}_3 = 0$$

or

$$v_{\circled１} - v_{\circled２} = 0$$

and

$$v_5 = r_m i_3$$

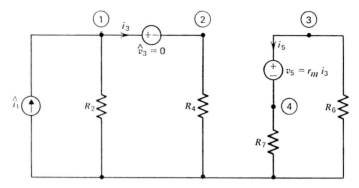

**Fig. 8-14  A network with a current-controlled voltage source analyzed in Example 8.6.**

or

$$v_{③} - v_{⑤} - r_m i_3 = 0$$

The six equations can then be solved simultaneously to yield values for the unknowns $v_{①}$, $v_{②}$, $v_{③}$, $v_{④}$, $i_3$, and $i_5$.  □

In the previous four examples the controlling branch was always shown as a zero-valued current source or a zero-valued voltage source. Sometimes it will be convenient to leave these zero-valued sources out of the network and consider the branch current or voltage associated with some nonsource element to be the controlling quantity. For this situation, an appropriate zero-valued source can always be inserted without altering the behavior of the network. In Figs. 8-15a to 8-15d we have redrawn the networks studied in Examples 8.3 to 8.6 without the zero-valued source. The student should convince himself that the behavior of these networks is not altered by removal of the zero-valued sources.

### Example 8.7

The network of Fig. 8-16 contains a gyrator. We can write the nodal equilibrium equations by inspection initially in terms of $i_3$ and $i_4$:

$$\frac{1}{R_2} v_{①} - i_3 = \hat{i}_1$$

and

$$\frac{1}{R_5} v_{②} + i_4 = 0$$

These two equations are in terms of four unknowns: $v_{①}$, $v_{②}$, $i_3$, and $i_4$. Two additional equations are supplied by the gyrator:

$$i_3 = \alpha v_4$$

$$= \alpha v_{②}$$

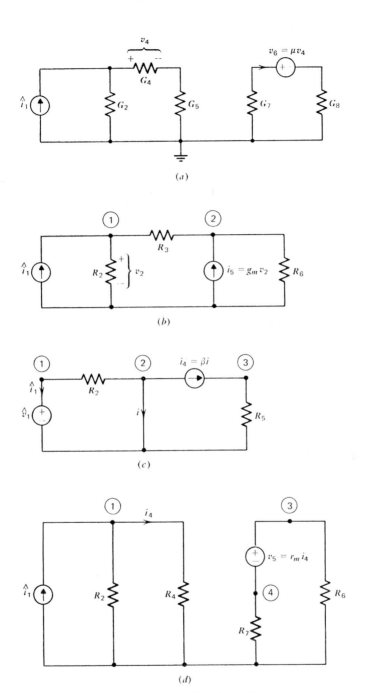

**Fig. 8-15** The networks of Examples 8.3 to 8.6 with the zero-valued sources removed.

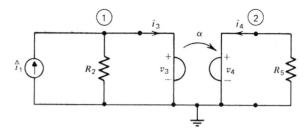

**Fig. 8-16   A network with a gyrator analyzed in Example 8.7.**

and

$$i_4 = -\alpha v_3$$
$$= -\alpha v_①$$

On substitution of these expressions into the nodal equilibrium, we have

and

$$\frac{1}{R_2} v_① + \alpha v_② = \hat{i}_1$$

$$-\alpha v_① + \frac{1}{R_5} v_② = 0 \quad \square$$

In summary, when transformers, controlled sources, or gyrators are present, we can write nodal equilibrium equations as always, introducing additional unknown branch currents and voltages when needed. The additional equations needed for solution are the unused branch relationships that are written in terms of the node voltages.

## 8-5  NODAL EQUILIBRIUM EQUATIONS FOR LINEAR NETWORKS CONTAINING TRANSFORMERS, CONTROLLED SOURCES, AND GYRATORS

The techniques used in the previous section to analyze simple networks that contain two-ports are easily generalized to larger more complex networks. Recall that the nodal equilibrium equations for any network that contains linear resistors, independent current sources, and independent voltage sources could easily be written by inspection and has the form

$$y_{11} v_① + y_{12} v_② + \cdots + y_{1m} v_⑩ = i_①$$
$$y_{21} v_① + y_{22} v_② + \cdots + y_{3m} v_⑩ = i_②$$
$$\cdots\cdots\cdots\cdots\cdots\cdots\cdots\cdots\cdots\cdots$$
$$y_{m1} v_① + y_{m2} v_② + \cdots + y_{mm} v_⑩ = i_⑩$$

where

$$m = n_n - 1$$

$y_{kk}$ = sum of all conductances connected between node (k) and all other nodes

$y_{jk}$ = sum of all conductances directly connected between nodes (j) and (k)

and

$i_{(k)}$ = sum of all independent current source currents entering node (k) plus the sum of all (unknown) independent voltage source currents entering node (k)

If there are $n_V$ independent voltage sources in the network, and additional $n_V$ equations would be required. These equations are the voltage source branch relationships in terms of the node voltages.

When dependent sources are present we can treat them as independent sources and write the nodal equilibrium equations as indicated. The branch relationships of the dependent sources, expressed in terms of the node voltages, can then be appended to the set of nodal equilibrium equations. For transformers and gyrators, we can initially treat the currents through the primary and secondary as unknowns and write the nodal equilibrium equations as above. Then the two transformer or gyrator branch relationships expressed in terms of the node voltages, can be appended.

Alternatively, when writing the nodal equilibrium equation for a node that has a two-port element connected to it, we can try to use the elements' description to express the current leaving the node through the element in terms of the node voltages. The second method will result in fewer equations to be solved. These alternative techniques for writing nodal equilibrium equations are best illustrated by two examples.

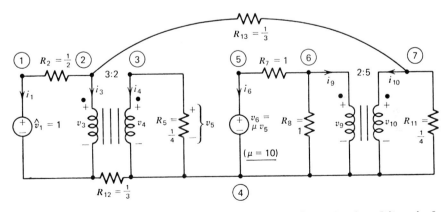

**Fig. 8-17   The network of Example 8.8 that contains several multiterminal elements.**

## Example 8.8

The network of Fig. 8-17 contains two transformers and a voltage controlled voltage source. Observe that the controlling branch voltages is $v_5$, the voltage across $R_5$. The nodal equilibrium equations are

node ①   $\displaystyle -\frac{1}{R_2}v_② + i_1$ $\displaystyle = -\frac{1}{R_2}\hat{v}_1$

node ②   $\displaystyle \left(\frac{1}{R_2}+\frac{1}{R_{13}}\right)v_② - \frac{1}{R_{13}}v_⑦ + i_3$ $\displaystyle = \frac{1}{R_2}\hat{v}_1$

node ③   $\displaystyle \frac{1}{R_5}v_③ - \frac{1}{R_5}v_④ + i_4$ $= 0$

node ④   $\displaystyle -\frac{1}{R_5}v_③ + \left(\frac{1}{R_5}+\frac{1}{R_8}+\frac{1}{R_{11}}+\frac{1}{R_{12}}\right)v_④$

$\displaystyle \qquad -\frac{1}{R_8}v_⑥ - \frac{1}{R_{11}}v_⑦ - i_4 - i_6 - i_9 - i_{10} = 0$

node ⑤   $\displaystyle \frac{1}{R_7}v_⑤ - \frac{1}{R_7}v_⑥ + i_6$ $= 0$

node ⑥   $\displaystyle -\frac{1}{R_8}v_④ - \frac{1}{R_7}v_⑤ + \left(\frac{1}{R_7}+\frac{1}{R_8}\right)v_⑥ \quad + i_9 = 0$

node ⑦   $\displaystyle -\frac{1}{R_{13}}v_② - \frac{1}{R_{11}}v_④ + \left(\frac{1}{R_{11}}+\frac{1}{R_{13}}\right)v_⑦ + i_{10} = 0$

These seven equations are expressed in terms of unknowns: $v_②$, $v_③$, $v_④$, $v_⑤$, $v_⑥$, $v_⑦$, $i_1$, $i_3$, $i_4$, $i_6$, $i_9$, and $i_{10}$. The five additional equations needed for solution are the four transformer relations; which can be written as:

$$v_② - \frac{3}{2}(v_③ - v_④) = 0$$

$$i_3 + \frac{2}{3}i_4 = 0$$

$$v_⑥ - v_④ - \frac{2}{5}(v_⑦ - v_④) = 0$$

$$i_9 + \frac{5}{2}i_{10} = 0$$

and the voltage controlled voltage source relation, which can be written as

$$(v_⑤ - v_④) - \mu(v_③ - v_④) = 0$$

(Observe that the branch relationships associated with the transformers and controlled source have been expressed in terms of the node voltages.) On substitution of the element values, the 12 simultaneous equations to be solved become:

$$- 2v_{②} + i_1 = -2$$

$$5v_{②} - 3v_{⑦} + i_3 = 2$$

$$4v_{③} - 4v_{④} + i_4 = 0$$

$$- 4v_{③} + 12v_{④} - v_{⑥} - 4v_{⑦} - i_4 - i_6 - i_9 - i_{10} = 0$$

$$v_{⑤} - v_{⑥} + i_6 = 0$$

$$- v_{④} - v_{⑤} + 2v_{⑥} + i_9 = 0$$

$$- 3v_{②} - 4v_{④} + 7v_{⑦} + i_{10} = 0$$

$$v_{②} - \frac{3}{2}v_{③} + \frac{3}{2}v_{④} = 0$$

$$i_3 + \frac{2}{3}i_4 = 0$$

$$- \frac{3}{5}v_{④} + v_{⑥} - \frac{2}{5}v_{⑦} = 0$$

$$i_9 + \frac{5}{2}i_{10} = 0$$

and

$$- 10v_{③} + 9v_{④} + v_{⑤} = 0$$

Use of GEP yields the solution $v_{②} = .476$, $v_{③} = .385$, $v_{④} = .068$, $v_{⑤} = 3.24$, $v_{⑥} = .204$, $v_{⑦} = .408$, $i_1 = -1.05$, $i_3 = .846$, $i_4 = -1.27$, $i_6 = -3.04$, $i_9 = 2.90$, and $i_{10} = -1.16$.   □

## Example 8.9

The network of Fig. 8-18 contains two voltage controlled current sources and one current controlled current source. Once again the zero-valued sources of the controlled source models are not explicitly shown. The nodal equilibrium equations for nodes ① and ② are easily written:

node ①        $$- \frac{1}{R_2} v_{②} + i_1 = - \frac{\hat{v}_1}{R_2}$$

and

node ②        $$\left( \frac{1}{R_2} + \frac{1}{R_3} + \frac{1}{R_4} \right) v_{②} - \frac{1}{R_4} v_{⑥} = \frac{\hat{v}_1}{R_2}$$

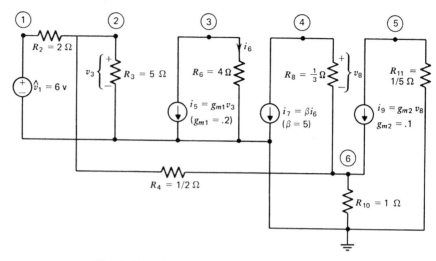

**Fig. 8-18   Network considered in Example 8.9.**

Node ③ has a controlled current source connected to it that is described by the relation

$$i_5 = g_{m1}v_3$$

or in terms of the node voltages

$$i_5 = g_{m1}v_②$$

Then the nodal equilibrium equation for node ③ is

*node ③*
$$g_{m1}v_② + \frac{1}{R_6}v_③ = 0$$

Node ④ also has a controlled current source connected to it. The controlled current source is described by

$$i_7 = \beta i_6$$

but

$$i_6 = \frac{1}{R_6}v_③$$

so that the controlled source relationship in terms of the node voltages is

$$i_7 = \frac{\beta}{R_6}v_③$$

and the nodal equilibrium equation is

*node ④*
$$\frac{\beta}{R_6}v_③ + \frac{1}{R_8}v_④ - \frac{1}{R_8}v_⑥ = 0$$

The relationship for the controlled current source connected to node ⑤ in terms of the node voltages is

$$i_9 = g_{m2}(v_④ - v_⑥)$$

so that the nodal equilibrium equati. 's

node ⑤ $$g_{m2}v_④ + \frac{1}{R_{11}}v_⑤ - g_{m2}v_⑥ = 0$$

Finally, the nodal equilibrium equation for node ⑥ is

node ⑥ $$-\frac{1}{R_4}v_② - \left[\frac{1}{R_8} + g_m\right]v_④ + \left(\frac{1}{R_8} + \frac{1}{R_{10}} + \frac{1}{R_4} + g_m\right)v_⑥ = 0$$

On substitution of the element values, the six nodal equilibrium equations become:

node ① $$-.5v_② + i_1 \qquad\qquad = -3$$

node ② $$2.7v_② - 2v_⑥ \qquad = 3$$

node ③ $$.2v_② + .25v_③ \qquad = 0$$

node ④ $$1.25v_③ + 3v_④ - 3v_⑥ = 0$$

node ⑤ $$.1v_④ + 5v_⑤ - .1v_⑥ = 0$$

and

node ⑥ $$-2v_② - 3.1v_④ + 6.1v_⑥ = 0$$

The solution of these equations is $i_1 = -.786$, $v_② = 4.43$, $v_③ = -3.54$, $v_⑨ = 6.0$, $v_⑤ = -.21$, and $v_⑥ = 4.48$. Observe that by expressing the two-port element branch relationships in terms of the node voltages it is easy to write the nodal equilibrium equations without resorting to the introduction of extra unknowns. □

## 8-6 TWO-PORT DESCRIPTIONS OF NETWORKS

In many networks only four terminals are really of interest: the two *input* terminals and the two *output* terminals. For such a network, the two-port parameters discussed in Section 8-2 provide a convenient method of describing network behavior. In order words, just as it was sometimes convenient to describe an arbitrary connection of resistors, transformers and controlled sources as a two terminal equivalent resistor, it might be convenient to describe a network comprised of resistors, transformers, and controlled sources as a two-port. Even more importantly, if a certain connection of elements, that can be modeled as a two-port, shows up in several places in the network,

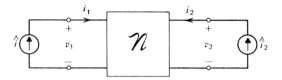

**Fig. 8-19    Characterizing a network using z-parameters.**

it is more convenient to use the two-port description rather than describing each element individually.

Consider the network $\mathcal{N}$ of Fig. 8-19. The port variables are $v_1$, $i_1$, and $v_2$, $i_2$. The currents $\hat{i}_1$ and $\hat{i}_2$ are two independent current cources external to the network $\mathcal{N}$. Suppose we wished to characterize $\mathcal{N}$ using z-parameters:

$$v_1 = z_{11}i_1 + z_{12}i_2$$

$$v_2 = z_{21}i_1 + z_{22}i_2$$

We must determine appropriate values for $z_{11}$, $z_{12}$, $z_{21}$, $z_{22}$. Observe that if $i_2$ were set to zero,

$$v_1|_{i_2=0} = z_{11}i_1$$

and

$$v_2|_{i_2=0} = z_{21}i_1$$

or

$$z_{11} = \frac{v_1}{i_1}\bigg|_{i_2=0}$$

and

$$z_{21} = \frac{v_2}{i_1}\bigg|_{i_2=0}$$

This situation can be realized if the current source $\hat{i}_2 = 0$, that is, open circuited. Observe if $\hat{i}_1 = 1$, then $i_1 = 1$ and

$$z_{11} = v_1\big|_{\substack{i_2=0 \\ i_1=1}}$$

and

$$z_{21} = v_2\big|_{\substack{i_2=0 \\ i_1=1}}$$

Thus the parameters $z_{11}$ and $z_{21}$ can be measured by open circuiting port 2 and applying a current (or voltage) excitation at port 1. Similarly, the parameters $z_{21}$ and $z_{22}$ can be measured by open circuiting port 1 (setting $\hat{i}_1 = 0$) and applying a current (or voltage) excitation at port 2:

$$z_{12} = \frac{v_1}{i_2}\bigg|_{i_1=0}$$

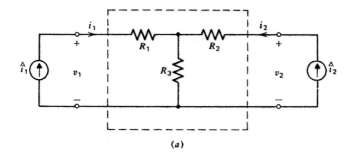

(a)

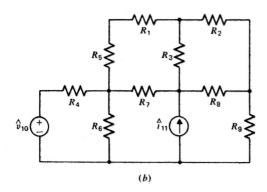

(b)

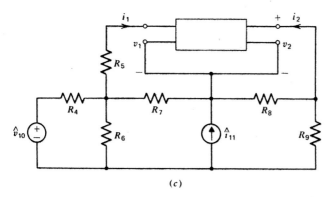

(c)

Fig. 8-20 (a) The network used in Example 8.10 to illustrate calculation of the z-parameters. (b) A network that contains a connection of resistors similar to the two-port of (a) is shown in (c) replaced by a two-port described by $v_1 = (R_1 + R_3)i_1 + R_3 i_2$ and $v_2 = R_3 i_2 + (R_2 + R_3)i_2$.

and

$$z_{22} = \left. \frac{v_2}{i_2} \right|_{i=0}$$

Since the $z$-parameters are measured by open circuiting one of the ports and have units of ohms (or impedance) they are sometimes called the *open circuit impedance parameters*. Observe that two separate analyses are required to determine the four $z$-parameters.

### Example 8.10

The network of Fig. 8-20a serves as an illustration of the procedure. To calculate $z_{11}$ and $z_{21}$ set $\hat{\imath}_2 = 0$ and $\hat{\imath}_1 = 1$ a. Then

$$v_1 = (R_1 + R_3)\hat{\imath}_1 = R_1 + R_3$$

and

$$v_2 = R_3\hat{\imath}_1 = R_3$$

so that

$$z_{11} = \left. \frac{v_1}{i_1} \right|_{i_2=0} = R_1 + R_3$$

and

$$z_{21} = \left. \frac{v_2}{i_1} \right|_{i_2=0} = R_3$$

To calculate $z_{12}$ and $z_{22}$ set $\hat{\imath}_1 = 0$ and $\hat{\imath}_2 = 1$ a. Then

$$v_1 = R_3\hat{\imath}_2 = R_3$$

and

$$v_2 = (R_2 + R_3)\hat{\imath}_2 = R_2 + R_3$$

so that

$$z_{12} = \left. \frac{v_1}{i_2} \right|_{i_1=0} = R_3$$

and

$$z_{22} = \left. \frac{v_2}{i_2} \right|_{i_1=0} = R_2 + R_3$$

The network $\mathcal{N}$ can be described by the two-port equations

$$v_1 = (R_1 + R_3)i_1 + R_3 i_2$$
$$v_2 = R_3 i_2 + (R_2 + R_3)i_2$$

If this two-port were embedded in some larger network, for example, as shown in Fig. 8-20b, it would not be necessary to consider each branch independently when writing node equations but rather treat the connection of elements as a two port as shown in Fig. 8-20c. The network of Fig. 8-20c although equivalent to the network of Fig. 8-20b has one less node. Observe

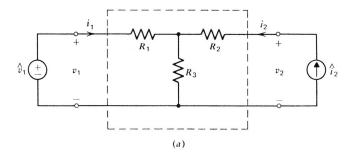

(a)

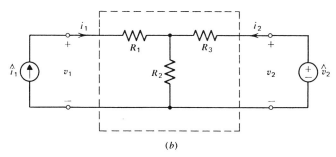

(b)

**Fig. 8-21** **A voltage source could also be used to calculate $z_{11}$ and $z_{21}$, and (b) $z_{12}$ and $z_{22}$.**

that replacing a connection of resistors by a two-port is only beneficial if we are unconcerned with the voltages and currents internal to the two-port.   □

It should be pointed out that independent voltage sources could also have been used to determine the $z$-parameters for the network of Fig. 8-20 as illustrated in Fig. 8-21a and 8-21b.

**Example 8.11**

In Fig. 8-21a, let $i_2 = 0$ and $v_1 = 1$. Then

$$i_1 = \frac{\hat{v}_1}{R_1 + R_3} = \frac{1}{R_1 + R_3}$$

and

$$v_2 = \frac{R_3}{R_1 + R_3} \hat{v}_1 = \frac{R_3}{R_1 + R_3}$$

from which we can find $z_{11}$ and $z_{21}$:

$$z_{11} = \left.\frac{v_1}{i_1}\right|_{i_2=0} = R_1 + R_2$$

$$z_{21} = \left.\frac{v_2}{i_1}\right|_{i_2=0} = R_3$$

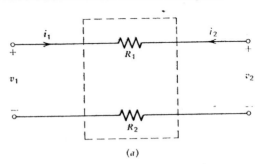

(a)

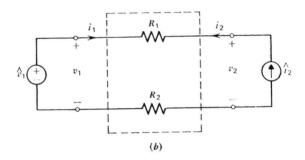

(b)

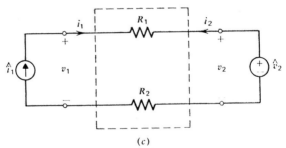

(c)

**Fig. 8-22** (*a*) A two-port network that does not possess a *z*-parameter description; (*b*) configuration used to attempt to find $z_{11}$ and $z_{21}$; (*c*) configuration used to attempt to find $z_{12}$ and $z_{22}$.

Similarly, $z_{12}$ and $z_{22}$ could be found from Fig. 8-21*b* by setting $i_1 = 0$ and $v_2 = 1$. $\square$

It is not always possible to describe two-port networks using *z*-parameters. For example, the network of Fig. 8-22*a* does not lend itself to a *z*-parameter description. We can try to find the *z*-parameters using the networks of Fig. 8-22*b* and *c*. In the network of Fig. 8-22*b*, let $\hat{v}_1 = 1$ and $\hat{i}_2 = 0$. Then

$$i_1 = 0$$

and

$$v_2 = \hat{v}_1 = 1$$

so that

$$z_{11} = \frac{v_1}{i_1}\bigg|_{i_2=0} = \frac{1}{0}$$

which is undefined. Similarly,

$$z_{21} = \frac{v_2}{i_1}\bigg|_{i_2=0} = \frac{1}{0}$$

which is also undefined. Furthermore, using the network of Fig. 8-22c we would find that $z_{12}$ and $z_{22}$ are undefined. This network does not have a z-parameter description. An alternative is to try the y-parameters:

$$i_1 = y_{11}v_1 + y_{12}v_2$$

$$i_2 = y_{21}v_1 + y_{22}v_2$$

$y_{11}$ and $y_{21}$ can be found by setting $v_2$ to zero:

$$y_{11} = \frac{i_1}{v_1}\bigg|_{v_2=0}$$

and

$$y_{21} = \frac{i_2}{v_1}\bigg|_{v_2=0}$$

$y_{12}$ and $y_{22}$ can be found by setting $v_1$ to zero:

$$y_{12} = \frac{i_2}{v_2}\bigg|_{v_1=0}$$

and

$$y_{22} = \frac{i_2}{v_2}\bigg|_{v_1=0}$$

Observe that two separate analyses are required to determine the four y-parameters. Since these parameters are found by forcing voltages to zero, that is, short circuiting one of the ports, they are referred to as *short circuit admittance parameters*. To determine the y-parameters for the two-port of Fig. 8-22a, we can use the configuration of Fig. 8-23. First, we set $\hat{v}_2 = 0$, then

$$i_1 = \frac{\hat{v}_1}{R_1 + R_2}$$

and

$$i_2 = -\frac{\hat{v}_1}{R_1 + R_2}$$

Since $v_1 = \hat{v}_1$:

$$y_{11} = \frac{i_1}{v_1}\bigg|_{v_2=0} = \frac{1}{R_1 + R_2}$$

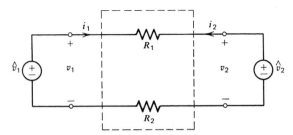

**Fig. 8-23   Network used to find y-parameters of the two-port of Fig. 8-14a.**

and

$$y_{21} = \frac{i_2}{v_1}\bigg|_{v_2=0} = -\frac{1}{R_1 + R_2}$$

Next set $\hat{v}_1 = 0$, so that

$$i_1 = \frac{\hat{v}_2}{R_1 + R_2}$$

and

$$i_1 = \frac{-\hat{v}_2}{R_1 + R_2}$$

Since $v_2 = \hat{v}_2$:

$$y_{12} = \frac{i_1}{v_2}\bigg|_{v_1=0} = \frac{-1}{R_1 + R_2}$$

and

$$y_{22} = \frac{i_2}{v_2}\bigg|_{v_1=0} = \frac{1}{R_1 + R_2}$$

The two-port description is

$$i_1 = \frac{1}{R_1 + R_2} v_1 - \frac{1}{R_1 + R_2} v_2$$

and

$$i_2 = \frac{1}{R_1 + R_2} v_1 + \frac{1}{R_1 + R_2} v_2$$

Observe that these equations are not linearly independent. The lack of independence of these equations is behind the fact that this network does not possess a $z$-parameter description. More will be said about the independence of port equations.

It is also possible for a network to possess a $z$-parameter description and not a $y$-parameter description. The network of Fig. 8-24 has the $z$-parameters:

$$z_{11} = R, \qquad z_{21} = R$$
$$z_{21} = R, \qquad z_{22} = R$$

but the $y$-parameters are undefined.

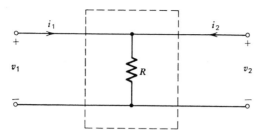

**Fig. 8-24  A network that possesses a $z$-parameter description but not a $y$-parameter description.**

So that the student does not believe that a network possesses only one type of description, the network of Fig. 8-20$a$ possesses a $y$-parameter description as well as the $z$-parameter description. It is left as an exercise for the student to show that the network can be described by

$$i_1 = \frac{R_2 + R_3}{R_1 R_2 + R_1 R_3 + R_2 R_3} v_1 - \frac{R_3}{R_1 R_2 + R_1 R_3 + R_2 R_3} v_2$$

$$i_2 = \frac{-R_3}{R_1 R_2 + R_1 R_3 + R_2 R_3} v_1 + \frac{R_1 + R_3}{R_1 R_2 + R_1 R_3 + R_2 R_3} v_2$$

Finally, it is interesting to observe that two ports can be modeled using controlled sources and resistors. Figure 8-25$a$, a model depicting the $z$-parameter description is derived by interpreting the equations

$$v_1 = z_{11} i_1 + z_{12} i_2$$

and

$$v_2 = z_{21} i_1 + z_{22} i_2$$

as loop equations. Figure 8-25$b$, a model depicting the $y$-parameter description is derived by interpreting the equations

$$i_1 = y_{11} v_1 + y_{12} v_2$$

and

$$i_2 = y_{21} v_1 + y_{22} v_2$$

as node equations. Figure 8-25$c$, a model depicting the hybrid parameter description is derived by interpreting the equation

$$v_1 = h_{11} i_1 + h_{12} v_2$$

as a loop equation and interpreting the equation

$$i_2 = h_{21} i_1 + h_{22} v_2$$

as a node equation. We leave it as an exercise for the student to derive models for the transmission inverse transmission, and inverse hybrid descriptions.

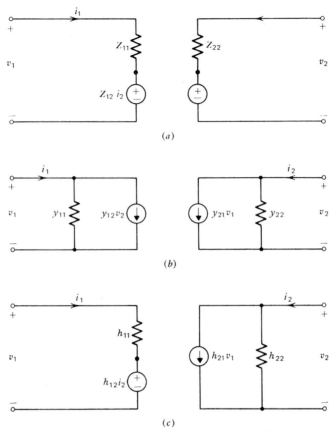

**Fig. 8-25** **Models depicting (*a*) the *z*-parameters, (*b*) the *y*-parameters, (*c*) the hybrid parameters.**

Various analysis schemes that are based on the use of two-port parameters can be derived. Some of these schemes have found great utility for special purpose analysis programs. The interested student is referred to the references.

## 8-7  SUMMARY

*Multiterminal resistive elements* were introduced in this chapter. At times it is convenient to consider the terminals in pairs. If the terminal pair is such that the current entering one terminal equals the current leaving the other terminal, it is called a *port*. *Two-port elements* are frequently encountered, and there are several methods of characterizing them, among these are the *open circuit impedance*, or *z-parameter* description, the *short circuit admittance*, or *y-parameter* description and the *hybrid*, or *h-parameter* description.

Some specific two-port elements are: the *ideal transformer*, described by

$$v_1 = \left(\frac{n_1}{n_2}\right) v_2$$

and

$$i_1 = -\left(\frac{n_2}{n_1}\right) i_2$$

where $n_1/n_2$ is the *turns ratio*; the *voltage controlled voltage source*, described by

$$v_2 = \mu v_1$$

and

$$i_1 = 0$$

where $\mu$ is the *voltage amplification factor*; the *voltage controlled current source*, described by

$$i_2 = g_m v_1$$

and

$$i_1 = 0$$

where $g_m$ is the *transconductance*; the *current controlled current source*, described by

$$i_2 = \beta i_1$$

and

$$v_1 = 0$$

where $\beta$ is the *current amplification factor*; the *current controlled voltage source*, described by

$$v_2 = r_m i_1$$

and

$$v_1 = 0$$

where $r_m$ is the *transresistance*; and the *gyrator*, described by

$$i_1 = \alpha v_2$$

and

$$i_2 = -\alpha v_1$$

where $\alpha$ is the *gyration ratio*. These two-ports are not physical elements but, they prove quite useful for modeling the behavior of physical devices.

The analysis schemes of previous chapters were extended to handle networks that contain multiterminal elements and finally the use of two-port descriptions to characterize entire networks was discussed.

## PROBLEMS

**8.1**     For each of the networks in Fig. P8.1, write a set of nodal equilibrium equations by inspection, using unknown currents if necessary for

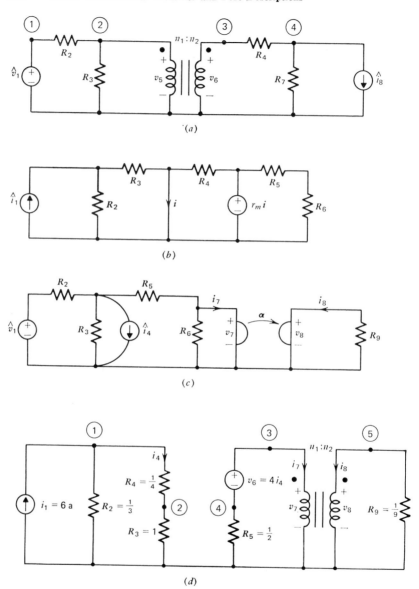

**Fig. P8.1**

the multiterminal elements. Append to this set the equations that describe the multiterminal element. Then by suitable manipulation reduce these equations to a set of $n_n - 1$ equations in terms of $n_n - 1$ unknowns.

**8.2**    Write the nodal equilibrium equations for the network for Fig. P8.2 and determine the value of $R_2$ for which $v_o/\hat{v}_1 = 10$.

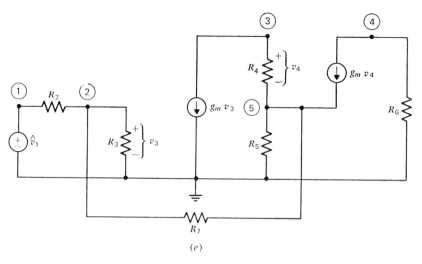

**Fig. P8.1** (*Continued*)

**8.3** Recall in the proof of Thévenin's and Norton's theorems the only restriction placed upon the network was that it be linear. Since the two-ports described here are linear, these theorems may still be applied. Find the Thévenin equivalent network as seen from terminals T-T' for the networks of Fig. P8.3. (*Note.* When determining the Thévenin or Norton equivalent resistance only the *independent* sources should be set to zero.)

**8.4** Determine the Norton equivalent for the networks of Fig. P8.4.

**8.5** Write the equations you would submit to GEP to find the Thévenin equivalent network associated with the network of Fig. P8.5.

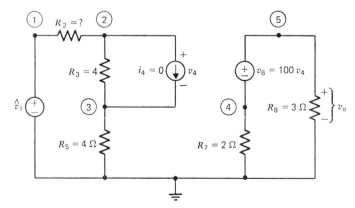

**Fig. P8.2**

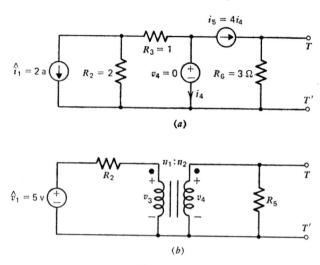

(a)

(b)

**Fig. P8.3**

**8.6**    A simplified small signal dc model for a transistor is shown in Fig. P8.6a. Using this model with $\alpha_F = .98$ determine the output voltage $v_o$ for the network of Fig. P8.6b.

**8.7**    Find $v_o$ in the network of Fig. P8.7. Use the transistor model of Fig. P8.6 with $\alpha_F = .98$.

**8.8**    Compute the power absorbed by each branch in the network of Fig. P8.2 and verify Tellegen's theorem.

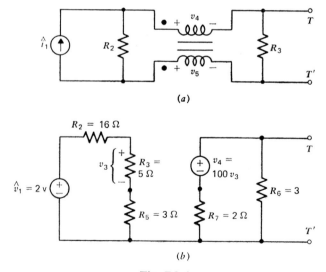

(a)

(b)

**Fig. P8.4**

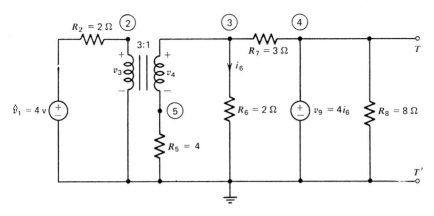

**Fig. P8.5**

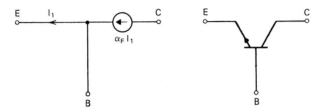

**Fig. P8.6a**

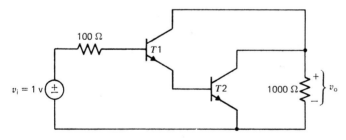

**Fig. P8.6b**

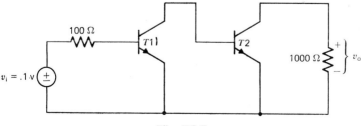

**Fig. P8.7**

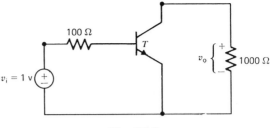

**Fig. P8.9**

**8.9**    The gain of a network can be found by dividing the magnitude of
the output by the magnitude of the input. Compare the gains of
the networks of Fig. P8.6$b$ and Fig. P8.7 with the single transistor
amplified stage shown in Fig. P8.9. Also compare the input impedance of the three networks. (The input impedance is $-v_i/i_i$.) What
conclusions can you draw concerning the operation of these circuits?

**8.10**    A more accurate small-signal dc transistor model is shown in Fig.
P8.10. Use this model for the transistor in network of Fig. P8.9 and
compute $v_o$. Let $r_\pi = 300\ \Omega$, $g_m = .15$, and $r_o = 10^5\ \Omega$.

**8.11**    Consider the gyrator-current source combination shown in Fig.
P8.11. Show that this device behaves as an independent voltage
source with voltage $\hat{\imath}$.

**8.12**    It can be seen from Prob. 8.11 that an alternate method for writing
node equations for networks that contain voltage sources is to
replace the voltage source with an equivalent gyrator-current source
combination as shown in Fig. P8.11. Node equations for this modified
network can easily be written because it contains only independent
current sources. Use this method to solve for the node voltages in
the networks of Fig. P8.12. Compare your results with those obtained in Prob. 4.3.

**8.13**    Find the $z$-parameters and $y$-parameters for each of the networks in
Fig. P8.13.

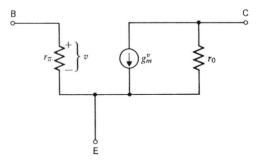

**Fig. P8.10**

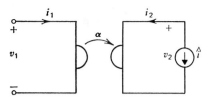

**Fig. P8.11**

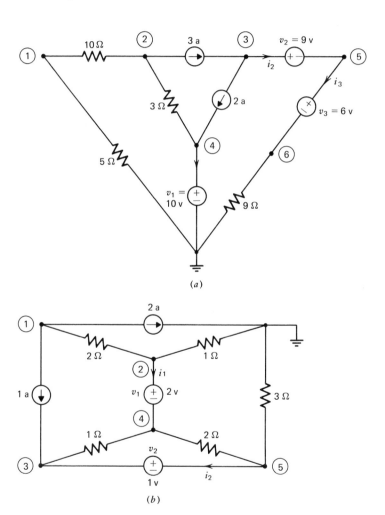

*(a)*

*(b)*

**Fig. P8.12**

271

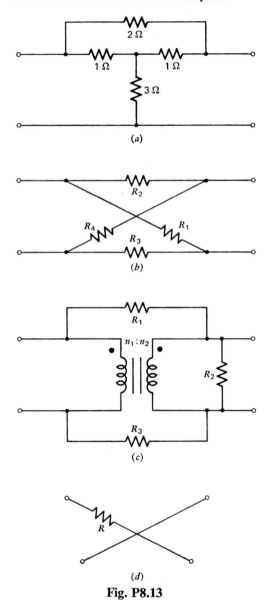

(a)

(b)

(c)

(d)

**Fig. P8.13**

**8.14** A hybrid description of a two-port network can be written as

$$v_1 = h_{11}i_1 + h_{12}v_2$$
$$i_2 = h_{21}i_1 + h_{22}v_2$$

Determine the hybrid description for the two-port network shown in Fig. P8.14.

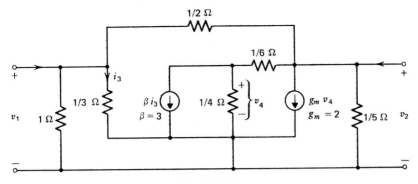

**Fig. P8.14**

**8.15**   Find the $y$-parameters associated with the networks of Fig. P8.15.

**8.16**   Consider the connection of two-ports shown in Fig. P8.16. The first
two-port is characterized by

$$i_1 = y_{11}v_1 + y_{12}v_2$$

$$i_2 = y_{21}v_1 + y_{22}v_2$$

and the second two-port is characterized by

$$\hat{i}_1 = \hat{y}_{11}\hat{v}_1 + \hat{y}_{12}\hat{v}_2$$

$$\hat{i}_2 = \hat{y}_{21}\hat{v}_1 + \hat{y}_{22}\hat{v}_2$$

Determine the $y$-parameters for the overall network. Can you extend
this result to arbitrary numbers of two-ports in parallel?

**8.17**   In the network of Fig. P8.17 each two-port is described by the equations

$$i_1 = 3v_1 + 5v_2$$

$$i_2 = 2v_1 - v_2$$

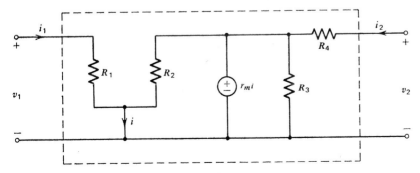

**Fig. P8.15**

**Fig. P8.16**

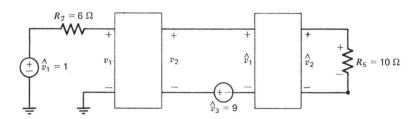

**Fig. P8.17**

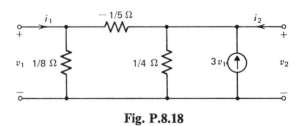

**Fig. P.8.18**

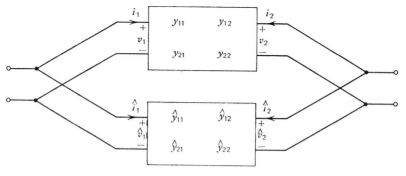

**Fig. P8.20**

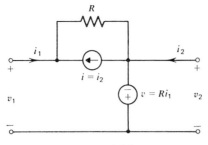

**Fig. P8.21**

Write the nodal equilibrium equations for this circuit and use GEP to determine the node voltages.

**8.18** Verify that the two-port of Fig. P8.18 is characterized by the two-port equations of Prob. 8.17. Replace each of the two-ports in the network of Prob. 8.17 with this model. Write the nodal equilibrium equations for this modified network. Use the GEP to solve for the node voltages and compare the results with those of Prob. 8.17.

**8.19** Using resistors and controlled sources find models for the transmission, inverse transmission and inverse hybrid descriptions.

**8.20** In the network of Fig. P8.20 what must the turns ratio of the transformer be in order that maximum power be delivered to $R_3$?

**8.21** Compute the $y$-parameters for the network of Fig. P8.21. What ideal element does this network behave like?

## CHAPTER 9

# Matrix Notation

In our work so far we have consistently dealt with sets of simultaneous algebraic equations. For instance, the nodal equilibrium equations for a network comprised of resistors and current sources can be written as

$$y_{11} v_{①} + y_{12} v_{②} + \cdots + y_{1m} v_{⑩} = i_{①}$$
$$y_{21} v_{①} + y_{22} v_{②} + \cdots + y_{2m} v_{⑩} = i_{②}$$
$$\cdots\cdots\cdots\cdots\cdots\cdots\cdots\cdots\cdots\cdots\cdots\cdots$$
$$y_{m1} v_{①} + y_{m2} v_{②} + \cdots + y_{mm} v_{⑩} = i_{⑩}$$

where $m = n_n - 1$,

$y_{kk}$ = sum of all conductances connected between node ⓚ and all other nodes

$y_{jk}$ = negative sum of all conductances directly connected between nodes ⓙ and ⓚ ($j \neq k$)

$i_{ⓚ}$ = sum of all current source currents entering node ⓚ

These equations may be written more concisely as

$$\sum_{k=1}^{n_n-1} y_{jk} v_{ⓚ} = i_{ⓙ}, \qquad j = 1, 2, \ldots, n_n - 1$$

but even this notation can become cumbersome. (Recall the proof of Tellegen's theorem in Chapter 6.) *Matrix notation* is a convenient shorthand to represent an array of numbers. In particular, the use of matrix notation allows us to represent simultaneous algebraic equations, and their manipulation, in a compact form. In this chapter we discuss matrices and how they are used in what we have done in Chapters 1 to 8.

## 9-1  NODAL EQUILIBRIUM EQUATIONS IN TERMS OF MATRICES

We start by reviewing the definition of a matrix.[1] A rectangular array of numbers having $n$ rows and $m$ columns is called an $(n \times m)$ *matrix*. When $n = m$ the matrix is called *square* and the number $n \ (= m)$ is its *order*. An example of a 2 × 3 matrix is

$$\begin{bmatrix} 3 & -2 & 6 \\ 6 & 1 & 4 \end{bmatrix}$$

An example of a 3rd order square matrix, or a 3 × 3 matrix is

$$\begin{bmatrix} 1 & 2 & 3 \\ 3 & 5 & 7 \\ 4 & 2 & 6 \end{bmatrix}$$

**Example 9.1**

We can think of a matrix as representing a table of numbers. For example, consider Table 9-1.

**Table 9.1**

| Nodal Equilibrium Equation for Node Number | Coefficient of Node Voltages | | |
|:---:|:---:|:---:|:---:|
| | $v_1$ | $v_2 \cdots v_m$ | |
| ① | $y_{11}$ | $y_{12}$ | $\cdots$ $y_{1m}$ |
| ② | $y_{21}$ | $y_{22}$ | $\cdots$ $y_{2m}$ |
| ③ | $y_{31}$ | $y_{32}$ | $\cdots$ $y_{3m}$ |
| $\vdots$ | $\vdots$ | $\vdots$ | |
| ⓜ $= n_n - 1$ | $y_{m1}$ | $y_{m2}$ | $\cdots$ $y_{mm}$ |

[1] Recall that matrices were first mentioned in Chapter 3 when we discussed Cramer's rule.

where we have catalogued the coefficients of the node voltages $v_①$, $v_②$, ... , $v_⓶$ that appear in the nodal equilibrium equations for nodes $①$, $②$, ... , $⓶ = n_n - 1$. This table may be represented by the matrix

$$
\begin{bmatrix}
y_{11} & y_{12} & \cdots & y_{1m} \\
y_{21} & y_{22} & \cdots & y_{2m} \\
\cdots & \cdots & \cdots & \cdots \\
y_{m1} & y_{m2} & \cdots & y_{mm}
\end{bmatrix}
$$

if we take the $l$th row to represent the coefficients of the nodal equilibrium equation associated with the $l$th node and the $k$th column to represent the coefficient of the $k$th node voltage $v_ⓚ$. $\square$

An $m \times n$ matrix is denoted by a capital (boldface) letter, for example, $\mathbf{A}$. The elements of $\mathbf{A}$ are written as $a_{lk}$ where the first subscript, $l$, denotes the row number and the second subscript, $k$, denotes the column number. We may define a matrix by all its elements as follows:

$$
\mathbf{A} =
\begin{bmatrix}
a_{11} & a_{12} & \cdots & a_{1n} \\
a_{21} & a_{22} & \cdots & a_{2n} \\
\cdots & \cdots & \cdots & \cdots \\
a_{m1} & a_{m2} & \cdots & a_{mn}
\end{bmatrix}.
$$

or symbolically by

$$
\mathbf{A} \equiv [a_{lk}]
$$

where the square brackets indicate the array, a typical element of which is $a_{lk}$. In what follows it will be convenient to denote the matrix of coefficients of the nodal equilibrium equations by $\mathbf{Y}$:

$$
\mathbf{Y} =
\begin{bmatrix}
y_{11} & y_{12} & \cdots & y_{1m} \\
y_{21} & y_{22} & \cdots & y_{2n} \\
\cdots & \cdots & \cdots & \cdots \\
y_{m1} & y_{m2} & \cdots & y_{mm}
\end{bmatrix}
$$

$\mathbf{Y}$ is referred to as the *nodal admittance matrix*, and can be written by inspection just as the nodal equilibrium equations can be written by inspection.

A matrix consisting of a single column, that is, an $n \times 1$ matrix, is called a *column matrix, column vector, n-vector,* or just *vector*. Vectors are usually denoted by lower case (boldface) letters, for example, $\mathbf{a}$, although capital boldface letters may also be used.

A vector may also be defined by its elements:

$$
\mathbf{a} = \begin{bmatrix} a_1 \\ a_2 \\ \cdot \\ \cdot \\ \cdot \\ a_p \end{bmatrix}
$$

It is convenient to denote the numbers that appear on the right hand side of a set of nodal equilibrium equations, $i_{①}, i_{②}, \ldots, i_{⑩}$ by the vector $\mathbf{I}$:

$$
\mathbf{I} = \begin{bmatrix} i_1 \\ i_2 \\ \cdot \\ \cdot \\ \cdot \\ i_m \end{bmatrix}
$$

$\mathbf{I}$ is called the *equivalent current vector*. We may also construct the *node voltage vector*

$$
\mathbf{v}_n = \begin{bmatrix} v_{①} \\ v_{②} \\ \cdot \\ \cdot \\ \cdot \\ v_{⑩} \end{bmatrix}
$$

The nodal equilibrium equations represent relationships between the nodal admittance matrix $\mathbf{Y}$, the node voltage vector $\mathbf{v}_n$ and the equivalent current source vector $\mathbf{I}$. We will indicate this relationship by

$$
\mathbf{Y}\mathbf{v}_n = \mathbf{I}
$$

To guarantee that this matrix equation is interpreted correctly we must define *matrix multiplication* precisely.

The *product* of the $(n \times m)$ matrix $\mathbf{A}$ and the $(m \times r)$ matrix $\mathbf{B}$ is the $(n \times r)$ matrix $\mathbf{C}$ where

$$
c_{lk} \equiv \sum_{j=1}^{m} a_{lj} b_{jk}
$$

for all $l = 1, 2, \ldots, n$ and for all $k = 1, 2, \ldots, r$.

**Example 9.2**

Suppose $A$ is $2 \times 2$ and $B$ is $2 \times 3$:

$$A = \begin{bmatrix} a_{11} & a_{12} \\ a_{21} & a_{22} \end{bmatrix} \qquad B = \begin{bmatrix} b_{11} & b_{12} & b_{13} \\ b_{21} & b_{22} & b_{23} \end{bmatrix}$$

then the product of $A$ and $B$ is the matrix $C$:

$$C = AB = \begin{bmatrix} a_{11}b_{11} + a_{12}b_{21} & a_{11}b_{12} + a_{12}b_{22} & a_{11}b_{13} + a_{12}b_{23} \\ a_{21}b_{11} + a_{22}b_{21} & a_{21}b_{12} + a_{22}b_{22} & a_{21}b_{13} + a_{22}b_{23} \end{bmatrix}$$

In other words, the $lk$th element of $C$, $c_{lk}$, is obtained by adding the products of the element in the first column of row $l$ in $A$ with the element in the first row of column $k$ in $B$, and the element in the second column of row $l$ in $A$ with the element in the second row of column $k$ in $B$ . . . , and the element in the $m$th column of row $l$ in $A$ with the element in the $m$th row of column $k$ in $B$.   □

**Example 9.3**

Consider the matrices

$$A = \begin{bmatrix} 1 & 2 \\ 1 & -1 \end{bmatrix}$$

and

$$B = \begin{bmatrix} 1 & 0 & -1 \\ 1 & -1 & 1 \end{bmatrix}$$

the product of $A$ and $B$ is the matrix $C$:

$$C = AB = \begin{bmatrix} 3 & -2 & 1 \\ 0 & 1 & -2 \end{bmatrix} \qquad □$$

**Example 9.4**

The product of the matrices

$$A = \begin{bmatrix} 1 & 2 & 3 & -1 \\ -2 & 1 & 3 & 5 \\ -1 & 3 & 1 & 2 \end{bmatrix}$$

and

$$B = \begin{bmatrix} 1 & 2 \\ -1 & 1 \\ 0 & 1 \\ 1 & 0 \end{bmatrix}$$

is the matrix **C**:

$$C = \begin{bmatrix} -2 & 7 \\ 2 & 0 \\ -2 & 2 \end{bmatrix} \quad \square$$

It is important to observe the order of multiplication of matrices as illustrated by the following example.

**Example 9.5**

Consider the two matrices

$$A = \begin{bmatrix} 1 & 3 \\ 2 & 1 \end{bmatrix}$$

and

$$B = \begin{bmatrix} 2 & 1 \\ 1 & 1 \end{bmatrix}$$

The matrix product **AB** is

$$AB = \begin{bmatrix} 5 & 4 \\ 5 & 3 \end{bmatrix}$$

while the matrix product **BA** is

$$BA = \begin{bmatrix} 4 & 7 \\ 3 & 4 \end{bmatrix}$$

Therefore

$$AB \neq BA \quad \square$$

Two ($n \times n$) matrices are said to *commute* if **AB** equals **BA.**

Observe that we cannot effect a multiplication of arbitrary matrices. The two matrices to be multiplied must be *conformable*, that is, the number of columns of the first must equal the number of rows of the second.

According to the above definitions, the set of nodal equilibrium equations

$$y_{11} v_{\textcircled{1}} + y_{12} v_{\textcircled{2}} + \cdots + y_{1m} v_{\textcircled{m}} = i_{\textcircled{1}}$$
$$y_{21} v_{\textcircled{1}} + y_{22} v_{\textcircled{2}} + \cdots + y_{2m} v_{\textcircled{m}} = i_{\textcircled{2}}$$
$$\cdots\cdots\cdots\cdots\cdots\cdots\cdots\cdots\cdots\cdots\cdots\cdots$$
$$y_{m1} v_{\textcircled{1}} + y_{m2} v_{\textcircled{2}} + \cdots + y_{mm} v_{\textcircled{m}} = i_{\textcircled{m}}$$

is adequately represented by the matrix equation

$$Y v_n = I$$

where

$$\mathbf{Y} \equiv \begin{bmatrix} y_{11} & y_{12} & \cdots & y_{1m} \\ y_{21} & y_{22} & \cdots & y_{2m} \\ \cdot & \cdot & \cdots & \cdot \\ y_{m1} & y_{m2} & \cdots & y_{mm} \end{bmatrix}$$

$$\mathbf{V}_n \equiv \begin{bmatrix} v_① \\ v_② \\ \cdot \\ \cdot \\ \cdot \\ v_m \end{bmatrix}$$

and

$$\mathbf{I} \equiv \begin{bmatrix} i_① \\ i_② \\ \cdot \\ \cdot \\ \cdot \\ i_⓶ \end{bmatrix}$$

In other words

$$\begin{bmatrix} y_{11} & y_{12} & \cdots & y_{1m} \\ y_{21} & y_{22} & \cdots & y_{2m} \\ \cdot & \cdot & \cdots & \cdot \\ y_{m1} & y_{m2} & \cdots & y_{mm} \end{bmatrix} \begin{bmatrix} v_① \\ v_② \\ \cdot \\ \cdot \\ v_⓶ \end{bmatrix} = \begin{bmatrix} i_① \\ i_② \\ \cdot \\ \cdot \\ i_⓶ \end{bmatrix}$$

## 9-2 USE OF MATRICES TO EXPRESS KCL, KVL, AND THE BRANCH RELATIONSHIPS

In Chapter 6 we found it convenient to express the $n_n - 1$ linearly independent KCL node equations in terms of the $n_b$ branch currents as

$$\sum_{k=1}^{n_b} a_{jk} i_k = 0 \qquad \text{for} \qquad j = 1, 2, \ldots, n_n - 1$$

where the coefficients $a_{jk}$ are defined as follows:

$$a_{jk} = \begin{cases} +1 \text{ if branch } k \text{ leaves node number } ⓙ \\ -1 \text{ if branch } k \text{ enters node number } ⓙ \\ 0 \text{ if branch } k \text{ does not touch node number } ⓙ \end{cases}$$

We define the *nodal incidence matrix* $\mathbf{A}$ as the matrix of coefficients $a_{jk}$

$$\mathbf{A} \equiv [a_{jk}]$$

and the *branch current vector*

$$\mathbf{i}_b = \begin{bmatrix} i_1 \\ i_2 \\ \cdot \\ \cdot \\ \cdot \\ i_{n_b} \end{bmatrix}$$

Then the KCL node equations can be expressed as

$$\mathbf{A}\mathbf{i}_b = \mathbf{0}$$

We use $\mathbf{0}$ to denote the *zero* or *null matrix* as a matrix of arbitrary dimensions and all elements equal to zero.

### Example 9.6

The nodal incidence matrix associated with the network graph of Fig. 9-1 is

$$\mathbf{A} = \begin{bmatrix} 1 & 1 & 0 & 0 & 0 & 0 & 0 & 0 \\ -1 & 0 & -1 & 1 & 0 & 0 & 0 & 0 \\ 0 & 0 & 0 & -1 & 1 & 1 & 0 & 0 \\ 0 & 0 & 0 & 0 & 0 & -1 & 0 & 1 \\ 0 & 0 & 1 & 0 & -1 & 0 & -1 & 0 \end{bmatrix}$$

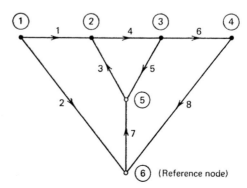

**Fig. 9-1   Network graph used in Example 9.5 and 9.6 to illustrate the nodal matrix incidence.**

The KCL equations are

$$\mathbf{A}\mathbf{i}_b = \mathbf{0}$$

where

$$\mathbf{i}_b = \begin{bmatrix} i_1 \\ i_2 \\ i_3 \\ i_4 \\ i_5 \\ i_6 \\ i_7 \\ i_8 \end{bmatrix}$$

since

$$\mathbf{A}\mathbf{i}_b = \mathbf{0}$$

represents

$$\begin{bmatrix} 1 & 1 & 0 & 0 & 0 & 0 & 0 & 0 \\ -1 & 0 & -1 & 1 & 0 & 0 & 0 & 0 \\ 0 & 0 & 0 & -1 & 1 & 1 & 0 & 0 \\ 0 & 0 & 0 & 0 & 0 & -1 & 0 & 1 \\ 0 & 0 & 1 & 0 & -1 & 0 & -1 & 0 \end{bmatrix} \begin{bmatrix} i_1 \\ i_2 \\ i_3 \\ i_4 \\ i_5 \\ i_6 \\ i_7 \\ i_8 \end{bmatrix} = \begin{bmatrix} 0 \\ 0 \\ 0 \\ 0 \\ 0 \\ 0 \\ 0 \\ 0 \end{bmatrix}$$

or when multiplied out

$$i_1 + i_2 = 0$$

$$-i_1 - i_3 + i_4 = 0$$

$$-i_4 + i_5 + i_6 = 0$$

$$-i_6 + i_8 = 0$$

and

$$i_3 - i_5 - i_7 = 0 \quad \square$$

Observe that the nodal incidence matrix completely characterizes the topology of the network—it describes how branches are interconnected. Given $\mathbf{A}$, one should be able to draw the network graph.

In Chapter 6 the KVL equations, in which all branch voltages are expressed in terms of node voltages was written as

$$v_k = \sum_{j=1}^{n_n-1} a_{jk} v_{ⓙ}, \qquad k = 1, 2, \ldots, n_b$$

We define the *branch voltage vector*

$$\mathbf{v}_b = \begin{bmatrix} v_1 \\ v_2 \\ \cdot \\ \cdot \\ \cdot \\ v_{n_b} \end{bmatrix}$$

and node voltage vector

$$\mathbf{v}_n = \begin{bmatrix} v_ⓛ \\ v_② \\ \cdot \\ \cdot \\ \cdot \\ v_{ⓝ_{n-1}} \end{bmatrix}$$

We wish to express KVL using matrix notation, but before proceeding we must define the *matrix transpose*. The *transpose* of an $(n \times m)$ matrix $\mathbf{B} = [b_{ij}]$ is the $(m \times n)$ matrix $\mathbf{B}' = [b_{ji}]$, that is, the rows and columns of $\mathbf{B}$ are interchanged to obtain $\mathbf{B}'$. For example, the transpose of the $(2 \times 3)$ matrix

$$\mathbf{B} = \begin{bmatrix} 1 & -3 & 2 \\ 9 & -4 & 0 \end{bmatrix}$$

is the $(3 \times 2)$ matrix.

$$\mathbf{B}' = \begin{bmatrix} 1 & 9 \\ -3 & -4 \\ 2 & 0 \end{bmatrix}$$

The transpose of the 3-vector

$$\mathbf{b} = \begin{bmatrix} -.5 \\ 15 \\ 3 \end{bmatrix}$$

is

$$\mathbf{b}' = [-.5 \quad 15 \quad 3]$$

Now the KVL equations can be written as

$$\mathbf{v}_b = \mathbf{A}'\mathbf{v}_n$$

**Example 9.7**

Observe that for the network of Fig. 9-1, KVL can indeed be expressed as

$$\mathbf{v}_b = \mathbf{A}'\mathbf{v}_n$$

since this matrix equation represents

$$
\begin{bmatrix} v_1 \\ v_2 \\ v_3 \\ v_4 \\ v_5 \\ v_6 \\ v_7 \\ v_8 \end{bmatrix}
=
\begin{bmatrix}
1 & -1 & 0 & 0 & 0 \\
1 & 0 & 0 & 0 & 0 \\
0 & -1 & 0 & 0 & 1 \\
0 & 1 & -1 & 0 & 0 \\
0 & 0 & 1 & 0 & -1 \\
0 & 0 & 1 & -1 & 0 \\
0 & 0 & 0 & 0 & -1 \\
0 & 0 & 0 & 1 & 0
\end{bmatrix}
\begin{bmatrix} v_① \\ v_② \\ v_③ \\ v_④ \\ v_⑤ \end{bmatrix}
$$

or when multiplied out

$$v_1 = v_① - v_②$$
$$v_2 = v_②$$
$$v_3 = -v_② + v_⑤$$
$$v_4 = v_② - v_③$$
$$v_5 = v_③ - v_⑤$$
$$v_6 = v_③ - v_④$$
$$v_7 = -v_⑤$$

and

$$v_8 = v_④ \quad \square$$

Finally, we consider the branch relationships. If the $k$th branch ($k = 1, 2, \ldots, n_b$) is a resistor of resistance $R_k$ then we can express its branch relationship as

$$i_k = \left(\frac{1}{R_k}\right) v_k$$

On the other hand, if the $k$th branch is a current source of value $i_k$ then the branch relationship is

$$i_k = \hat{i}_k$$

All the branch relationships can be expressed in matrix form as

$$\mathbf{i}_b = \mathcal{Y}\mathbf{v}_b + \mathcal{I}$$

where $\mathcal{Y} = [y_{jk}]$ is an $n_b \times n_b$ square matrix such that $y_{jk} = 0$ for $j \neq k$ and

$$y_{kk} = \begin{cases} 1/R_k \text{ if the } k\text{th branch is a resistor of value } R_k \\ 0 \text{ if the } k\text{th branch is a current source} \end{cases}$$

and $\mathcal{I} = [i_k]$ is an $n_b$-vector such that

$$i_k = \begin{cases} 0 \text{ if the } k\text{th branch is a resistor} \\ \hat{i}_k \text{ if the } k\text{th branch is a current source of value } i_k \end{cases}$$

The matrix $\mathcal{Y}$ is a *diagonal* matrix. An $n \times n$ square matrix $\mathbf{B} = [b_{ij}]$ is called *diagonal* if $b_{ij} = 0$ for $i \neq j$ and is sometimes denoted by the diagonal $(b_{11}, b_{22}, \ldots, b_{nn})$.

**Example 9.8**

The branch relationships for the network of Fig. 9-2 are

$$i_1 = \frac{1}{R_1} v_1$$

$$i_2 = \frac{1}{R_2} v_2$$

$$i_3 = -\hat{i}_3$$

$$i_4 = \frac{1}{R_4} v_4$$

$$i_5 = \frac{1}{R_5} v_5$$

$$i_6 = \frac{1}{R_6} v_6$$

$$i_7 = \frac{1}{R_7} v_7$$

and

$$i_8 = \hat{i}_8$$

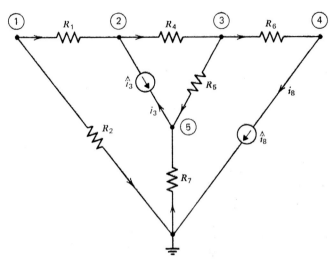

**Fig. 9-2    Network used in Example 9.7 to illustrate the derivation of nodal equilibrium equations using matrices. The reference directions are the same as used for the graph of Fig. 9-1.**

In matrix form, these equations become

$$\mathbf{i}_b = \mathscr{Y}\mathbf{v}_b + \mathscr{I}$$

where

$$
\mathscr{Y} =
\begin{bmatrix}
\dfrac{1}{R_1} & 0 & 0 & 0 & 0 & 0 & 0 & 0 \\
0 & \dfrac{1}{R_2} & 0 & 0 & 0 & 0 & 0 & 0 \\
0 & 0 & 0 & 0 & 0 & 0 & 0 & 0 \\
0 & 0 & 0 & \dfrac{1}{R_4} & 0 & 0 & 0 & 0 \\
0 & 0 & 0 & 0 & \dfrac{1}{R_5} & 0 & 0 & 0 \\
0 & 0 & 0 & 0 & 0 & \dfrac{1}{R_6} & 0 & 0 \\
0 & 0 & 0 & 0 & 0 & 0 & \dfrac{1}{R_7} & 0 \\
0 & 0 & 0 & 0 & 0 & 0 & 0 & 0
\end{bmatrix}
$$

or $\mathscr{Y} = \text{diag}\left(\dfrac{1}{R_1}, \dfrac{1}{R_2}, 0, \dfrac{1}{R_4}, \dfrac{1}{R_5}, \dfrac{1}{R_6}, \dfrac{1}{R_7}, 0\right)$

and

$$\mathscr{I} = \begin{bmatrix} 0 \\ 0 \\ -\hat{\imath}_3 \\ 0 \\ 0 \\ 0 \\ 0 \\ \hat{\imath}_8 \end{bmatrix}$$

## 9-3  DERIVATION OF NODAL EQUILIBRIUM EQUATIONS AND PROOF OF TELLEGEN'S THEOREM USING MATRICES

It is instructive to reconsider the derivation of the nodal equilibrium equations for networks comprised of resistors and current sources in terms of matrices.

Until now we have expressed the KCL equations as

$$\mathbf{A}\mathbf{i}_b = \mathbf{0}$$

the KVL equations as

$$\mathbf{v}_b = \mathbf{A}'\mathbf{v}_n$$

and the branch relationships as

$$\mathbf{i}_b = \mathscr{Y}\mathbf{v}_b + \mathscr{I}$$

We can substitute the branch relationships

$$\mathbf{i}_b = \mathscr{Y}\mathbf{v}_b + \mathscr{I}$$

into the KCL equations

$$\mathbf{A}\mathbf{i}_b = \mathbf{0}$$

to obtain

$$\mathbf{A}(\mathscr{Y}\mathbf{v}_b + \mathscr{I}) = \mathbf{0}$$

Because matrix multiplication is distributive (to be shown later) this expression becomes

$$\mathbf{A}\mathscr{Y}\mathbf{v}_b + \mathbf{A}\mathscr{I} = \mathbf{0}$$

Finally, we can use the KVL equations

$$\mathbf{v}_b = \mathbf{A}'\mathbf{v}_n$$

to obtain

$$\mathbf{A}\mathscr{Y}\mathbf{A}'\mathbf{v}_n + \mathbf{A}\mathscr{I} = \mathbf{0}$$

or

$$A\mathscr{Y} A' \mathbf{v}_n = -A\mathscr{I}$$

On comparison of the above expression with the form found earlier

$$\mathbf{Yv}_n = \mathbf{I}$$

we recognize that

$$\mathbf{Y} = A\mathscr{Y} A'$$

and

$$\mathbf{I} = -A\mathscr{I}$$

We see that matrix multiplication arises naturally in the solution of sets of simultaneous algebraic equations. In the product

$$A\mathscr{Y}$$

$A$ is said to be *postmultiplied* by $\mathscr{Y}$ and $\mathscr{Y}$ is said to be *premultiplied* by $A$.

### Example 9.9

As an illustration of the manipulation of matrix equations in the derivation of nodal equilibrium equations, consider the network of Fig. 9-3. The nodal incidence matrix is

$$A = \begin{bmatrix} 1 & -1 & -1 & 0 & 0 & 0 \\ 0 & 1 & 0 & 1 & 1 & 0 \\ 0 & 0 & 1 & 0 & -1 & -1 \end{bmatrix}$$

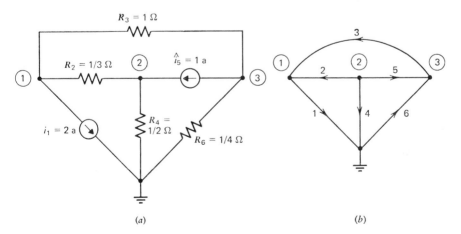

(a)                              (b)

Fig. 9-3  (a) A simple network used to demonstrate nodal analysis using matrices, and (b) its graph.

Therefore, the KCL node equations are

$$
\begin{bmatrix}
1 & -1 & -1 & 0 & 0 & 0 \\
0 & 1 & 0 & 1 & 1 & 0 \\
0 & 0 & 1 & 0 & -1 & -1
\end{bmatrix}
\begin{bmatrix}
i_1 \\ i_2 \\ i_3 \\ i_4 \\ i_5 \\ i_6
\end{bmatrix} = 0
$$

and the KVL equations are

$$
\begin{bmatrix}
v_1 \\ v_2 \\ v_3 \\ v_4 \\ v_5 \\ v_6
\end{bmatrix}
=
\begin{bmatrix}
1 & 0 & 0 \\
-1 & 1 & 0 \\
-1 & 0 & 1 \\
0 & 1 & 0 \\
0 & 1 & -1 \\
0 & 0 & -1
\end{bmatrix}
\begin{bmatrix}
v_① \\ v_② \\ v_③
\end{bmatrix}
$$

The branch relationships are

$$
\begin{bmatrix}
i_1 \\ i_2 \\ i_3 \\ i_4 \\ i_5 \\ i_6
\end{bmatrix}
=
\begin{bmatrix}
0 & 0 & 0 & 0 & 0 & 0 \\
0 & \dfrac{1}{R_2} & 0 & 0 & 0 & 0 \\
0 & 0 & \dfrac{1}{R_3} & 0 & 0 & 0 \\
0 & 0 & 0 & \dfrac{1}{R_4} & 0 & 0 \\
0 & 0 & 0 & 0 & 0 & 0 \\
0 & 0 & 0 & 0 & 0 & \dfrac{1}{R_6}
\end{bmatrix}
\begin{bmatrix}
v_1 \\ v_2 \\ v_3 \\ v_4 \\ v_5 \\ v_6
\end{bmatrix}
+
\begin{bmatrix}
\hat{\imath}_1 \\ 0 \\ 0 \\ 0 \\ -\hat{\imath}_5 \\ 0
\end{bmatrix}
$$

so that

$$
\mathscr{Y} = \mathrm{diag}\left(0, \frac{1}{R_2}, \frac{1}{R_3}, \frac{1}{R_4}, 0, \frac{1}{R_6}\right)
$$

and

$$
\mathscr{I}' = [i_1 \quad 0 \quad 0 \quad 0 \quad -\hat{\imath}_5 \quad 0]
$$

From the above we find that

$$\mathbf{A}\mathscr{Y} = \begin{bmatrix} 0 & -\dfrac{1}{R_2} & -\dfrac{1}{R_3} & 0 & 0 & 0 \\[2ex] 0 & \dfrac{1}{R_2} & 0 & \dfrac{1}{R_4} & 0 & 0 \\[2ex] 0 & 0 & \dfrac{1}{R_3} & 0 & 0 & -\dfrac{1}{R_6} \end{bmatrix}$$

so that

$$\mathbf{A}\mathscr{Y}\mathbf{A}' = \begin{bmatrix} \dfrac{1}{R_2}+\dfrac{1}{R_3} & -\dfrac{1}{R_2} & -\dfrac{1}{R_3} \\[2ex] -\dfrac{1}{R_2} & \dfrac{1}{R_2}+\dfrac{1}{R_4} & 0 \\[2ex] -\dfrac{1}{R_3} & 0 & \dfrac{1}{R_3}+\dfrac{1}{R_6} \end{bmatrix}$$

and

$$\mathbf{A}\mathscr{Y} = \begin{bmatrix} \hat{\imath}_1 \\ -\hat{\imath}_5 \\ \hat{\imath}_5 \end{bmatrix}$$

The nodal equilibrium equations are

$$\mathbf{A}\mathscr{Y}\mathbf{A}'\mathbf{v}_n = -\mathbf{A}\mathscr{I}$$

or

$$\begin{bmatrix} \dfrac{1}{R_2}+\dfrac{1}{R_3} & -\dfrac{1}{R_2} & -\dfrac{1}{R_3} \\[2ex] -\dfrac{1}{R_2} & \dfrac{1}{R_2}+\dfrac{1}{R_4} & 0 \\[2ex] -\dfrac{1}{R_3} & 0 & \dfrac{1}{R_3}+\dfrac{1}{R_6} \end{bmatrix} \begin{bmatrix} v_{①} \\ v_{②} \\ v_{③} \end{bmatrix} = -\begin{bmatrix} \hat{\imath}_1 \\ -\hat{\imath}_5 \\ \hat{\imath}_5 \end{bmatrix}$$

When multiplied out these equations are

$$\left(\frac{1}{R_2}+\frac{1}{R_3}\right)v_{①} - \frac{1}{R_2}v_{②} - \frac{1}{R_3}v_{③} = -\hat{\imath}_1$$

$$-\frac{1}{R_2}v_{①} + \left(\frac{1}{R_2}+\frac{1}{R_4}\right)v_{②} = \hat{\imath}_5$$

$$-\frac{1}{R_3}v_{①} + \left(\frac{1}{R_3}+\frac{1}{R_6}\right)v_{③} = -\hat{\imath}_5 \quad \square$$

It is clear from the above example, that the matrix $\mathbf{Y} = \mathbf{A}\mathscr{Y}\mathbf{A}'$ and the vector $-\mathbf{A}\mathscr{I}$ can easily be written by inspection.

A rather short, but elegant, proof of Tellegen's theorem is possible by using matrix notation. Recall that Tellegen's theorem states that if $v_k$ is the $k$th branch voltage of one network and $i_k$ is the $k$th branch current of a second network that has the same topology as the first, then

$$\sum_{k=1}^{n_b} v_k i_k = 0$$

Let $\mathbf{v}_b$ denote the branch voltage vector for network one, that is,

$$\mathbf{v}_b = \begin{bmatrix} v_1 \\ v_2 \\ . \\ . \\ . \\ v_{n_b} \end{bmatrix}$$

and $\mathbf{i}_b$ the branch current for network two, that is,

$$\mathbf{i}_b = \begin{bmatrix} i_1 \\ i_2 \\ . \\ . \\ . \\ i_{n_b} \end{bmatrix}$$

Tellegen's theorem can be written as

$$\mathbf{v}_b'\mathbf{i}_b = (v_1 \; v_2 \cdots v_{n_b}) \begin{bmatrix} i_1 \\ i_2 \\ . \\ . \\ . \\ i_{n_b} \end{bmatrix}$$

$$= \sum_{k=1}^{n_b} v_k i_k = 0$$

To prove that

$$\mathbf{v}_b'\mathbf{i}_b = 0$$

we note that KVL can be written as

$$\mathbf{v}_b = \mathbf{A}'\mathbf{v}_n$$

where $\mathbf{v}_n$ is the node voltage vector for network one and $\mathbf{A}$ is the nodal incidence matrix associated with both networks one and two, since both networks have the same topology. Therefore

$$\mathbf{v}'_b\mathbf{i}_b = (\mathbf{A}'\mathbf{v}_n)'\mathbf{i}_b$$

$$= \mathbf{v}'_n\mathbf{A}\mathbf{i}_b$$

$$= 0$$

because KCL for the second network is

$$\mathbf{A}\mathbf{i}_b = 0$$

## 9-4  SOME PROPERTIES OF MATRICES

It is now convenient to define in detail some of the operations that may be performed on matrices. The student should constantly bear in mind that the operations to be discussed are defined, so that when they are performed on matrices they correspond to the operations as if they were performed on sets of simultaneous equations.

Given the $(n \times m)$ matrix $\mathbf{A}$ and the scalar $\alpha$ (a scalar is any single number), the product of $\alpha\mathbf{A}$ is the $(n \times m)$ matrix $\mathbf{C}$ where

$$c_{lk} = \alpha a_{lk}$$

for all $l = 1, 2, \ldots, n$ and all $k = 1, 2, \ldots, m$. For example, the product of the matrix

$$\begin{bmatrix} 1 & 5 \\ -2 & 7 \end{bmatrix}$$

by the scalar 3 is

$$3\begin{bmatrix} 1 & 5 \\ -2 & 7 \end{bmatrix} = \begin{bmatrix} 3 & 15 \\ -6 & 21 \end{bmatrix}$$

Two $(m \times n)$ matrices

$$\mathbf{A} = [a_{lk}]$$
$$\mathbf{B} = [b_{lk}]$$

are said to be *equal* if and only if

$$a_{lk} = b_{lk}$$

for all $l = 1, 2, \ldots, m$ and all $k = 1, 2, \ldots, n$. (*Note.* Matrix equality may only be defined for matrices of equal size). The *sum* of two $(m \times n)$ matrices

$$\mathbf{A} = [a_{lk}]$$

and

$$\mathbf{B} = [b_{lk}]$$

is the $(m \times n)$ matrix

$$C = A + B$$

where

$$c_{lk} = a_{lk} + b_{lk}$$

for all $l = 1, 2, \ldots, m$, and all $k = 1, 2, \ldots, n$.

Observe the following identities:

$$A \cdot 1 = 1 \cdot A = A$$

for any square matrix $A$; and

$$A + 0 = 0 + A = A$$

for any $(m \times n)$ matrix $A$.

It is left as an exercise for the student to show that the following rules are satisfied by matrix operations.

1. *Distributive rules:*

$$(A + B)C = AC + BC$$

where $A$ and $B$ are $(n \times m)$ and $C$ is $(m \times r)$;

$$(A(B + C)) = AB + AC$$

where $A$ is $(n \times m)$ and $B$ and $C$ are $(m \times r)$;

$$(\alpha + \beta)A = \alpha A + \beta A$$

where $\alpha$ and $\beta$ are scalars;

$$\alpha(A + B) = \alpha A + \alpha B$$

where $\alpha$ is a scalar and $A$ and $B$ are $(n \times m)$

2. *Associative rules:*

$$\alpha A = A\alpha \quad \text{or} \quad (\alpha A)B = \alpha AB = A(\alpha B)$$

where $A$ is $(n \times m)$, $B$ is $(m \times r)$ and $\alpha$ is a scalar;

$$A(BC) = (AB)C$$

where $A$ is $(n \times m)$, $B$ is $(m \times r)$, $C$ is $(r \times s)$;

$$(A + B) + C = A + (B + C)$$

where $A$, $B$, and $C$ are $(n \times m)$.

3. *Commutative rules:*

$$A + B = B + A$$

where $A$ and $B$ are $(n \times m)$

In the above **1** denotes the identity matrix, that is, a diagonal matrix with all elements equal to unity.

## 9-5   THE MATRIX INVERSE

Consider the two simultaneous algebraic equations in two unknowns

$$a_{11}x_1 + a_{12}x_2 = b_1$$

$$a_{21}x_1 + a_{22}x_2 = b_2$$

or in matrix form

$$\mathbf{Ax} = \mathbf{b}$$

where

$$\mathbf{A} = \begin{bmatrix} a_{11} & a_{12} \\ a_{21} & a_{22} \end{bmatrix} \qquad \mathbf{x} = \begin{bmatrix} x_1 \\ x_2 \end{bmatrix} \qquad \text{and} \qquad \mathbf{b} = \begin{bmatrix} b_1 \\ b_2 \end{bmatrix}$$

Using simple algebraic techniques it is easy to find expressions for $x_1$ and $x_2$ in terms of $b_1$ and $b_2$

(9.1a)
$$x_1 = \frac{a_{22}}{a_{11}a_{22} - a_{12}a_{21}} b_1 - \frac{a_{12}}{a_{11}a_{22} - a_{12}a_{21}} b_2$$

(9.1b)
$$x_2 = \frac{-a_{21}}{a_{11}a_{22} - a_{12}a_{21}} b_1 + \frac{a_{11}}{a_{11}a_{22} - a_{12}a_{21}} b_2$$

if the quantity $(a_{11}a_{22} - a_{12}a_{21})$ is nonzero. The quantity $(a_{11}a_{22} - a_{12}a_{21})$, as discussed in Chapter 3, is known as the *determinant* of the matrix $\mathbf{A}$ and is denoted by det $(\mathbf{A})$, that is,

$$\det(\mathbf{A}) = a_{11}a_{22} - a_{12}a_{21}$$

Thus we can say that the set of simultaneous equations represented by

(9.2)
$$\mathbf{Ax} = \mathbf{b}$$

can be solved for $\mathbf{x}$ if det $(\mathbf{A})$ is nonzero; that is, if they are "determinant." The solution of the matrix equation

$$\mathbf{Ax} = \mathbf{b}$$

(9.1) can be written in matrix form as

(9.3)
$$\mathbf{x} = \mathbf{Wb}$$

where

$$\mathbf{W} \equiv \frac{1}{\det(\mathbf{A})} \begin{bmatrix} a_{22} & -a_{12} \\ -a_{21} & a_{22} \end{bmatrix}$$

We can combine (9.2) and (9.3) to yield

$$\mathbf{AWb} = \mathbf{Ax} = \mathbf{b}$$

or

$$\mathbf{WAx} = \mathbf{Wb} = \mathbf{x}$$

For these equalities to hold, we must have

$$\mathbf{AW} = \mathbf{1}$$

or

$$\mathbf{WA} = \mathbf{1}$$

The matrix $\mathbf{W}$ is called the *inverse* of $\mathbf{A}$ and is denoted by $\mathbf{A}^{-1}$. Therefore, if

$$\mathbf{Ax} = \mathbf{b}$$

and if $\mathbf{A}$ is nonsingular, that is,

$$\det{(\mathbf{A})} \neq 0$$

then

$$\mathbf{x} = \mathbf{A}^{-1}\mathbf{b}$$

To be somewhat more general, suppose we are given a set of $n$ independent algebraic equations in terms of $n$ unknowns

$$\sum_{k=1}^{n} a_{jk}x_k = b_j \qquad j = 1, 2, \ldots, n$$

where the $a_{jk}$ and $b_j$ are known constants and the $x_k$ are unknown. These equations can be written in matrix form as

$$\begin{bmatrix} a_{11} & a_{12} & \cdots & a_{1k} & \cdots & a_{1n} \\ a_{21} & a_{22} & \cdots & a_{2k} & \cdots & a_{in} \\ \cdots & \cdots & \cdots & \cdots & \cdots & \cdots \\ a_{j1} & a_{j2} & \cdots & a_{jk} & \cdots & a_{jn} \\ \cdots & \cdots & \cdots & \cdots & \cdots & \cdots \\ a_{n1} & a_{n2} & \cdots & a_{nk} & \cdots & a_{nn} \end{bmatrix} \begin{bmatrix} x_1 \\ x_2 \\ \cdot \\ x_k \\ \cdot \\ x_n \end{bmatrix} = \begin{bmatrix} b_1 \\ b_2 \\ \cdot \\ b_j \\ \cdot \\ b_n \end{bmatrix}$$

or

$$\mathbf{Ax} = \mathbf{b}$$

Cramer's rule can be employed to find values for the $x_k$'s:

$$x_k = \frac{1}{\Delta} \sum_{j=1}^{n} b_j \Delta_{kj}, \qquad k = 1, 2, \ldots, n$$

where $\Delta = \det{(\mathbf{A})} \neq 0$ and $\Delta_{kj}$ is the cofactor of $a_{kj}$. In practice the use of Cramer's rule is computationally inefficient but it suffices for the present. (Observe that $\det{(\mathbf{A})} \neq 0$ because the equations were assumed to be independent. We will return to this point later.) We rewrite the above equation as

$$\sum_{j=1}^{n} \frac{\Delta_{kj}}{\Delta} b_j = x_k \qquad k = 1, 2, \ldots, n$$

or in matrix form

$$\mathbf{W}\mathbf{b} = \mathbf{x}$$

where we have defined

$$\mathbf{W} = \left[\frac{\Delta_{kj}}{\Delta}\right]$$

that is, the elements of $\mathbf{W}$ are

$$w_{jk} = \frac{\Delta_{kj}}{\Delta}$$

Now observe that

$$\mathbf{A}\mathbf{W}\mathbf{b} = \mathbf{A}\mathbf{x}$$

$$= \mathbf{b} = \mathbf{1} \cdot \mathbf{b}$$

so that

$$\mathbf{A}\mathbf{W} = \mathbf{1}$$

Similarly, since

$$\mathbf{x} = \mathbf{W}\mathbf{b}$$

we have

$$\mathbf{x} = \mathbf{W}\mathbf{A}\mathbf{b}$$

so that

$$\mathbf{W}\mathbf{A} = \mathbf{1}$$

Any matrix $\mathbf{W}$ such that $\mathbf{W}\mathbf{A} = \mathbf{A}\mathbf{W} = \mathbf{1}$, called the *inverse of* $\mathbf{A}$ and denoted by $\mathbf{A}^{-1}$. Therefore

$$\mathbf{A}^{-1} = \left[\frac{\Delta_{kj}}{\Delta}\right]$$

and

$$\mathbf{A}\mathbf{A}^{-1} = \mathbf{A}^{-1}\mathbf{A} = \mathbf{1}$$

Moreover, observe that $\mathbf{A}^{-1}$ exists if and only if det $(\mathbf{A}) \neq 0$, that is, $\mathbf{A}$ is *nonsingular*.

### Example 9.10

As an example of how matrix inversion can be used to solve for the node voltages of a network, consider the network of Fig. 9-3. The nodal equilibrium equations are

$$4v_{①} - 3v_{②} - v_{③} = -2$$
$$-3v_{①} + 5v_{②} \qquad = 1$$
$$- v_{①} \qquad + 5v_{③} = -1$$

Or, in matrix form:

$$\begin{bmatrix} 4 & -3 & -1 \\ -3 & 5 & 0 \\ -1 & 0 & 5 \end{bmatrix}\begin{bmatrix} v_{①} \\ v_{②} \\ v_{③} \end{bmatrix} = \begin{bmatrix} -2 \\ 1 \\ -1 \end{bmatrix}$$

The solution can be found by inverting the matrix

$$\mathbf{Y} = \begin{bmatrix} 4 & -3 & -1 \\ -3 & 5 & 0 \\ -1 & 0 & 5 \end{bmatrix}$$

The determinant of $\mathbf{Y}$ is

$$\Delta = \det(\mathbf{Y}) = -150$$

The cofactors of $\mathbf{Y}$ are

$$\Delta_{11} = 25, \qquad \Delta_{12} = 15, \qquad \Delta_{13} = 5$$
$$\Delta_{21} = 15, \qquad \Delta_{22} = -21, \qquad \Delta_{23} = 3$$
$$\Delta_{31} = 5, \qquad \Delta_{32} = 3, \qquad \Delta_{33} = -29$$

so that the inverse of $Y$ is

$$\mathbf{Y}^{-1} = \frac{-1}{150} \begin{bmatrix} 25 & 15 & 5 \\ 15 & -21 & 3 \\ 5 & 3 & -29 \end{bmatrix}$$

Observe that

$$\mathbf{YY}^{-1} = 1$$

Then the solution is

$$\begin{bmatrix} v_① \\ v_② \\ v_③ \end{bmatrix} = \frac{-1}{150} \begin{bmatrix} 25 & 15 & 5 \\ 15 & -21 & 3 \\ 5 & 3 & -29 \end{bmatrix} \begin{bmatrix} -2 \\ 1 \\ -1 \end{bmatrix}$$

or $v_① = 4/15$, $v_② = 54/150$, and $v_③ = -21/150$. □

We leave it as an exercise for the student to show that for two matrices $\mathbf{A}$ and $\mathbf{B}$:

1. $(\mathbf{A} + \mathbf{B})' = \mathbf{A}' + \mathbf{B}'$ where $\mathbf{A}$ and $\mathbf{B}$ are $(n \times m)$.
2. $(\mathbf{AB})' = \mathbf{B}'\mathbf{A}'$ where $\mathbf{A}$ is $(n \times m)$ and $\mathbf{B}$ is $(m \times r)$.

If $\mathbf{A}$ and $\mathbf{B}$ are nonsingular, square matrices, then

3. $(\mathbf{AB})^{-1} = \mathbf{B}^{-1}\mathbf{A}^{-1}$
4. $(\mathbf{A}^{-1})^{-1} = \mathbf{A}$
5. $(\mathbf{A}^{-1})' = (\mathbf{A}')^{-1}$

Until now we have relied on the use of Gaussian elimination for the solution of sets of simultaneous algebraic equations. Even though we have introduced Cramer's rule and matrix inversion we will continue to use Gaussian elimination, with possibly some modifications to be discussed, for solution of equations of the form

$$\mathbf{Ax} = \mathbf{b}$$

We will use the inverse $\mathbf{A}^{-1}$ as a convenient notation to denote the solutions of matrix equations:

$$\mathbf{x} = \mathbf{A}^{-1}\mathbf{b}$$

but for the present we will *not* actually compute $\mathbf{A}^{-1}$ and perform the indicated multiplication.

## 9-6  SUMMARY

In this chapter we introduced matrix notation as a convenient shorthand to represent arrays of numbers. The nodal *admittance matrix, equivalent current source vector* and *node voltage vector* were defined and used to represent the nodal equilibrium equations and the concept of *matrix multiplication* was defined.

We defined a nodal *incidence matrix* that characterized the topology of a graph and used it to express the KCL node equations and KVL loop equations. The branch relationships of a network could also be simply expressed in terms of matrices. We then illustrated the derivation of the nodal equilibrium equations in terms of matrices. Some basic matrix manipulations such as *transposition* and *inversion*, as well as some of the properties of matrix addition and multiplication were discussed.

## PROBLEMS

**9.1**    Evaluate the matrix products

(a)
$$\begin{bmatrix} 1 & -1 \\ -1 & 0 \\ 1 & 1 \end{bmatrix} \begin{bmatrix} -1 & 2 \\ 1 & 1 \end{bmatrix}$$

(b)
$$\begin{bmatrix} 1 & 2 & 1 \\ 4 & 6 & 1 \\ 1 & 1 & 0 \end{bmatrix} \begin{bmatrix} 6 & 0 & -2 \\ 2 & 2 & 1 \\ 3 & 1 & 1 \end{bmatrix}$$

(c)    $\mathbf{a'b}$ and $\mathbf{ab'}$ where

$$\mathbf{a} = \begin{bmatrix} a_1 \\ a_2 \\ \cdot \\ \cdot \\ \cdot \\ a_n \end{bmatrix} \qquad \mathbf{b} = \begin{bmatrix} b_1 \\ b_2 \\ \cdot \\ \cdot \\ \cdot \\ b_n \end{bmatrix}$$

**9.2**    Determine which of the following pairs of matrices commute

(a)
$$\mathbf{A} = \begin{bmatrix} \lambda_1 & 0 \\ 0 & \lambda_2 \end{bmatrix} \quad \text{and} \quad \mathbf{B} = \begin{bmatrix} 1 & 2 \\ 0 & 1 \end{bmatrix}$$

(b)
$$\mathbf{A} = \begin{bmatrix} 1 & 2 \\ 0 & 1 \end{bmatrix} \quad \text{and} \quad \mathbf{B} = \begin{bmatrix} 1 & 0 \\ 1 & 1 \end{bmatrix}$$

(c)
$$\mathbf{A} = \begin{bmatrix} 1 & 1 & 1 \\ 0 & 1 & 1 \\ 0 & 0 & 1 \end{bmatrix} \quad \text{and} \quad \mathbf{B} = \begin{bmatrix} 1 & 1 & 0 \\ 0 & 1 & 0 \\ 0 & 1 & 1 \end{bmatrix}$$

**9.3**    Determine the nodal incidence matrix for the network graphs shown in Fig. P9.3.

**9.4**    Given the following nodal incidence matrices construct the appropriate network graph. (*Hint.* First determine which branches are connected to the datum node and their orientation.)

(a)
$$\mathbf{A} = \begin{bmatrix} -1 & +1 & 0 & 0 & -1 & -1 \\ 1 & -1 & 0 & 0 & 0 & 0 \\ 0 & 0 & 1 & 1 & 1 & 0 \end{bmatrix}$$

(b)
$$\mathbf{A} = \begin{bmatrix} 1 & 0 & 0 & 0 & 1 & 1 & 0 & 0 & 0 \\ -1 & 0 & 1 & 1 & 0 & 0 & 0 & 0 & 0 \\ 0 & 1 & 0 & -1 & -1 & 0 & 0 & 0 & 0 \\ 0 & -1 & 0 & 0 & 0 & 0 & -1 & -1 & 0 \\ 0 & 0 & 0 & 0 & 0 & -1 & 1 & 0 & 1 \end{bmatrix}$$

(c)
$$\mathbf{A} = \begin{bmatrix} 1 & 1 & -1 & 0 & 0 & 0 & 0 \\ 0 & -1 & 0 & 1 & -1 & 0 & 1 \\ 0 & 0 & 1 & 0 & 1 & 1 & 0 \end{bmatrix}$$

**9.5\***    In addition to the nodal incidence matrix, other topological matrices can be defined and used to characterize a network graph. We can define the *mesh incidence matrix* **M**—with elements $m_{jk}$ such that

$$m_{jk} = \begin{cases} +1 & \text{if branch } k \text{ is in mesh } j \text{ and has the the same orientation} \\ -1 & \text{if branch } k \text{ is in mesh } j \text{ and has the opposite orientation} \\ 0 & \text{if branch } k \text{ is not in mesh } j \end{cases}$$

Find the mesh incidence matrix **M** for the graphs of Fig. P9.5 and the meshes shown. Let $\mathbf{v}_b$ denote the branch voltage vector and

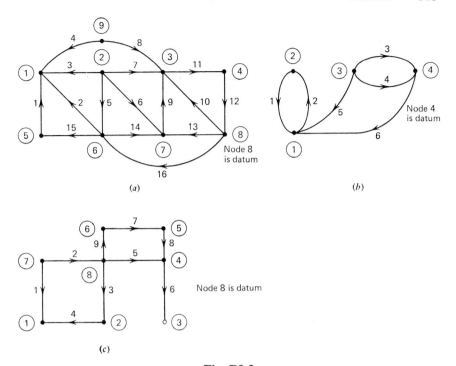

**Fig. P9.3**

verify that $\mathbf{Mv}_b = \mathbf{0}$ are in fact Kirchhoff's voltage law loop equations obtained in Prob. 2.14.

**9.6**     We can define the fundamental cutset matrix $\mathbf{Q}$ with elements $q_{jk}$ such that

$$q_{jk} = \begin{cases} +1 & \text{if branch } k \text{ is in the fundamental cutset} \\ & \text{defined by tree branch } j \text{ and has the} \\ & \text{same orientation} \\ -1 & \text{if branch } k \text{ is in the fundamental cutset} \\ & \text{defined by tree branch } j \text{ and has the} \\ & \text{opposite orientation} \\ 0 & \text{if branch } k \text{ is not in the fundamental} \\ & \text{cutset defined by loop } j \end{cases}$$

Note that there will be a different $\mathbf{Q}$ for different trees. Find one fundamental cutset matrix $\mathbf{Q}$ for the graph of Fig. P9.6$a$ for each of the trees shown in Figs. P9.6$b$ to P9.6$d$. Verify that $\mathbf{Qi}_b = \mathbf{0}$ are the same fundamental cutset equations obtained in Prob. 2.8.

**9.7**     Using the ideas of Probs. 9.5 and 9.6 define a fundamental loop matrix $\mathbf{B}$. Using this definition find a fundamental loop matrix for

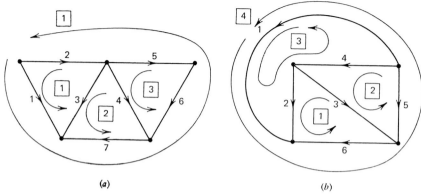

(a)                                            (b)

**Fig. P9.5**

the graph of Fig. P9.6a for the trees of Fig. P9.6b to P9.6d. Verify that $\mathbf{Bv}_b = \mathbf{0}$ are the same fundamental loop equations obtained in Prob. 2.8.

**9.8**    For each of the networks in Fig. P9.8 write the KCL, KVL, and branch relationships in matrix form. By manipulating the appropriate matrices derive the nodal equilibrium equations. Compare your results with those obtained in Prob. 3.5.

**9.9**    Use matrix methods to find the nodal equilibrium equations for the networks of Fig. P9.9.

**9.10**   Prove that matrices obey the distributive laws on page 296.

**9.11**   Prove that matrices obey the associative laws on page 296.

**9.12**   Prove that

$$(\mathbf{A} + \mathbf{B})' = \mathbf{A}' + \mathbf{B}'$$

and

$$(\mathbf{AB})' = \mathbf{B}'\mathbf{A}'$$

**9.13**   Given the matrix

$$\mathbf{A} = \begin{bmatrix} -1 & 0 & 1 \\ 2 & 1 & 0 \\ 1 & 2 & -1 \end{bmatrix}$$

find

(a) $\mathbf{A}'$
(b) det $(\mathbf{A})$
(c) $\mathbf{A}^{-1}$

**9.14**   Given the matrices

$$\mathbf{A} = \begin{bmatrix} -3 & 1 & 2 \\ 6 & 6 & 4 \\ -2 & 6 & 3 \end{bmatrix} \quad \mathbf{B} = \begin{bmatrix} 4 & 2 & 1 \\ 1 & -7 & -1 \\ 2 & 8 & 6 \end{bmatrix} \quad \mathbf{C} = \begin{bmatrix} 2 & -1 & 8 \\ 0 & 2 & 1 \\ 3 & 8 & 4 \end{bmatrix}$$

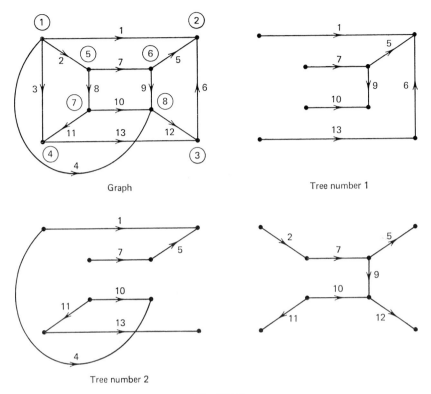

Graph

Tree number 1

Tree number 2

**Fig. P9.6**

find

    (a) the determinant of **A**, **B**, and **C**
    (b) **B**$^{-1}$, if it exists and **B**′
    (c) **A** · **C** and **C** · **A**
    (d) **A(B + C)** and **AB + AC**
    (e) **[AB]**$^{-1}$ and **(AB)**′

**9.15**    Write the nodal equilibrium equations for the network of Fig. P9.15 and solve for the node voltages using matrix inversion.

**9.16**    Suppose we wished to find the inverse of some matrix **A**. Let **B** denote this inverse. Then

$$\mathbf{AB} = \mathbf{1}$$

We can use GEP to solve for the columns of **B**, since we can consider each column of **B** to be the solution of a set of algebraic equations of the form

$$\mathbf{Ab}_k = \mathbf{e}_k$$

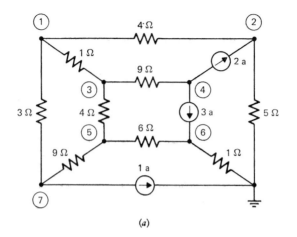

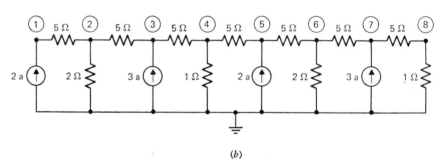

(b)

**Fig. P9.8**

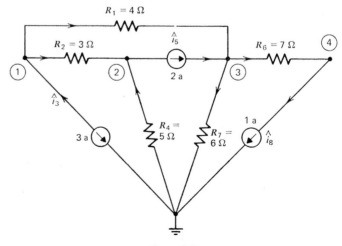

**Fig. P9.9**

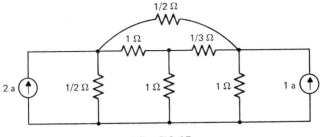

**Fig. P9.15**

where $b_k$ is the $k$th column of **B** and

$$\mathbf{e}_k = \begin{bmatrix} 0 \\ 0 \\ \cdot \\ \cdot \\ 1 \\ \cdot \\ \cdot \\ 0 \\ 0 \end{bmatrix} \leftarrow k\text{th row}$$

Thus, to find $\mathbf{A}^{-1}$ we must solve $n$ sets of simultaneous equations. Use this technique to find the inverse of the following matrices.

$$\mathbf{A} = \begin{bmatrix} -1 & 0 & 1 \\ 2 & 1 & 0 \\ 1 & 2 & -1 \end{bmatrix} \qquad \mathbf{B} = \begin{bmatrix} 4 & 2 & 1 \\ 1 & -7 & -1 \\ 2 & 8 & 6 \end{bmatrix}$$

Compare your results with those obtained in Probs. 9.13 and 9.14.

**9.17**   Use GEP and the method outlined in Prob. 9.16 to find the inverse of the matrix.

$$\mathbf{A} = \begin{bmatrix} 1 & 2 & 3 & 0 & 3 \\ 2 & 1 & 1 & 5 & 1 \\ 4 & 4 & 2 & 1 & 1 \\ 1 & 1 & 5 & 3 & 1 \\ 1 & 2 & 2 & 1 & 10 \end{bmatrix}$$

**9.18**   Write a subroutine that will invert a matrix. Use the method described in Prob. 9.16 and the subroutine GAUSS.

## CHAPTER 10

# DC Analysis of Transistor Networks

In most present-day analog circuits such as amplifiers, the bipolar transistor is the most common physical device. It is therefore worthwhile to consider the analysis of networks containing transistors in some depth. We can view the transistor as a nonlinear three terminal device. Characterization of transistor behavior is somewhat more complex than characterization of the behavior of elements studied previously. In fact we will find it convenient to think of the transistor as an interconnection of some of the elements already introduced. In other words we will interpret or *model* the equations that describe a transistor by connections of ideal elements. The model that results can be viewed as a nonlinear two-port element.

## 10-1 THE *npn* BIPOLAR TRANSISTOR MODEL

The circuit symbol of an *npn* bipolar transistor is shown in Fig. 10-1. The three accessible terminals are labeled the emitter, denoted by $E$, the base, denoted by $B$, and the collector, denoted by $C$. A simple description of the dc behavior of a transistor was given by Ebers and Moll[1]

$$(10.1a) \quad i_E = -I_{ES}[\exp\left(-v_{EB}/v_T\right) - 1] + \alpha_R I_{CS}[\exp\left(-v_{CB}/v_T\right) - 1]$$

[1] J. J. Ebers and J. L. Moll, "Large-signal Behavior of Junction Transistors," *Proc. IRE*, Vol. 42, pp. 1761–1772, Dec. 1954.

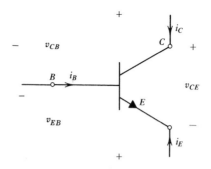

**Fig. 10-1   Circuit symbol of an *npn* bipolar transistor. The three accessible terminals are the emitter *E*, the base *B*, and the collector *C*.**

and

(10.1b)   $i_C = \alpha_F I_{ES}[\exp{(-v_{EB}/v_T)} - 1] - I_{CS}[\exp{(-v_{CB}/v_T)} - 1]$

where $I_{ES}$ is the saturation current of the emitter-base junction, $I_{CS}$ is the saturation current of the collector-base junction, $\alpha_R$ is the reverse, or collector-to-emitter, current gain, $\alpha_F$ is the forward, or emitter-to-collector, current gain, and $v_T = q/kT$ is the thermal voltage. A detailed explanation of the operation of a transistor and the derivation of the Ebers-Moll equations are beyond the scope of this text.

We desire to model the transistor with some interconnection of elements that have been introduced previously. Since the transistor is characterized by two equations with $v_{EB}$ and $v_{CB}$ as the independent variables and $i_E$ and $i_C$ as the dependent variables, we should expect to model it by some form of two-port, as indicated in Fig. 10-2a. We recognize that the term

$$I_{ES}(e^{-v_{EB}/v_T} - 1)$$

is the current that flows through an ideal diode whose branch voltage is $-v_{EB}$ and saturation current $I_{ES}$. This current will be denoted by $i_{DE}$:

(10.2)                    $i_{DE} = I_{ES}(e^{-v_{EB}/v_T} - 1)$

Similarly, the term

$$I_{CS}(e^{-v_{CB}/v_T} - 1)$$

is the current that flows through an ideal diode whose branch voltage is $-v_{CB}$ and saturation current $I_{CS}$. This current will be denoted by $i_{DC}$:

(10.3)                    $i_{DC} = I_{CS}(e^{-v_{CB}/v_T} - 1)$

Expressions (10.1a) and (10.1b) can be rewritten as

$$i_E = -i_{DE} + \alpha_R i_{DC}$$

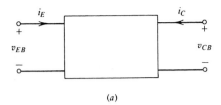

(a)

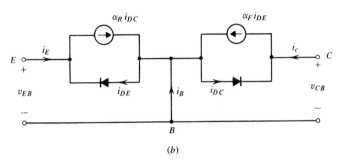

(b)

**Fig. 10-2**   (a) **A two-port may be used to model a transistor. In particular, (b) is the *Ebers-Moll model* of the dc behavior of transistors.**

and

$$i_C = \alpha_F i_{DE} - i_{DC}$$

These equations can be modeled using the connection of current controlled current sources and ideal diodes shown in Fig. 10-2b. Finally, we observe that

(10.4) $$i_E + i_B + i_C = 0$$

and

(10.5) $$v_{EB} + v_{CE} - v_{CB} = 0$$

The Ebers–Moll model does not describe all the properties of transistors. For instance, the emitter, base, and collector all have some ohmic resistance associated with them. To account for this, resistors are sometimes added to the basic model as shown in Fig. 10-3. For our present purposes the simplest model will suffice.

The Ebers–Moll equations are a special form of the general nonlinear two-port equations

(10.6a) $$i_1 = f_1(v_1, v_2)$$

and

(10.6b) $$i_2 = f_2(v_1, v_2)$$

where the voltages $v_1$ and $v_2$ and the currents $i_1$ and $i_2$ are defined in Fig. 10-4. The functions $f_1$ and $f_2$ are nonlinear functions of the voltages $v_1$ and $v_2$.

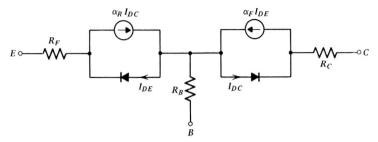

**Fig. 10-3    The Ebers-Moll model with resistors added to account for emitter, base, and collector resistance.**

**Fig. 10-4    A nonlinear two-port element.**

## 10-2    THE LINEARIZED EQUIVALENT TRANSISTOR MODEL

Given the Ebers–Moll model of Fig. 10-2, we are in a position to perform a dc analysis of transistor networks. All we need do is replace each transistor by the Ebers–Moll model. We then have a network composed of resistors, ideal diodes, and current controlled current sources that can be analyzed with the techniques discussed in Chapters 7 and 8. There are two points, however, which justify further consideration of this approach to analysis of transistor networks. First, the controlling currents of the controlled current sources are currents through nonlinear diodes, not linear elements as was the case in Chapter 8. Second, since transistors themselves are usually more prevalent in modern day circuits than diodes, it pays to consider it as an element in its own right. To this end we wish to develop the linearized equivalent model for a transistor. Just as was the case for diode networks, the analysis of transistor networks proceeds by replacing each transistor by its linearized equivalent model, and iteratively analyzing the resulting linear network until convergence has been obtained.

At this point there are two approaches we may take. We can either start with the Ebers–Moll model of Fig. 10-2$b$, and consider the linearization of each circuit element, or we can begin with the Ebers–Moll equations (10.1) and perform a Taylor series in terms of the variables $v_{EB}$ and $v_{CB}$. In this section we will use the first approach. The second approach, which will lead to the same results, but is manipulatively more complex, will be considered in Sections 10-4 and 10-5.

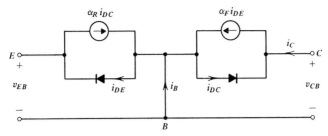

**Fig. 10-5   The Ebers–Moll model.**

Consider again the Ebers–Moll model that is redrawn in Fig. 10-5. The only nonlinear elements of the model are the two diodes $DE$ and $DC$ with branch relationships

$$i_{DE} = I_{ES}(e^{-v_{EB}/v_T} - 1)$$

and

$$i_{DC} = I_{CS}(e^{-v_{CB}/v_T} - 1)$$

The linearized models associated with each of these diodes (see Chapter 7) is obtained by performing a Taylor series expansion of the diode branch relationship about the $l$th estimate of the diode voltage and retaining only first-order terms. These models are shown in Figs. 10-6a and 10-6b where the superscripts $l$ and $l + 1$ denote the $l$th and $(l + 1)$st estimates of the voltages or currents (*Note*. The voltage across the diodes are $-v_{EB}$ and $-v_{CB}$) and:

(10.7a) $\qquad G_E^l = \dfrac{I_{ES}}{v_T} e^{-v_{EB}^l/v_T}$

(10.7b) $\qquad \hat{i}_{DE}^l = I_{ES}(e^{-v_{EB}^l/v_T} - 1) + \left( \dfrac{I_{ES}}{v_T} e^{-v_{EB}^l/v_T} \right) v_{EB}^l$

$\qquad\qquad = I_{ES}\left[ e^{-v_{EB}^l/v_T}\left( 1 + \dfrac{v_{EB}^l}{v_T} \right) - 1 \right]$

(10.8a) $\qquad G_C^l = \dfrac{I_{CS}}{v_T} e^{-v_{CB}^l/v_T}$

and

(10.8b) $\qquad \hat{i}_{DC}^l = I_{CS}(e^{-v_{CB}^l/v_T} - 1) + \left( \dfrac{I_{CS}}{v_T} e^{-v_{CB}^l/v_T} \right) v_{CB}^l$

$\qquad\qquad = I_{CS}\left[ e^{-v_{CB}^l/v_T}\left( 1 + \dfrac{v_{CB}^l}{v_T} \right) - 1 \right]$

The linearized transistor model is obtained by replacing the diodes in the Ebers–Moll model with their linearized models, as shown in Fig. 10-7. Thus

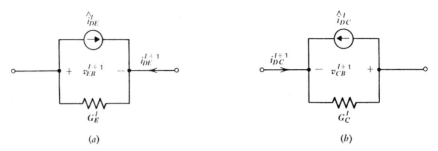

**Fig. 10-6   The linearized models associated with the diodes in the Ebers-Moll model.**

the linearized transistor model is described by the relationships

(10.9a)     $$i_E^{l+1} = -[i_{DE}^l - G_E^l v_{EB}^{l+1}] + \alpha_R i_{DC}^{l+1}$$

and

(10.9b)     $$i_C^{l+1} = \alpha_F i_{DE}^{l+1} - [i_{DC}^l - G_C^l v_{DB}^{l+1}]$$

Since

$$i_{DE}^{l+1} = i_{DE}^l - G_E^l v_{EB}^{l+1}$$

and

$$i_{DC}^{l+1} = i_{DC}^l - G_C^l v_{CB}^{l+1}$$

Equation (10.9) can be written as

(10.10a)     $$i_E^{l+1} = -[i_{DE}^l - G_E^l v_{EB}^{l+1}] + \alpha_R[i_{DC}^l - G_C^l v_{CB}^{l+1}]$$

and

(10.10b)     $$i_C^{+1} = \alpha_F[i_{DE}^l - G_E^l v_{EB}^{l+1}] - [i_{DC}^l - G_C^l v_{CB}^{l+1}]$$

When analyzing networks that contain several nonlinear diodes or transistors, the Newton–Raphson iteration may not converge if we work directly

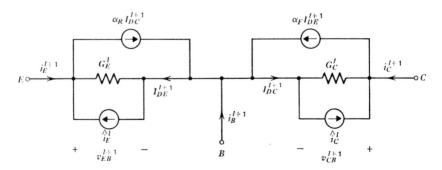

**Fig. 10-7   Linearized transistor model.**

with the linearized diode model of Chapter 7 or the linearized transistor model derived above. The reason for this situation is that the linearization procedure may require evaluation of an exponential at such a large value of voltage so as to cause an overflow in the computer. Alternatively, the resistance used in the linearized model may be so large as to cause problems when Gaussian elimination is used to solve the nodal equilibrium equations in the linearized network. In order to avoid these problems we must modify the linearized diode and, hence, linearized transistor models slightly. Before presenting an example of the analysis of transistor networks, we digress briefly to discuss this modification.

## 10-3  MODIFICATION OF LINEARIZED DIODE AND TRANSISTOR MODELS TO AID IN CONVERGENCE

As indicated briefly in the last section, direct use of the linearized models for diodes and transistors may cause ill conditioning. This problem can be avoided by modifying the linearized models.

The two regions of operation with which we are particularly concerned is the heavily forward biased diode and the reversed biased diode. First, we recognize that the largest positive voltage that can appear across the $p$–$n$ junction of a diode is about .7 volts. Therefore, we limit the largest positive voltage across any $pn$ junction to be .8 volts, which allows for some additional flexibility. Second, for all practical purposes, we can think of a reversed biased $pn$ junction as a large valued (open circuit) resistor. In order to avoid any discontinuity in the assumed diode characteristic, we choose the value of this resistor to be equal to the inverse of the slope of the diode's $i$–$v$ curve at zero bias. In other words, given a diode described by

$$i_D = I_s(e^{v_D/v_T} - 1)$$

the value of the resistor that represents a reversed biased diode is

$$R = \left(\frac{I_s}{v_T}\right)^{-1}$$

Since, at room temperature $v_T \simeq .025$ volts,

$$R = \frac{.025}{I_s}$$

For the transistor circuits we will be concerned with, three situations can arise:[2]

1. The transistor is in the *normal region*, that is, the emitter-base junction is forward biased and the collector-base junction is reversed biased.

---

[2] A fourth region, the *inverse region*, is also possible where the collector-base junction is forward biased and the emitter-base junction is reverse biased. However this situation occurs infrequently.

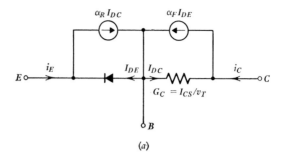

(a)

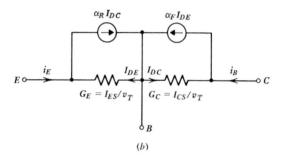

(b)

**Fig. 10-8  Modification of Ebers-Moll model for (a) normal transistor operation; and for (b) cutoff transistor operation.**

2. The transistor is in the *cut-off region*, that is, both the emitter-base and collector-base junctions are reversed biased.
3. The transistor is in the *saturation region*, that is, both the emitter-base and collector-base junctions are forward biased.

From discussion above, convergence of the Newton-iteration can be sped up and numerical ill conditioning reduced if modifications of the Ebers Moll model and, hence, the linearized model is used for transistors operating in the normal and cutoff mode, as shown in Fig. 10-8. The linearized models associated with each of these modified Ebers–Moll models is shown in Fig. 10-9.

One additional problem can arise if the Newton–Raphson scheme is implemented as originally described. It is possible for iterations to oscillate about the correct solution. To avoid this situation, we can "damp" the change indicated by the algorithm. In otherwords, if the $l$th estimate of the diode voltage is $v_D^l$ and the $(l+1)$st estimate found by analyzing the linearized network is $v_D^{l+1}$, we can use

(10.11)    $$\hat{v}_D^{l+1} = v_D^l + \gamma(v_D^{l+1} - v_D^l) = (1 - \gamma)v_D^l + \gamma v_D^{l+1}$$

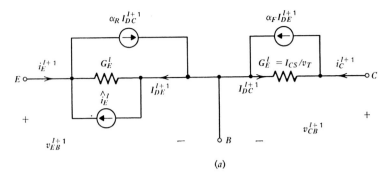

(a)

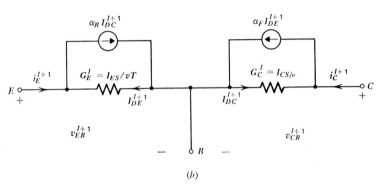

(b)

**Fig. 10-9** Modified linearized Ebers-Moll models for (a) normal operation and (b) cutoff operation.

where $0 \leq \gamma \leq 1$, as the voltage to update the linearized models, instead of $v_D^{l+1}$. Observe that when $\gamma = 1$, $\hat{v}_D^{l+1} = v_D^{l+1}$, and when $\gamma = 0$, $\hat{v}_D^{l+1} = v_D^l$. These techniques are illustrated in the next example.

**Example 10.1**

Consider the network of Fig. 10-10a. Using the linearized version of the Ebers–Moll mode to represent each transistor, we can redraw the network as shown in Fig. 10-10b. The nodal equilibrium equations of the linearized network are:

$$-\frac{1}{R_1} v_{②}^{l+1} - \frac{1}{R_2} v_{④}^{l+1} - i_3^{l+1} = -\left(\frac{1}{R_1} + \frac{1}{R_2}\right)\hat{v}_3$$

$$\left(\frac{1}{R_1} + G_{C1}^l\right)v_{②}^{l+1} - \alpha_{F1}G_{E1}^l v_{⑤}^{l+1}$$

$$= -\alpha_{F1}i_{DE1}^l + i_{DC1}^l + \frac{1}{R_1}\hat{v}_3 - (\alpha_{F1}G_{E1}^l - G_{C1}^l)\hat{v}_1$$

$$G_{C1}^l(\alpha_{R1} - 1)v_{\textcircled{2}}^{l+1} + G_{E1}^l(\alpha_{F1} - 1)v_{\textcircled{5}}^{l+1} + i_1^{l+1}$$

$$= -i_{DE1}^l(1 - \alpha_{F1}) - i_{DC1}^l(1 - \alpha_{R1}) - [G_{C1}^l(1 - \alpha_{R1}) + G_{E1}^l(1 - \alpha_{F1})]\hat{v}_1$$

$$\left(\frac{1}{R_2} + G_{C2}^l\right)v_{\textcircled{4}}^{l+1} - \alpha_{F2}G_{E2}^l v_{\textcircled{5}}^{l+1} = -\alpha_{F2}i_{DE2}^l + i_{DC2}^l + \frac{1}{R_2}\hat{v}_3$$

$$- \alpha_{R1}G_{C1}^l v_{\textcircled{2}}^{l+1} - \alpha_{R2}G_{C2}^l v_{\textcircled{4}}^{l+1} + \left(G_{E1}^l + G_{E2}^l + \frac{1}{R_3}\right)v_{\textcircled{5}}^{l+1}$$

$$= -\alpha_{R1}i_{DC1}^l - \alpha_{R2}i_{DC2}^l + i_{DE1}^l + i_{DE2}^l - (\alpha_{R1}G_{C1}^l - G_{E1}^l)\hat{v}_1 + \frac{1}{R_3}\hat{v}_2$$

$$- \frac{1}{R_3}v_{\textcircled{5}}^{l+1} + i_2^{l+1} = -\frac{1}{R_3}\hat{v}_2$$

where we have used the fact that

$$v_{\textcircled{1}} = \hat{v}_3$$
$$v_{\textcircled{3}} = \hat{v}_1$$

and

$$v_{\textcircled{6}} = \hat{v}_2$$

The unknowns are $v_{\textcircled{2}}^{l+1}$, $v_{\textcircled{4}}^{l+1}$, $v_{\textcircled{5}}^{l+1}$, $i_2^{l+1}$, and $i_3^{l+1}$. The model parameters $G_{E1}^l$, $G_{C1}^l$, $i_{DE1}^l$, $i_{DC1}^l$, $G_{E2}^l$, $G_{C2}^l$, $i_{DE2}^l$, $i_{DC2}^l$, are found in terms of

$$v_{EB1}^l = v_{\textcircled{5}}^l - \hat{v}_1$$
$$v_{CB1}^l = v_{\textcircled{2}}^l - \hat{v}_1$$
$$v_{EB2}^l = v_{\textcircled{5}}^l$$

and

$$v_{CB2}^l = v_{\textcircled{4}}^l$$

using Equations (10.7) and (10.8)

A computer program that iteratively solves these equations is shown in Fig. 10-11a. Several points are worthy of note. Observe first that after reading in the network data the iteration loop begins with two calls to the subroutine TRAN, that is, once for each transistor before calling the subroutine that sets up the nodal equilibrium equations, NODEQN. Subroutine TRAN computes the model parameters for the linearized transistor model by first determining which region the transistor is operating and then using the appropriate model (see Figs. 10-7 and 10-9). After calling GAUSS and printing the results of the current iteration, the subroutine CNVG is called to check for convergence. The parameter KNVG is set to true if convergence has occurred, that is, if the current estimated values of the emitter-base and collector-base voltages

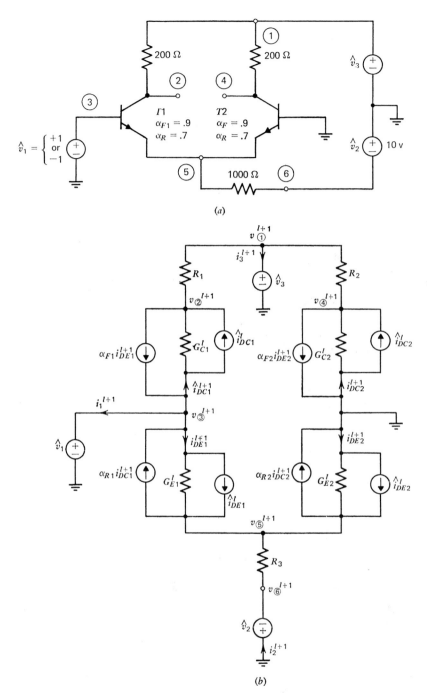

(a)

(b)

**Fig. 10-10** (a) Network used in Example 10.2; (b) the linearized equivalent network.

```
C     CHAPTER 10 EXAMPLE 10.1
C
C     DIMENSIONS
C
      IMPLICIT REAL*8 (A-H,O-Z)
      LOGICAL*1 KNVG
      DIMENSION Y(6,6),CIN(6)
C
C     INITIALIZE ITERATION COUNTER AND CONVERGENCE DETECTOR
C
    4 I=1
      KNVG=.FALSE.
C
C     READ IN ELEMENT VALUES, TRANSISTOR PARAMETERS AND
C     INITIAL GUESSES FOR VEB1,VCB1,VEB2,VCB2,AND GAMMA
C
      CALL READ(V1,V2,V3,R1,R2,R3,ALF1,ALFR1,ALF2,ALFR2,VEB1,VCB1,
     +VEB2,VCB2,GAMMA)
C
C     CALCULATE TRANSISTOR MODEL PARAMETERS
C
    1 CALL TRAN(GE1,GC1,CIE1,CIC1,VEB1,VCB1)
      CALL TRAN(GE2,GC2,CIE2,CIC2,VEB2,VCB2)
C
C     SET UP AND SOLVE NODE EQUATIONS
C
      CALL NODEQN(V1,V2,V3,R1,R2,R3,ALF1,ALFR1,ALF2,ALFR2,GE1,GC1,CIE1,
     +CIC1,GE2,GC2,CIE2,CIC2,Y,CIN)
      CALL GAUSS(Y,CIN,6,6)
C
C     CALCULATE OUTPUT VOLTAGES AND NEW ESTIMATE OF
C     TRANSISTOR JUNCTION VOLTAGES
C
      V1OUT=CIN(1)
      V2OUT=CIN(2)
      VEB1T=CIN(3)-V1
      VCB1T=CIN(1)-V1
      VEB2T=CIN(3)
      VCB2T=CIN(2)
C
C     PRINT SOLUTION AND TERMINATE PROGRAM IF KNVG=TRUE
C
    3 CALL PRINT(V1OUT,V2OUT,VEB1T,VCB1T,VEB2T,VCB2T,I,KNVG)
      IF(KNVG) GO TO 4
C
C     CHECK FOR CONVERGENCE
C
      CALL CNVG(VEB1T,VCB1T,VEB2T,VCB2T,VEB1O,VCB1O,VEB2O,VCB2O,KNVG)
C
C     IF CONVERGENCE HAS OCCURRED PRINT FINAL SOLUTION
C
      IF(KNVG) GO TO 3
C
```

**Fig. 10-11** (*a*) **The program used to analyze the network of Fig. 10-7a.**

(denoted by VEB1T, VCB2T, etc.) are within 1 percent of the previous, or "old" estimated values (denoted by VEB1O, VCB2O, etc.). If convergence has not occurred the new estimates are used to update the model parameters. Finally, observe that a damping factor GAMMA is used to damp the Newton iteration, that is, the updated estimates of $v_{EB1}$, $v_{CB1}$, $v_{EB2}$, and $v_{CB2}$ are obtained by using Equation (10.11).

The results of the program are shown in Fig. 10-11d. Observe from Fig. 10-11b and 10-11c that the same results are obtained for $\gamma = 1$ and $\gamma = .5$.

$\square$

## 10-4* TAYLOR SERIES EXPANSION OF A FUNCTION OF TWO VARIABLES

In the next section we will discuss the derivation of the linearized transistor model directly from the Ebers–Moll equations. Before proceeding with this

```
C
C      TERMINATE PROGRAM IF CONVERGENCE HAS NOT OCCURRED
C      AFTER 20 ITERATIONS
C
    2 IF(I.GE.50) GO TO 4
C
C      CONVERGENCE HAS NOT OCCURRED. INCREASE ITERATION
C      COUNTER, STORE NEW SOLUTION,COMPUTE PREDICTED
C      VOLTAGES AND ITERATE
C
      I=I+1
      VEB1O=VEB1T
      VCB1O=VCB1T
      VEB2O=VEB2T
      VCB2O=VCB2T
      VEB1=VEB1+GAMMA*(VEB1T-VEB1)
      VCB1=VCB1+GAMMA*(VCB1T-VCB1)
      VEB2=VEB2+GAMMA*(VEB2T-VEB2)
      VCB2=VCB2+GAMMA*(VCB2T-VCB2)
      GO TO 1
      END



      SUBROUTINE READ(V1,V2,V3,R1,R2,R3,ALF1,ALFR1,ALF2,ALFR2,VEB1,VCB1,
     +VEB2,VCB2,GAMMA)
      IMPLICIT REAL*8 (A-H,O-Z)
      READ(5,1) V1,V2,V3,R1,R2,R3,ALF1,ALFR1,ALF2,ALFR2,VEB1,VCB1,VEB2,
     +VCB2,GAMMA
      PRINT 2,V1,VEB1,VEB2,VCB1,VCB2,GAMMA
    1 FORMAT(6G12.5)
    2 FORMAT('ORESULTS FOR EX. 10.1 WITH INPUT VOLTAGE V1=',G12.5//
     +' INITIAL GUESS',8X,'VEB1=',G12.5,5X,'VEB2=',G12.5/22X,'VCB1=',
     +G12.5,5X,'VCB2=',G12.5//' GAMMA=',G12.5)
      RETURN
      END
```

**Fig. 10-11a (Continued)**

```
      SUBROUTINE NODEQN(V1,V2,V3,R1,R2,R3,ALF1,ALFR1,ALF2,ALFR2,GE1,GC1,
     +CIE1,CIC1,GE2,GC2,CIE2,CIC2,Y,CIN)
      IMPLICIT REAL*8 (A-H,O-Z)
      DIMENSION Y(6,6),CIN(6)
C
C     INITIALIZE COEFFICIENT ARRAYS
C
      DO 1 J=1,6
      CIN(J)=0.DO
      DO 1 K=1,6
    1 Y(J,K)=0.DO
C
C     SET UP NODAL EQUILIBRIUM COEFFICIENTS
C
      Y(1,1)=-1./R1
      Y(1,2)=-1./R2
      Y(1,6)=1.
      CIN(1)=(-(1./R1)-(1./R2))*V3
      Y(2,1)=(1./R1)+GC1
      Y(2,3)=-ALF1*GE1
      CIN(2)=-(ALF1*CIE1)+CIC1+(1./R1)*V3-(ALF1*GE1-GC1)*V1
      Y(3,1)=GC1*(ALFR1-1.)
      Y(3,3)=GE1*(ALF1-1.)
      Y(3,4)=1.
      CIN(3)=CIC1*(ALFR1-1.)+CIE1*(ALF1-1.)-(GC1*(1.-ALFR1)+GE1*(1.-
     +ALF1))*V1
      Y(4,2)=(1./R2)+GC2
      Y(4,3)=-ALF2*GE2
      CIN(4)=-(ALF2*CIE2)+(CIC2)+(1./R2)*V3
      Y(5,1)=-ALFR1*GC1
      Y(5,2)=-ALFR2*GC2
      Y(5,3)=GE1+GE2+(1./R3)
      CIN(5)=CIE1+CIE2-(ALFR1*CIC1)-(ALFR2*CIC2)-(ALFR1*GC1-GE1)*V1
     ++(1./R3)*V2
      Y(6,3)=-1./R3
      Y(6,5)=1.
      CIN(6)=-(1./R3)*V2
      RETURN
      END



      SUBROUTINE CNVG(VEB1T,VCB1T,VEB2T,VCB2T,VEB1O,VCB1O,VEB2O,VCB2O,
     +KNVG)
      IMPLICIT REAL*8(A-H,O-Z)
      LOGICAL*1 KNVG
      ERR1=DABS((VEB1T-VEB1O)/VEB1T)
      ERR2=DABS((VCB1T-VCB1O)/VCB1T)
      ERR3=DABS((VEB2T-VEB2O)/VEB2T)
      ERR4=DABS((VCB2T-VCB2O)/VCB2T)
      TOL=.0001
      IF(ERR1.LT.TOL.AND.ERR2.LT.TOL.AND.ERR3.LT.TOL.AND.ERR4.LT.
     +TOL) KNVG=.TRUE.
      RETURN
      END
```

**Fig. 10-11a** (*Continued*)

```fortran
      SUBROUTINE TRAN(GE,GC,CIE,CIC,VEB,VCB)
      IMPLICIT REAL*8 (A-H,O-Z)
      REAL*8 IES,ICS
      DATA IES/1.D-13/,ICS/1.D-13/,VT/.025/
C
C     PLACE LOWER BOUND ON JUNCTION VOLTAGES
C
      VEB=DMAX1(VEB,-.8D0)
      VCB=DMAX1(VCB,-.8D0)
C
C     TEST FOR NORMAL, CUTOFF, SATURATION OR REVERSE MODE OF OPERATION
C     AND SET UP APPROPRIATE MODEL
C
      IF(VEB.LT.O..AND.VCB.GT.O.) GO TO 1
      IF(VEB.GT.O..AND.VCB.GT.O.) GO TO 2
      IF(VEB.LT.O..AND.VCB.LT.O.) GO TO 3
C
C     REVERSE MODE
C
      CIE=0.
      CIC=ICS*((DEXP(-VCB/VT)*(1.+(VCB/VT)))-1.)
      GE=IES/VT
      GC=ICS*DEXP(-VCB/VT)/VT
      RETURN
C
C     NORMAL MODE
C
    1 CIE=IES*((DEXP(-VEB/VT)*(1.+(VEB/VT)))-1.)
      CIC=0.
      GE=IES*DEXP(-VEB/VT)/VT
      GC=ICS/VT
      RETURN
C
C     CUTOFF MODE
C
    2 CIE=0.
      CIC=0.
      GE=IES/VT
      GC=ICS/VT
      RETURN
C
C     SATURATION MODE
C
    3 CIE=IES*((DEXP(-VEB/VT)*(1.+(VEB/VT)))-1.)
      CIC=ICS*((DEXP(-VCB/VT)*(1.+(VCB/VT)))-1.)
      GC=ICS*DEXP(-VCB/VT)/VT
      GE=IES*DEXP(-VEB/VT)/VT
      RETURN
      END

      SUBROUTINE PRINT(V1OUT,V2OUT,VEB1T,VCB1T,VEB2T,VCB2T,I,KNVG)
      IMPLICIT REAL*8 (A-H,O-Z)
      LOGICAL*1 KNVG
C
C     IF FIRST ITERATION (I=1) PRINT HEADINGS
C
```

**Fig. 10-11a** (*Continued*)

```
      IF(I.GT.1) GO TO 4
      PRINT 5
    5 FORMAT(/' ITERATION NO.',8X,'V1OUT',12X,'V2OUT',13X,'VEB1',13X,
     +'VCB1',13X,'VEB2',13X,'VCB2')
      GO TO 1
C
C     PRINT FINAL SOLUTION IF KNVRG=1
C
    4 IF(.NOT.KNVG) GO TO 1
      PRINT 3,V1OUT,V2OUT
      RETURN
    1 PRINT 2,I,V1OUT,V2OUT,VEB1T,VCB1T,VEB2T,VCB2T
    2 FORMAT(5X,I4,10X,G12.5,5X,G12.5,5X,G12.5,5X,G12.5,5X,G12.5,5X,
     +G12.5)
    3 FORMAT(/' FINAL SOLUTION'/' V1OUT=',G12.5,5X,' V2OUT=',G12.5)
      RETURN
      END
```

**Fig. 10-11a (Continued)**

however, it is convenient to digress briefly to discuss the Taylor series expansion of a function of two variables. Let $f$ be a function of two variables $x_1$ and $x_2$, that is,

$$f = f(x_1, x_2)$$

The Taylor series expansion of $f$ about the point $(x_1{}^*, x_2{}^*)$ is

$$(10.12) \quad f(x_1, x_2) = f(x_1{}^*, x_2{}^*) + \frac{\partial f}{\partial x_1}\bigg|_{x_1{}^*, x_2{}^*}(x_1 - x_1{}^*) + \frac{\partial f}{\partial x_2}\bigg|_{x_1{}^*, x_2{}^*}(x_2 - x_2{}^*)$$

$$+ \frac{1}{2}\frac{\partial^2 f}{\partial x_1{}^2}\bigg|_{x_1{}^*, x_2{}^*}(x_1 - x_1{}^*)^2 + \frac{1}{2}\frac{\partial^2 f}{\partial x_2{}^2}\bigg|_{x_1{}^*, x_2{}^*}(x_2 - x_2{}^*)^2$$

$$+ \frac{\partial^2 f}{\partial x_1 \partial x_2}\bigg|_{x_1{}^*, x_2{}^*}(x_1 - x_1{}^*)(x_2 - x_2{}^*) + \cdots$$

This expression can be written more succinctly in terms of matrices. We make the following definitions:

$$\mathbf{x} \equiv (x_1, x_2)'$$

$$\mathbf{x}^* \equiv (x_1{}^*, x_2{}^*)'$$

$$\Delta \mathbf{x} \equiv \mathbf{x} - \mathbf{x}^*$$

$$\nabla f \equiv \left(\frac{\partial f}{\partial x_1}, \frac{\partial f}{\partial x_2}\right)$$

RESULTS FOR EX. 10.1 WITH INPUT VOLTAGE V1= 1.0000

INITIAL GUESS    VEB1=-0.50000    VEB2=-0.50000
               VCB1= 10.000     VCB2= 10.000

GAMMA= 1.0000

| ITERATION NO. | V1OUT | V2OUT | VEB1 | VCB1 | VEB2 | VCB2 |
|---|---|---|---|---|---|---|
| 1 | 9.1079 | 9.4572 | -3.0287 | 8.1079 | -2.0287 | 9.4572 |
| 2 | -28418. | 28436. | -1.2750 | -28419. | -0.27502 | 28436. |
| 3 | 0.22534 | 10.000 | -0.77479 | -0.77466 | 0.22521 | 10.000 |
| 4 | 0.25127 | 10.000 | -0.74923 | -0.74873 | 0.25077 | 10.000 |
| 5 | 0.27889 | 10.000 | -0.72269 | -0.72111 | 0.27731 | 10.000 |
| 6 | 0.31175 | 10.000 | -0.69324 | -0.68825 | 0.30676 | 10.000 |
| 7 | 0.36581 | 10.000 | -0.65393 | -0.63419 | 0.34607 | 10.000 |
| 8 | 0.63400 | 10.000 | -0.56300 | -0.36600 | 0.43700 | 10.000 |
| 9 | 8.1864 | 10.000 | -0.95258 | 7.1864 | 0.47422D-01 | 10.000 |
| 10 | 8.1595 | 10.000 | -0.77503 | 7.1595 | 0.22497 | 10.000 |
| 11 | 8.1550 | 10.000 | -0.75012 | 7.1550 | 0.24988 | 10.000 |
| 12 | 8.1506 | 10.000 | -0.72536 | 7.1506 | 0.27464 | 10.000 |
| 13 | 8.1462 | 10.000 | -0.70101 | 7.1462 | 0.29899 | 10.000 |
| 14 | 8.1420 | 10.000 | -0.67772 | 7.1420 | 0.32228 | 10.000 |
| 15 | 8.1383 | 10.000 | -0.65708 | 7.1383 | 0.34292 | 10.000 |
| 16 | 8.1356 | 10.000 | -0.64205 | 7.1356 | 0.35795 | 10.000 |
| 17 | 8.1343 | 10.000 | -0.63524 | 7.1343 | 0.36476 | 10.000 |
| 18 | 8.1341 | 10.000 | -0.63413 | 7.1341 | 0.36587 | 10.000 |
| 19 | 8.1341 | 10.000 | -0.63411 | 7.1341 | 0.36589 | 10.000 |

FINAL SOLUTION
V1OUT= 8.1341         V2OUT= 10.000

**Fig. 10-11b  The output for an undamped Newton iteration.**

RESULTS FOR EX. 10.1 WITH INPUT VOLTAGE V1= 1.0000

INITIAL GUESS    VEB1=-0.50000    VEB2=-0.50000
              VCB1= 10.000     VCB2= 10.000

GAMMA= 0.50000

| ITERATION NO. | V1OUT | V2OUT | VEB1 | VCB1 | VEB2 | VCB2 |
|---|---|---|---|---|---|---|
| 1 | 9.1079 | 9.4572 | -3.0287 | 8.1079 | -2.0287 | 9.4572 |
| 2 | -28418. | 28436. | -1.2750 | -28419. | -0.27502 | 28436. |
| 3 | 0.22529 | 11.155 | -0.77485 | -0.77471 | 0.22515 | 11.155 |
| 4 | 0.23821 | 10.000 | -0.76209 | -0.76179 | 0.23791 | 10.000 |
| 5 | 0.25136 | 10.000 | -0.74919 | -0.74864 | 0.25081 | 10.000 |
| 6 | 0.26496 | 10.000 | -0.73604 | -0.73504 | 0.26396 | 10.000 |
| 7 | 0.27934 | 10.000 | -0.72245 | -0.72066 | 0.27755 | 10.000 |
| 8 | 0.29512 | 10.000 | -0.70808 | -0.70488 | 0.29192 | 10.000 |
| 9 | 0.31364 | 10.000 | -0.69231 | -0.68636 | 0.30769 | 10.000 |
| 10 | 0.33813 | 10.000 | -0.67385 | -0.66187 | 0.32615 | 10.000 |
| 11 | 0.37912 | 10.000 | -0.64953 | -0.62088 | 0.35047 | 10.000 |
| 12 | 0.49662 | 10.000 | -0.60954 | -0.50338 | 0.39046 | 10.000 |
| 13 | 2.2773 | 10.000 | -0.52872 | 1.2773 | 0.47128 | 10.000 |
| 14 | 8.1534 | 10.000 | -0.74107 | 7.1534 | 0.25893 | 10.000 |
| 15 | 8.1362 | 10.000 | -0.64563 | 7.1362 | 0.35437 | 10.000 |
| 16 | 8.1353 | 10.000 | -0.64036 | 7.1353 | 0.35964 | 10.000 |
| 17 | 8.1347 | 10.000 | -0.63703 | 7.1347 | 0.36297 | 10.000 |
| 18 | 8.1343 | 10.000 | -0.63527 | 7.1343 | 0.36473 | 10.000 |
| 19 | 8.1342 | 10.000 | -0.63451 | 7.1342 | 0.36549 | 10.000 |
| 20 | 8.1342 | 10.000 | -0.63423 | 7.1342 | 0.36577 | 10.000 |
| 21 | 8.1341 | 10.000 | -0.63414 | 7.1341 | 0.36586 | 10.000 |
| 22 | 8.1341 | 10.000 | -0.63412 | 7.1341 | 0.36588 | 10.000 |

FINAL SOLUTION
V1OUT= 8.1341       V2OUT= 10.000

Fig. 10-11c   The output for a damped Newton iteration with $\delta = .5$ and $V1 = 1$.

RESULTS FOR EX. 10.1 WITH INPUT VOLTAGE V1= -1.0000

INITIAL GUESS
VEB1=-0.50000    VEB2=-0.50000
VCB1= 10.000    VCB2= 10.000

GAMMA= 0.50000

| ITERATION NO. | V1OUT | V2OUT | VEB1 | VCB1 | VEB2 | VCB2 |
|---|---|---|---|---|---|---|
| 1 | 9.5288 | 9.1795 | -1.8239 | 10.529 | -2.8239 | 9.1795 |
| 2 | 28436. | -28418. | -0.27501 | 28437. | -1.2750 | -28418. |
| 3 | 11.155 | -0.77466 | 0.22519 | 12.155 | -0.77481 | -0.77466 |
| 4 | 10.000 | -0.76168 | 0.23800 | 11.000 | -0.76200 | -0.76168 |
| 5 | 10.000 | -0.74843 | 0.25096 | 11.000 | -0.74904 | -0.74843 |
| 6 | 10.000 | -0.73466 | 0.26424 | 11.000 | -0.73576 | -0.73466 |
| 7 | 10.000 | -0.71997 | 0.27806 | 11.000 | -0.72194 | -0.71997 |
| 8 | 10.000 | -0.70361 | 0.29282 | 11.000 | -0.70718 | -0.70361 |
| 9 | 10.000 | -0.68389 | 0.30932 | 11.000 | -0.69068 | -0.68389 |
| 10 | 10.000 | -0.65649 | 0.32931 | 11.000 | -0.67069 | -0.65649 |
| 11 | 10.000 | -0.60549 | 0.35735 | 11.000 | -0.64265 | -0.60549 |
| 12 | 10.000 | -0.41169 | 0.40972 | 11.000 | -0.59028 | -0.41169 |
| 13 | 10.000 | 5.6878 | 0.44433 | 11.000 | -0.55567 | 5.6878 |
| 14 | 10.000 | 8.3240 | 0.31113 | 11.000 | -0.68887 | 8.3240 |
| 15 | 10.000 | 8.3139 | 0.36706 | 11.000 | -0.63294 | 8.3139 |
| 16 | 10.000 | 8.3138 | 0.36794 | 11.000 | -0.63206 | 8.3138 |
| 17 | 10.000 | 8.3137 | 0.36827 | 11.000 | -0.63173 | 8.3137 |
| 18 | 10.000 | 8.3137 | 0.36838 | 11.000 | -0.63162 | 8.3137 |
| 19 | 10.000 | 8.3137 | 0.36841 | 11.000 | -0.63159 | 8.3137 |

FINAL SOLUTION
V1OUT= 10.000    V2OUT= 8.3137

**Fig. 10-11d** The output for a damped Newton iteration with $\delta = .5$ and $V1 = -1$.

and

$$\nabla^2 f \equiv \begin{bmatrix} \dfrac{\partial^2 f}{\partial x_1{}^2} & \dfrac{\partial^2 f}{\partial x_1\, \partial x_2} \\[2ex] \dfrac{\partial^2 f}{\partial x_2\, \partial x_1} & \dfrac{\partial^2 f}{\partial x_2{}^2} \end{bmatrix}$$

We may now write

$$f(x_1, x_2) = f(\mathbf{x})$$

and the Taylor series expansion of $f(\mathbf{x})$ about the point $\mathbf{x}^*$ is

(10.13)     $$f(\mathbf{x}) = f(\mathbf{x}^*) + \nabla f|_{\mathbf{x}^*} \Delta \mathbf{x} + \frac{1}{2} (\Delta \mathbf{x})' \nabla^2 f|_{\mathbf{x}^*} (\Delta \mathbf{x}) + \cdots$$

To see that (10.12) and (10.13) are identical observe that

$$\nabla f|_{\mathbf{x}^*} \Delta \mathbf{x} = \begin{bmatrix} \dfrac{\partial f}{\partial x_1}\Big|_{x_1{}^*,\, x_2{}^*}, & \dfrac{\partial f}{\partial x_2}\Big|_{x_1{}^*,\, x_2{}^*} \end{bmatrix} \begin{bmatrix} x_1 - x_1{}^* \\ x_2 - x_2{}^* \end{bmatrix}$$

$$= \dfrac{\partial f}{\partial x_1}\Big|_{x_1{}^*,\, x_2{}^*} (x_1 - x_1{}^*) + \dfrac{\partial f}{\partial x_2}\Big|_{x_1{}^*,\, x_2{}^*} (x_2 - x_2{}^*)$$

and

$$\frac{1}{2} (\Delta \mathbf{x})' \nabla^2 f|_{\mathbf{x}^*} (\Delta x)$$

$$= \frac{1}{2} (x_1 - x_1{}^*,\, x_2 - x_2{}^*) \begin{bmatrix} \dfrac{\partial^2 f}{\partial x_1{}^2}\Big|_{x_1{}^*,\, x_2{}^*} & \dfrac{\partial^2 f}{\partial x_1\, \partial x_2}\Big|_{x_1{}^*,\, x_2{}^*} \\[2ex] \dfrac{\partial^2 f}{\partial x_2\, \partial x_1}\Big|_{x_1{}^*,\, x_2{}^*} & \dfrac{\partial^2 f}{\partial x_2{}^2}\Big|_{x_1{}^*,\, x_2{}^*} \end{bmatrix} \begin{bmatrix} x_1 - x_1{}^* \\ x_2 - x_2{}^* \end{bmatrix}$$

$$= \frac{1}{2} \dfrac{\partial^2 f}{\partial x_1{}^2}\Big|_{x_1{}^*,\, x_2{}^*} (x_1 - x_2)^2 + \dfrac{\partial^2 f}{\partial x_1\, \partial x_2}\Big|_{x_1{}^*,\, x_2{}^*} (x_1 - x_1{}^*)(x_2 - x_2{}^*)$$

$$+ \frac{1}{2} \dfrac{\partial^2 f}{\partial x_2{}^2}\Big|_{x_1{}^*,\, x_2{}^*} (x_2 - x_2{}^*)^2$$

since

$$\frac{\partial^2 f}{\partial x_1\, \partial x_2} = \frac{\partial^2 f}{\partial x_2\, \partial x_1}$$

For our purposes we will only be concerned with the first two terms of the Taylor series expansion. Thus we will approximate $f(x_1, x_2)$ by

$$f(\mathbf{x}) \approx f(\mathbf{x}^*) + \nabla f|_{\mathbf{x}^*} \Delta \mathbf{x}$$

or

$$f(x_1, x_2) \simeq f(x_1{}^*, x_2{}^*) + \frac{\partial f}{\partial x_1}\bigg|_{x_1{}^*, x_2{}^*} (x_1 - x_1{}^*) + \frac{\partial f}{\partial x_2}\bigg|_{x_1{}^*, x_2{}^*} (x_2 - x_2{}^*)$$

**Example 10.2**

Consider the expansion of the function

$$f(x_1, x_2) = x_1{}^2 + x_1 x_2 + \frac{1}{2} x_2{}^3$$

about the point $x_1 = 3$, $x_2 = 2$ using the first two terms of a Taylor series:

$$f(x_1, x_2) \simeq f(3, 2) + \frac{\partial f}{\partial x_1}\bigg|_{3, 2} (x_1 - 3) + \frac{\partial f}{\partial x_2}\bigg|_{3, 2} (x_2 - 2)$$

First we evaluate $f(3, 2)$:

$$f(3, 2) = (3)^2 + (3)(2) + \frac{1}{2}(2)^3 = 19$$

then $\dfrac{\partial f}{\partial x_1}\bigg|_{3, 2}$ and $\dfrac{\partial f}{\partial x_2}\bigg|_{3, 2}$

$$\frac{\partial f}{\partial x_1}\bigg|_{3, 2} = 2x_1 + x_2\big|_{3, 2}$$

$$= 2(3) + 2 = 8$$

$$\frac{\partial f}{\partial x_2}\bigg|_{3, 2} = x_1 + \frac{3}{2} x_2{}^2\big|_{3, 2}$$

$$= 3 + \frac{3}{2}(2)^2 = 9$$

Therefore

$$f(x_1, x_2) \simeq 19 + 8(x_1 - 3) + 9(x_2 - 2)$$

$$\simeq -23 + 8x_1 + 9x_2$$

Suppose we desired $f(x_1, x_2)$ for the point $x_1 = 3.1$ and $x_2 = 1.9$. The exact value of $f$ is

$$f(3.1, 1.9) = (3.1)^2 + (3.1)(1.9) + \frac{1}{2}(1.9)^3 = 18.9285$$

whereas using the approximation

$$\hat{f}(3.1, 1.9) = -23 + 8(3.1) + 9(1.9) = 18.9$$

As should be expected, the further we move from the point at which our approximation was made the poorer it becomes. For instance, at the point

$x_1 = 3.5$ and $x_2 = 1.5$, the exact value of $f$ is

$$f(3.5, 1.5) = (3.5)^2 + (3.5)(1.5)^{\cdot} + \frac{1}{2}(1.5)^3 = 19.1875$$

while the approximate value of $f$ is

$$\hat{f}(3.5, 1.5) = -23 + 8(3.5) + 9(1.5) = 17.5. \quad \square$$

## 10-5*  AN ALTERNATE DERIVATION OF THE LINEARIZED TRANSISTOR MODEL

When diodes are present in a network we will find the node voltages by iteratively analyzing the linearized version of the network. In particular, at each iteration the diode was replaced by a conductor in parallel with a current source. At the $l$th iteration, the conductor has a value equal to the slope of the tangent to the diode characteristic curve about the $l$th estimate of the diode voltage (operating point). The current source has a value equal to the intercept of this tangent with the $i$-axis. The above scheme for analyzing networks that contain diodes, where diodes are replaced by linearized equivalent models was used in Section 10-2 to derive a linearized transistor model. In this section we derive a linearized model directly from the Ebers–Moll equations (10.1). The reason for presenting this approach is that it is useful for determining linearized models for general nonlinear two-ports that cannot be modelled by appropriate two terminal elements.

Let $i_E^{l+1}$, $i_C^{l+1}$, $v_{EB}^{l+1}$, and $v_{CB}^{l+1}$ denote the $(l + 1)$st estimates of the currents and voltages $i_E$, $i_C$, $v_{EB}$, and $v_{CB}$, respectively, of a transistor. Under the assumption that these estimates are in fact the actual currents and voltages, from Equation (10.1) we have

$$i_E^{l+1} = -I_{ES}\left[\exp\left(-\frac{v_{EB}^{l+1}}{v_T}\right) - 1\right] + \alpha_R I_{CS}\left[\exp\left(-\frac{v_{CB}^{l+1}}{v_T}\right) - 1\right]$$

and

$$i_C^{l+1} = \alpha_F I_{CS}\left[\exp\left(-\frac{v_{EB}^{l+1}}{v_T}\right) - 1\right] - I_{CS}\left[\exp\left(-\frac{v_{CB}^{l+1}}{v_T}\right) - 1\right]$$

Since the $(l + 1)$st estimates of $i_E$, $i_C$, $v_{EB}$, and $v_{CB}$ are obtained from the $l$th estimates by

$$i_E^{l+1} = i_E^l + \Delta i_E^l$$

$$i_C^{l+1} = i_C^l + \Delta i_C^l$$

$$v_{EB}^{l+1} = v_{EB}^l + \Delta v_{EB}^l$$

$$v_{CB}^{l+1} = v_{CB}^l + \Delta v_{CB}^l$$

the following relationships exists:

$$i_E^{l+1} = -I_{ES}\left\{\exp\left[-\left(\frac{v_{EB}^l + \Delta v_{EB}^l}{v_T}\right)\right] - 1\right\}$$

$$+ \alpha_R I_{CS}\left\{\exp\left[-\left(\frac{v_{CB}^l + \Delta v_{CB}^l}{v_T}\right)\right] - 1\right\}$$

and

$$i_C^{l+1} = \alpha_F I_{ES}\left\{\exp\left[-\left(\frac{v_{EB}^l + \Delta v_{EB}^l}{v_T}\right)\right] - 1\right\}$$

$$- I_{CS}\left\{\exp\left[-\left(\frac{v_{CB}^l + \Delta v_{CB}^l}{v_T}\right)\right] - 1\right\}$$

Expanding these expressions about $v_{EB}^l$ and $v_{CB}^l$ using the first two terms of a Taylor series yields

$$i_E^{l+1} = -I_{ES}(e^{-v_{EB}^l/v_T} - 1) + \frac{I_{ES}}{v_T} e^{-v_{EB}^l/v_T}(v_{EB}^{l+1} - v_{EB}^l)$$

$$+ \alpha_R I_{CS}(e^{-v_{CB}^l/v_T} - 1) - \alpha_R \frac{I_{CS}}{v_T} e^{-v_{CB}^l/v_T}(v_{CB}^{l+1} - v_{CB}^l)$$

and

$$i_C^{l+1} = \alpha_F I_{ES}(e^{-v_{EB}^l/v_T} - 1) - \alpha_F \frac{I_{ES}}{v_T} e^{-v_{EB}^l/v_T}(v_{EB}^{l+1} - v_{EB}^l)$$

$$- I_{CS}(e^{-v_{EB}^l/v_T} - 1) + \frac{I_{CS}}{v_T} e^{-v_{EB}^l/v_T}(v_{EB}^{l+1} - v_{CB}^l)$$

After grouping terms that are known at the $(l + 1)$st iteration we have

$$i_E^{l+1} = \left\{I_{ES}\left[1 - e^{-v_{EB}^l/v_T}\left(1 + \frac{v_{EB}^l}{v_T}\right)\right] - \alpha_R I_{CS}\left[1 - e^{-v_{CB}^l/v_T}\left(1 + \frac{v_{CB}^l}{v_T}\right)\right]\right\}$$

$$+ \left(\frac{I_{ES}}{v_T} e^{-v_{EB}^l/v_T}\right)v_{EB}^{l+1} - \left(\alpha_R \frac{I_{CS}}{v_T} e^{-v_{CB}^l/v_T}\right)v_{CB}^{l+1}$$

and

$$i_C^{l+1} = \left\{I_{CS}\left[1 - e^{-v_{CB}^l/v_T}\left(1 + \frac{v_{CB}^l}{v_T}\right)\right] - \alpha_F I_{ES}\left[1 - e^{-v_{EB}^l/v_T}\left(1 + \frac{v_{EB}^l}{v_T}\right)\right]\right\}$$

$$- \left(\alpha_F \frac{I_{ES}}{v_T} e^{-v_{EB}^l/v_T}\right)v_{EB}^{l+1} + \left(\frac{I_{CS}}{v_T} e^{-v_{BC}^l/v_T}\right)v_{CB}^{l+1}$$

For manipulative ease, we define the following quantities:

$$i_{DE}^l \equiv I_{ES}\left[e^{-v_{EB}^l/v_T}\left(1 + \frac{v_{EB}^l}{v_T}\right) - 1\right]$$

$$i_{DC}^l \equiv I_{CS}\left[1 - e^{-v_{CB}^l/v_T}\left(1 + \frac{v_{CB}^l}{v_T}\right) - 1\right]$$

$$G_E^l \equiv \frac{I_{ES}}{v_T}\, e^{v_{EB}^l/v_T}$$

$$G_C^l \equiv \frac{I_{CS}}{v_T}\, e^{-v_{CB}^l/v}$$

as was done in Equation (10.7). Then we have

$$i_E^{l+1} = -(i_{DE}^l - G_E^l v_{EB}^{l+1}) + \alpha_R(i_{DC}^l - G_C^l v_{CB}^{l+1})$$

$$i_C^{l+1} = \alpha_F(i_{DE}^l - G_E^l v_{EB}^{l+1}) - (i_{DC}^l - G_C^l v_{CB}^{l+1})$$

which are identical to Equations 10.10, as expected.

### Example 10.3

To illustrate this approach for finding the linearized equivalent model for an arbitrary nonlinear two-port, consider a two-port element characterized by the relationships

$$v_1 = i_1^2 + i_1 v_2 + v_2^3$$

and

$$i_2 = i_1 + 2i_1 v_2^2 + v_2$$

Let the superscript $(l + 1)$ denote the $(l + 1)$st estimate of the voltages and currents than are assumed to be the actual voltages and current. Taking a Taylor series expansion about the $l$th estimate and retaining only first-order terms, we have

$$v_1^{l+1} = (i_1^l)^2 + i_1^l v_2^l + (v_2^l)^3 + (2i_1^l + v_2^l)[i_1^{l+1} - i_1^l]$$
$$+ [i_1^l + 3(v_2^l)^2][v_2^{l+1} - v_2^l]$$

and

$$i_2^{l+1} = i_1^l + 2i_1^l(v_2^l)^2 + v_2^l + (1 + 2(v_2^l)^2)[i_1^{l+1} - i_1^l]$$
$$+ (4i_1^l v_2^l + 1)[v_2^{l+1} - v_2^l]$$

Combining terms yields the expressions

$$v_1^{l+1} = [-(i_1^l)^2 + i_1^l v_2^l - 2(v_2^l)^3] + [2i_1^l + v_2^l]i_1^{l+1} + [i_1^l + 3(v_2^l)^2]v_2^{l+1}$$

and

$$i_2^{l+1} = [-4i_1^l(v_2^l)^2] + [1 + 2(v_2^l)^2]i_1^{l+1} + [4i_1^l v_2^{l+1} + 1]v_2^{l+1}$$

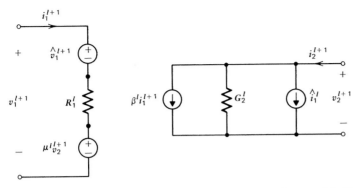

**Fig. 10-12** The linearized model of the nonlinear two-port discussed in Example 10.3.

A linear model that is characterized by these expressions is shown in Fig. 10.12 where

$$\hat{v}_1^l = [-(i_1^l)^2 + i_1^l v_2^l - 2(v_2^l)^3]$$

$$R_1^l = [2i_1^l + v_2^l]$$

$$\mu^l = [i_1^l + 3(v_2^l)^2]$$

$$\beta^l = [1 + 2(v_2^l)^2]$$

$$G_2 = [4i_1^l v_2^l + 1]$$

and

$$\hat{i}_2^l = [-4i_1^l(v_2^l)^2] \quad \square$$

## 10-6 SMALL-SIGNAL ANALYSIS AND ALTERNATE REPRESENTATIONS

In Section 7-7 we discussed the small-signal analysis of networks containing diodes. Small-signal analysis is useful if once a solution has been obtained for a nonlinear network it is desired to find the solution for the same network after a small perturbation of the excitation has occurred. In particular, for transistors we assume that the operating point $i_E$, $i_C$, $v_{EB}$, and $v_{CB}$ has been found. For small perturbations of the excitations the operating point becomes $i_E + \Delta i_E$, $i_C + \Delta i_C$, $v_{EB} + \Delta v_{EB}$, and $v_{CB} + \Delta v_{CB}$. Rather than undertake a complete nonlinear analysis to determine the exact operating point, we assume superposition holds and solve for the changes $\Delta i_E$, $\Delta i_C$, $\Delta v_{EB}$, and $\Delta v_{CB}$ from the linear small-signal network, that is, the original network with all nonlinear elements linearized about their operating points and excited only by the perturbed signals. We found in the Sections 10-2 and 10-5 that $\Delta i_E$,

$\Delta i_C$, $\Delta v_{EB}$, and $\Delta v_{CB}$ are related by

$$\Delta i_E = \left[\frac{I_{ES}}{v_T} e^{-v_{EB}/v_T}\right]\Delta v_{EB} - \left[\alpha_R \frac{I_{CS}}{v_T} e^{-v_{CB}/v_T}\right]\Delta v_{CB}$$

and

$$\Delta i_C = -\left[\alpha_F \frac{I_{ES}}{v_T} e^{-v_{EB}/v_T}\right]\Delta v_{EB} + \left[\frac{I_{CS}}{v_T} e^{-v_{CB}/v_T}\right]\Delta v_{CB}$$

Observe that $v_{EB}$ and $v_{CB}$ are quantities we know from the original nonlinear analysis. We can interpret the small-signal model of a transistor as a two-port whose $y$-parameters are given by

$$y_{11} = \frac{I_{ES}}{v_T} e^{-v_{EB}/v_T}$$

$$y_{22} = \frac{I_{CS}}{v_T} e^{-v_{CB}/v_T}$$

$$y_{12} = -\alpha_R \frac{I_{CS}}{v_T} e^{-v_{CB}/v_T} = -\alpha_R y_{22}$$

and

$$y_{21} = -\alpha_F \frac{I_{ES}}{v_T} e^{-v_{EB}/v_T} = -\alpha_F y_{11}$$

The small-signal transistor model is then characterized by

$$\Delta i_E = y_{11}\,\Delta v_{EB} + y_{12}\,\Delta v_{CB}$$

and

$$\Delta i_C = y_{21}\,\Delta v_{EB} + y_{22}\,\Delta v_{CB}$$

or

$$\Delta i_E = y_{11}\,\Delta v_{EB} - \alpha_R y_{22}\,\Delta v_{CB}$$

and

$$\Delta i_C = -\alpha_F y_{11}\,\Delta v_{EB} + y_{22}\,\Delta v_{CB}$$

These equations are modeled by the network of Fig. 10-13. This representation is sometimes referred to as the *T-model*.

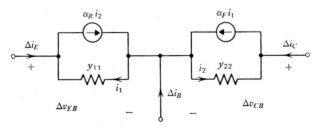

**Fig. 10-13   The *T*-model, a small-signal transistor model.**

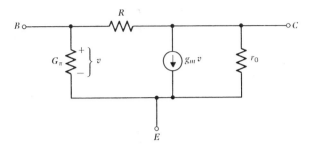

**Fig. 10-14    The $\pi$-model—a more common small-signal model of a transistor.**

## Example 10.4

A much more common form of the transistor small-signal model is the $\pi$-model of Fig. 10-14. This model is very convenient and a modified form of it arises quite frequently when studying the ac-steady state behavior of transistor networks. The $\pi$-model is derived from the $T$-model as follows. First the $T$-model is redrawn as shown in Fig. 10-15a, which is equivalent to the network of Fig. 10-15b, since

$$\Delta i_B = -y_{22}\,\Delta v_{CB} - y_{11}\,\Delta v_{EB} - \alpha_F i_1 + \alpha_R i_2$$

$$\Delta i_C = y_{22}\,\Delta v_{EB} + \alpha_F i_1$$

and

$$\Delta i_E = y_{11}\,\Delta v_{EB} + \alpha_R i_2$$

in both networks. However, since

$$i_1 = -y_{11}\,\Delta v_{EB}$$

and

$$\alpha_F i_1 = -\alpha_F y_{11}\,\Delta v_{BE}$$

the network of Fig. 10-15b is equivalent to the network of Fig. 10-15c, which in turn is equivalent, as far as terminal behavior, to the network of Fig. 10-15d. Repeating these steps we can show that the network of Fig. 10-15d is equivalent to the network of Fig. 10-15e. Now define

$$v = -\Delta v_{EB}$$

so that

$$\Delta v_{CB} = \Delta v_{CE} - v$$

Then we have

$$-\alpha_R y_{22}\,\Delta v_{CB} = \alpha_R y_{22} v - \alpha_R y_{22}\,\Delta v_{CE}$$

and

$$-\alpha_F y_{11}\,\Delta v_{EB} = \alpha_F y_{11} v$$

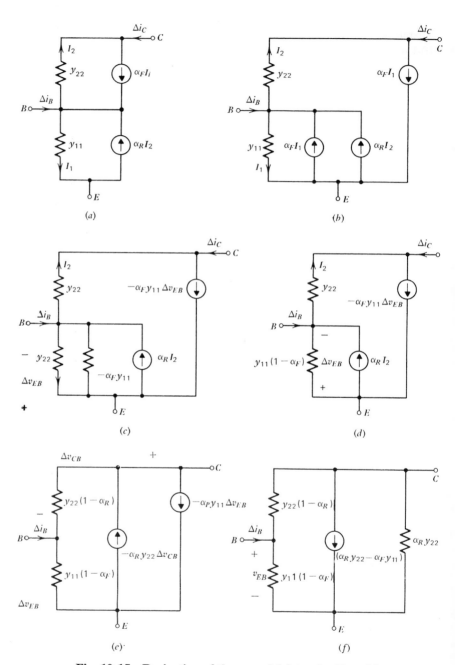

**Fig. 10-15  Derivation of the $\pi$-model from the $T$-model.**

Therefore, the network of Fig. 10-15e is equivalent to the network of Fig. 10-15f, which is the $\pi$-model with

$$G = y_{11}(1 - \alpha_F), \quad R = \frac{1}{[y_{22}(1 - \alpha_R)]}$$

$$g_m = \alpha_R y_{22} - \alpha_F y_{11} \quad \text{and} \quad r_0 = \alpha_R y_{22} \quad \square$$

## Example 10.5

The network of Fig. 10-16a is a differential amplifier and is similiar to the network of Fig. 10-10, which we analyzed in Example 10.1. The output of the differential amplifier is $v_0 = v_② - v_④$. Suppose we wished to determine the output for the set of inputs $\hat{v}_1 = -.1, -.05, -.01, -.005, -.004, -.003, -.002, -.001, .001, .002, .003, .004, .005, .01, .05,$ and $.1$ volts. Clearly we could determine this information by running the program of Fig. 10-11 for each of these voltages. The results of this approach are displayed in Table 10-1. (*Note.* These results were obtained using an initial guess of $V_{EB1} = V_{EB2} = -.5$ and $V_{CB1} = V_{CB2} = 10.0$, and GAMMA $= .5$.) Alternatively we could perform a nonlinear analysis for $\hat{v}_1 = 0.0$ to find the dc operating point and analyze the small signal network (shown in Fig. 10-16) under the excitations of $\Delta v_1 = -.1, -.05, -.01, -.005, -.004, -.003, -.002, -.001, .001, .002, .003, .004, .005, .01, .05,$ and $.1$ and add the results to the corresponding results of the original nonlinear (large signal) analysis. Note that since $v_② = v_④$ for $\hat{v}_1 = 0$, the output of the small signal circuit

$$\Delta v_o = \Delta v_② - \Delta v_④$$

should be approximately equal to the output of the large signal network. Of course we would expect that this approximation should get poorer as $\hat{v}_1$ gets larger. (Why?) The advantage of the second approach, of course, is that there is no iterative procedure involved with each analysis as there is using the first approach. Therefore, the amount of computer time required should be drastically reduced.

It is convenient in what follows to denote the small signal quantities $\Delta v_1$, $\Delta i_{DE1}$, etc by capital letters, that is,

$$\Delta v_1 \equiv V_1$$

$$\Delta i_{DE1} \equiv I_{DE1}$$

etc. (This type of notation is common in the literature.) The nodal equilibrium equations for the small-signal network are:

$$-\frac{1}{R_1} V_② - \frac{1}{R_2} V_④ + I_3 = -\left(\frac{1}{R_1} + \frac{1}{R_2}\right)\hat{V}_3$$

$$\left(\frac{1}{R_1} + G_{C1}\right)V_② - \alpha_F G_{E1}V_⑤ = \frac{1}{R_1}\hat{V}_3 - (\alpha_{F1}G_{E1} - G_{C1})\hat{V}_1$$

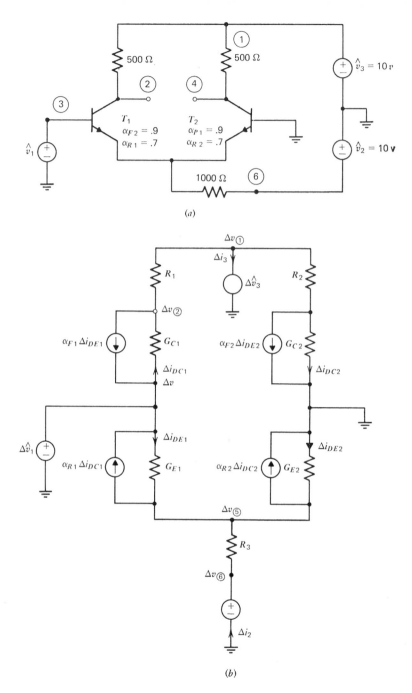

**Fig. 10-16** (*a*) The differential amplifier analyzed in Example 10.5. (*b*) The small-signal circuit associated with the nonlinear circuit of (*a*). The conductances $G_{E1}$, $G_{C1}$, $G_{E2}$, and $G_{C2}$ are equal to the associated conductances of the linearized transistor model at convergence.

338

Table 10–1. Results of large signal analysis in Example 10.5.

| $\hat{v}_1$ | $v_②$(V1OUT) | $v_④$(V2OUT) | $v_o = v_② - v_④$ |
|---|---|---|---|
| −.1 | 9.92 | 5.86 | 4.06 |
| −.05 | 9.50 | 6.29 | 3.21 |
| −.01 | 8.30 | 7.47 | .83 |
| −.005 | 8.099 | 7.678 | .421 |
| −.004 | 8.057 | 7.720 | .337 |
| −.003 | 8.015 | 7.762 | .253 |
| −.002 | 7.973 | 7.804 | .169 |
| −.001 | 7.9306 | 6.8461 | .0845 |
| .0 | 7.89 | 7.89 | .0 |
| .001 | 7.8459 | 7.9303 | −.0844 |
| .002 | 7.804 | 7.972 | −.168 |
| .003 | 7.761 | 8.014 | −.253 |
| .004 | 7.719 | 8.056 | −.0337 |
| .005 | 7.677 | 8.098 | −0.421 |
| .01 | 7.47 | 8.30 | −.83 |
| .05 | 6.27 | 9.49 | −3.22 |
| .1 | 5.82 | 9.97 | −4.10 |

$$G_{C1}(\alpha_{R1} - 1)V_② + G_{E1}(\alpha_{F1} - 1)V_⑤ + I_1$$

$$= \left[ G_{C1}(1 - \alpha_{R1}) + G_{E1}(1 - \alpha_{F1}) \right]\hat{V}_1$$

$$\left(\frac{1}{R_2} + G_{C2}\right)V_④ - \alpha_{F2}G_{E2}V_⑤ = \frac{1}{R_2}\hat{V}_3$$

$$- \alpha_{R1}G_{C1}V_② - \alpha_{R2}G_{C2}V_④ + \left(G_{E1} + G_{E2} + \frac{1}{R_3}\right)V_⑤$$

$$= -(\alpha_{R1}G_{C1} - G_{E1})\hat{V}_1 + \frac{1}{R_3}\hat{V}_2$$

$$-\frac{1}{R_3}V_⑤ + I_2 = -\frac{1}{R_3}\hat{V}_2$$

where we have used the fact that

$$V_① = \hat{V}_3$$

$$V_③ = \hat{V}_1$$

and

$$V_⑥ = \hat{V}_2$$

Observe the similarity between these equations and the nodal equilibrium equations obtained for the linearized network in Example 10.1.

We wish to solve this set of equations for $\hat{V}_2 = \hat{V}_3 = 0$, and $\hat{V}_1$ equal to each of the $\Delta\hat{v}_1$ given above. A program that can be used for this purpose is shown in Fig. 10-17. Observe the differences between this program and the program used to solve the original nonlinear network given in Fig. 10-11. In particular, the present program has no interation loop in the main program; the transistor subroutine is used to supply only the small signal model

```
C      CHAPTER 10 EXAMPLE 10.5
C
C      DIMENSIONS
C
       IMPLICIT REAL*8 (A-H,O-Z)
       DIMENSION Y(6,6),CIN(6)
C
C      READ IN ELEMENT VALUES, TRANSISTOR PARAMETERS, AND
C      DC OPERATING POINTS: VEB1, VCB1, VEB2, VCB2
C      NOTE: WHEN K=1 ONLY NEW SMALL SIGNAL INPUT IS READ
C
       K=0
     4 CALL READ(V1,V2,V3,R1,R2,R3,ALF1,ALFR1,ALF2,ALFR2,VEB1,VCB1,
      +VEB2,VCB2,K)
       K=1
C
C      CALCULATE SMALL SIGNAL TRANSISTOR PARAMTERS
C
     1 CALL TRAN(GE1,GC1,CIE1,CIC1,VEB1,VCB1)
       CALL TRAN(GE2,GC2,CIE2,CIC2,VEB2,VCB2)
C
C      MODIFY VALUES RETURNED FROM TRAN FOR SMALL SIGNAL CASE
C
       CIE1=0
       CIC1=0
       CIE2=0
       CIC2=0
C
C      SET UP AND SOLVE NODE EQUATIONS
C
       CALL NODEQN(V1,V2,V3,R1,R2,R3,ALF1,ALFR1,ALF2,ALFR2,GE1,GC1,CIE1,
      +CIC1,GE2,GC2,CIE2,CIC2,Y,CIN)
       CALL GAUSS(Y,CIN,6,6)
C
C      CALCULATE SMALL SIGNAL OUTPUT VOLTAGES AND
C      TRANSISTOR JUNCTION VOLTAGES
C
       V1OUT=CIN(1)
       V2OUT=CIN(2)
       VEB1T=CIN(3)-V1
       VCB1T=CIN(1)-V1
       VEB2T=CIN(3)
       VCB2T=CIN(2)
C
C      PRINT SMALL SIGNAL SOLUTION
C
     3 CALL PRINT(V1OUT,V2OUT,VEB1T,VCB1T,VEB2T,VCB2T)
       GO TO 4
       END
```

**Fig. 10-17   Program used for small-signal analysis in Example 10.5.**

```fortran
      SUBROUTINE READ(V1,V2,V3,R1,R2,R3,ALF1,ALFR1,ALF2,ALFR2,VEB1,VCB1,
     +VEB2,VCB2,K)
      IMPLICIT REAL*8 (A-H,O-Z)
      IF(K.EQ.1) GO TO 4
      READ(5,1) V1,V2,V3,R1,R2,R3,ALF1,ALFR1,ALF2,ALFR2,VEB1,VCB1,VEB2,
     +VCB2
      GO TO 5
    4 READ(5,1) V1
    5 PRINT 2,V1,VEB1,VEB2,VCB1,VCB2
    1 FORMAT(6G12.5)
    2 FORMAT('ORESULTS FOR EX. 10.5 WITH INPUT VOLTAGE V1=',G12.5//
     +' DC OPERATING POINT',/8X,'VEB1=',G12.5,5X,'VEB2=',G12.5,5X,
     +'VCB1=',G12.5,5X,'VCB2=',G12.5)
      RETURN
      END




      SUBROUTINE NODEQN(V1,V2,V3,R1,R2,R3,ALF1,ALFR1,ALF2,ALFR2,GE1,GC1,
     +CIE1,CIC1,GE2,GC2,CIE2,CIC2,Y,CIN)
      IMPLICIT REAL*8 (A-H,O-Z)
      DIMENSION Y(6,6),CIN(6)
C
C     INITIALIZE COEFFICIENT ARRAYS
C
      DO 1 J=1,6
      CIN(J)=0.D0
      DO 1 K=1,6
    1 Y(J,K)=0.D0
C
C     SET UP NODAL EQUILIBRIUM COEFFICIENTS
C
      Y(1,1)=-1./R1
      Y(1,2)=-1./R2
      Y(1,6)=1.
      CIN(1)=(-(1./R1)-(1./R2))*V3
      Y(2,1)=(1./R1)+GC1
      Y(2,3)=-ALF1*GE1
      CIN(2)=-(ALF1*CIE1)+CIC1+(1./R1)*V3-(ALF1*GE1-GC1)*V1
      Y(3,1)=GC1*(ALFR1-1.)
      Y(3,3)=GE1*(ALF1-1.)
      Y(3,4)=1.
      CIN(3)=CIC1*(ALFR1-1.)+CIE1*(ALF1-1.)-(GC1*(1.-ALFR1)+GE1*(1.-
     +ALF1))*V1
      Y(4,2)=(1./R2)+GC2
      Y(4,3)=-ALF2*GE2
      CIN(4)=-(ALF2*CIE2)+(CIC2)+(1./R2)*V3
      Y(5,1)=-ALFR1*GC1
      Y(5,2)=-ALFR2*GC2
      Y(5,3)=GE1+GE2+(1./R3)
      CIN(5)=CIE1+CIE2-(ALFR1*CIC1)-(ALFR2*CIC2)-(ALFR1*GC1-GE1)*V1
     ++(1./R3)*V2
      Y(6,3)=-1./R3
      Y(6,5)=1.
      CIN(6)=-(1./R3)*V2
      RETURN
      END
```

**Fig. 10-17 (*Continued*)**

```fortran
      SUBROUTINE TRAN(GE,GC,CIE,CIC,VEB,VCB)
      IMPLICIT REAL*8 (A-H,O-Z)
      REAL*8 IES,ICS
      DATA IES/1.D-13/,ICS/1.D-13/,VT/.025/
C
C     PLACE LOWER BOUND ON JUNCTION VOLTAGES
C
      VEB=DMAX1(VEB,-.8D0)
      VCB=DMAX1(VCB,-.8D0)
C
C     TEST FOR NORMAL, CUTOFF, SATURATION OR REVERSE MODE OF OPERATION
C     AND SET UP APPROPRIATE MODEL
C
      IF(VEB.LT.0..AND.VCB.GT.0.) GO TO 1
      IF(VEB.GT.0..AND.VCB.GT.0.) GO TO 2
      IF(VEB.LT.0..AND.VCB.LT.0.) GO TO 3
C
C     REVERSE MODE
C
      CIE=0.
      CIC=ICS*((DEXP(-VCB/VT)*(1.+(VCB/VT)))-1.)
      GE=IES/VT
      GC=ICS*DEXP(-VCB/VT)/VT
      RETURN
C
C     NORMAL MODE
C
    1 CIE=IES*((DEXP(-VEB/VT)*(1.+(VEB/VT)))-1.)
      CIC=0.
      GE=IES*DEXP(-VEB/VT)/VT
      GC=ICS/VT
      RETURN
C
C     CUTOFF MODE
C
    2 CIE=0.
      CIC=0.
      GE=IES/VT
      GC=ICS/VT
      RETURN
C
C     SATURATION MODE
C
    3 CIE=IES*((DEXP(-VEB/VT)*(1.+(VEB/VT)))-1.)
      CIC=ICS*((DEXP(-VCB/VT)*(1.+(VCB/VT)))-1.)
      GC=ICS*DEXP(-VCB/VT)/VT
      GE=IES*DEXP(-VEB/VT)/VT
      RETURN
      END

      SUBROUTINE PRINT(V1OUT,V2OUT,VEB1T,VCB1T,VEB2T,VCB2T)
      IMPLICIT REAL*8 (A-H,O-Z)
      PRINT 3
    3 FORMAT(/' V1OUT',12X,'V2OUT',13X,'VEB1',13X,'VCB1',
     +13X,'VEB2',13X,'VCB2')
    1 PRINT 2,V1OUT,V2OUT,VEB1T,VCB1T,VEB2T,VCB2T
    2 FORMAT(1X,G12.5,5X,G12.5,5X,G12.5,5X,G12.5,5X,G12.5,5X,
     +G12.5)
      RETURN
      END
```

**Fig. 10-17 (*Continued*)**

Table 10-2. Results of small signal analysis in Example 10.5

| $\hat{V}_1 = \hat{v}_1 - .0$ | $V_② = v_② - 7.89$ | $V_④ = v_④ - 7.88$ | $V_0 = V_② - V_④$ | $\dfrac{V_0 - v_0}{v_0} \times 100\%$ |
|---|---|---|---|---|
| −.1 | 4.24 | −4.24 | 8.45 | 180% |
| −.05 | 2.12 | −2.11 | 4.23 | 31% |
| −.01 | .424 | −.421 | .845 | 1.8% |
| −.005 | .212 | −.211 | .423 | .47% |
| −.004 | .169 | −.169 | .338 | .3% |
| −.003 | .127 | −.126 | .253 | 0% |
| −.002 | .0847 | −.0843 | .169 | 0% |
| −.001 | .0424 | −.0421 | .0845 | 0% |
| .0 | .0 | .0 | .0 | 0% |
| .001 | −.0424 | .0421 | −.0845 | .1% |
| .002 | −.0847 | .0843 | −.169 | .6% |
| .003 | −.127 | .126 | −.253 | 0% |
| .004 | −.169 | .169 | −.338 | .3% |
| .005 | −.212 | .211 | −.423 | .47% |
| .01 | −.424 | .421 | −.845 | 1.8% |
| .05 | −2.12 | 2.11 | −4.23 | 31% |
| .1 | −4.24 | 4.21 | −8.45 | 106% |

parameters at a given dc operating point, the dc sources are set equal to zero; however, the subroutine NODEQN remains the same. The results of running this program for the different values of $\hat{V}_1$ are given in Table 10-2. The percent deviate between these approximate results and the actual results listed in Table 10-1, that is

$$\frac{V_0 - v_0}{v_0} \times 100\%$$

is also given. Observe how the deviation between the approximate and true results increases as $|\hat{V}_1|$ increases. If we would accept an error less than 5 percent as good engineering judgment then the small signal analysis would be valid for all $\Delta \hat{v}_1 < .02$ volts.  □

## 10-7  SUMMARY

In this Chapter we discussed the dc analysis of networks that contain transistors. The equations that describe a transistor can be *modeled* by a nonlinear two-port that is made up of ideal circuit elements. Analysis then proceeds in the usual fashion by either linearizing the nonlinear two-port, or, alternatively, linearizing the nonlinear two-terminal elements within the model.

In order to speed up convergence of the iterative analysis procedure we found it convenient to specialize the *Ebers–Moll* transistor model to each of the four possible regions of transistor operation: normal, cutoff, saturation, and inverse. Moreover, to prevent oscillation a *damped* Newton–Raphson iteration scheme was discussed.

Finally the dc *small-signal analysis* of networks that contain transistors was described and a dc small-signal transistor model derived.

## PROBLEMS

**10.1**    Use ideal elements to model the two-ports described by the following equations

(a) $i_1 = 3v_1 + 4(e^{v_2/v_T} - 1)$

   $i_2 = v_1 + (e^{v_2/v_T} - 1)$

(b) $v_1 = i_1 - 10^{-6}(e^{-v_2/v_T} - 1)$

   $i_2 = 20i_1 + 10^{-6}(e^{-v_2/v_T} - 1)$

**10.2**    Find linearized equivalent models for each of the following nonlinear two-ports:

(a) $i_1 = -e^{v_1} + e^{-v_2}$

   $i_2 = v_1^2 + v_2$

(b) $i_1 = -e^{-v_1} + i_2 v_1$

   $v_2 = v_1^2 + i_2^2 + v_1 i_2$

(c) $v_1 = \cos i_1 + i_1 i_2 + i_1^2$

   $v_2 = e^{i_1} + i_2$

**10.3**    Find the small signal models associated with each of the nonlinear elements in Prob. 10.2.

**10.4**    In Example 10.1 the network of Fig. 10-10 was analyzed using the program of Fig. 10-11. Suppose now that a 500 Ω resistor is connected between nodes 2 and 4 as shown in Fig. P10.4. Modify the subroutine NODEQN in the program of Fig. 10-11 and analyze this modified network for $\hat{v}_1 = +1$ and $\hat{v}_1 = -1$. (*Note.* Make changes directly in NODEQN using the 500 Ω value—do not bother modifying the subroutine READ, etc.)

**10.5**    The circuit of Fig. P10.5 is called a Darlington pair and was analyzed using a very simple transistor model in Prob. 8.6. (a) Replace the transistors by their linearized models and write the nodal equilibrium equations for this network. (b) Modify the

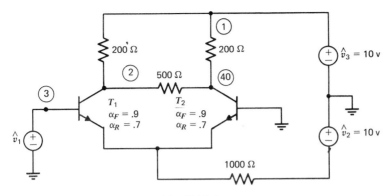

**Fig. P10.4**

program of Fig. 10-11 so that it can be used to analyze this network. (*Note.* Only the subroutine NODEQN needs to be altered.) Using this program analyze the network for $\hat{v}_1 = 1$ v.

**10.6**  Using the program written in Prob. 10.5 to determine the input impedence and current gain of the Darlington pair in Fig. P10.5.

**10.7**  The network of Fig. P10.7 is called a Schmitt trigger. (a) Replace the transistors with their linearized models and write the nodal equilibrium equations for the linearized network. (b) Modify the program of Fig. 10-11 so that it can be used to analyze this network.

**10.8**  We can study the operation of a Schmitt trigger by using the program written in Prob. 10.7. Set $\hat{v}_1 = -10$ v and analyze the network. Observe that transistor $T1$ is "off," since $v_2 \simeq v_1$, and $T_2$ is "on," since $v_4 \simeq v_5$. Increment $\hat{v}_1$ several times and reanalyze the network. Observe that when $v_1 \simeq v_5$, the circuit "triggers" and $T_1$ turns on, that is, $v_2 \simeq v_5$, and $T_2$ turns off, that is, $v_4 \simeq v_1$.

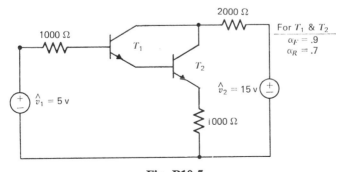

**Fig. P10.5**

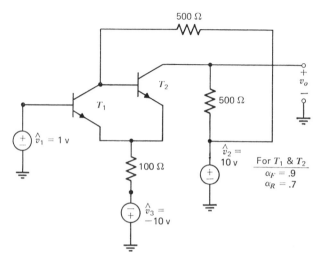

**Fig. P10.7**

**10.9**   An alternative description of transistor behavior is given by the so-called *transport model* equations.

$$i_E = -\left(1 + \frac{1}{\beta_F}\right)I_{SS}(e^{-v_{EB}/v_T} - 1) + I_{SS}(e^{-v_{CB}/v_T} - 1)$$

$$i_C = I_{SS}(e^{-v_{EB}/v_T} - 1) - \left(1 + \frac{1}{\beta_R}\right)I_{SS}(e^{-v_{CB}/v_T} - 1)$$

(a) Derive a model that is comprised of ideal diodes and controlled sources that can be used to represent these equations.

(b) Find the linearized transport model that can be used in an iterative analysis program.

(c) Find the small signal transport model.

**10.10**   Using the linearized transport model derived in Prob. 10.9b, write a subroutine similiar to SUBROUTINE TRAN of Fig. 10-11a, which can be used in an iterative analysis program.

**10.11**   Use the programs of Fig. 10-10 and 10-17 to verify the results of Example 10.5 and Tables 10-1 and 10-2.

**10.12**   (a) Use the program of Prob. 10.4 to find the $v_{②}$ and $v_{④}$ for the following values of $\hat{v}_1$: $-.1$, $-.01$, $-.005$, $-.001$, $0.0$, $.001$, $.005$, $.01$, and $.1$.

(b) Modify this program to perform a small signal analysis and find $\Delta v_{②}$ and $\Delta v_{④}$ for the values $\Delta \hat{v}_1 = -.1$, $-.01$, $-.005$, $-.001$, $.001$, $.005$, $.01$, and $.1$.

   (c) Compute the approximate large signal values for $v_②$ and $v_④$ for each of the inputs $\hat{v}_1$ by adding the large signal value obtained in (a) for $\hat{v}_1 = 0.0$ to the values obtained in (b).

   (d) Compare the approximate results of (c) with the exact results of (a).

**10.13**    Modify the program of Prob. 10.5 to perform small signal analysis and repeat Prob. 10.12.

# CHAPTER 11

# Analysis of First-Order Networks

At this point the only elements we have considered have been those whose branch relationships have been algebraic equations. Such an element is termed *instantaneous* or *memoryless* because a sudden change in its branch voltage corresponds to a similar change in its branch voltage. In this chapter we introduce two elements, the capacitor and inductor, which have the ability to store energy and whose branch relationships are differential equations.

## 11-1 DEFINITION OF ELEMENTS

The linear *capacitor* is an element described by the branch relationship

$$i(t) = \frac{d}{dt} [C(t)v(t)]$$

and is denoted by the symbol of Fig. 11-1a. The element $C$ is the *capacitance* in farads. For the present we assume the capacitor to be *time-invariant*, that is, its value is independent of time, so the branch relationship becomes

$$i(t) = C \frac{d}{dt} v(t)$$

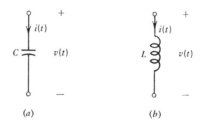

**Fig. 11-1** (*a*) **The circuit symbol for a capacitor.** (*b*) **The circuit symbol for an inductor.**

An alternate form of this branch relationship emerges upon integration of both sides of this expression

$$v(t) = v(t_0) + \frac{1}{C} \int_{t_0}^{t} i(\tau) \, d\tau$$

where $t_0$ denotes some *initial time*. The voltage across the capacitor at $t_0$ is $v(t_0)$. More precisely, we may relate voltage across a capacitor to the charge, $q$, (measured in coulombs) stored on the capacitor:

$$q(t) = Cv(t)$$

Then

$$v(t_0) = \frac{1}{C} q(t_0)$$

where $q(t_0)$ represents the initial stored charge of a capacitor. Of course,

$$i(t) = \frac{d}{dt} q(t)$$

To better motivate the concept of initial stored charge, consider a capacitor that was built at the time of creation ($t = -\infty$) that has no stored charge. Then

$$v(t) = \frac{1}{C} \int_{-\infty}^{t} i(\tau) \, d\tau \qquad \text{for all } t > -\infty$$

For any time $t > t_0$ we can write

$$v(t) = \frac{1}{C} \int_{-\infty}^{t} i(\tau) \, d\tau$$

$$= \frac{1}{C} \int_{-\infty}^{t_0} i(\tau) \, d\tau + \frac{1}{C} \int_{t_0}^{t} i(\tau) \, d\tau$$

Thus the initial stored charge is

$$q(t_0) = \int_{-\infty}^{t_0} i(\tau) \, d\tau$$

and the initial voltage is

$$v(t_0) = \frac{1}{C} \int_{-\infty}^{t} i(\tau) \, d\tau$$

We will usually let the initial time be zero, $t_0 = 0$, so that

$$v(t) = v(0) + \frac{1}{C} \int_{0}^{t} i(\tau) \, d\tau$$

The linear *inductor* is an element described by the branch relationship

$$v(t) = \frac{d}{dt} [L(t)i(t)]$$

where $L$ is the *inductance* in henrys. The inductor symbol is shown in Fig. 11-1$b$. For the present only time-invariant inductors are considered so the branch relationships becomes

$$v(t) = L \frac{d}{dt} i(t)$$

An alternate characterization is obtained by integrating both sides of this expression:

$$i(t) = i(t_0) + \frac{1}{L} \int_{t_0}^{t} v(\tau) \, d\tau$$

where $i(t_0)$ is the initial current and can be related to the initial stored flux. Observe that since flux $\phi$ and current are related by

$$\phi(t) = Li(t)$$

then

$$i(t_0) = \frac{1}{L} \phi(t_0)$$

We could go through an argument similar to that given for the capacitor to show that if the inductor was created at time $t = -\infty$ with no initial stored flux, then

$$i(t_0) = \frac{1}{L} \int_{-\infty}^{t_0} v(\tau) \, d\tau$$

If $t_0 = 0$ then the inductance branch relationship is

$$i(t) = i(0) + \frac{1}{L} \int_{0}^{t} v(\tau) \, d\tau$$

In the sequel we will always assume the initial time to be zero ($t_0 = 0$) unless otherwise specified.

There are two situations concerning the behavior of capacitors and inductors that deserve mention. First, observe that if the initial stored energy of the capacitor is zero (we say the capacitor is initially *relaxed*) then it behaves initially (at time $t_0$) as if it was a zero-valued voltage source or short circuit, while if the initial stored energy of the inductor is zero it behaves initially as if it was a zero-valued current source or an open circuit. The second situation arises when these elements experience constant, or dc, excitations. Observe that if the voltage across a capacitor remains constant, the current becomes zero so that it behaves as if it were a zero-valued current source or open circuit while if the current through an inductor remains constant, the voltage across it becomes zero so that it behaves as if it were a zero-valued voltage source, or a short circuit.

## 11-2 THE FIRST-ORDER NETWORK

In this section we consider the analysis of several networks that contain a single energy storage element. Such networks will usually be characterized by a first-order differential equation and are called *first-order networks*.

### Example 11.1

Consider first the $RC$ (for resistor-capacitor) network of Fig. 11-2. The single nodal equilibrium equation is obtained from the KCL node equation

$$i_1(t) + i_2(t) + i_3(t) = 0$$

the branch relationships

$$i_1(t) = \hat{i}_1(t)$$

$$i_2(t) = \frac{1}{R_2} v_2(t)$$

$$i_3(t) = C_3 \frac{dv_3}{dt}$$

and the KVL equations

$$v_1(t) = v_{①}(t)$$
$$v_2(t) = v_{①}(t)$$
$$v_3(t) = v_{①}(t)$$

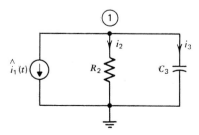

**Fig. 11-2   A simple $RC$ network.**

In addition, the initial capacitance voltage is $v_3(0)$. The nodal equilibrium equation is the first-order differential equation

$$\frac{1}{R_2} v_① + C_3 \frac{dv_①}{dt} = -\hat{\imath}_1$$

which can also be written as

$$\frac{dv_①}{dt} = -\frac{1}{R_2 C_3} v_① - \frac{1}{C_3} \hat{\imath}_1$$

To determine the unique solution of this differential equation, the initial condition $v_①(0)$ must be known. (We take $t_0 = 0$.) But

$$v_①(t) = v_3(t)$$

so that

$$v_①(0) = v_3(0)$$

that is, the initial condition is obtained from the initial capacitance voltage. (We defer discussion of the solution of first-order differential equations to the next section.)    □

## Example 11.2

Now consider the *RL* (for resistor-inductor) network of Fig. 11-3a. Suppose we use mesh analysis. The KVL mesh equation is

$$v_1(t) - v_2(t) - v_3(t) = 0$$

The KCL equations in terms of the mesh current $i_▯(t)$ are

$$i_1(t) = -i_▯(t)$$
$$i_2(t) = i_▯(t)$$

and

$$i_3(t) = i_▯(t)$$

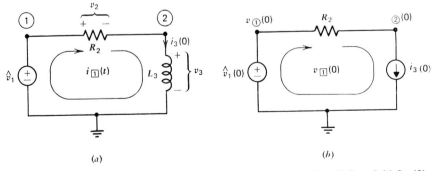

(a)    (b)

**Fig. 11-3**    (a) A simple *RL* network studied in Examples 11.2 and 11.3; (b) The initial condition network associated with (a).

and the branch relationships are

$$v_1(t) = \hat{v}_1(t)$$

$$v_2(t) = R_2 i_2(t)$$

and

$$v_3(t) = L_3 \frac{di_3(t)}{dt}$$

Substitution of the branch relationships into the KVL mesh equation and the KCL equations into this result yields the first-order differential equation

$$\hat{v}_1(t) - R_2 i_{\boxed{1}}(t) - L_3 \frac{di_{\boxed{1}}(t)}{dt} = 0$$

or

$$\frac{di_{\boxed{1}}(t)}{dt} = -\frac{R_2}{L_3} i_{\boxed{1}}(t) + \frac{1}{L_3} \hat{v}_1(t)$$

The initial condition required for solution of this differential equation is $i_{\boxed{1}}(0)$, which is easily determined from the known initial inductance current

$$i_{\boxed{1}}(0) = i_3(0)$$ □

## Example 11.3

It is also worthwhile to consider the nodal analysis of the network of Fig. 11-3a. To this end we express the KCL node equations as

$$i_1(t) + i_2(t) = 0$$

and

$$-i_2(t) + i_3(t) = 0$$

the KVL equations as:

$$v_1(t) = v_{①}(t)$$

$$v_2(t) = v_{①}(t) - v_{②}(t)$$

and

$$v_3(t) = v_{②}(t)$$

and the branch relationships as

$$v_1(t) = \hat{v}_1(t)$$

$$i_2(t) = v_2(t)/R$$

and

$$i_3(t) = i_3(0) + \int_0^t \frac{1}{L_3} v_3(\tau)\, d\tau$$

The nodal equilibrium equation for node ② becomes

$$\frac{1}{R_2} v_{②}(t) + i_3(0) + \int_0^t \frac{1}{L_3} v_{②}(\tau)\, d\tau = \frac{1}{R_2} \hat{v}_1(t)$$

We seek a differential equation. Upon differentiation, this equation becomes

$$\frac{1}{R_2} \frac{dv_{②}(t)}{dt} + \frac{1}{L_3} v_{②}(t) = \frac{1}{R_2} \frac{d\hat{v}_1(t)}{dt}$$

or

$$\frac{dv_{②}(t)}{dt} = -\frac{R_2}{L_3} v_{②}(t) + \frac{d\hat{v}_1(t)}{dt}$$

Observe that $i_3(0)$ is a constant so that $di_3(0)/dt = 0$. Furthermore, since $\hat{v}_1(t)$ is a known function of time, $d\hat{v}_1(t)/dt$ is also known and may be thought of as merely being another forcing function.

The initial condition required for solution of this equation is $v_{②}(0)$. To find $v_{②}(0)$ we consider the *initial condition network* of Fig. 11-3b. The initial condition network is a dc resistive network that is obtained from the original network by letting $t = 0$ and replacing the inductor by a current source whose value is the initial inductance current $i_3(0)$. (If $i_3(0) = 0$, then the inductor is replaced by a zero-valued current source, that is, an open circuit.) This network is easily solved for $v_{②}(0)$:

$$v_{②}(0) = \hat{v}_1(0) - R_2 i_3(0)$$

Thus the initial condition needed for the solution of the differential equation that arises from nodal analysis is found in terms of the initial value of the input voltage source and the initial inductance current. □

### Example 11.4

The mesh equation of Fig. 11-4a is

$$\hat{v}_1(t) - R_2 i_{①}(t) - v_3(0) - \int_0^t \frac{1}{C_3} i_{①}(\tau)\, d\tau = 0$$

(a)    (b)

**Fig. 11-4** (*a*) **Network analyzed in Example 11.4.** (*b*) **The initial condition network.**

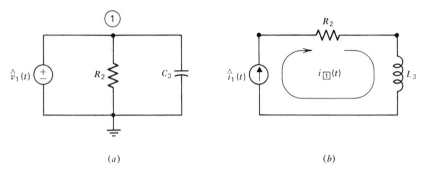

(a)                                                        (b)

**Fig. 11-5  Two networks that contain a single energy storage element but which are not first-order networks.**

or, on differentiation

$$\frac{di_{\boxed{1}}(t)}{dt} = -\frac{1}{R_2 C_3} i_{\boxed{1}}(t) + \frac{1}{R_2} \frac{d\hat{v}_1(t)}{dt}$$

The initial condition required for solution of this differential equation is $i_{\boxed{1}}(0)$ which can be found from the initial condition network of Fig. 11-4b. Observe that the capacitor has been replaced by a voltage source whose value equals the initial capacitance voltage. (If the initial capacitance voltage was zero, a zero-valued voltage source, or short circuit, would be used.) We find that

$$i_{\boxed{1}}(0^+) = \frac{(\hat{v}_1(0) - v_3(0^+))}{R_2} \qquad \Box$$

Before we leave this section, we should observe that not all networks which contain a single energy storage element are first-order networks. Consider the networks of Fig. 11-5. In the first network, the capacitor voltage is specified by the voltage source and in the second network the inductor current is specified by the current source. Meaningful first-order differential equations cannot be written in either case. In the first network this situation arises because the voltage source and capacitor form a loop, so that the capacitor voltage is not independent. In the second network, the current source and inductor form a cutset, so that the inductor current is not independent.

## 11-3  SOLUTION OF FIRST-ORDER DIFFERENTIAL EQUATIONS

The first-order differential equations that resulted in Examples 11.1, 11.2, 11.3, and 11.4 are of the form of the general first-order differential equation

(11.1)                          $$\frac{dx(t)}{dt} = ax(t) + bw(t)$$

where $x(t)$ is either a node voltage or mesh current, $a$ and $b$ are constants related to the network elements, and $w(t)$ is a forcing function related to the input. By finding the solution of the general first-order differential equation (11.1) we also find the solution of the differential equations of Examples 11.1 to 11.4. In this section we present a general method for solution of (11.1), which is not dependent on the particular form of the forcing function $w(t)$. Many other methods of solution exist; those students interested in these methods should consult the appropriate references.

The first step in finding the solution of the first-order differential equation

$$\frac{dx(t)}{dt} = ax(t) + bw(t)$$

is to find the solution to the *homogeneous equation*. The homogeneous equation results when the forcing function is set to zero:

$$\frac{dx(t)}{dt} = ax(t)$$

A solution of the homogeneous equation is easily found by the method of *separation of variables*.[1] First recognize that the homogeneous equation can be written as

$$\frac{dx_h(t)}{x_h(t)} = a \, dt$$

The subscript $h$ is used to denote the homogeneous solution. On integration from $\tau$ to $t$ we have

$$\ln \frac{x_h(t)}{x_h(\tau)} = \int_\tau^t a \, d\xi$$

or

$$x_h(t) = \exp \left( \int_\tau^t a \, d\xi \right) x_h(\tau)$$

Given an initial condition at $t_0$, this equation becomes

$$x_h(t) = e^{a(t-t_0)} x_h(t_0)$$

The second step is to solve for the particular solution. The method of *variation of parameters*,[1] although unmotivated is to be employed. Since $x_h(t)$ is the homogeneous solution, it satisfies the homogeneous equation

$$\frac{dx_h(t)}{dt} = ax_h(t)$$

and, at $t_0$ it equals the initial condition

$$x_h(t_0) = x_0$$

[1] I. S. Sokolnikoff and R. M. Redheffer, *Mathematics of Physics and Modern Engineering* McGraw-Hill, 1958, pp. 17–29.

It is relatively simple to show that the solution to the homogeneous equation for a given initial condition is unique provided that

$$a < \infty$$

and, moreover, that if the solution vanishes at any finite time it must vanish for all finite time. These properties ensure that $x_h(t)$ is nonzero for all finite time provided only that $x_h(t_c)$ is nonzero. We try, as the particular solution,

$$x_p(t) = x_h(t)\eta(t)$$

where $\eta(t)$ is at this point, an arbitrary function of time, and $x_h(t)$ is a *non-trivial* (nonzero) solution of the homogeneous equation. The subscript $p$ denotes the particular solution. Substitution of the particular solution into the original differential equation yields

$$x_h(t)\dot{\eta}(t) = bw(t)$$

or

$$\dot{\eta}(t) = [x_h(t)]^{-1}bw(t)$$

Observe that $[x_h(t)]^{-1}$ exists for all finite time since $x_h(t)$ is a *nontrivial solution* to the homogeneous equation. [i.e., $x_h(t) \neq 0$, for all time]. On integration and multiplication by $x_h(t)$ this expression becomes

$$x_p(t) = x_h(t)\eta(t) = \int_{t_0}^t x_h(t)[x_h(\tau)]^{-1}bw(\tau)\,d\tau$$

[*Note.* We have now let $\eta(t_0) = 0$]. Since $x_h(t)$ is the solution to the homogeneous equation,

$$x_h(t) = e^{a(t-\tau)}x_h(\tau)$$

the particular solution may be written as

$$x_p(t) = \int_{t_0}^t e^{a(t-\tau)}bw(\tau)\,d\tau$$

The complete solution is the sum of the homogeneous and particular solutions:

(11.2)     $$x(t) = e^{a(t-t_0)}x(t_0) + \int_{t_0}^t e^{a(t-\tau)}bw(\tau)\,d\tau$$

where $t \geq t_0$.

Observe that if the input, $w(t)$, is set to zero for all time, then the response, $x(t)$ is due solely to the initial condition and equals the homogeneous response:

$$x(t) = e^{a(t-t_0)}x(t_0)$$

This response is also termed the *zero-input response*. If the initial condition, $x(t_0)$, is set to zero, the response $x(t)$ is due solely to the input and equals the particular solution:

$$x(t) = \int_{t_0}^t e^{a(t-\tau)}b(\tau)w(\tau)\,d\tau$$

Since the initial condition is sometimes referred to as the *initial state* of the network, this response is also called the *zero-state response*. Thus the total, or *complete*, response equals the sum of the *zero-input response* and the *zero-state response*.

## Example 11.5

The nodal equilibrium equation for the network of Fig. 11-2 was found in Example 11.1 to be

$$\frac{dv_①}{dt} = -\frac{1}{R_2C_3}v_① - \frac{1}{C_3}\hat{\imath}_1$$

with $v_①(0) = v_3(0)$.
   The homogeneous equation is

$$\frac{dv_①}{dt} = -\frac{1}{R_2C_3}v_①$$

and the homogeneous solution is

$$v_{①h}(t) = e^{-(1/R_2C_3)t}v_①(0)$$

The particular solution is

$$v_{①p}(t) = -\int_0^t e^{-(1/R_2C_3)(t-\tau)}\frac{1}{C_3}\hat{\imath}_1(\tau)\,d\tau$$

Therefore, the complete solution is

$$v_①(t) = e^{-(1/R_2C_3)t}v_3(0) - \frac{1}{C_3}\int_0^t e^{(-1/R_2C_3)(t-\tau)}\hat{\imath}_1(\tau)\,d\tau$$

for $t \geq 0$ where $v_3(0)$ has been substituted for $v_1(0)$.
   In the previous example the quantity $R_2C_3$ has units of time since

$$e^{(-1/R_2C_3)t}$$

must be dimensionless. The quantity $R_2C_3$ is called the *time-constant* of the network and is a measure of how fast the zero-input response decays. To see this let $\tau = R_2C_3$, then the zero-input response is

$$v_1(t) = e^{-t/\tau}v_3(0)$$

When $t = \tau$,

$$v_1(\tau) = e^{-1}v_3(0)$$

$$\simeq .37v_3(0)$$

In other words, at the end of one time constant the voltage decreases to about 37 percent of its initial value. Also observe that

$$v_1(2\tau) = e^{-2}v_3(0)$$

$$= .14v_3(0)$$

$$v_1(3\tau) = e^{-3}v_3(0)$$

$$= .05v_3(0)$$

and

$$v_1(4\tau) = e^{-4}v_3(0)$$

$$= .018v_4(0)$$

so that after about four time constants, the voltage can be thought of as zero, although strictly speaking, zero is reached only when $t$ goes to infinity.

The time constant is a useful concept when comparing one response with another. Suppose $R = 10\ \Omega$ and $C = 10$ pF, then $\tau = 10^{-10}$ seconds. Then the response would decay to 37 percent of its initial value after $10^{-10}$ seconds. However if $R = 100$ K $\Omega$ and $C = 10$ pF, then $\tau = 1$ second and the response would decrease to 37 percent of its initial value after 1 second. We can see, therefore, that the larger the time constant the slower the decay.

Since the time constant is a property of the network itself, that is, it is not dependent on any input, it is called a *natural property*. Furthermore, since frequency is the reciprocal of time, $1/\tau$ is also called the *natural frequency* of the network.

**Example 11.6**

The solution of the nodal equilibrium equation obtained in Example 11.3 for the network of Fig. 11-3,

$$\frac{dv_{②}(t)}{dt} = \frac{-R_2}{L_3}v_{②}(t) + \frac{d\hat{v}_1(t)}{dt}$$

where $v_{②}(0) = \hat{v}_1(0) - R_2 i_3(0)$ is

$$v_{②}(t) = [e^{-R_2 t/L_3}](\hat{v}_1(0) - R_2 i_3(0)) + \int_0^t e^{-R_2(t-\tau)/L_3}\frac{d\hat{v}_1(\tau)}{d\tau}\,d\tau$$

As will be seen, it may be undesirable to consider $d\hat{v}_1(t)/dt$ as an input. For such a situation, we can integrate the above expression by parts to yield

$$v_{②}(t) = [e^{-R_2 t/L_3}][\hat{v}_1(0) - R_2 i_3(0)] + [e^{-R_2(t-\tau)/L_3}\hat{v}_1(\tau)]_0^t$$

$$- \int_0^t \frac{R_2}{L_3} e^{-R_2(t-\tau)/L_3}\hat{v}_1(\tau)\,d\tau$$

$$= e^{-R_2 t/L_3}[-R_2 i_3(0)] + \hat{v}_1(t) - \int_0^t \frac{R_2}{L_3} e^{-R_2(t-\tau)/L_3}\hat{v}_1(\tau)\,d\tau$$

The quantity $L_3/R_2$ has units of time and is called the *time-constant* of this network. Like the time-constant of the network of Fig. 11-2, it measures the rate of decay of the zero-input response.  ☐

### Example 11.7

Consider the network of Fig. 11-6a. The two nodal equilibrium equations are

$$node \ ① \quad -\frac{1}{3}v_{②} + i_1 \quad = -\frac{1}{3}\hat{v}_1(t)$$

$$node \ ② \quad \frac{1}{3}v_{②} + 4\frac{dv_{②}}{dt} = \frac{1}{3}\hat{v}_1(t)$$

The second equation can be put into the form of the general first-order differential equation (11.1):

$$\frac{dv_{②}}{dt} = -\frac{1}{12}v_{②} + \frac{1}{12}\hat{v}_1(t)$$

and, therefore, the solution is of the form of (11.2):

$$v_{②}(t) = e^{-(1/12)t}v_{②}(0) + \frac{1}{12}\int_0^t e^{-(1/12)(t-\tau)}\hat{v}_1(\tau)\,d\tau$$

Note that $v_{②}(t) = v_3(t)$, so that the initial node voltage $v_{②}(0)$ equals the initial capacitance voltage $v_3(0)$.

Suppose $v_3(0) = 1$ and consider the zero-input response

$$v_{②}(t) = e^{-(1/12)t}$$

that is plotted in Fig. 11-6b. At time $t = 0$, $v_{②}(t) = 1$, which is the initial capacitance voltage. As $t$ increases, $v_{②}(t)$ decreases exponentially to zero. At $t = 12$ seconds, which is the time constant of the network, the response has decayed to a value of $1/e \approx .37$. In other words, the response has decayed approximately 63 percent of its initial value during this time.

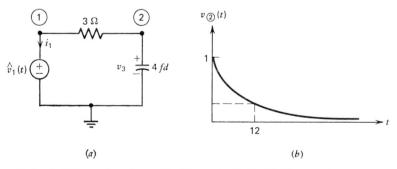

(a)                                                                 (b)

**Fig. 11-6**   (a) The network used in Example 11.7. (b) Its zero-input response.

Now suppose the input was

$$\hat{v}_1(t) = \begin{cases} 2 & t \geq 0 \\ 0 & t < 0 \end{cases}$$

Then the response becomes

$$v_{\textcircled{2}}(t) = e^{-(1/12)t} + \frac{1}{6} \int_0^t e^{-(1/12)(t-\tau)} \, d\tau$$

$$= e^{-(1/12)t} + 2[1 - e^{-(1/12)t}]$$

$$= 2 - e^{-(1/12)t}$$

for $t \geq 0$. Observe that the response may be separated into two parts: that which decreases to zero as $t$ increases, this part is termed the *transient response*, and the part that remains constant as $t$ increases, this part is termed the *steady-state part*. The transient part of $v_2(t)$ is

$$v_2(t)\big|_{tr} = e^{-(1/12)t}$$

and the steady-state part is

$$v_2(t)\big|_{ss} = 2$$

We will discuss transient and steady-state responses in more detail later.

Before leaving this example, we observe that once $v_{\textcircled{2}}(t)$ is known, it can be substituted into the nodal equilibrium equation for node $\textcircled{1}$ to yield a solution for $i_1(t)$:

$$i_1(t) = \frac{1}{3} e^{-(1/12)t} v_3(0) + \frac{1}{36} \int_0^t e^{-(1/12)(t-\tau)} \hat{v}_1(\tau) \, d\tau - \frac{1}{3} \hat{v}_1(t) \quad \square$$

The following example illustrates how the techniques described above can be used to solve more complex networks that contain a single energy storage element.

**Example 11.8**

The nodal equilibrium equations for the network of Fig. 11-7a are

node $\textcircled{1}$ $\qquad -2v_{\textcircled{2}}(t) + i_1(t) \qquad\qquad\qquad = -2\hat{v}_1(t)$

node $\textcircled{2}$ $\qquad 5v_{\textcircled{2}}(t) + \frac{1}{2} \int_0^t [v_{\textcircled{2}}(\tau) - v_{\textcircled{3}}(\tau)] \, d\tau = -i_4(0)$

node $\textcircled{3}$ $\qquad v_{\textcircled{3}}(t) + \frac{1}{2} \int_0^t [v_{\textcircled{3}}(\tau) - v_{\textcircled{2}}(\tau)] \, d\tau = \quad i_4(0) - \hat{i}_6(t)$

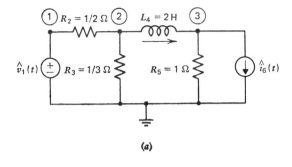

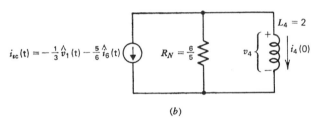

(b)

Fig. 11-7   (a) The network used in Example 11.8. (b) An equivalent network.

where we have used the inductance branch relationship

$$i_4(t) = i_4(0) + \frac{1}{2} \int_0^t v_4(\tau)\, d\tau$$

$$= i_4(0) + \frac{1}{2} \int_0^t [v_{\textcircled{2}}(\tau) - v_{\textcircled{3}}(\tau)]\, d\tau$$

Rather than trying to solve the simultaneous algebraic-integral equations at this time, we may obtain a solution by first finding the Norton equivalent as seen by the inductor $L_4$. The student should verify that the Norton equivalent seen by $L_4$ is the one shown in Fig. 11-7b. The nodal equilibrium equation in terms of $v_4$ is

$$\frac{dv_4(t)}{dt} = -\frac{R_N}{L_4} v_4(t) - \frac{R_N}{L_4} \frac{di_{sc}(t)}{dt}$$

with $v_4(0) = -R_N[\hat{i}_{so}(0) + i_4(0)]$
The solution is

$$v_4(t) = e^{-(R_N/L_4)t} v_4(0) - \frac{R_N}{L_4} \int_0^t e^{-(R_N/L_4)(t-\tau)} \frac{d\hat{i}_{sc}(\tau)}{d\tau}\, d\tau$$

or

$$v_4(t) = -e^{-(6/10)t} \frac{6}{5} [\hat{i}_{so}(0) + \hat{i}_4(0)] + \frac{3}{5} \int_0^t e^{-(6/10)(t-\tau)} \frac{1}{3} \left[ \frac{d\hat{v}_1(\tau)}{d\tau} + \frac{5}{6} \frac{d\hat{i}_6(\tau)}{d\tau} \right] d\tau$$

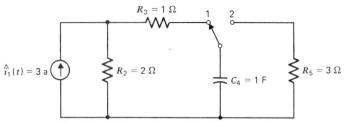

**Fig. 11-8    Network used in Example 11.9.**

Once $v_4(t)$ is known, $v_{②}(t)$, $v_{③}(t)$, and $i_1(t)$ can be determined by considering the original nodal equilibrium equations.    □

The following example will illustrate the importance of calculating initial conditions.

**Example 11.9**

Consider the circuit of Fig. 11-8. The switch has been in position 1 for a long period of time and at time $t = 0$ is switched to position 2. We wish to determine the current that flows in the 3 Ω resistor. To find the initial capacitance voltage, we recognize that at the instant just prior to the switch flipping positions, denoted by $t = 0-$, the capacitor is essentially an open circuit with no current flowing through it. Therefore, $v_4(0-) = R_2 i_1 = 6$ v. After $t = 0$, the current $i_5(t)$ is governed by the equations

$$i_5(t) = \frac{1}{R_5} v_4(t)$$

and

$$C_4 \dot{v}_4(t) = -\frac{1}{R_5} v_4(t)$$

or

$$v_4(t) = e^{-(1/C_4 R_5)t} v_4(0^+)$$

where $t = 0^+$ denotes the time instant just after the switch changes positions. We will see in the next chapter that the voltage across a capacitor cannot change instantaneously so that $v_4(0^+) = v_4(0^-)$ and, therefore,

$$v_4(t) = 6e^{-1/3t}$$

Finally, we have

$$i_s(t) = 2e^{-1/3t}    □$$

## 11-4    SUMMARY

Two energy storage elements were introduced in this chapter: the linear, time invariant *capacitor*, described by

$$i(t) = C \frac{dv(t)}{dv}$$

where $C$ is the capacitance in farads, and the linear time-invariant *inductor*, described by

$$v(t) = L\frac{di(t)}{dt}$$

where $L$ is the inductance in henrys.

*First-order networks* are networks describable by first-order differential equations. Usually a network that contains a single capacitor or inductor is a first-order network. The general form of a *first-order differential equation* is

$$\dot{x}(t) = ax(t) + bw(t)$$

where $w(t)$ is the *input*, or *forcing function*. The *complete solution* of this equation is

$$x(t) = e^{at}x(0) + \int_0^t e^{a(t-\tau)}bw(\tau)\,d\tau$$

which is the sum of the *homogeneous solution*

$$x_h(t) = e^{at}x(0)$$

and the *particular solution*

$$x_p(t) = \int_0^t e^{a(t-\tau)}bw(\tau)\,d\tau$$

The *zero-input response* is determined from the complete solution by setting the input $w(t)$ to zero. The *zero-state response* is determined from the complete response by setting the initial condition $x(0)$ to zero and finally, the *time constant* is a measure of the speed at which the zero-input response of a network decays.

## PROBLEMS

**11.1**   (a) Using the definition of linearity, show that the branch relationship

$$i(t) = \frac{d}{dt}[C(t)v(t)]$$

does indeed describe a linear element.

(b) Show that in order for a capacitor to be both linear and time-invariant its branch relationship must be of the form

$$i(t) = C\frac{d}{dt}v(t)$$

**11.2**   Show that $n$ capacitors with capacitances $C_1, C_2, \ldots, C_n$ connected in parallel behave as a single capacitor with capacitance $C_1 + C_2 \cdots + C_n$.

**11.3**    Show that $n$ capacitors with capacitances $C_1, C_2, \ldots, C_n$ connected in series behave as a single capacitor with capacitance

$$\frac{1}{\left(\dfrac{1}{C_1} + \dfrac{1}{C_2} + \cdots + \dfrac{1}{C_n}\right)}$$

**11.4**    Consider $n$ inductors $L_1, L_2, \ldots, L_n$. How do these inductors behave if they are connected in parallel? If they are in series?

**11.5**    Find the voltages or currents indicated in the networks of Fig. P11.5 by first writing a first-order differential equation and then finding its solution.

**11.6**    Solve the following differential equations:

(a) $\dfrac{dv}{dt} + 2v(t) = 0,\ v(0) = 2$

(b) $\dfrac{dv}{dt} + 3v(t) = 3,\ v(0) = 0$

(c) $2\dfrac{dv}{dt} + 5v(t) = 2e^{-3t},\ v(0) = 0$

(d) $\dfrac{dv}{dt} + 3v(t) = 2 \cos 4t,\ v(0) = 1$

**11.7**    Determine the capacitor voltage in the network of Fig. P11.7 by first determining the Norton equivalent seen by the capacitor and then writing a first order differential equation. What is the $RC$ time constant?

**11.8**    Find the inductance current in the network of Fig. P11.8. (*Hint.* Use a Thévenin equivalent and write a loop equation.) What is the circuit's time constant?

**11.9**    What are the zero-state response and zero-input response associated with the networks considered in Prob. 11.5.

**11.10**    The switch in the network of Fig. P11.10 has been in position 1 for a long period of time. At time $t = 0$ the switch is moved to position 2. Find the voltage $v_5(t)$ for $t \geq 0$.

**11.11**    In the circuit of Fig. P11.11 the switch has been closed for $t < 0$. The switch abruptly opens at $t = 0$. Find and sketch $v_1(t)$.

**11.12**    Find the current $i_4(t)$ in the network of Fig. P11.12a if the voltage $\hat{v}_1(t)$ is the delayed rectangular pulse shown in Fig. P11.12b.

**11.13**    Find the value of $L_4$ in the network of Fig. P11.13 such that $v_5(t)$ reaches 90 percent of its final value at $t = 1$ sec.

**11.14**    Determine, and sketch, the voltage $v_6(t)$ in the network of Fig. P11.14.

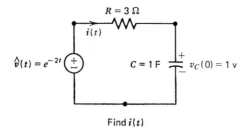

Find $i(t)$

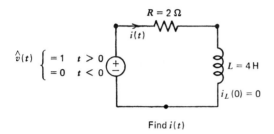

Find $i(t)$

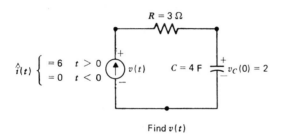

Find $v(t)$

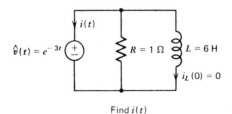

Find $i(t)$

**Fig. P11.5**

**11.15**   Find the value of $\hat{v}_1$ and $R_4$ in the network of Fig. P11.15 so that 2 milliseconds after closing $S_1$ $v_3(t) = 12$ volts; once $v_3(t)$ reaches this value it is to be held there by closing $S_2$.

**11.16**   The switch in the network of Fig. P11.16 has been closed for a long time and all transients have vanished. The switch is suddenly opened at time $t = 0$. Find and sketch $v_6(t)$.

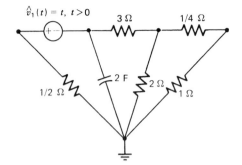

**Fig. P11.7**

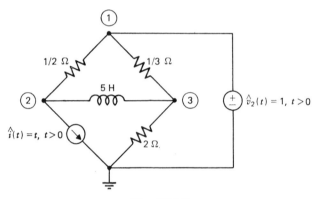

**Fig. P11.8**

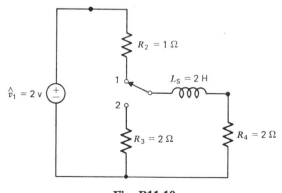

**Fig. P11.10**

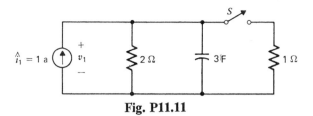

**Fig. P11.11**

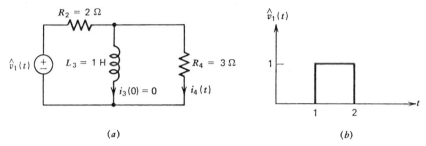

(a)                                               (b)

**Fig. P11.12**

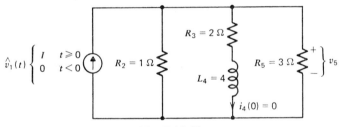

**Fig. P11.13**

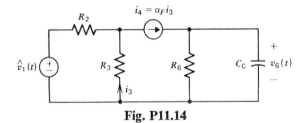

**Fig. P11.14**

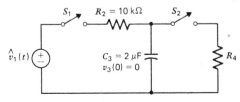

**Fig. P11.15**

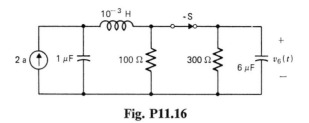

**Fig. P11.16**

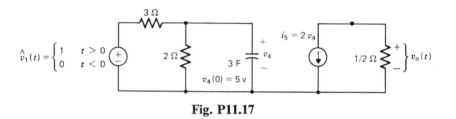

$$\hat{v}_1(t) = \begin{cases} 1 & t > 0 \\ 0 & t < 0 \end{cases}$$

**Fig. P11.17**

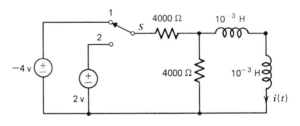

**Fig. P11.18**

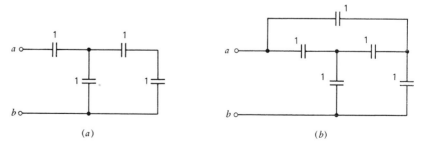

(a)                                    (b)

**Fig. P11.19**

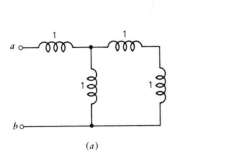

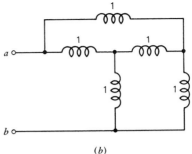

(a)                                    (b)

**Fig. P11.20**

**11.17**   Find $v_0(t)$ in the network of Fig. P11.17. What is the zero-state response? What is zero-input response?

**11.18**   The switch in the network of Fig. P11.18 has been in position 1 for a long time. At time $t = 0$ the switch is thrown to position 2. Determine $i(t)$ for all time.

**11.19**   Find equivalent capacitors that may be used to represent the capacitor networks at terminals $a$ and $b$ for each of the capacitor networks of Fig. P11.19.

**11.20**   Find equivalent inductors that may be used to represent the inductor networks at terminals $a$ and $b$ for each of the inductor networks of Fig. P11.20.

## CHAPTER 12

# Singularity Functions

In this chapter we describe a class of very useful functions for investigating the behavior of circuits. These functions are related to one another through differentiation and integration. After these functions are introduced, various manipulations involving them will be discussed. Finally, some network examples will be presented.

The reader is cautioned that some of the signals to be discussed are not functions in the strict sense. Therefore various phrases, such as *singularity functions*, *generalized functions*, or *distributions* are often used. However, a detailed discussion of this point is beyond the scope of this text. It suffices that what follows is mathematically correct. The interested student should consult one of the many references on the subject.[1]

## 12-1 THE UNIT STEP

The first function we wish to consider is the unit step shown in Fig. 12-1a. The unit step is denoted by $u(t)$ and defined by

$$(12.1) \qquad u(t) = \begin{cases} 1 & \text{for} \quad t > 0 \\ 0 & \text{for} \quad t < 0 \end{cases}$$

[1] For example, see S. W. Director and R. A. Rohrer, *Introduction to System Theory*, McGraw-Hill, 1972.

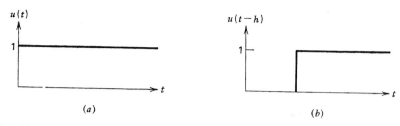

**Fig. 12-1** (*a*) **The unit step** $u(t)$**.** (*b*) **A unit step shifted by** $h$ **seconds is repre-sented by** $u(t - h)$**.**

that is, it has a value of zero for all time less than zero and has a value of one for all time greater than zero. There is an abrupt change, or *discontinuity* at $t = 0$ and the unit step is not well defined at this point. At times it will be convenient to use a unit step that is shifted in time, such as the one shown in Fig. 12-1*b*. This step can be represented by the shifted unit step function $u(t - h)$ where

$$u(t - h) = \begin{cases} 1 & \text{for} \quad t - h > 0 \quad \text{that is, } t > h \\ 0 & \text{for} \quad t - h < 0 \quad \text{that is, } t < h \end{cases}$$

In general, the shifted unit step function, $u(t - h)$, has a value of zero when-ever its argument $(t - h)$ is less than zero, that is, $t < h$, and has a value of one whenever its argument is greater than zero, that is, $t > h$.

A sequence of unit steps can be used to mathematically represent a wide variety of waveforms, as illustrated in the following examples.

### Example 12.1

A square pulse of unit height and width $T$ shown in Fig. 12-2*a* can be represented by the unit steps shown in Fig. 12-2*b*. Symbolically, we can write

$$p(t) = u(t) - u(t - T)$$

The shifted pulse shown in Fig. 12-2*c* can be expressed as

$$p_h(t) = p(t - h) = u(t - h) - u(t - h - T) \quad \square$$

### Example 12.2

The waveform of Fig. 12-3*a* can be represented by the series of step func-tions shown in Fig. 12-3*b*. Symbolically we have

$$f(t) = 2u(t) + u(t - t_1) - 2u(t - t_2) + 5u(t - t_3) - 6u(t - t_4)$$

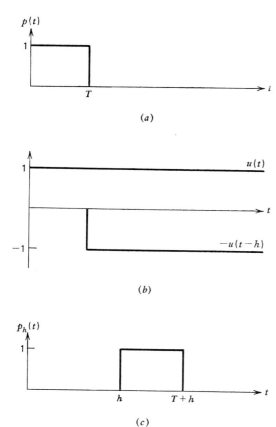

Fig. 12-2  (a) A pulse of width $T$ unit height is equal to the sum of the unit step and the sifted unit step shown in (b). (c) A shifted unit pulse.

Observe that we could have also expressed $f(t)$ as

$$f(t) = \begin{cases} 0 & t < 0 \\ 2 & 0 < t < t_1 \\ 3 & t_1 < t < t_2 \\ 1 & t_2 < t < t_3 \\ 6 & t_3 < t < t_4 \\ 0 & t > t_4 \end{cases}$$

Clearly the first expression is more compact.  ☐

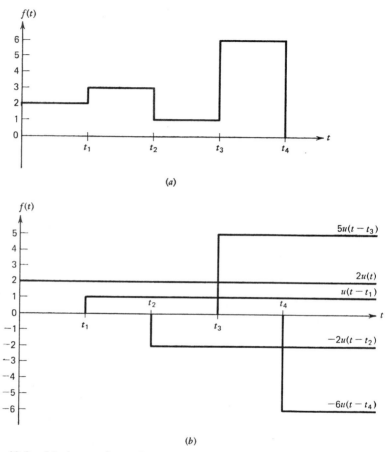

(a)

(b)

**Fig. 12-3** (*a*) **A waveform that can be represented by the sequence of step functions shown in** (*b*).

## 12-2 THE UNIT RAMP AND UNIT PARABOLIC FUNCTIONS

The unit step can be used to generate a whole series of unit functions by repeated integration. In order to establish this relationship it is convenient to designate the unit step by $u_0(t)$.[2] The *unit ramp* is designated by $u_{-1}(t)$ and is obtained by integrating the unit step:

$$u_{-1}(t) = \int_{-\infty}^{t} u_0(\tau)\, d\tau$$

(12.2)

$$= \begin{cases} 0, & \text{for} \quad t < 0 \\ t, & \text{for} \quad t > 0 \end{cases}$$

[2] We will use both $u_0(t)$ and $u(t)$ interchangeably to denote the unit step.

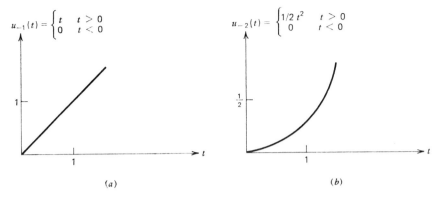

Fig. 12-4   (a) The unit ramp; and (b) the unit parabolic.

Similarly the *unit parabolic* is designated by $u_{-2}(t)$ and is obtained by integrating the unit ramp:

$$u_{-2}(t) = \int_{-\infty}^{t} u_{-1}(\tau) \, d\tau$$

(12.3)

$$= \begin{cases} 0 & \text{for} \quad t < 0 \\ \dfrac{1}{2} t^2 & \text{for} \quad t > 0 \end{cases}$$

These functions are illustrated in Fig. 12-4.

It should be clear that this procedure could be repeated indefinitely with

(12.4)       $$u_{-(n+1)}(t) = \int_{-\infty}^{t} u_{-n}(\tau) \, d\tau$$

or

(12.5)       $$u_{-n}(t) = \frac{d}{dt} u_{-(n+1)}(t)$$

As was the case for these unit step, the unit functions can be used to describe more complicated waveforms.

**Example 12.3**

The waveform of Fig. 12-5a can be represented by the series of unit functions shown in Fig. 12-5b. Symbolically we have

$$f(t) = 3u_0(t) + u_{-1}(t - 1) - u_{-1}(t - 2) - 2u_{-2}(t - 4) - 4u_0(t - 6) +$$
$$2u_{-2}(t - 4)u_0(t - 6) + 6u_{-1}(t - 6) - 9u_{-1}(t - 7)$$

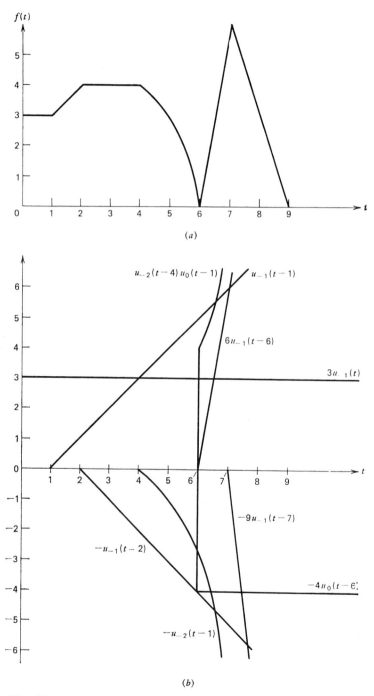

**Fig. 12-5  Decomposition of a waveform into singularity functions.**

which could also have been written as

$$f(t) = \begin{cases} 0 & t < 0 \\ 3 & 0 < t < 1 \\ 2 + t & 1 < t < 2 \\ 4 & 2 < t < 4 \\ -t^2 + 8t - 12 & 4 < t < 6 \\ -36 + 6t & 6 < t < 7 \\ 27 - 3t & 7 < t < 9 \quad \square \end{cases}$$

## 12-3  THE UNIT IMPULSE

Given that the unit ramp is the derivative of the unit parabolic and that the unit step is the derivative of the unit ramp, a logical question is—what is the derivative of the unit step? Formally, we may write

(12.6) $$u_1(t) = \frac{d}{dt} u_0(t)$$

Before proceeding it is worthwhile to show that this situation can arise during analysis of networks and, therefore, it is appropriate to consider the derivative of the unit step. To see this, consider the network of Fig. 12-6. This network was considered in Examples 11.3 and 11.6. The nodal equilibrium equation associated with node ② was found to be

$$\frac{dv_②(t)}{dt} = -\frac{R_2}{L_3} v_②(t) + \frac{d\hat{v}_1(t)}{dt}$$

Suppose that the input was the unit step, that is,

$$\hat{v}_1(t) = u(t)$$

Then the response, $v_②(t)$ depends upon the derivative of the unit step, $du(t)/dt$.

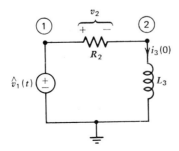

**Fig. 12-6   Circuit used to motivate the unit impulse.**

The difficulty with expression (12.6) is that $u(t)$ is a discontinuous function, therefore, its derivative does not exist in the ordinary way. In order to be able to handle this and other situations mathematically, the *unit impulse* is defined. (Some other names given to the unit impulse in the literature include the *delta function*, *impulse function*, and *dirac delta function*.) We could simply define the unit impulse as the derivative of the unit step however, a more precise definition follows.

Since the unit impulse appears so frequently, it is given a special symbol $\delta(t)$. In our previous notation $\delta(t) = u_1(t)$. The unit impulse is defined by its behavior upon integration. In particular, it is defined by the integral equation:

(12.6)
$$f(0) = \int_{-\infty}^{\infty} f(t)\, \delta(t)\, dt$$

where $f(t)$ is any ordinary function.[3] In other words, the integral of the product of a unit impulse and an ordinary function equals the value of the function when its argument is set to zero. Several examples illustrate the interpretation of the impulse function.

**Example 12.4**

If
$$f(t) = \cos t$$
then $f(0) = 1$ so that
$$\int_{-\infty}^{\infty} \cos t\, \delta(t)\, d\tau = 1$$

Similarly, if
$$f(t) = e^{-3t} \sin(6t + 30°)$$
then $f(0) = .5$ so that
$$\int_{-\infty}^{\infty} e^{-3t} \sin(6t + 30°)\, \delta(t)\, dt = .5$$

Finally, if $f(t)$ is constant, for example if $f(t) = 2$, then $f(0) = 2$ so that
$$\int_{-\infty}^{\infty} 2\delta(t)\, dt = 2 \quad \square$$

**Example 12.5**

The unit impulse is often used in an expression without an associated integral, for instance
$$g(t) = 2\delta(t) + \cos t$$

[3] Actually $f(t)$ can be more general than indicated here but for our purposes this statement suffices.

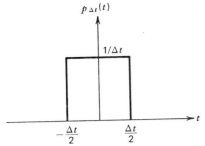

**Fig. 12-7  A sequence of pulses which behave like a unit impulse as $\Delta t$ approaches zero.**

However, in the strict mathematical sense this expression only has meaning when interpreted in terms of integration. Given an ordinary function $f(t)$ then

$$\int_{-\infty}^{\infty} f(t)g(t) \, dt = 2f(0) + \int_{-\infty}^{\infty} f(t) \cos t \, dt \quad \square$$

A unit impulse shifted in time is also useful and may be defined as follows

$$f(\tau) = \int_{-\infty}^{\infty} f(t) \, \delta(t - \tau) \, dt$$

in other words, the integral of the product of a function and a unit impulse equals the value of the function when its argument is chosen so that the argument of the unit impulse equals zero.

**Example 12.6**

Given

$$f(t) = \cos t$$

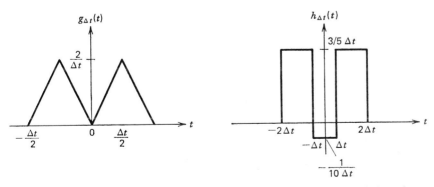

**Fig. 12-8  Two sequences of pulses which also behave like the unit impulse as $\Delta t$ approaches zero.**

then

$$\int_{-\infty}^{\infty} \cos t \, \delta \left( \frac{t - \pi}{2} \right) dt = \cos \frac{\pi}{2} = 0$$

Given

$$f(t) = e^{6t} \cos (3t)$$

then

$$\int_{-\infty}^{\infty} [e^{6t} \cos 3t] \, \delta \left( \frac{t - \pi}{3} \right) dt = -e^{2\pi} \quad \square$$

### Example 12.7

The unit impulse is also useful in modifying expressions. Consider the two equations

$$\frac{dx(t)}{dt} = ax(t) + bw(t)$$

and

$$y(t) = cx(t) + dw(t)$$

The solution to the first-order differential equation is

$$x(t) = e^{at}x(0) + \int_{0}^{t} e^{a(t-\tau)}bw(\tau) \, d\tau$$

Substitution of this expression into the algebraic expression above yields

$$y(t) = ce^{at}x(0) + c\int_{0}^{t} e^{a(t-\tau)}bw(\tau) \, d\tau + dw(t)$$

or

$$y(t) = ce^{at}x(0) + \int_{0}^{t} [e^{a(t-\tau)}cb + d\delta(t - \tau)w(\tau) \, d\tau]$$

because

$$\int_{0}^{t} d\delta(t - \tau)w(\tau) \, d\tau = dw(t) \quad \square$$

It is important for the student to have some physical feeling for the unit impulse. Towards this end we first show that the unit impulse as defined above is, indeed, equal to the derivative of the unit step. What follows serves as justification of this fact but is not meant to be a rigorous proof. We wish to show that

$$\delta(t) = \frac{du(t)}{dt}$$

Since the impulse function can only be interpreted in the strict sense in terms of integration, first multiply $du(t)/dt$ by $f(t)$, where $f(t)$ is some ordinary function, and integrate:

$$\int_{-\infty}^{\infty} \frac{du(t)}{dt} f(t) \, dt$$

On integration by parts, we have

$$\int_{-\infty}^{\infty} \frac{du(t)}{dt} f(t)\, dt = [u(t)f(t)]_{-\infty}^{\infty} - \int_{-\infty}^{\infty} u(t) \frac{df(t)}{dt}\, dt$$

$$= f(\infty) - \int_{0}^{\infty} \frac{df(t)}{dt}\, dt$$

$$= f(\infty) - [f(\infty) - f(0)]$$

$$= f(0)$$

where it is assumed $f(\infty)$ exists, and is finite. Since

$$\int_{-\infty}^{\infty} \left[ \frac{du(t)}{dt} \right] f(t)\, dt = f(0)$$

is the definition of the impulse function, we must have

(12.7)
$$\delta(t) = \frac{du(t)}{dt}$$

In a strict sense it is meaningless to consider the "value" of the unit impulse at a particular instant of time. However, by so doing we can start to get a physical feeling for the unit impulse. The unit step is defined as

$$u(t) = \begin{cases} 1, & t > 0 \\ 0, & t < 0 \end{cases}$$

Therefore, except for the point $t = 0$, the unit step is constant. Now, since

$$\delta(t) = \frac{du(t)}{dt}$$

we conclude that

$$\delta(t) = 0 \qquad \text{for all} \qquad t \neq 0$$

that is, the unit impulse is nonzero only at a single point in time. By letting $f(t) = 1$ in the defining expression

$$\int_{-\infty}^{\infty} f(t)\, \delta(t)\, dt = f(0)$$

we can establish that

$$\int_{-\infty}^{\infty} \delta(t)\, dt = 1$$

in other words, the unit impulse has unit area. Moreover, all this area is concentrated about the point $t = 0$ because

$$\int_{-\infty}^{\infty} \delta(t) \, dt = \int_{-\infty}^{0^-} \delta(t) \, dt + \int_{0^-}^{0^+} \delta(t) \, dt + \int_{0^+}^{\infty} \delta(t) \, dt$$

$$= \int_{0^-}^{0^+} \delta(t) \, dt = 1$$

where $0^-$ is used to designate an instant before true 0 and $0^+$ is used to designate an instant after true 0.

The unit impulse may be viewed as the limiting case of the sequence of pulses defined by

$$p_{\Delta t}(t) = \begin{cases} \dfrac{1}{\Delta t}, & \dfrac{-\Delta t}{2} < t < \dfrac{\Delta t}{2} \\ 0, & \text{elsewhere} \end{cases}$$

and shown in Fig. 12-7. The area under these pulses is unity. As $\Delta t$ approaches zero the pulses become narrower and narrower, however the area remains equal to unity. To see that these pulses behave like an impulse in the limit as $\Delta t \to 0$, observe that if $f(t)$ is an ordinary function

$$\lim_{\Delta t \to 0} \int_{-\infty}^{\infty} p_{\Delta t}(t) f(t) \, dt = \lim_{\Delta t \to 0} \int_{-\Delta t/2}^{\Delta t/2} \frac{f(t)}{\Delta t} \, dt \approx \lim_{\Delta t \to 0} \frac{f(\Delta t)}{\Delta t} \Delta t = f(0)$$

Therefore, on integration, the sequence of pulses $p_{\Delta t}(t)$ behave like the unit impulse.

The student should note that it is folly to think of $\delta(t)$ of having a particular value at time $t = 0$, since it is possible to show that the sequences of pulses shown in Fig. 12-8, among others will also behave like a unit impulse in the limit. Observe that

$$\lim_{\Delta t \to 0} p_{\Delta t}(0) = \infty$$

$$\lim_{\Delta t \to 0} g_{\Delta t}(0) = 0$$

and

$$\lim_{\Delta t \to 0} h_{\Delta t}(0) = -\infty$$

Two properties of the unit impulse are easily shown.

1. *Sifting Property*: If $f(t)$ is continuous at $t = \tau$, then

(12.8) $$\int_{-\infty}^{\infty} f(t) \, \delta(t - \tau) \, dt = f(\tau)$$

To prove this relationship, we make a change of variables. Let

$$\sigma = t - \tau$$

then the left-hand side of (12.8) becomes

$$\int_{-\infty}^{\infty} f(\tau + \sigma)\, \delta(\sigma)\, d\sigma = f(\tau)$$

2. *Time Scaling*

(12.9)
$$\int_{-\infty}^{\infty} f(t)\, \delta(at)\, dt = \frac{1}{|a|} f(0)$$

This property is also shown using a change of variables. Let

$$\tau = at$$

then the left-hand side of (12.9) becomes

$$\frac{1}{a} \int_{-\infty}^{\infty} f\left(\frac{\tau}{a}\right) \delta(\tau)\, d\tau = \frac{1}{a} f(0), \qquad \text{if} \qquad a > 0$$

and

$$-\frac{1}{a} \int_{-\infty}^{\infty} f\left(\frac{\tau}{a}\right) \delta(\tau)\, d\tau = -\frac{1}{a} f(0), \qquad \text{if} \qquad a < 0$$

Therefore, (12.9) follows.

### Example 12.8

The following integrals illustrate further the proper interpretation of expressions involving unit impulses:

(a) $\displaystyle\int_{0^-}^{\infty} \delta(t)\, \cos{(\omega t - 3)}\, dt = \cos{(-3)}$

(b) $\displaystyle\int_{0^+}^{\infty} \delta(t)\, \cos{(\omega t - 3)}\, dt = 0 \qquad (t = 0 \text{ not within limits of integral})$

(c) $\displaystyle\int_{-\infty}^{\infty} \delta\left(\frac{t}{2} - 1\right)(t^2 + 2)\, dt = 12$

$$\left(\text{make a change of variable, let } \tau = \frac{t}{2} - 1\right) \quad \square$$

### Example 12.9

Consider again the nodal equilibrium equation obtained in Example 11.3:

$$\frac{dv_{②}(t)}{dt} = \frac{R_2}{L_3} v_{②}(t) + \frac{d\hat{v}_1(t)}{dt}$$

with $v_{②}(0^+) = \hat{v}_1(0^+) - R_2 i_3(0)$. Suppose $R_2 = 2\,\Omega$, $L_3 = 3$ H, $i_3(0) = 1$ a and that $\hat{v}_1(t) = u(t)$, so that $\hat{v}_1(0^+) = 1$. The solution of the above equation

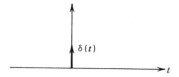

Fig. 12-9  Symbolic representation of the unit impulse.

is

$$v_{②}(t) = -e^{-2/3t} + \int_{0^+}^{t} e^{-2/3(t-\tau)}\,\delta(\tau)\,d\tau$$

$$= -e^{-2/3t}, \qquad t > 0$$

because the point $\tau = 0$ is not within the limits of integration $0^+$ and $t$. Observe that we have indicated that this solution is only valid for $t > 0$.  □

As in the case of the other unit functions, it is sometimes convenient to display the unit impulse graphically. The symbol of a unit impulse is shown in Fig. 12-9.

## 12-4  RESPONSE TO SINGULARITY FUNCTIONS

In this section we will study the behavior of simple first-order networks that are excited by unit steps and unit impulses. Three important theorems are introduced to aid us in this study.

### Theorem  12.1

If the current through a capacitor remains finite, the voltage across it cannot change instantaneously and, if the voltage across an inductor remains finite, the current through it cannot change instantaneous.

### Proof

Consider a capacitor and assume that the voltage across it $v_C(t)$ changed instantaneously, that is, $v_C(t) = u(t)$. Then

$$i_C(t) = C\frac{dv_C}{dt} = C\delta(t)$$

which is not finite. This contradiction establishes the theorem. A similar argument can be made for inductors.  □

This theorem actually forms the theoretical basis for the analysis of networks that contain switches.

### Example  12.10

Consider the network of Fig. 12-10. The switch $S$ is in position 1 for a long time. At time $t = 0$ the switch moves (instantaneously) to position 2. Suppose

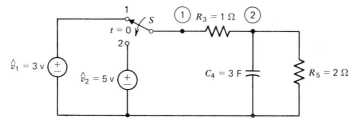

**Fig. 12-10    Network studied in Example 12.10.**

we wish to determine the capacitor voltage for $t \geq 0$. The initial condition is recognized from Theorem 12.1, $v_4(0^+) = v_4(0^-)$, that is the voltage across the capacitor just prior to the switch changing positions must be equal to the voltage across the capacitor just after the switch changes positions. Since the switch has been in position 1 for a long time, all transients have died down and the capacitor appears as an open circuit. Therefore, at time $t = 0^-$

$$v_4(0^-) = \frac{\hat{v}_1 R_5}{R_3 + R_5} = 2$$

For $t > 0$, the nodal equilibrium equation for node ② is

$$C_4 \dot{v}_{②}(t) + \left(\frac{1}{R_3} + \frac{1}{R_5}\right) v_{②}(t) = \frac{1}{R_3} \hat{v}_2(t)$$

or, on substitution of element values, and recognizing that $v_{②} = v_4$:

$$3\dot{v}_4 + \frac{3}{2} v_4 = 5$$

The solution of this equation is

$$v_4(t) = e^{-(1/2)t} v_4(0^+) + \int_{0^+}^{t} e^{-(1/2)(t-\tau)} \left(\frac{5}{3}\right) d\tau$$

$$= \frac{10}{3} - \frac{4}{3} e^{-(1/2)t}$$

for $t > 0$. Observe that as $t \to \infty$, $v_4(t) \to 10/3$ as can be seen by again replacing $C_4$ by an open circuit.    ☐

**Theorem   12.2**

A unit impulse of current flowing through a capacitor causes an instantaneous change of $1/C$ volts across it and a unit impulse of voltage across an inductor causes an instantaneous change of $1/L$ amperes through it.

## Proof

Consider an initially uncharged capacitor, so that

$$v_C(t) = \frac{1}{C} \int_{-\infty}^{t} i_C(\tau) \, d\tau$$

If an impulse of current passes through the capacitor at time $t = 0$, the voltage becomes

$$v_C(t) = \frac{1}{C} \int_{-\infty}^{t} \delta(\tau) \, d\tau = \frac{1}{C} u(t)$$

A similar proof establishes the theorem for inductors.    □

Theorem 12.2 forms the basis for using impulses to generate initial conditions. In particular, an impulse of current applied by a current source in parallel with a capacitor will generate an initial condition of voltage on the capacitor and, an impulse of voltage applied by a voltage source in series with an inductor will generate an initial condition of current in the inductor.

### Example 12.11

In Example 12.10 it was found that the initial condition for the network of Fig. 12-10 was $v_4(0^+) = 2$. From Theorem 12.2 and the previous discussion, this initial condition could also be established by putting a current source in parallel with $C_4$, which has a value of

$$\hat{i}_1(t) = 6\delta(t)$$

Then if $v_4(0^-) = 0$

$$v_4(0^+) = \frac{6}{C_4} = 2 \text{ volts}$$

The appropriately modified network, for $t > 0$, is shown in Fig. 12-11. To verify this result, observe that the nodal equilibrium equation for node ② is

$$3\dot{v}_② + \frac{3}{2} v_② = 5 + 6\delta(t)$$

The solution of this equation is

$$v_②(t) = e^{-(1/2)t} v_②(0^+) + \int_0^t e^{-(1/2)(t-\tau)} \left[ \frac{5}{3} + 2\delta(\tau) \right] d\tau$$

Since $v_②(t) = v_4(t)$, and $v_②(0^+) = 0$, this equation becomes

$$v_4(t) = \frac{10}{3} [1 - e^{-(1/2)t}] + 2e^{-(1/2)t}$$

$$= \frac{10}{3} - \frac{4}{3} e^{-(1/2)t}$$

which is the same expression derived in Example 12.10.    □

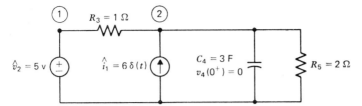

**Fig. 12-11** **An alternate form of the network of Fig. 12-10 for $t > 0$.**

It should be apparent from this example that the response of a given network due to sources that apply impulses can be thought of as the response of the same network that results from initial conditions that have been fixed by some other means. This observation leads us to consider the *impulse response* of a network. The impulse response is the response of a network that has no initial stored energy to a unit impulse input. Since the impulse response is identical with the response of a network excited only by initial conditions, it is a *natural* or intrinsic property of the network. Moreover, it can be shown that in a linear network there is a relationship between the impulse response, usually denoted by $h(t)$, and the response obtained by any other waveform. For example, the *step response*, denoted by $s(t)$ is equal to the integral of the impulse responses, just as the unit step is the integral of the unit impulse. Because of this relationship, which will be discussed later, the impulse response can be thought of as completely characterizing a linear network which has no initial stored energy.

**Example 12.12**

Consider the network of Fig. 12-12 which has no initial stored energy, that is, $v_3(0^-) = 0$. Let the response be the voltage source current $i_1(t)$. The pertinent equations are

$$-\frac{1}{R_2} v_{\textcircled{2}} + i_1(t) = -\frac{1}{R_2} \hat{v}_1(t)$$

$$\frac{1}{R_2} v_{\textcircled{2}}(t) + C_3 \frac{dv_{\textcircled{2}}(t)}{dt} = \frac{1}{R_2} \hat{v}_1(t)$$

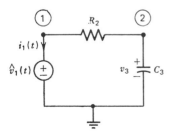

**Fig. 12-12** **Network of Example 12.12.**

from which we find

$$(12.10) \quad i_1(t) = \frac{1}{R_2}\left[ e^{-(1/R_2C_3)t}v_{\textcircled{2}}(0) + \int_0^t e^{-(1/R_2C_3)(t-\tau)}\frac{1}{R_2C_3}\hat{v}_1(\tau)\,d\tau \right] - \frac{1}{R_2}\hat{v}_1(t)$$

To compute the step response we let $\hat{v}_1(t) = u(t)$ and evaluate the initial condition at $t = 0^-$. Since the network is initially relaxed, $v_2(0^-) = v_3(0^-) = 0$, and the step response is

$$i_1(t) = s(t) = \frac{1}{R_2{}^2C_3} e^{-(1/R_2C_3)t} \int_0^t e^{(1/R_2C_3)\tau}\,d\tau - \frac{1}{R_2} u(t) \qquad \text{for} \qquad t > 0$$

$$= -\frac{1}{R_2} e^{-t/R_2C_3} \qquad \text{for} \qquad t > 0$$

which can be written as

$$s(t) = -\left[ \frac{1}{R_2} e^{-t/R_2C_3} \right] u(t)$$

The impulse response is found from (12.10) by setting $\hat{v}_1(t) = \delta(t)$. We have

$$h(t) = \frac{1}{R_2}\left[ e^{-(1/R_2C_3)t}v_{\textcircled{2}}(0^-) + \int_0^t e^{-(1/R_2C_3)(t-\tau)}\frac{1}{R_2C_3}\delta(\tau)\,d\tau \right] - \frac{1}{R_2}\delta(t)$$

$$= \frac{1}{R_2{}^2C_3} e^{-(1/R_2C_3)t} - \frac{1}{R_2}\delta(t)$$

for $t > 0$, or

$$h(t) = \frac{1}{R_2{}^2C_3} e^{-(1/R_2C_3)t}u(t) - \frac{1}{R_2}\delta(t)$$

Observe that

$$h(t) = \frac{d}{dt} s(t)$$

because $e^{-t/R_2C_3}\delta(t) = \delta(t)$. $\square$

### Theorem 12.3

The voltage across a capacitor and the current through an inductor must remain finite.

### Proof

Consider a capacitor. Recall from Chapter 2 that the power absorbed by an element is given by

$$p_C(t) = v_C(t)i_C(t)$$

Since power is the time rate of change of energy we have that the energy stored by an initially relaxed capacitor is

$$e_C(t) = \int_{-\infty}^{t} v_C(\tau) i_C(\tau) \, d\tau$$

but

$$i_C(t) = C \frac{dv_C(t)}{dt}$$

so that

$$e_C(t) = C \int_{-\infty}^{t} v_C(\tau) \frac{dv_C(\tau)}{d\tau} \, d\tau = \frac{1}{2} C v_C^2(t)$$

Since the energy stored by a capacitor must remain finite, so must its voltage.

A similar argument holds for an inductor. Observe that the energy stored by an initially relaxed inductor is

$$e_L(t) = \int_{-\infty}^{t} v_L(\tau) i_L(\tau) \, d\tau = \frac{1}{2} L i_L^2(t) \quad \square$$

## Example 12.13

The network of Fig. 12-13 was considered in Example 11.9. Switch $S$ has been in position 1 for a long time and is moved instantaneously to position 2 at $t = 0$. At time $t = 0^-$

$$v_4(0^-) = 6 \text{ volts}$$

so that the initial energy stored in the capacitor is

$$e_{C_4}(0^-) = \frac{1}{2} C_4 v_4^2(0^-)$$

$$= 18 \text{ joules}$$

For $t > 0$,

$$v_4(t) = 6e^{-1/3t}$$

so that

$$e_{C_4}(t) = \frac{1}{2} C_4 v_4^2(t)$$

$$= 18e^{-2/3t} \qquad t > 0$$

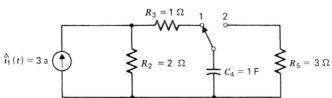

Fig. 12-13   Network studied in Example 12.13.

Observe that as $t \to \infty$ the energy stored by the capacitor decreases to zero, as expected.  □

## 12-5  SUMMARY

A class of useful functions known as *singularity functions* was introduced in this chapter. The most useful of these functions are the *unit step* and *unit impulse*. These functions are all related to one another through differentiation and integration.

A number of useful manipulations involving the unit step and unit impulse were presented. Moreover, by using these functions we were able to prove three useful theorems concerning the behavior of capacitors and inductors. Finally, the concept of an *impulse response* was introduced.

## PROBLEMS

**12.1**    Write mathematical expressions for each of the waveforms of Fig. P12.1.

**12.2**    Sketch the following functions:

(a) $tu_0(t)$

(b) $tu_0(t - 1)$

(c) $(t - 1)u_0(t - 1)$

(d) $-tu_0(-t)$

(e) $u_0(t - 1)u_0(t - 2)$

(f) $u_0(t - 1) - u_0(t - 2)$

**12.3**    Sketch the following functions:

(a) $u_{-1}(-t + 3)$

(b) $u_{-1}(3t + 1)$

(c) $u_{-1}(t - 1) - 2u_{-1}(t - 2)$

(d) $u_{-1}(t)u_0(t - 1)$

(e) $u_{-1}(-2t - 1)u_0(t - 1)$

(f) $e^{-\alpha t}u_0(t - 1)$

(g) $\cos \beta t u_{-1}(t)$

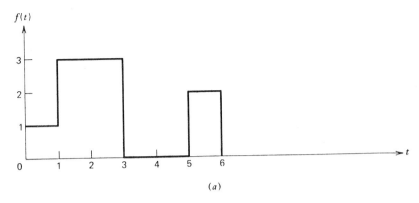

(a)

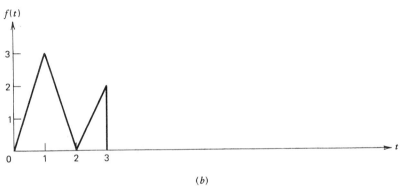

(b)

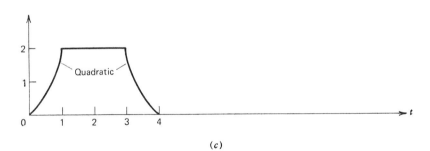

(c)

**Fig. P12.1**

**12.4**    Evaluate the following integrals:

(a) $\displaystyle\int_0^\infty \delta(t - 2)t^2 \, dt$

(b) $\displaystyle\int_0^\infty \delta(2t - 1)t \, dt$

(c) $\displaystyle\int_0^\infty \delta(t - 4)e^{-j\omega t} \, dt$

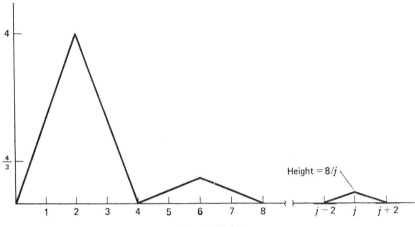

Fig. P12.11

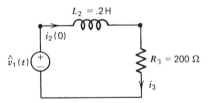

Fig. P12.12

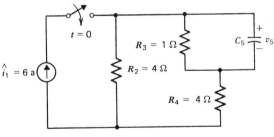

Fig. P12.13

Determine the current $i_3(t)$ if $i_2(0) = 2$ ma. Is there a value of $\phi$ such that there is no transient response?

**12.13** The switch in the network of Fig. P12.13 is closed at $t = 0$. It is found that $v_5(0^+) = 0$ and that $\dot{v}_5(0^+) = 6$. Determine the value of $C$.

## CHAPTER 13

# Numerical Solution of First-Order Networks

In Chapter 11 we introduced two circuit elements, the capacitor and the inductor, whose branch relationships were differential equations. In order to solve for the node voltages of a network that contains a single energy storage element, we had to solve a first-order differential equation of the form

$$\dot{x}(t) = ax(t) + bw(t)$$

We have already discussed the analytic solution of first-order differential equations. In this chapter we introduce some basic methods for the numerical solution of differential equations, and discuss their use for the analysis of first-order networks.

The student should bear in mind that if a first-order network is encountered, the easiest way to solve it would probably be analytically. However, it is convenient to consider the numerical solution of first-order networks as a means to motivate the more involved numerical solution of higher-order networks.

## 13-1 NUMERICAL INTEGRATION: THE FORWARD EULER INTEGRATION ALGORITHM

Consider the first-order differential equation

(13.1) $$\dot{y}(t) = f(y(t), z(t), t)$$

where $f$ is a function of the independent variable $t$, the dependent variable $y$ and some forcing function $z(t)$. Observe that the first-order differential equation we have studied previously:

$$\dot{x}(t) = ax(t) + bw(t)$$

is of the form of (13.1) under the following identifications:

$$y(t) = x(t)$$
$$z(t) = w(t)$$

and

$$f(x(t), w(t), t) = ax(t) + bw(t)$$

In what follows it will be notationally convenient at times to drop the time arguments of $y$ and $z$. Therefore, (13.1) will also be written as

$$(13.2) \qquad \dot{y} = f(y, z, t)$$

Time arguments will only be used if they are needed for clarity. Initially we will study the unforced case so that (13.2) becomes

$$(13.3) \qquad \dot{y} = f(y, t)$$

Ideally we wish to determine the solution to (13.3) for all instants of time $t$. However, since we will use a numerical scheme that is to be implemented on the digital computer, we will actually only be able to evaluate the solution to (13.3) for a finite set of time "points." These time points are denoted by $t_0, t_1, \ldots, t_{k-1}, t_k, t_{k+1}, \ldots, t_N$, where $t_0$ is the initial time and $t_N$ the final time of interest. These points are sometimes referred to as *sample points*. For the present we will assume that the time points are equally spaced so that

$$t_k - t_{k-1} = h$$

a constant, for all $k = 1, 2, \ldots, N$. Observe that

$$t_1 = t_0 + h$$
$$t_2 = t_1 + h = t_0 + 2h$$

and, in general

$$t_k = t_{k-1} + h = t_0 + kh$$

for $k = 1, 2, \ldots, N$. If the initial time is $t_0 = 0$, as it usually is, then

$$t_k = kh$$

The spacing between sample points, $h$, is called the *step size*.

Now consider integration of both sides of (13.3) between the time points $t_k$ and $t_{k+1}$:

$$\int_{t_k}^{t_{k+1}} \dot{y}(\tau) \, d\tau = \int_{t_k}^{t_{k+1}} f(y(\tau), \tau) \, d\tau$$

where $\tau$ is used as the dummy argument of integration. The left-hand side represents an integral of a total differential so that this expression may be re-written as

$$y(t_{k+1}) - y(t_k) = \int_{t_k}^{t_{k+1}} f(y(\tau), \tau) \, d\tau$$

or

(13.4) $$y(t_{k+1}) = y(t_k) + \int_{t_k}^{t_{k+1}} f(y(\tau), \tau) \, d\tau$$

On the computer, only a finite set of values of $f(y, t)$ can be stored. Suppose these values are the values that $f(y, t)$ takes on at the sample points $t_k (k = 0, 1, \ldots, N)$. Let $g(t)$ be a function that equals $f(y, t)$ at the sample points, that is,

$$g(t_k) = f(y(t_k), t_k)$$

but has some other specified form between the sample points. For example, if $f(y, t)$ is the function shown in Fig. 13-1a and if $g(t)$ is assumed to be constant between sample points, then $g(t)$ is the piecewise-constant function shown in Fig. 13-1c.

The integral

$$\int_{t_k}^{t_{k+1}} f(y(\tau), \tau) \, d\tau$$

in (13.4) equals the area under the curve $f(y, t)$ between $t_k$ and $t_{k+1}$. The *forward Euler* integration algorithm approximates this area by the area of a rectangle whose height is $f(y(t_k), t_k)$ and width is $h$. If $f(y, t)$ is the curve depicted in Fig. 13.1a then the forward Euler scheme essentially approximates $f(y, t)$ by the piecewise constant curve of Fig. 13.1c. The approximation improves, of course, as $h$ decreases. We have

$$\int_{t_k}^{t_{k+1}} f(y(\tau), \tau) \, d\tau \simeq h[f(y(t_k), t_k)]$$

so that (13.4) becomes

(13.5) $$y(t_{k+1}) \simeq y(t_k) + h[f(y(t_k), t_k)]$$

To further ease notational complexity, we define

$$y^k \equiv y(t_k)$$

and

$$f^k \equiv f(y(t_k), t_k)$$

Therefore, (13.5) can be written most succinctly as

(13.6) $$y^{k+1} = y^k + hf^k$$

Expression (13.6) in known as a *recursion* relation because it is used itera-tively or "recursively" to determine the solution $y(t_k)$, that is, given $y_0$ and

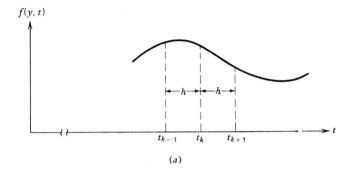

(a)

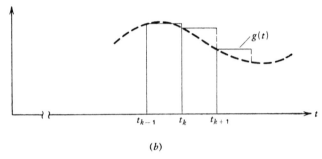

(b)

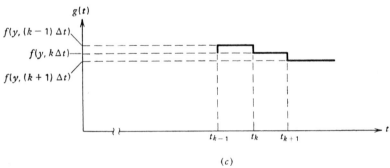

(c)

**Fig. 13-1** **Approximation of a continuous curve by a piecewise constant curve.**

$f_0$, $y_1$ is found, then $y_2$ from $y_1$ and $f_1$, etc. A recursion relation, such as (13.6), is also called a *difference equation*. Since the forward Euler scheme approximates the area under a curve by the area under rectangles, it is also referred to as *rectangular integration*.

**Example 13.1**

As an illustration of the Forward Euler integration scheme, consider the solution of the differential equation

$$\dot{y}(t) = t \qquad y(0) = 0$$

The forward Euler recursion relationship (13.6) is

$$y^{k+1} = y^k + hf^k$$

But

$$f(y, t) = t$$

so that

$$f^k = f(y(t_k), t_k) = t_k$$

and the recursion relationship becomes

$$y^{k+1} = y^k + ht_k$$
$$= y^k + kh^2$$

since $t_k = kh$ ($t_0 = 0$). For a step size of $h = .1$, we have (with $y^0 = 0$):

$$y^1 = \hat{y}(.1) = y^0 + (0)(.1)^2$$
$$= 0$$
$$y^2 = \hat{y}(.2) = y^1 + 1(.1)^2$$
$$= .01$$
$$y^3 = \hat{y}(.3) = y^2 + 2(.1)^2$$
$$= .03$$

etc. (The "hat" is used to distinguish the numerical solution from the analytical solution.)

It is interesting to compare the numerical solution with the analytical solution. From Chapter 11, the analytic solution of the differential equation

$$\dot{y}(t) = t \qquad y(0) = 0$$

is

$$y(t) = y(0) + \int_0^t \tau \, d\tau$$
$$= \frac{1}{2} t^2$$

Therefore, the exact solution at the specified sample points is

$$y(.1) = .005$$
$$y(.2) = .02$$
$$y(.3) = .045$$

or, in general,

$$y(kh) = \frac{1}{2}(kh)^2 = .05(k)^2$$

Observe that the numerical solution does not equal the analytic solution. However, for a step size of $h = .05$ the recursion relation yields

$$y^1 = \hat{y}(.05) = 0$$

and

$$y^2 = \hat{y}(.1) = .0025$$

Observe that the numerical solution for a step size of .05 is closer to the actual solution at $t = .1$ than was the numerical solution for a step size of .1. This result agrees with our intuition that the smaller the step size the better the approximation of the area under a rectangle to the area under the curve being integrated. A more detailed investigation of the correspondence between the step size and the accuracy of the numerical solution will be undertaken in the sequel. Table 13-1 compares the numerical and analytic solutions for two step sizes over the time interval $0 \leq t \leq 1$   □

**Example 13.2**

In Example 11.7 the solution of the first-order differential equation

$$\dot{v}_{②} = \frac{-1}{12} v_{②} + \frac{1}{12} \hat{v}_1(t)$$

with $\hat{v}_1(t) = 2u(t)$ and $v_{②}(0) = 1$ was found to be

$$v_{②}(t) = 2 - e^{-(1/12)t}$$

**Table 13-1**   Comparison of the actual and the forward Euler solution of the differential equation $\dot{y} = t$ for the step sizes $h = .1$ and $h = .05$

| | | Computed Solution | | | |
|---|---|---|---|---|---|
| | Actual Solution | $h = .1$ | | $h = .05$ | |
| $t$ | $(t)^2/2$ | $\hat{y}$ | Error (Magnitude) | $\hat{y}$ | Error |
| .0 | .0 | .0 | .0 | .0 | .0 |
| .1 | .005 | .0 | .005 | .0025 | .0025 |
| .2 | .020 | .01 | .010 | .015 | .005 |
| .3 | .045 | .03 | .015 | .0325 | .0075 |
| .4 | .080 | .06 | .020 | .07 | .0100 |
| .5 | .125 | .10 | .025 | .1125 | .0125 |
| .6 | .180 | .15 | .030 | .165 | .0150 |
| .7 | .245 | .21 | .035 | .2275 | .0175 |
| .8 | .320 | .28 | .040 | .30 | .0200 |
| .9 | .405 | .36 | .045 | .3825 | .0225 |
| 1.0 | .500 | .45 | .050 | .475 | .0250 |

for $t > 0$. Let us find the solution numerically using the forward Euler method. The recursion relation for the differential equation

$$\dot{y} = f(y, t)$$

is

$$y^{k+1} = y^k + hf(y_k, t_k)$$

For the case on hand, $y = v_{\textcircled{2}}$ and

$$f(y, t) = f(v_{\textcircled{2}}, t)$$

$$= -\frac{1}{12} v_{\textcircled{2}}(t) + \frac{1}{6} u(t)$$

Therefore, the recursion relation becomes

$$v_{\textcircled{2}}^{k+1} \equiv v_{\textcircled{2}}^k + h\left[ -\frac{1}{12} v_{\textcircled{2}}^k + \frac{1}{6} u(t_k) \right]$$

$$\left( 1 - \frac{h}{12} \right) v_{\textcircled{2}}^k + \frac{h}{6}$$

provided $t_k > 0$. A simple computer program for evaluating this recursion relationship for various values of $h$ is shown in Fig. 13-2. The program terminates and reads a new step size when the time $T$ exceeds the final time of interest $TF$. A summary of the results of this program for various values of $h$ along with the exact values of $v_{\textcircled{2}}$ is given in Table 13-2. Observe that as the step size increases so does the error. In fact, for step sizes of $h = 25.$ and $h = 50.$ the numerical solution diverges rather drastically from the actual solution. □

A physical interpretation of the forward Euler method for solving the differential equation

$$\dot{y} = f(y, t)$$

can be made as follows. The recursion relationship (13.6)

$$y^{k+1} = y^k + hf^k$$

can be written as

$$f^k = \frac{y^{k+1} - y^k}{h}$$

but

$$f^k = \dot{y}^k$$

so we conclude that we are using as an approximation for the derivative of $y(t)$ at time $t_k$:

$$\dot{y}^k = \frac{y^{k+1} - y^k}{h}$$

```
C     CHAPTER 13 EXAMPLE 13.2
C
C     READ TIME STEP AND FINAL TIME AND PRINT HEADINGS
C
    1 READ 10,H,TF
      PRINT 14,H
      PRINT 13
C
C     SETUP INITIAL VALUES
C
      T=0.
      V2=1.
C
C     PRINT T AND V2
C
    3 PRINT 11,T,V2
C
C     CHECK T AND TF
C
      IF(T.GT.TF) GO TO 1
C
C     INCREMENT TIME AND COMPUTE V2
C
      T=T+H
      V2=(1.-H/12.)*V2+H/6.
      GO TO 3
   10 FORMAT(2G10.3)
   11 FORMAT(10X,G12.5,6X,G12.5)
   13 FORMAT(12X,'TIME',18X,'V2'/)
   14 FORMAT(//' RESULTS FOR EXAMPLE 13.2 WITH A STEP SIZE=',G12.5//)
      END
```

```
----- DATA FOR PROGRAM OF EXAMPLE 13.2 -----
    1.       1000.
   20.       1000.
   25.       1000.
   50.       1000.
```

**Fig. 13-2   Computer program that implements the forward Euler integration algorithm to solve $\dot{v}_{\textcircled{2}} = -1/12v_{\textcircled{2}} + 1/12\hat{v}_1(t)$.**

In other words, the forward Euler method approximates the slope of the computed solution at time $t_k$ with that of the actual solution at time $t_k$ as illustrated in Fig. 13-3. The recursion relationship (13.6) is sometimes written as

$$(13.7) \qquad\qquad y^{k+1} = y^k + h\dot{y}^k.$$

## 13-2   THE BACKWARD EULER INTEGRATION ALGORITHM

We will see later that in many situations if the forward Euler method is used and if the step size is too large the approximate solution generated by the

**Table 13-2** Comparison of the results of using the forward Euler integration scheme for the solution of the differential equation $\dot{v}_{②} = -\frac{1}{12}v_{②} + \frac{1}{6}u(t)$ for various step sizes

| t, sec | Actual $v_{②}$ | Computed $v_{②}$ | | | |
|---|---|---|---|---|---|
| | | $h = 1$ | $h = 20$ | $h = 25$ | $h = 50$ |
| 0 | 1.0 | 1.0 | 1.0 | 1.0 | 1.0 |
| 1. | 1.079 | 1.083 | | | |
| 10. | 1.565 | 1.581 | | | |
| 20. | 1.811 | 1.824 | 2.666 | | |
| 25. | 1.875 | 1.886 | | 3.083 | |
| 50. | 1.984 | 1.987 | | 0.826 | 5.166 |
| 100. | 1.999 | 1.999 | 2.131 | 0.622 | −8.027 |
| 200. | 2.000 | 1.999 | 1.982 | 0.102 | −98.556 |
| 300. | 2.000 | 1.999 | 2.002 | −0.613 | −1006.35 |
| 400. | 2.000 | 1.999 | 1.999 | −1.599 | −10109.5 |
| 500. | 2.000 | 1.999 | 2.000 | −2.957 | −101394. |
| 1000. | 2.000 | 2.000 | 2.000 | −22.574 | −0.102E11 |

recursion relationship will diverge from the actual solution of the differential equation. In this section we introduce an alternate integration method that does not have this problem.

We are interested in solving the differential equation

$$\dot{y}(t) = f(y(t)), t)$$

and displaying the values of $y(t)$ at the equally spaced time points $t_0, t_1,$

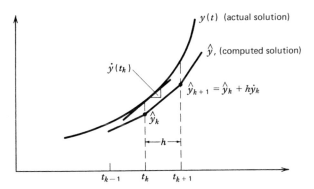

**Fig. 13-3** Physical interpretation of the forward Euler Method; at each approximate solution point the extrapolation to the next point is linearly based on an approximation to the slope of the actual solution at the point.

$t_2, \ldots$ . Integrating this equation between two time points $t_k$ and $t_{k+1}$ yields

$$(13.8) \qquad y(t_{k+1}) = y(t_k) + \int_{t_k}^{t_{k+1}} f(y(\tau), \tau)\, d\tau$$

The forward Euler method resulted when the area under $f(y, t)$ between $t_k$ and $t_{k+1}$ was approximated by the area under a rectangle with height $f^k = f(y(t_k), t_k)$. Another way of expressing this statement is to say that $f(y, t)$ is approximated in the interval $t_k \leq t < t_{k+1}$ by a piecewise constant function equal to $f^k = f(y(t_k), t_k)$.

The *backward Euler* method results if $f(y, t)$ is approximated in the interval $t_k < t \leq t_{k+1}$ by a piecewise constant function equal to $f^{k+1} = f(y(t_{k+1}), t_{k+1})$. In other words, the area under the curve $f(y, t)$ between the time points $t_k$ and $t_{k+1}$ is approximated by the area under a rectangle of height $f^{k+1} = f(y(t_{k+1}, t_{k+1})$. This situation is illustrated in Fig. 13-4. With this approximation, expressions (13.8) becomes

$$(13.9) \qquad y^{k+1} = y^k + hf^{k+1}$$

for $k = 0, 1, 2, \ldots$ . Observe that the time point $t_{k+1} = (k + 1)h$ appears on both the left and right hand sides of this expression. This type of integration scheme is called an *implicit method*. By way of contrast, in the forward Euler method, the time point $t_{k+1}$ only appeared on the left hand side and, therefore, is called an *explicit method*. We will discuss the relative merits of implicit and explicit integration schemes later.

### Example 13.3

As an example of the use of the backward Euler method, consider again the differential equation

$$\dot{y}(t) = t$$

that was solved by the forward Euler method in Example 13.1. The recursion relationship (13.9) becomes

$$y^{k+1} = y^k + ht_{k+1}$$
$$= y^k + h((k + 1)h)$$

The results for step sizes of .1 and .05 are summarized in Table 13-3. Observe that the errors are comparable with those obtained with the forward Euler method, as should be expected because in both instances a piecewise constant function is being used to approximate a first-order curve.   □

### Example 13.4

Consider again the first-order differential equation of Examples 11.7 and 13.2:

$$\dot{v}_{\circled{2}} = -\frac{1}{12} v_{\circled{2}} + \frac{1}{12} \hat{v}_1(t)$$

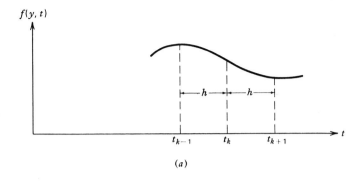

$(a)$

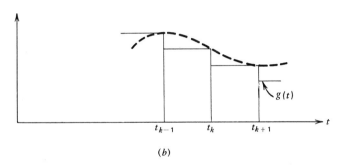

$(b)$

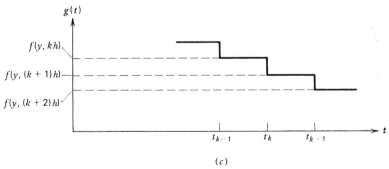

$(c)$

**Fig. 13-4   Approximation of a curve using a piecewise constant curve defined by the future time point.**

with $\hat{v}_1(t) = 2u(t)$ and $v_2(0) = 1$. Application of the backward Euler method yields the recursion relation

$$v_{\circled{2}}^{k+1} = v_{\circled{2}}^{k} + h\left[-\frac{1}{12}v_{\circled{2}}^{k+1} + \frac{1}{6}u(t_{k+1})\right]$$

or

$$v_{\circled{2}}^{k+1} = \left(1 + \frac{h}{12}\right)^{-1}\left(v_{\circled{2}}^{k} + \frac{h}{6}\right)$$

**Table 13-3** Comparison of the actual and backward Euler solutions of the differential equation $\dot{y}(t) = t$ for the step sizes $h = .1$ and $h = .05$

| $t$ | Actual Solution $\dfrac{t^2}{2}$ | Computed Solution $h = .1$ | Computed Solution $h = .05$ |
|---|---|---|---|
| 0 | 0.0 | 0.0 | 0.0 |
| .1 | .005 | .01 | .0075 |
| .2 | .02 | .03 | .025 |
| .3 | .045 | .06 | .0525 |
| .4 | .08 | .10 | .09 |
| .5 | .125 | .15 | .137 |
| .6 | .18 | .21 | .195 |
| .7 | .245 | .28 | .262 |
| .8 | .32 | .36 | .340 |
| .9 | .405 | .45 | .427 |
| 1.0 | .500 | .55 | .525 |

for $t_k > 0$. A simple program to evaluate this recursion relation is shown in Fig. 13-5. The results are summarized in Table 13-4.

Recall that this same problem was solved using the forward Euler integration scheme in Example 13.2. Comparison of those results, Table 13-2, with the ones obtained here is interesting. We observe that even for a large step size, like $h = 50$, the backward-Euler solution approaches the true solution while the forward Euler solution diverges from the true solution. □

A physical interpretation of the backward Euler method can easily be made. The recursion relationship (13.9)

$$y^{k+1} = y^k + hf^{k+1}$$

can be written as

$$f^{k+1} = \frac{y^{k+1} - y^k}{h}$$

Since

$$f^{k+1} = \dot{y}^{k+1}$$

then

$$\dot{y}^{k+1} = \frac{y^{k+1} - y^k}{h}$$

In other words, the backward Euler method approximates the slope of the computed solution at time $t_k$ with that of the actual solution at time $t_{k+1}$ as illustrated in Fig. 13-6.

```
C     CHAPTER 13 EXAMPLE 13.4
C
C     READ TIME STEP AND FINAL TIME AND PRINT HEADINGS
C
    1 READ 10,H,TF
      PRINT 14,H
      PRINT 13
C
C     SETUP INITIAL VALUES
C
      T=0.
      V2=1.
C
C     PRINT T AND V2
C
    3 PRINT 11,T,V2
C
C     CHECK T AND TF
C
      IF(T.GT.TF) GO TO 1
C
C     INCREMENT TIME AND COMPUTE V2
C
      T=T+H
      V2=(V2+H/6.)/(1.+H/12.)
      GO TO 3
   10 FORMAT(2G10.3)
   11 FORMAT(10X,G12.5,6X,G12.5)
   13 FORMAT(12X,'TIME',18X,'V2'/)
   14 FORMAT(//' RESULTS FOR EXAMPLE 13.4 WITH A STEP SIZE=',G12.5//)
      END
```

```
----- DATA FOR PROGRAM OF EXAMPLE 13.4 -----
 1.         1000.
20.         1000.
25.         1000.
50.         1000.
```

**Fig. 13-5  Computer program that implements the backward Euler integration algorithm to solve** $\dot{v}_{②} = -1/12 v_{②} + 1/12 \hat{v}_1(t)$.

It is important that the student be aware of the major difference between the forward and backward Euler methods. Consider the differential equation

$$\dot{y}(t) = f(y(t), t)$$

The forward Euler method yields the *explicit* recursion relationship

$$y^{k+1} = y^k + hf(y_k, t_k)$$

so that given $y^k$ the solution at time $t_{k+1}$, namely $y^{k+1}$, is *explicitly* available. The backward Euler method yields the *implicit* recursion relationship

$$y^{k+1} = y^k + hf(y_{k+1}, t_{k+1})$$

**Table 13-4** Comparison of the results of using the backward Euler integration scheme for the solution of the differential equation $\dot{v}_{②} = -\frac{1}{12}v_{②} + \frac{1}{6}$ for various step sizes

| | | Computed $v_2$ | | | |
|---|---|---|---|---|---|
| $t$, sec | Actual $v_{②}$ | $h = 1$ | $h = 20$ | $h = 25$ | $h = 50$ |
| 0 | 1.0 | 1.0 | 1.0 | 1.0 | 1.0 |
| 1. | 1.079 | 1.076 | | | |
| 10. | 1.565 | 1.550 | | | |
| 20. | 1.811 | 1.798 | 1.625 | | |
| 25. | 1.875 | 1.864 | | 1.675 | |
| 50. | 1.984 | 1.981 | | 1.894 | 1.806 |
| 100. | 1.999 | 1.999 | 1.992 | 1.988 | 1.962 |
| 200. | 2.000 | 1.999 | 1.999 | 1.999 | 1.998 |
| 300. | 2.000 | 1.999 | 2.000 | 2.000 | 1.999 |
| 400. | 2.000 | 1.999 | 2.000 | 2.000 | 2.000 |
| 500. | 2.000 | 1.999 | 2.000 | 2.000 | 2.000 |
| 1000. | 2.000 | 2.000 | 2.000 | 2.000 | 2.000 |

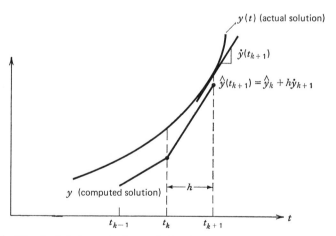

**Fig. 13-6** Physical interpretation of the backward Euler method; at each approximate solution point the extrapolation to the next point is linearly based on an approximation to the slope of the actual solution at the *next* point.

so that given $y^k$ the solution at $t_{k+1}$ (i.e. $y^{k+1}$) is *implicitly* available since $y^{k+1}$ appears on both sides of this equation. If the function $f(y, t)$ is a linear function of $y$ then the implicit equation may always be rewritten as an explicit equation. On the other hand, if $f(y, t)$ is a nonlinear function of $y$ then an analytic solution for $y^{k+1}$ usually does not exist and some additional numerical method, such as the Newton-Raphson method, may have to be employed. For example, suppose

$$\dot{y} = e^y$$

then the backward Euler recursion relationship becomes

$$y^{k+1} = y^k + h e^{y^{k+1}}$$

This equation cannot be solved analytically for $y^{k+1}$.

Before leaving this section, observe that the methods we have described for the numerical solution of the differential equation

$$\dot{y}(t) = f(y(t), t)$$

are also applicable for the solution of the more general differential equation

$$y(t) = f(y(t), z(t), t)$$

where $z(t)$ is some forcing function. In particular, we have the following recursion relationships:

$$y^{k+1} = y^k + h f(y^k, z^k, t_k)$$

for the forward Euler method; and

$$y^{k+1} = y^k + h f(y^{k+1}, z^{k+1}, t_{k+1})$$

for the backward Euler method.


## 13-3  USE OF NUMERICAL INTEGRATION IN THE ANALYSIS OF FIRST-ORDER NETWORKS

We are now in a position to solve first-order networks using the numerical techniques discussed in sections 13-1 and 13-2. In Chapter 11 it was shown that the nodal equilibrium equations for a first-order network can be manipulated into the form

(13.10)                     $$\dot{x}(t) = ax(t) + bw(t)$$

where $x(t)$ is a node voltage and $w(t)$ some function of the sources of the network. We wish to show that if forward or backward Euler numerical integration methods are used to solve (13.10), the recursion relationship that results is identical to one that would result if the numerical integration method was applied to the energy storage element branch relationship first and then

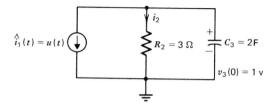

**Fig. 13-7  A simple $RC$ network.**

the nodal equilibrium equations were written. The latter procedure will ultimately lead to a rather general transient analysis scheme.

The student should recall that a similar situation arose during the study of nonlinear networks. Equivalent results were obtained if the nodal equilibrium equations were written first and then solved by the Newton-Raphson algorithm, or if the Newton-Raphson algorithm was used to linearize the nonlinear branch relationship first and then the nodal equilibrium equations were written. The second approach led to the general analysis scheme based on linearized equivalent models.

For motivational purposes refer to the network of Fig. 13-7. The nodal equilibrium equation is

$$\frac{1}{R_2} v_{①}(t) + C_3 \dot{v}_{①}(t) = \hat{\imath}_1(t)$$

which can also be written as

(13.11) $$\dot{v}_{①}(t) = -\frac{1}{R_2 C_3} v_{①}(t) - \frac{1}{C_3} \hat{\imath}_1(t)$$

Application of the backward Euler algorithm yields the recursion relationship

(13.12) $$v_{①}^k = v_{①}^k + h\left(-\frac{1}{R_2 C_3} v_{①}^{k+1} - \frac{1}{C_3} \hat{\imath}_1^{k+1}\right)$$

which can be rewritten as

$$v_{①}^{k+1} = \left(1 + \frac{h}{R_2 C_3}\right)^{-1}\left(v_{①}^k - \frac{h}{C_3} \hat{\imath}_1^{k+1}\right)$$

or, after substitution of the element value

(13.13) $$v_{①}^{k+1} = \left(1 + \frac{h}{6}\right)^{-1}\left(v_{①}^k - \frac{h}{6}\right)$$

The initial condition needed for this equation is

$$v_{①}(0) = v_3(0) = 1$$

```
4 READ (5,1) H,TF
1 FORMAT(2G10.3)
  T=0.
  V1=1.
3 PRINT 2,T,V1
2 FORMAT(' TIME=',G10.3,2X,'V1=',G10.3)
  IF(T.GT.TF) GO TO 4
  T=T+H
  V1=(V1-H/2)/(1+H/6)
  GO TO 3
  END
```

**Fig. 13-8    A simple computer program to evaluate the recursion relationship (13.13).**

A computer program that can be used to solve (13.13) for various step sizes is shown in Fig. 13-8. Results from this program are plotted in Fig. 13-9.

It is possible to make a meaningful physical interpretation of the numerical integration process. This interpretation will lead to a general analysis procedure. We start by rewriting (13.12) as

$$(13.14) \qquad \left(\frac{1}{R_2} + \frac{C_3}{h}\right) v_{①}^{k+1} = -i_1^{k+1} + \frac{C_3}{h} v_3^k$$

where we have used the fact that

$$v_{①}^k = v_3^k$$

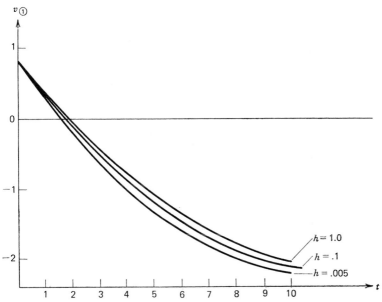

**Fig. 13-9    The solution of (13.11) predicted by the computer program of Fig. 13-8.**

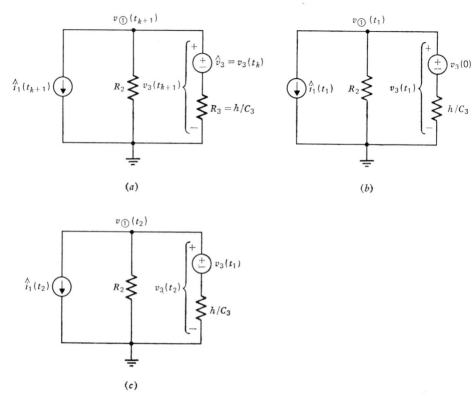

**Fig. 13-10**    **Network interpretation of the recursion relationship (13.14) (a) for all $k$; (b) for $k = 0$, (c) for $k = 1$.**

At time $t_{k+1}$, $v_3(t_k)$ is known and we can interprete this expression as the nodal equilibrium equation of the dc network of Fig. 13-10a. Then the solution of (13.14) for $k = 0$:

$$\left(\frac{1}{R_2} + \frac{C_3}{h}\right) v_{\text{①}}^1 = -\hat{i}_1^1 + \frac{C_3}{h} v_3^0$$

is equivalent to the analysis of the network of Fig. 13-10b. When $v_{\text{①}}^1$ is found (13.14) can be used with $k = 1$ to determine $v_{\text{①}}^2$:

$$\left(\frac{1}{R_2} + \frac{C_3}{h}\right) v_{\text{①}}^2 = -\hat{i}_1^2 + \frac{C_3}{h} v_3^1$$

which is equivalent to the analysis of the network of Fig. 13-10c. Thus the iterative solution of the recursion relation (13.12) parallels the repeated analysis of the network of Fig. 13-10a.

We can make a closer tie between the networks of Fig. 13-10 and the original network of Fig. 13-7. First, recognize that since the node voltage $v_{\text{①}}$

equals the capacitor voltage $v_3$,

$$(13.15) \qquad v_1^k = v_3^k$$

Next, observe that the capacitor current $i_3$ is related to the source current $\hat{i}_1$ and resistor current $i_2$ by

$$i_3(t) = -\hat{i}_1(t) - i_2(t)$$

so that

$$i_3(t) = -\hat{i}_1(t) - \frac{1}{R_2} v_①(t)$$

or, more specifically,

$$(13.16) \qquad i_3^k = -\hat{i}_1^k - \frac{1}{R_2} v_①^k$$

Comparison of expressions (13.12), (13.15), and (13.16) indicates that the following relationship must hold between the capacitor's voltage $v_3$ and current $i_3$ at time $t_{k+1}$:

$$(13.17) \qquad v_3^{k+1} = v_3^k + \frac{h}{C_3} i_3^{k+1}$$

This expression may be viewed as the branch relationship of a voltage source of value $v_3^k$ in series with a resistor of value $h/C_3$. These elements are exactly what appears in place of the capacitor in Fig. 13-10$a$. Finally, observe that (13.17) is the expression that would result if the backward Euler method was applied directly to the capacitor branch relationship

$$(13.18) \qquad \dot{v}_3 = \frac{1}{C_3} i_3$$

We call (13.17) the *discretized* form of (13.18) or the *discretized capacitor branch relationship* under the backward Euler integration scheme. The model used to represent the discretized branch relationship is called the *companion model*.[1] The *companion network* is the original network with the capacitor replaced by its companion model and all voltages and currents evaluated at time $t_{k+1}$.

In summary, analysis of the network of Fig. 13-7 over some interval of time can be viewed as a series of dc analyses of a network similar to the original network except that the capacitor has been replaced by its companion model. The companion model arises from a network interpretation of the recursion relation obtained from application of the backward Euler

---

[1] This phraseology was introduced by D. A. Calahan in *Computer Aided Network Design*, McGraw-Hill, 1972.

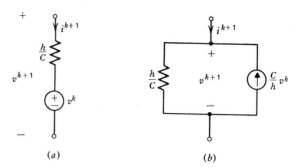

**Fig. 13-11  Two equivalent backward Euler companion models for a capacitor.**

method to the capacitor branch relationship

$$\dot{v} = \frac{1}{C}\, i$$

The recursion relation is

$$v^{k+1} = v^k + \frac{h}{C}\, i^{k+1}$$

which may be interpreted as a voltage source of value $v^k$ in series with a resistor of value $h/C$, as shown in Fig. 13-11. The voltage across this series combination is $v^{k+1}$ and the current through it is $i^{k+1}$. Since, for the most part we are interested in the nodal equilibrium equations of a network, it is more convenient to perform a current source conversion on this companion model. The companion model then takes on the form shown in Fig. 13-11b: a current source of value $Cv^k/h$ in parallel with a resistor of value $h/C$. This new model could have also been derived from the equations that follow. From the previous section, we know that the backward Euler method can be viewed as the following approximation of the derivative of $v$ at time $t_{k+1}$:

$$\dot{v}^{k+1} = \frac{v^{k+1} - v^k}{h}$$

Therefore, the capacitor branch relationship

$$i^{k+1} = C\dot{v}^{k+1}$$

becomes

$$i^{k+1} = \frac{C}{h}\, v^{k+1} - \frac{C}{h}\, v^k$$

which is a description of the model shown in Fig. 13-11b.

**Example 13.5**

Consider the numerical solution of the network of Fig. 13-12a. The companion network associated with this network for the backward Euler method

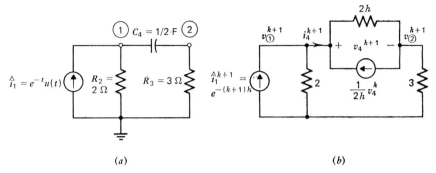

**Fig. 13-12**  (*a*) The network of Example 13.5 and (*b*) its companion network.

is shown in Fig. 13-12*b*. The nodal equilibrium equations for the companion network are

$$\left(\frac{1}{2} + \frac{1}{2h}\right)v_{①}^{k+1} - \frac{1}{2h}v_{②}^{k+1} = e^{-(k+1)h} + \frac{1}{2h}v_4^k$$

and

$$-\frac{1}{2h}v_{①}^{k+1} + \left(\frac{1}{3} + \frac{1}{2h}\right)v_{②}^{k+1} = -\frac{1}{2h}v_4^k$$

with $v_4^k = v_{①}^k - v_{②}^k$. For $k = 0$, $v_4^0 = 0$ so that everything on the right-hand side is known. Given a step size $h$ these equations can be solved to yield values for $v_{①}^1$ and $v_{②}^1$. Then $v_4^1$ can be computed, the right hand side updated and the equations solved again to yield values for $v_{①}^2$ and $v_{②}^2$. This process is repeated over the time interval of interest. A computer program to solve these equations recursively is shown in Fig. 13-13*a*. The program's output is shown in Fig. 13-13*b*.  □

The companion model we have found above emerges from a network interpretation of the recursion relationship which results from the application of the backward Euler method to the capacitor branch relationship. If the forward Euler method is to be employed for the analysis of the network of Fig. 13-7, a different companion model would be used.

Consider the analysis of the network of Fig. 13-7 using the forward Euler method. Application of the forward Euler method to the capacitor branch relationship

$$\dot{v}_3 = \frac{1}{C_3} i_3$$

yields the discretized branch relationship

(13.19)
$$v_3^{k+1} = v_3^k + \frac{h}{C_3} i_3^k$$

```
C      CHAPTER 13 EXAMPLE 13.5
C
C      DIMENSIONS
C
       IMPLICIT REAL*8 (A-H,O-Z)
       DIMENSION Y(2,2),CIN(2)
C
C      READ IN ELEMENT VALUES,TIME STEP,FINAL TIME AND
C         INITIAL CAPACITANCE VOLTAGE(V4I)
C
     1 CALL READ (R2,R3,C4,V4I,H,TF)
C
C      SET UP AND SOLVE NODE EQUATIONS
C
       T=0.
     2 CALL NODEQN (R2,R3,C4,V4I,V4,H,T,Y,CIN)
       CALL GAUSS (Y,CIN,2,2)
       CALL PRINT (CIN,H,T)
       V4=CIN(1)-CIN(2)
C
C      IF T IS LESS THAN FINAL TIME (TF) INCREMENT T
C
       IF(T.GT.TF) GO TO 1
       T=T+H
       GO TO 2
       END



       SUBROUTINE READ (R2,R3,C4,V4I,H,TF)
       IMPLICIT REAL*8 (A-H,O-Z)
       READ(5,1) R2,R3,C4,V4I,H,TF
     1 FORMAT(6G10.0)
       RETURN
       END



       SUBROUTINE PRINT (CIN,H,T)
       IMPLICIT REAL*8 (A-H,O-Z)
       DIMENSION CIN(2)
       IF(T.NE.0.) GO TO 1
       PRINT 2,H
     1 PRINT 3,T,(CIN(J),J=1,2)
     2 FORMAT(///' RESULTS FOR EXAMPLE 13.5 USING A STEP SIZE OF',G12.5//
      +12X,'TIME',18X,'V1',18X,'V2')
     3 FORMAT(10X,G12.5,8X,G12.5,10X,G12.5)
       RETURN
       END
       SUBROUTINE NODEQN (R2,R3,C4,V4I,V4,H,T,Y,CIN)
       IMPLICIT REAL*8 (A-H,O-Z)
       DIMENSION Y(2,2),CIN(2)
C
C      DETERMINE SOURCE VALUE
C
       CI1=DEXP(-T)
C
C      AT TIME T=0 SOLVE INITIAL CONDITION NETWORK
C
```

**Fig. 13-13***a*  **Computer program used to recursively solve the nodal equilibrium equations for the companion network of Fig. 13-12.**

```
         IF(T.GT.O.) GO TO 1
C
C        INITIAL CONDITION NETWORK
C
         Y(1,1)=1./R2
         Y(1,2)=1./R3
         CIN(1)=CI1
         Y(2,1)=1.
         Y(2,2)=-1.
         CIN(2)=V4I
         RETURN
C
C        UPDATE COMPANION MODEL
C
       1 CI4=(1./(2.*H))*V4
         CR4=2.*H
C
C        COMPANION NETWORK
C
         Y(1,1)=(1./R2)+(1./CR4)
         Y(1,2)=-1./CR4
         CIN(1)=CI1+CI4
         Y(2,1)=-1./CR4
         Y(2,2)=(1./CR4)+(1./R3)
         CIN(2)=-CI4
         RETURN
         END
```

```
----- DATA FOR PROGRAM OF EXAMPLE 13.5 -----
         2.          3.          .5          0.          .8          3.
         2.          3.          .5          0.          .4          3.
         2.          3.          .5          0.          .2          3.
         2.          3.          .5          0.          .1          3.
         2.          3.          .5          0.          .05         2.5
```

**Fig. 13-13a (*Continued*)**

which at time $t_{k+1}$ can be modelled by a voltage source with value $v_3^k + (h/C_3)i_3^k$. The network that results when the capacitor is replaced by this companion model is shown in Fig. 13-14. The node voltage $v_{\text{①}}^{k+1}$ is

$$(13.20) \qquad v_{\text{①}}^{k+1} = v_3^k + \left(\frac{h}{C_3}\right)i_3^k$$

Observe that since

$$v_3(t) = v_{\text{①}}(t)$$

and

$$i_3(t) = -\hat{i}_1(t) - i_2(t)$$

$$= -\hat{i}_1(t) - \frac{1}{R_2}v_{\text{①}}(t)$$

RESULTS FOR EXAMPLE 13.5 USING A STEP SIZE OF 0.80000

| TIME | V1 | V2 |
|------|------|------|
| 0.0 | 1.2000 | 1.2000 |
| 0.80000 | 0.62634 | 0.40848 |
| 1.6000 | 0.34745 | 0.84517D-01 |
| 2.4000 | 0.20613 | -0.37044D-01 |
| 3.2000 | 0.13051 | -0.73478D-01 |

RESULTS FOR EXAMPLE 13.5 USING A STEP SIZE OF 0.40000

| TIME | V1 | V2 |
|------|------|------|
| 0.0 | 1.2000 | 1.2000 |
| 0.40000 | 0.87835 | 0.69343 |
| 0.80000 | 0.65254 | 0.36918 |
| 1.2000 | 0.49238 | 0.16501 |
| 1.6000 | 0.37744 | 0.39531D-01 |
| 2.0000 | 0.29386 | -0.34778D-01 |
| 2.4000 | 0.23219 | -0.76137D-01 |
| 2.8000 | 0.18600 | -0.96574D-01 |
| 3.2000 | 0.15085 | -0.10399 |

RESULTS FOR EXAMPLE 13.5 USING A STEP SIZE OF 0.20000

| TIME | V1 | V2 |
|------|------|------|
| 0.0 | 1.2000 | 1.2000 |
| 0.20000 | 1.0310 | 0.90970 |
| 0.40000 | 0.88903 | 0.67741 |
| 0.60000 | 0.76947 | 0.49223 |
| 0.80000 | 0.66851 | 0.34523 |
| 1.0000 | 0.58299 | 0.22916 |
| 1.2000 | 0.51033 | 0.13809 |
| 1.4000 | 0.44840 | 0.67195D-01 |
| 1.6000 | 0.39543 | 0.12551D-01 |
| 1.8000 | 0.34996 | -0.29043D-01 |
| 2.0000 | 0.31079 | -0.60184D-01 |
| 2.2000 | 0.27693 | -0.82984D-01 |
| 2.4000 | 0.24754 | -0.99154D-01 |
| 2.6000 | 0.22193 | -0.11008 |
| 2.8000 | 0.19954 | -0.11689 |
| 3.0000 | 0.17989 | -0.12048 |
| 3.2000 | 0.16258 | -0.12158 |

RESULTS FOR EXAMPLE 13.5 USING A STEP SIZE OF 0.10000D 00

| TIME | V1 | V2 |
|------|------|------|
| 0.0 | 1.2000 | 1.2000 |
| 0.10000D 00 | 1.1136 | 1.0440 |
| 0.20000 | 1.0344 | 0.90453 |
| 0.30000 | 0.96174 | 0.77984 |
| 0.40000 | 0.89497 | 0.66851 |
| 0.50000 | 0.83360 | 0.56919 |
| 0.60000 | 0.77716 | 0.48070 |
| 0.70000 | 0.72520 | 0.40195 |
| 0.80000 | 0.67735 | 0.33197 |
| 0.90000 | 0.63323 | 0.26986 |

Fig. 13-13*b*    Output of the program in Fig 13-13a.

| | | |
|---|---|---|
| **1.0000** | **0.59253** | **0.21484** |
| 1.1000 | 0.55496 | 0.16618 |
| 1.2000 | 0.52023 | 0.12324 |
| 1.3000 | 0.48811 | 0.85427D-01 |
| 1.4000 | 0.45838 | 0.52216D-01 |
| 1.5000 | 0.43084 | 0.23131D-01 |
| 1.6000 | 0.40530 | -0.22590D-02 |
| 1.7000 | 0.38159 | -0.24341D-01 |
| 1.8000 | 0.35957 | -0.43464D-01 |
| 1.9000 | 0.33910 | -0.59943D-01 |
| 2.0000 | 0.32004 | -0.74060D-01 |
| 2.1000 | 0.30229 | -0.86072D-01 |
| 2.2000 | 0.28574 | -0.96208D-01 |
| 2.3000 | 0.27030 | -0.10467 |
| 2.4000 | 0.25587 | -0.11166 |
| 2.5000 | 0.24239 | -0.11732 |
| 2.6000 | 0.22976 | -0.12182 |
| 2.7000 | 0.21794 | -0.12529 |
| 2.8000 | 0.20686 | -0.12785 |
| 2.9000 | 0.19646 | -0.12961 |
| 3.0000 | 0.18669 | -0.13067 |
| 3.1000 | 0.17751 | -0.13111 |

RESULTS FOR EXAMPLE 13.5 USING A STEP SIZE OF 0.50000D-01

| TIME | V1 | V2 |
|---|---|---|
| 0.0 | 1.2000 | 1.2000 |
| 0.50000D-01 | 1.1564 | 1.1191 |
| 0.10000D 00 | 1.1146 | 1.0426 |
| 0.15000 | 1.0746 | 0.97021 |
| 0.20000 | 1.0363 | 0.90180 |
| 0.25000 | 0.99951 | 0.83714 |
| 0.30000 | 0.96427 | 0.77604 |
| 0.35000 | 0.93049 | 0.71832 |
| 0.40000 | 0.89810 | 0.66380 |
| 0.45000 | 0.86704 | 0.61233 |
| 0.50000 | 0.83724 | 0.56374 |
| 0.55000 | 0.80865 | 0.51788 |
| 0.60000 | 0.78121 | 0.47462 |
| 0.65000 | 0.75487 | 0.43383 |
| 0.70000 | 0.72959 | 0.39537 |
| 0.75000 | 0.70532 | 0.35912 |
| 0.80000 | 0.68201 | 0.32498 |
| 0.85000 | 0.65961 | 0.29282 |
| 0.90000 | 0.63810 | 0.26256 |
| 0.95000 | 0.61743 | 0.23408 |
| 1.0000 | 0.59756 | 0.20730 |
| 1.0500 | 0.57846 | 0.18213 |
| 1.1000 | 0.56009 | 0.15848 |
| 1.1500 | 0.54242 | 0.13627 |
| 1.2000 | 0.52543 | 0.11543 |
| 1.2500 | 0.50908 | 0.95889D-01 |
| 1.3000 | 0.49335 | 0.77570D-01 |
| 1.3500 | 0.47821 | 0.60412D-01 |
| 1.4000 | 0.46363 | 0.44353D-01 |
| 1.4500 | 0.44958 | 0.29334D-01 |
| 1.5000 | 0.43606 | 0.15300D-01 |
| 1.5500 | 0.42303 | 0.21976D-02 |
| 1.6000 | 0.41048 | -0.10024D-01 |
| 1.6500 | 0.39837 | -0.21411D-01 |

**Fig. 13-13b (Continued)**

| 1.7000 | 0.38671 | -0.32011D-01 |
|--------|---------|--------------|
| 1.7500 | 0.37546 | -0.41865D-01 |
| 1.8000 | 0.36461 | -0.51015D-01 |
| 1.8500 | 0.35414 | -0.59499D-01 |
| 1.9000 | 0.34404 | -0.67354D-01 |
| 1.9500 | 0.33429 | -0.74615D-01 |
| 2.0000 | 0.32488 | -0.81315D-01 |
| 2.0500 | 0.31579 | -0.87486D-01 |
| 2.1000 | 0.30702 | -0.93157D-01 |
| 2.1500 | 0.29854 | -0.98357D-01 |
| 2.2000 | 0.29035 | -0.10311 |
| 2.2500 | 0.28243 | -0.10745 |
| 2.3000 | 0.27478 | -0.11139 |
| 2.3500 | 0.26738 | -0.11496 |
| 2.4000 | 0.26022 | -0.11817 |
| 2.4500 | 0.25330 | -0.12106 |
| 2.5000 | 0.24660 | -0.12364 |
| 2.5500 | 0.24011 | -0.12593 |

**Fig. 13-13b (Continued)**

(13.20) can be written as

$$v_①^{k+1} = v_①^k + \frac{h}{C_3}\left(-\frac{1}{R_2}v_1^k - i_1^k\right)$$

which, of course, is exactly the expression that would result if the forward Euler method were used directly on (13.11), the nodal equilibrium equation for the network of Fig. 13-7.

The numerical solution of first-order networks which contain an inductor proceeds in the same manner as the numerical solution of first-order networks which contain a capacitor. First, we apply the numerical integration scheme to be used directly to the inductor branch relationship

(13.21)
$$\frac{di(t)}{dt} = \frac{1}{L}v(t)$$

to arrive at the discretized inductor branch relationship. In particular, the discretized inductor branch relationship is

(13.22)
$$i^{k+1} = i^k + \frac{h}{L}v^k$$

for the forward Euler method; and

(13.23)
$$i^{k+1} = i^k + \frac{h}{L}v^{k+1}$$

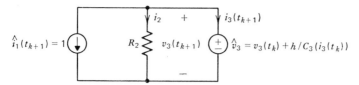

**Fig. 13-14   The network of Fig. 12-10 with the capacitor replaced by its forward Euler companion model.**

for the backward Euler method. Second, the companion model is found by making a network interpretation of the discretized branch relationships. The companion model at time $t_{k+1}$ for the forward Euler method, (13.22), is a current source of value $i^k + (h/L)v^k$. The companion model at time $t_{k+1}$ for the backward Euler method, (13.33), is a current source of value $i^k$ in parallel with a resistor of value $L/h$. These models are shown in Fig. 13-15. Repeated analysis of the companion network yields the solution of the original network.

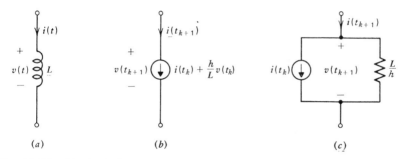

**Fig. 13-15   (a) An inductor and the companion models for (b) the forward Euler method; (c) the backward Euler method.**

## 13-4   THE SUBROUTINE PLOT

Since visual displays are so useful when trying to understand new concepts we introduce, at this point, a subroutine that can produce plots on the line printer. Because this book is not a text on computer programing, it is not our intention to discuss the logic behind the plot subroutine; rather, we merely describe its use. The interested student should be able to comprehend how this program works by studying the listing in Fig. 13-16.

The subroutine PLOT is capable of printing a two dimensional graph of two variables. It is called from another program by the statement

$$\text{CALL PLOT (X, Y, NPTS, NPLOT, IT)}$$

```
C                        SUBROUTINE 'PLOT'.
C
C
C       PLOT IS A SUBROUTINE FOR PRINTING A TWO DIMENSIONAL
C       GRAPH OF TWO VARIABLES.
C
C
C       PLOT IS CALLED FROM ANOTHER SUBPROGRAM BY
C       USING THE STATEMENT
C
C              CALL PLOT(X,Y,NPTS,NPLOT,IT)
C
C       X IS A SINGLE SUBSCRIPTED REAL*8 ARRAY,DIMENSIONED
C       NPTS IN THE CALLING PROGRAM. X FORMS THE AXIS OF
C       ABSCISSAS OF THE GRAPH. Y IS A DOUBLE SUBSCRIPTED
C       REAL*8 ARRAY DIMENSIONED NPLOT X NPTS IN THE CALLING
C       PROGRAM. Y FORMS THE AXIS OF ORDINATES OF THE GRAPH.
C       NPTS IS THE NUMBER OF POINTS TO BE PLOTTED FOR EACH
C       PLOT ON THE GRAPH. NPLOT IS THE NUMBER OF PLOTS
C       TO BE PLACED ON ONE GRAPH. NPLOT CANNOT EXCEED 2.
C       'IT' IS AN INTEGER VARIABLE THAT DETERMINES WHETHER
C       THE AXIS OF ABSCISSAS (X AXIS) IS TO BE SCALED
C       LINEARLY OR LOGARITHMICALLY. FOR 'IT' EQUAL TO 1,
C       THE AXIS OF ABSCISSAS(X AXIS) IS LOGARITHMIC.
C       FOR 'IT' NOT EQUAL 1, THE AXIS OF ABSCISSAS (X AXIS)
C       IS LINEAR.
C
C
        SUBROUTINE PLOT(X,Y,NPTS,NPLOT,IT)
        IMPLICIT REAL*8(A-H,O-Z)
        REAL*8 X(NPTS),Y(NPLOT,NPTS),XSCAL(11)
        LOGICAL*1 BL,$PT,BLANK,DIV,BAR,MINUS,PLUS,LINE(101),SYMB(2)
        DATA BLANK/' '/,BAR/'I'/,MINUS/'-'/,PLUS/'+'/,
       1SYMB/'X','O'/
C
C       INITIAL DATA SET UP --- FIND MAXIMUM AND MINIMUM VALUES OF X AND Y
C
        IF(NPLOT.GE.1.AND.NPLOT.LE.2) GO TO 1
        PRINT 2
2       FORMAT(' NUMBER OF PLOTS NOT BETWEEN 1 AND 2')
        GO TO 100
1       YMAX=Y(1,1)
        YMIN=YMAX
        DO 3 I=1,NPLOT
        DO 3 J=1,NPTS
        YT=Y(I,J)
        IF(YT.GT.YMAX) YMAX=YT
        IF(YT.LT.YMIN) YMIN=YT
3       CONTINUE
        XMAX=X(1)
        XMIN=XMAX
        DO 4 I=1,NPTS
        XT=X(I)
        IF(XT.GT.XMAX) XMAX=XT
        IF(XT.LT.XMIN) XMIN=XT
4       CONTINUE
        YR=YMAX-YMIN
        XR=XMAX-XMIN
```

**Fig. 13-16   Listing of the subroutine PLOT.**

```
      IF(IT.NE.1) GO TO 5
      DFL=(DLOG10(XMAX)-DLOG10(XMIN))/100.DO
      DF=10.**((DLOG10(XMAX)-DLOG10(XMIN))/10.DO)
C
C     SET UP EACH LINE OF GRAPH
C
5     IYS=0
      DO 70 IYT=1,51
      IYA=52-IYT
      IYS=IYS+1
      BL=BLANK
      DIV=BAR
      $PE=.FALSE.
      IF(IYS.NE.1) GO TO 30
      BL=MINUS
      DIV=PLUS
      $PT=.TRUE.
      YSCAL=YMAX-(IYT-1)*YR/50.
30    DO 35 IXA=1,101
35    LINE(IXA)=BL
      DO 40 IXA=1,101,10
40    LINE(IXA)=DIV
C
C     INSERT DATA POINTS ALONG GRAPH LINE
C
50    DO 60 L=1,NPLOT
      DO 60 M=1,NPTS
      IY=50.*(Y(L,M)-YMIN)/YR+1.4999
      IF(IT.NE.1) GO TO 58
      IX=(DLOG10(X(M))-DLOG10(XMIN))/DFL+1.49999
      GO TO 59
58    IX=100.*(X(M)-XMIN)/XR+1.49999
59    IF(IY.NE.IYA) GO TO 60
      LINE(IX)=SYMB(L)
60    CONTINUE
      IF($PT) PRINT 1000,YSCAL,LINE
      IF(.NOT.$PT) PRINT 1100,LINE
      IF(IYS.EQ.5) IYS=0
70    CONTINUE
C
C     PRINT X-AXIS SCALE VALUES
C
80    DO 90 IXM=1,11
      IF(IT.NE.1) GO TO 89
      XSCAL(IXM)=XMIN*DF**(IXM-1)
      GO TO 90
89    XSCAL(IXM)=XMIN+(IXM-1)*XR/10.
90    CONTINUE
      PRINT 1200, XSCAL(1),XSCAL(3),XSCAL(5),XSCAL(7),XSCAL(9),
     1    XSCAL(11),XSCAL(2),XSCAL(4),XSCAL(6),XSCAL(8),XSCAL(10)
1000  FORMAT(' ',5X,1PG11.4,1X,101A1)
1100  FORMAT(' ',17X,101A1)
1200  FORMAT('0',3X,1P6G20.3/13X,5G20.3)
100   RETURN
      END
```

**Fig. 13-16** (*Continued*)

where

X      is a single subscripted, double precision, array dimensioned NPTS in the calling program that forms the abscissa of the graph;

Y      is a double subscripted, double precision, array dimensioned (NPLOT, NPTS) in the calling program; Y(1, NPTS) contains the ordinates of the first graph to be plotted and Y(2, NPTS) contains the ordinates of the second graph to be plotted;

NPTS      is the number of points to be plotted;

NPLOT      is the number of plots (1 or 2) to be placed on the graph;

and

IT      is used to determine if the abscissa is to be plotted linearly or logarithmically (for use in frequency response plots to be discussed later.) If IT $= 1$ a logarithmic scale is used; if IT $\neq 1$ a linear scale is used.

The following example illustrates the use of PLOT.


**Example 13.6**

The companion network associated with the network of Fig. 13-17$a$ for the forward Euler method is shown in Fig. 13-17$b$. The nodal equilibrium equations for the companion network are

$$3v_{①}^{k+1} - 2v_{②}^{k+1} = \sin{(k+1)}h - \left( i_3^k + \frac{h}{L}v_3^k \right)$$

and

$$-2v_{①}^{k+1} + 2\frac{1}{3}v_{②}^{k+1} = i_3^k + \frac{h}{L}v_3^k$$

where $v_3^k = v_{①}^k - v_{②}^k$ and $i_3^k = i_3^{k-1} + (h/L)v_3^{k-1}$ for $k \geq 1$. For the case of $k = 0$, the right-hand side of these equations involvesthe term $i_3^0 + (h/L)v_3^0$. The term $i_3^0$ is given as the initial condition. The term $v_3^0$ must be found from the initial condition network shown in Fig. 13-17$c$. A program which solves these equations that calls PLOT to plot the response is given in Fig. 13-18$a$. The results are shown in Fig. 13-18$b$.   □

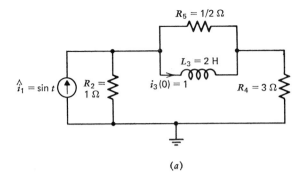

(a)

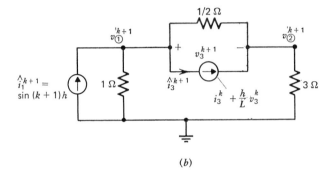

(b)

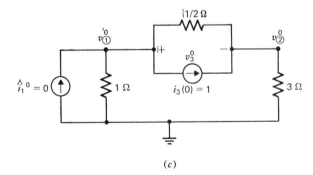

(c)

**Fig. 13-17**  (a) The network of Example 13.6 and (b) the companion network for the forward Euler method. (c) The initial condition network.

```
C     CHAPTER 13 EXAMPLE 13.6
C
C     DIMENSIONS
C
      IMPLICIT REAL*8 (A-H,O-Z)
      DIMENSION Y(2,2),CIN(2),XPLOT(50),YPLOT(2,50)
C
C     READ IN ELEMENT VALUES, TIME STEP, FINAL TIME AND
C         INITIAL INDUCTANCE CURRENT(CI3I)
C
    4 CALL READ (R2,EL3,R4,R5,CI3I,H,TF)
C
C     SET UP AND SOLVE NODE EQUATIONS
C
      NPTS=0
      T=0.
    2 NPTS=NPTS+1
      XPLOT(NPTS)=T
      CALL NODEQN (R2,EL3,R4,R5,CI3I,CI3,V3,H,T,Y,CIN)
      CALL GAUSS (Y,CIN,2,2)
C
C     STORE SOLUTION FOR PLOTTING
C
      YPLOT(1,NPTS)=CIN(1)
      YPLOT(2,NPTS)=CIN(2)
      CALL PRINT (CIN,H,T)
      V3=CIN(1)-CIN(2)
C
C     IF T LESS THAN FINAL TIME (TF) INCREMENT T
C
      IF(T.GT.TF) GO TO 1
      T=T+H
      GO TO 2
    1 PRINT 3,H
      CALL PLOT(XPLOT,YPLOT,NPTS,2,0)
    3 FORMAT('1',38X,'GRAPH OF V1 AND V2 VS TIME USING A STEP SIZE OF
     +,G12.5//53X,'V1="X"',10X,'V2="O"'//)
      GO TO 4
      END



      SUBROUTINE READ (R2,EL3,R4,R5,CI3I,H,TF)
      IMPLICIT REAL*8 (A-H,O-Z)
      READ(5,1) R2,EL3,R4,R5,CI3I,H,TF
    1 FORMAT(7G10.0)
      RETURN
      END
```

**Fig. 13-18a** Computer program used to recursively solve the nodal equilibrium equations for the companion network of Fig. 13-17b.

```
      SUBROUTINE NODEQN (R2,EL3,R4,R5,CI3I,CI3,V3,H,T,Y,CIN)
      IMPLICIT REAL*8 (A-H,O-Z)
      DIMENSION Y(2,2),CIN(2)
C
C     DETERMINE SOURCE VALUE
C
      CI1=DSIN(T)
C
C     AT TIME T=0 SOLVE INITIAL CONDITION NETWORK
C
      IF(T.GT.0.) GO TO 1
C
      CI3=CI3I
      GO TO 2
C
C     UPDATE COMPANION MODEL
C
    1 CI3=CI3+(H/EL3)*V3
C
C     COMPANION NETWORK
C
    2 Y(1,1)=(1./R2)+(1./R5)
      Y(1,2)=-1./R5
      CIN(1)=CI1-CI3
      Y(2,1)=-1./R5
      Y(2,2)=(1./R4)+(1./R5)
      CIN(2)=CI3
      RETURN
      END



      SUBROUTINE PRINT (CIN,H,T)
      IMPLICIT REAL*8 (A-H,O-Z)
      DIMENSION CIN(2)
      IF(T.NE.0.) GO TO 1
      PRINT 2,H
    1 PRINT 3,T,(CIN(J),J=1,2)
    2 FORMAT(' RESULTS FOR EXAMPLE 13.6 USING A STEP SIZE OF',G12.5///
     +12X,'TIME',18X,'V1',18X,'V2')
    3 FORMAT(10X,G12.5,8X,G12.5,10X,G12.5)
      RETURN
      END
```

----- DATA FOR PROGRAM OF EXAMPLE 13.6 -----

|    |    |    |    |    |    |    |
|----|----|----|----|----|----|----|
| 1. | 2. | 3. | .5 | 1. | .1 | 3. |
| 1. | 2. | 3. | .5 | 1. | .2 | 3. |
| 1. | 2. | 3. | .5 | 1. | .4 | 3. |
| 1. | 2. | 3. | .5 | 1. | .8 | 3. |

**Fig. 13-18a** (*Continued*)

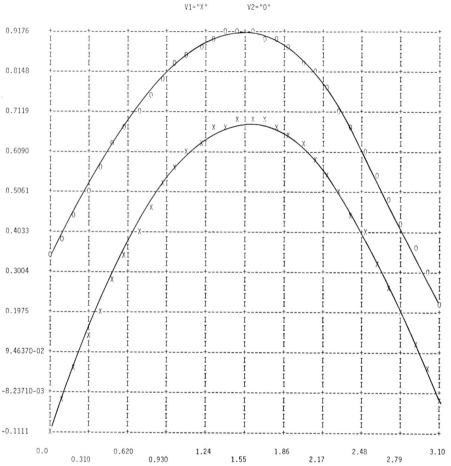

Fig. 13-18*b*   **Results of program in (*a*).**

## 13-5  SUMMARY

We presented two numerical methods that can be used to solve the first-order differential equation

$$\dot{y}(t) = f(y(t), v(t), t)$$

These methods are the *forward Euler method*, expressed by the *recursion relationship*

$$y^{k+1} = y^k + hf^k$$

and the *backward Euler method* expressed by the recursion relationship

$$y^{k+1} = y^k + hf^{k+1}$$

The forward Euler method is an example of an *explicit method* while the backward Euler method is an example of an *implicit method*.

The numerical solution of first-order networks is most easily undertaken in terms of *companion models*. The companion model for capacitors and inductors arise by making a network interpretation of the recursion relationship which results on application of the numerical integration technique to the branch relationship. Analysis of first-order networks is thereby transformed into the repeated analysis of a dc resistive network.

## PROBLEMS

**13.1**    Write the recursion relationship that would result from using the forward and backward Euler methods to numerically solve each of the following differential equations.

(a) $\dot{y}(t) = e^{-t}y(t)$

(b) $\dot{x}(t) = -3x(t) + 4e^{-3t}$

(c) $\dot{x}(t) = 3tx(t) + t$

**13.2**    Use the forward Euler integration scheme to solve the following differential equations. Use a step size of $h = .1$ and find the solution in the interval $0 \leq t \leq .3$. Using a step size of $h = 3$, find the solution in the interval $0 \leq t \leq 9$. (Carry out the recursive process by hand.) Compare your results in each case with the exact solution obtained in Prob. 11.6.

(a) $\dfrac{dv}{dt} + 2v(t) = 0, \qquad v(0) = 2$

(b) $\dfrac{dv}{dt} + 3v(t) = 3, \qquad v(0) = 0$

**13.3**    Repeat Prob. 13.2 using the backward Euler method.

**13.4**    Write simple computer programs such as the one shown in Fig. 13-2 to solve the differential equation of Prob. 13.1b with $x(0) = 1$ using the forward Euler method. Use the program to find $x(t)$ for $0 \leq t \leq 6$ for several different step sizes. Calculate the analytic solution and compare the numerical solution with it.

**13.5**    Repeat Prob. 13.4 using the backward Euler method.

**13.6**    Using the forward Euler integration scheme, write a simple program and use it to solve the differential equations

(a) $\dot{x}(t) = -x(t) + t, \; x(0) = 1$

(b) $\dot{x}(t) = -3tx(t), \; x(0) = 2$

in the interval $0 \leq t \leq 5$ using a step size of $h = .1$.

**13.7**    Repeat Prob. 13.6 using the backward Euler method.

**13.8**     If we were to write a computer program based upon mesh analysis, a companion model that had an independent voltage source would be more convenient to use than one with an independent current source. Derive companion models for capacitors and inductors which have voltage sources.

**13.9**     Use the program of Fig. 13-13a to study the relationship between the time constant of a network and its zero-state response. To do this, let $v_4(0) = 0$ and $\hat{i}_1(t) = u(t)$. Use the program to compare the response, $v_2(t)$, for the following values of $R_2$, $C_4$, and $R_3$:

(a) $R_2 = 3\ \Omega$,     $C_4 = \dfrac{1}{2}\,\mathrm{F}$,     $R_3 = 3\ \Omega$

(b) $R_2 = 1\ \Omega$,     $C_4 = \dfrac{1}{2}\,\mathrm{F}$,     $R_3 = 3\ \Omega$

(c) $R_2 = 2\ \Omega$,     $C_4 = 1\mathrm{F}$,     $R_3 = 3\ \Omega$

(d) $R_2 = 2\ \Omega$,     $C_4 = \dfrac{1}{4}\,\mathrm{F}$,     $R_3 = 3\ \Omega$

(e) $R_2 = 2\ \Omega$,     $C_4 = \dfrac{1}{2}\,\mathrm{F}$,     $R_3 = 4\ \Omega$

(f) $R_2 = 2\ \Omega$,     $C_4 = \dfrac{1}{2}\,\mathrm{F}$,     $R_3 = 2\ \Omega$

How does changes in $R_2$, $C_4$, and $R_3$ effect the time constant?

**13.10**     Determine the response $v_2(t)$, if a three ohm resistor is placed across the network of Fig. 13-12. (*Hint*. Modify the programs of Fig. 13-13a and use the subroutine PLOT.)

**13.11**     Use the program of Fig. 13-18a to plot the step response of the network of Fig. 13-17.

**13.12**     Consider again the networks of Prob. 11.5. Replace each energy storage element by the companion modal associated with the forward Euler integration scheme. Use these networks to determine the response for a time interval equal to twice the network's time constant using a step size equal to one half of the time constant, a step size equal to the time constant, and a step size equal to twice the time constant. Compare these results with the actual solutions obtained in Prob. 11.5.

**13.13**     Repeat Prob. 13.12 using the backward Euler method.

**13.14**     Use the backward Euler companion model to transform the networks of Probs. 11.7 and 11.8 into companion networks and write the nodal equilibrium equations for each of the companion networks. Write a computer program which calls GAUSS to solve the nodal equilibrium equations numerically and PLOT to plot the capacitor voltage or inductor current for various step sizes. Find an appropriate step size in each case.

## CHAPTER 14

# Time Domain Analysis of Higher-Order Circuits

In Chapter 11 we considered the analysis of first-order networks. Such analysis is called *time-domain analysis* because time is the independent variable. We now are ready to discuss the time-domain analysis of general *RLC* circuits. In this chapter we present the analytic solution and start by considering networks that contain two energy storage elements.

## 14-1  THE RLC NETWORK

Consider the *RLC* network of Fig. 14-1. The single independent *KCL* node equation is

$$-i_1(t) + i_2(t) + i_3(t) + i_4(t) = 0$$

On substitution of the branch relationships expressed in terms of the node voltage $v_①$, we have the nodal equilibrium equation

$$C_3 \frac{dv_①(t)}{dt} + \frac{1}{R_2} v_①(t) + i_4(0) + \int_0^t \frac{1}{L_4} v_①(\tau) \, d\tau + \hat{\imath}_1(t)$$

This integrodifferential equation must be solved for $v_①(t)$. Differentiating both sides yields the second order differential equation

(14.1) $$C_3 \frac{d^2 v_①(t)}{dt^2} + \frac{1}{R_2} \frac{dv_①(t)}{dt} + \frac{1}{L_4} v_①(t) = \frac{d\hat{\imath}_1(t)}{dt}$$

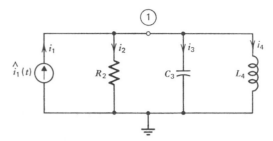

**Fig. 14-1   A second-order *RLC* network.**

This second-order differential equation is a special form of the more general constant-coefficient second-order differential equation

(14.2)
$$a_2 \frac{d^2x(t)}{dt^2} + a_1 \frac{dx(t)}{dt^2} + a_0 x(t) = f(t)$$

For ease of manipulation we introduce the differential operator notation:

$$D \equiv \frac{d}{dt}$$

$$D^2 \equiv \frac{d^2}{dt^2}$$

and

$$D^k \equiv \frac{d^k}{dt^k}, \qquad \text{for all positive integers } k$$

We also define the *integral operator*

$$D^{-1} = \int_0^t \cdot \, dt$$

The general second-order differential equation written in terms of these operators is

$$a_2 D^2 x(t) + a_1 D x(t) + a_0 x(t) = f(t)$$

or

(14.3)
$$(a_2 D^2 + a_1 D + a_0)x(t) = f(t)$$

because

$$a_2 D^2 x(t) = a_2 \frac{d^2 x(t)}{dt}$$

and

$$a_1 D x(t) = a_1 \frac{dx(t)}{dt}$$

These operators can be used to express the capacitor branch relationship as:

$$i(t) = C Dv(t)$$

or

$$v(t) = v(0) + \frac{1}{CD} i(t)$$

and the inductor branch relationship as

$$v(t) = L Di(t)$$

or

$$i(t) = i(0) + \frac{1}{LD} v(t)$$

The operators $D$ and $1/D$ are *linear operators* and obey the usual distributive, commutative, and associative laws. Some useful relations are shown below:

$$D[x_1(t) + x_2(t)] = Dx_1(t) + Dx_2(t)$$

$$\frac{1}{D} [x_1(t) + x_2(t)] = \frac{1}{D} x_1(t) + \frac{1}{D} x_2(t)$$

$$D[ax(t)] = a[Dx(t)], \qquad a \text{ is a scalar}$$

$$\frac{1}{D} [ax(t)] = a\left(\frac{1}{D} x(t)\right), \qquad a \text{ is a scalar}$$

$$D[Dx(t)] = D^2 x(t) \equiv \frac{d^2 x(t)}{dt^2}$$

$$D^m x(t) = \frac{d^m x(t)}{dt^m}$$

$$[D^m + D^n + D]x(t) = D^m x(t) + D^n x(t) + Dx(t)$$

and

$$D^m \left[\frac{1}{D_n} x(t)\right] = D^{m-n} x(t)$$

Just as in the first-order case, the solution of the second-order differential equation is the sum of the *homogeneous solution* and the *particular solution*. The homogeneous equation is (14.3) with $f(t)$ set to zero:

(14.4)     $$(a_2 D^2 + a_1 D + a_0)x(t) = 0$$

Observe that the homogeneous solution must be such that the sum of it, its first derivative, and its second derivative, each multiplied by a constant must be zero. We are led, therefore, to assume that the homogeneous solution is of the form

$$x_H(t) = ce^{st}$$

where $c$ and $s$ are some finite constants. (These constants may be complex.) Substitution of $x_H$ into the homogeneous equation yields

$$c(a_2 s^2 + a_1 s + a_0)e^{st} = 0$$

Therefore, $x_H(t)$ is indeed the homogeneous solution if $s$ is chosen to satisfy the *characteristic equation*

$$(a_2 s^2 + a_1 s + a_0)e^{st} = 0$$

in other words, $s = s_1$ or $s = s_2$ where $s_1$ and $s_2$ are the roots of the characteristic equation, that is,

$$s_1 = \frac{-a_1 + (a_1{}^2 - 4a_0 a_2)^{1/2}}{2a_2}$$

and

$$s_2 = \frac{-a_1 - (a_1{}^2 - 4a_0 a_2)^{1/2}}{2a_2}$$

The roots of the characteristic equation are referred to as the *natural frequencies*, *critical frequencies*, or *eigenvalues* of the network.

Thus, there are two solutions to the homogeneous equation, $c_1 e^{s_1 t}$ and $c_2 e^{s_2 t}$. Since the sum of these solutions is also a solution of the homogeneous equation we have that

(14.5)
$$x_H(t) = c_1 e^{s_1 t} + c_2 e^{s_2 t}$$

where $c_1$ and $c_2$ are constants that depend on the two initial conditions. If the two roots of the characteristic equation are equal, that is, $s_1 = s_2$, then (14.5) reduces to

$$x_H(t) = k_1 e^{s_1 t}$$

and since only one arbitrary constant is present this expression cannot be the general solution to (14.4). In order to remedy this situation, we try

$$x_H(t) = \xi(t)e^{s_1 t}$$

where $\xi(t)$ is an arbitrary function of time, as the homogeneous solution. Substitution of this expression into the homogeneous equation yields

(14.6)    $$[a_2 D^2 + (2a_2 s_1 + a_1)D + (a_2 s_1{}^2 + a_1 s_1 + a_0)]\xi(t)e^{s_1 t} = 0$$

But

$$a_2 s_1{}^2 + a_1 s_1 + a_0 = 0$$

and because $s_1$ is a multiple root of the characteristic equation

$$(a_1{}^2 - 4a_0 a_2)^{1/2} = 0$$

so that

$$s_1 = \frac{-a_1}{2a_2}$$

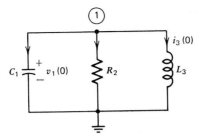

**Fig. 14-2   A second-order unforced *RLC* network.**

or

$$2a_2s_1 + a_1 = 0$$

Thus (14.6) becomes

$$a_2 D^2 \xi(t)e^{s_1t} = 0$$

Therefore,

$$x_H(t) = \xi(t)_1 e^{s_1t}$$

is a solution of the homogeneous equation if

$$\frac{d^2\xi(t)}{dt^2} = 0$$

Certainly if we choose

$$\xi(t) = \hat{k}_1 + \hat{k}_2 t$$

where $\hat{k}_1$ and $\hat{k}_2$ are constants, this requirement is met. Hence, if the characteristic equation has two coincident roots, the solution to the homogeneous equation is

$$x_H(t) = c_1 e^{s_1t} + c_2 t e^{s_1t}$$

**Example 14.1**

Consider the unforced *RLC* network of Fig. 14-2. The nodal equilibrium equation is

$$\left(C_1 D + \frac{1}{R_2} + \frac{1}{L_3 D}\right)v_①(t) + i_3(0) = 0$$

or on differentiation

(14.7) $$\left(C_1 D^2 + \frac{1}{R_2} D + \frac{1}{L_3}\right)v_①(t) = 0$$

Suppose the element values are $C_1 = 1/2$F. $R_2 = 1/3 \ \Omega$ and $L_3 = 1/4$H. Then the characteristic equation is

$$\frac{1}{2}s^2 + 3s + 4 = 0$$

and the roots are $s_1 = -4$ and $s_2 = -2$

The solution becomes

(14.8)
$$v_①(t) = c_1 e^{-2t} + c_2 e^{-4t}$$

To find the constants $c_1$ and $c_2$ we need two initial conditions:

$$v_1(0) \quad \text{and} \quad \frac{dv_1(t)}{dt}\bigg|_{t=0}$$

that is, the initial node voltage and the derivative of the node voltage evaluated at time $t = 0$. (*Note.* We will use $\dfrac{dv_1(0)}{dt}$ to denote $\dfrac{dv_①(t)}{dt}\bigg|_{t=0}$.) However, the two initial conditions usually specified for this network are the initial capacitance voltage, $v_1(0)$, and the initial inductance current, $i_3(0)$. Thus we must determine $v_①(0)$ and $dv_①(0)/dt$ in terms of $v_3(0)$ and $i_4(0)$. First, we observe that

$$v_①(t) = v_1(t)$$

so that

(14.9)
$$v_①(0) = v_1(0)$$

Next, we recognize that since

$$i_1(t) = C_1 \frac{dv_2(t)}{dt}$$

$$= C_1 \frac{dv_①(t)}{dt}$$

then

$$\frac{dv_①(0)}{dt} = \frac{1}{C_1} i_1(0)$$

But, from *KCL*

$$i_1(0) = -i_2(0) - i_3(0)$$

$$= -\frac{1}{R_2} v_1(0) - i_3(0)$$

so that

(14.10)
$$\frac{dv_①(0)}{dt} = \frac{1}{C_1}\left[-\frac{1}{R_2} v_1(0) - i_3(0)\right] = -24$$

(In general, determination of the initial conditions needed for solution of the differential equations in terms of the initial capacitance voltages and inductance currents is somewhat tricky. This topic is discussed in more detail later.)

We can now solve for the constants $c_1$ and $c_2$ of (14.8). Suppose that $v_1(0) = 1$ v and $i_3(0) = 2$ a. From (14.8), (14.9), and (14.10) we have

$$v_1(0) = c_1 + c_2 = 1$$

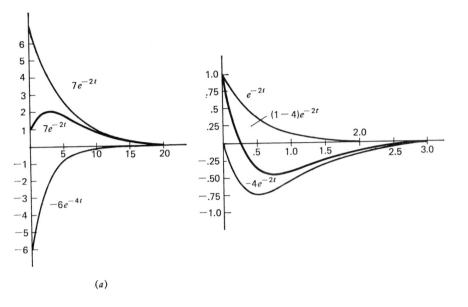

(a)

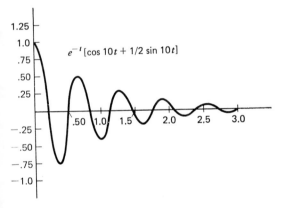

(c)

**Fig. 14-3** **Sketch of the voltage $v_①(t)$ in the network of Fig. 14-2 for (a)**
$C_1 = 1/2F$, $R_2 = 1/3\Omega$, and $L_3 = 1/4H$; (b) $C_1 = 1F$, $R_2 = 1/4\Omega$, and
$L_3 = 1/4H$; and (c) $C_1 = 1/2F$, $R_2 = 1\Omega$, and $L_3 = 1/50H$.

and

$$\frac{dv_1(0)}{dt} = -2c_1 - 4c_2 = -10$$

Thus $c_1 = 7$ and $c_2 = -6$ and the solution becomes

$$v_①(t) = 7e^{-2t} - 6e^{-4t}$$

which is a decaying exponential as illustrated in Fig. 14-3a.

Now consider the case where $C_1 = 1$ F, $R_2 = 1/4 \, \Omega$ and $L_3 = 1/4$ H. The characteristic equation associated with (14.7) becomes

$$s^2 + 4s + 4 = 0$$

and the roots are $s_1 = s_2 = -2$. Therefore, the solution takes on the form

$$v_①(t) = c_1 e^{-2t} + c_2 t e^{-2t}$$

Using the same initial conditions as before and (14.9) and (14.10) we can find $c_1$ and $c_2$:

$$v_①(0) = c_1 = 1$$

and

$$\frac{dv_①(0)}{dt} = -2c_1 + c_2 = -6$$

Therefore, $c_1 = 1$ and $c_2 = -4$ and the solution becomes

$$v_①(t) = (1 - 4t)e^{-2t}$$

A sketch of this response is shown in Fig. 14-3$b$.

Finally, consider the situation where $C_1 = 1/2$ F, $R_2 = 1 \, \Omega$ and $L_3 = 1/50$ H. Then the characteristic equation associated with (14.7) is

$$\tfrac{1}{2}s^2 + s + 50 = 0$$

and the roots are complex: $s_1 = -1 + 10j$ and $s_2 = -1 - 10j$ and the solution is

$$v_①(t) = c_1 e^{(-1+j10)t} + c_2 e^{(-1-j10)t}$$

Once again, using the initial conditions of $v_1(0) = 1$ v and $i_3(0) = 2$ a and expressions (14.9) and (14.10) we have

$$v_①(0) = c_1 + c_2 = 1$$

and

$$\frac{dv_①(0)}{dt} = (-1 + j10)c_1 + (-1 - j10)c_2 = -6$$

so that $c_1 = (1/2 + j1/4)$ and $c_2 = (1/2 - 1/4j)$. The solution is

$$v_①(t) = \left(\frac{1}{2} + j\frac{1}{4}\right)e^{(-1+j10)t} + \left(\frac{1}{2} - j\frac{1}{4}\right)e^{(-1-j10)t}$$

which is a damped sinusoid as sketched in Fig. 14-3$c$. To see this more clearly we may manipulate the solution as follows

$$v_①(t) = e^{-t}(\tfrac{1}{2}e^{j10t} + \tfrac{1}{2}e^{-j10t} + j\tfrac{1}{4}e^{j10t} - j\tfrac{1}{4}e^{-j10t})$$

Invoking Euler's identity:

$$e^{j\theta} = \cos \theta + j \sin \theta$$

and collecting terms we can rewrite the solution as

$$v_1(t) = e^{-t}\left(\cos 10t + \frac{1}{2} \sin 10t\right)$$

which is a damped sinusoid.  $\square$

## 14-2  THE FORCED RESPONSE

It is convenient at this point to introduce the "dot" convention for representing derivatives. One dot over the variable is used to indicate the first derivative, two dots the second derivative, etc. Thus we have

$$\dot{x}(t) \equiv \frac{dx(t)}{dt} = Dx(t)$$

$$\ddot{x}(t) \equiv \frac{d^2x(t)}{dt^2} = D^2x(t)$$

etc. In the sequel, we will interchangeably use differential operators and the dot convention—whichever is the most convenient at the time.

We are now ready to determine the particular solution of the general second-order differential equation (14-2):

$$\ddot{x}(t) + a_1\dot{x}(t) + a_0 = f(t)$$

The particular solution is found by variation of parameters as was done for the first-order case. First, we rewrite the homogeneous solution as

$$x_H(t) = c_1\lambda_1(t) + c_2\lambda_2(t)$$

where

$$\lambda_1(t) = e^{s_1 t}$$

and

$$\lambda_2(t) = e^{s_2 t}$$

if $s_1 \neq s_2$, or

$$\lambda_1 = e^{s_1 t}$$

and

$$\lambda_2 = te^{s_1 t}$$

if $s_1 = s_2$. Then we assume that the particular solution is of the form

$$x_P(t) = \eta_1(t)\lambda_1(t) + \eta_2(t)\lambda_2(t)$$

where $\eta_1(t)$ and $\eta_2(t)$ are unknown functions of time. Differentiation of $x_P(t)$ yields

$$\dot{x}_P = (\eta_1\dot{\lambda}_1 + \eta_2\dot{\lambda}_2) + (\dot{\eta}_1\lambda_1 + \dot{\eta}_2\lambda_2)$$

Since $\eta_1$ and $\eta_2$ are as yet undefined, let us choose them so that

(14.12)                     $$\dot{\eta}_1\lambda_1 + \dot{\eta}_2\lambda_2 = 0$$

then

$$\ddot{x}_P = \eta_1\ddot{\lambda}_1 + \eta_1\ddot{\lambda}_2 + \dot{\eta}_1\dot{\lambda}_1 + \dot{\eta}_2\dot{\lambda}_2$$

Substitution of $x_P$, $\dot{x}_P$, and $\ddot{x}_P$ into the original differential equation (14.11) yields

$$(a_2\ddot{\lambda}_1 + a_1\dot{\lambda}_1 + a_0\lambda_1)\eta_1 + (a_2\ddot{\lambda}_2 + a_1\dot{\lambda}_2 + a_0\lambda_2)\eta_2 + \dot{\lambda}_1\dot{\eta}_1 + \dot{\lambda}_2\dot{\eta}_2 = f(t)$$

But since $\lambda_1$ and $\lambda_2$ are solutions of the homogeneous equation, the expression reduces to

$$\dot{\lambda}_1\dot{\eta}_1 + \dot{\lambda}_2\dot{\eta}_2 = f(t)$$

This expression and (14.12) may be solved simultaneously (using Cramer's rule) to yield

(14.13a)                     $$\dot{\eta}_1 = (\dot{\lambda}_1 - \lambda_1\lambda_2^{-1}\dot{\lambda}_2)^{-1}f(t)$$

and

(14.13b)                     $$\dot{\eta}_2 = (\dot{\lambda}_2 - \lambda_2\lambda_1^{-1}\dot{\lambda}_1)^{-1}f(t)$$

For the case of distinct roots, the previous expressions reduce to

$$\dot{\eta}_1(t) = \frac{e^{-s_1 t}}{s_1 - s_2} f(t)$$

and

$$\dot{\eta}_2(t) = \frac{-e^{-s_1 t}}{s_1 - s_2} f(t)$$

so that

$$\eta_1(t) = \int_{t_0}^{t} \frac{e^{-s_1\tau}}{s_1 - s_2} f(\tau)\, d\tau$$

and

$$\eta_2(t) = \int_{0}^{t} \frac{-e^{-s_2\tau}}{s_1 - s_2} f(\tau)\, d\tau$$

Thus, the particular solution for the case of distinct roots is

(14.14)          $$x_P(t) = \int_{0}^{t} \frac{1}{s_1 - s_2} (e^{s_1(t-\tau)} - e^{s_2(t-\tau)}) f(\tau)\, d\tau$$

If the characteristic equation has repeated roots, then equations (14.13) become

$$\dot{\eta}_1 = -t e^{-s_1 t} f(t)$$

and

$$\dot{\eta}_2 = e^{-s_1 t} f(t)$$

so that the particular solution is

(14.15)
$$x_P(t) = \int_0^t e^{s_1(t-\tau)}(t - \tau)f(\tau)\,d\tau$$

Therefore, the complete solution of the second-order differential equation (14.11) is

(14.16)
$$x(t) = c_1 e^{s_1 t} + c_2 e^{s_2 t} + \int_0^t \frac{1}{s_1 - s_2}[e^{s_1(t-\tau)} - e^{s_2(t-\tau)}]f(\tau)\,d\tau$$

if $s_1 \neq s_2$ and

(14.17)
$$x(t) = (c_1 + c_2 t)e^{s_1 t} + \int_0^t e^{s_1(t-\tau)}(t - \tau)f(\tau)\,d\tau$$

if $s_1 = s_2$.

It remains to determine the constants $c_1$ and $c_2$. These constants are dependent on the initial conditions $x(0)$ and $\dot{x}(0)$. (We use the shorthand notation $\dot{x}(0)$ to denote the derivative of $x(t)$ evaluated at time $t = 0$, that is,

$$\dot{x}(0) = \left.\frac{dx(t)}{dt}\right|_{t=0}.$$ Given the solution, (14.16) or (14.17), if $t$ is set to zero it should equal $x(0)$. Similarly if the solution is differentiated and $t$ is set to zero, the result should equal $\dot{x}(0)$. Therefore, we have two expressions in terms of the two unknowns $c_1$ and $c_2$, which can be solved to yield values for $c_1$ and $c_2$.

## Example 14.2

The nodal equilibrium equation for the network of Fig. 14-4 is

$$\ddot{v}_①(t) + 5\dot{v}_①(t) + 6v_①(t) = 3\delta(t)$$

The homogeneous equation is

$$\ddot{v}_①(t) + 5\dot{v}_①(t) + 6v_①(t) = 0$$

so that the homogeneous solution is

$$v_{①_H}(t) = c_1 e^{-3t} + c_2 e^{-2t}$$

Since the roots of the characteristic equation are distinct, the particular solution is of the form of (14.14). In particular

$$v_{①_P}(t) = -\int_0^t (e^{-3(t-\tau)} - e^{-2(t-\tau)})3\delta(\tau)\,d\tau$$

$$= -3(e^{-3t} - e^{-2t})$$

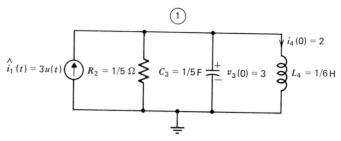

**Fig. 14-4** Second-order network used in Example 14-2 to illustrate the solution of second-order differential equations.

The complete solution is

$$v_①(t) = c_1 e^{-3t} + c_2 e^{-2t} - 3(e^{-3t} - e^{-2t})$$

The initial conditions $v_①(0)$ and $\dot{v}_①(0)$ can be expressed in terms of $\hat{i}_1(0)$, $v_3(0)$, and $i_4(0)$ as follows

$$v_①(0) = v_3(0) = 3$$

and

$$v_①(0) = \frac{1}{C_3}\left[\hat{i}_1(0) - \frac{1}{R_2}v_3(0) - i_4(0)\right]$$
$$= -70$$

But, from the solution obtained

$$v_①(0) = c_1 + c_2$$

and

$$\dot{v}_①(t) = -3c_1 e^{-3t} - 2c_2 e^{-2t} + 9e^{-3t} - 6e^{-2t}$$

so that

$$\dot{v}_①(0) = -3c_1 - 2c_2 + 3$$

Therefore,

$$c_1 + c_2 = 3$$

and

$$-3c_1 - 2c_2 + 3 = -70$$

which can be solved simultaneously to yield $c_1 = 67$ and $c_2 = -64$. Thus the complete solution is

$$v_①(t) = 64e^{-3t} - 61e^{-2t} \quad \square$$

The following examples illustrate that analysis of networks that contain two energy storage elements, that is, either two capacitors, two inductors, or one of each, *usually* requires solution of second-order differential equations.

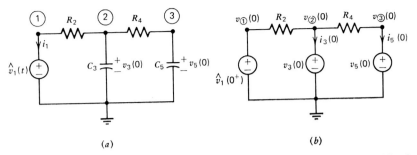

**Fig. 14-5** **Second-order network analyzed in Example 14.3 and (b) the initial condition network.**

## Example 14.3

The three nodal equilibrium equations for the network of Fig. 14-5a can be written by inspection:

$$\text{node } \textcircled{1} \qquad i_1 - \frac{1}{R_2} v_{\textcircled{2}} = -\frac{1}{R_2} \hat{v}_1$$

$$\text{node } \textcircled{2} \qquad \left(\frac{1}{R_4} + \frac{1}{R_2}\right) v_{\textcircled{2}} + C_3 \frac{dv_{\textcircled{2}}}{dt} - \frac{1}{R_4} v_3 = \frac{1}{R_2} \hat{v}_1$$

$$\text{node } \textcircled{3} \qquad -\frac{1}{R_4} v_{\textcircled{2}} + \frac{1}{R_4} v_{\textcircled{3}} + C_5 \frac{dv_{\textcircled{3}}}{dt} = 0$$

For the present we ignore the equation for node $\textcircled{1}$. Observe that use of the differential operator $D$ simplifies the equations for node $\textcircled{2}$ and node $\textcircled{3}$:

(14.18) $$\left(\frac{1}{R_2} + \frac{1}{R_4} + C_3 D\right) v_{\textcircled{2}} - \frac{1}{R_4} v_{\textcircled{3}} = \frac{1}{R_2} \hat{v}_1$$

(14.19) $$-\frac{1}{R_4} v_{\textcircled{2}} + \left(\frac{1}{R_4} + C_5 D\right) v_{\textcircled{3}} = 0$$

To solve this *set of simultaneous differential equations* we will manipulate them so as to derive two differential equations, each in terms of a single node voltage. We can eliminate $v_{\textcircled{1}}$ from (14.18) by recognizing that from (14.19)

$$v_{\textcircled{2}} = R_4\left(\frac{1}{R_4} + C_5 D\right) v_{\textcircled{3}}$$

and

$$Dv_{\textcircled{2}} = R_4\left(\frac{1}{R_4} D + C_5 D^2\right) v_{\textcircled{3}}$$

Substitution of these expressions into (14.18) yields

$$\left(\frac{1}{R_2} + \frac{1}{R_4}\right) R_4 \left(\frac{1}{R_4} + C_5 D\right) v_{(3)} + C_3 R_4 \left(\frac{1}{R_4} D + C_5 D^2\right) v_{(3)} - \frac{1}{R_4} v_{(3)} = \frac{1}{R_2} \hat{v}_1$$

or

$$(14.20) \quad C_3 R_4 C_5 \ddot{v}_{(3)} + \left(\left(\frac{1}{R_2} + \frac{1}{R_4}\right) R_4 C_5 + C_3\right) \dot{v}_{(3)} + \frac{1}{R_2} v_{(3)} = \frac{1}{R_2} \hat{v}_1$$

a second-order differential equation in terms of $v_{(3)}$.

A second-order differential equation in terms of $v_{(2)}$ can be found in a similar manner. From (14.18) we have

$$v_{(3)} = R_4 \left(\frac{1}{R_2} + \frac{1}{R_4} + C_3 D\right) v_{(2)} - \frac{R_4}{R_2} \hat{v}_1$$

and

$$D v_{(3)} = R_4 \left(\frac{1}{R_2} + \frac{1}{R_4} D + C_3 D^2\right) v_{(2)} - \frac{R_4}{R_2} D \hat{v}_1$$

On substitution of these expressions into (14.19), we have

$$- \frac{1}{R_4} v_{(2)} + \left(\frac{1}{R_2} + \frac{1}{R_4} + C_3 D\right) v_{(2)} + R_4 \left(\frac{C_5}{R_2} + \frac{C_5}{R_4} D + C_5 C_3 D^2\right) v_{(2)}$$

$$- \frac{R_4 C_5}{R_2} D \hat{v}_1 - \frac{1}{R_2} \hat{v}_1 = 0$$

or

$$(14.21) \qquad C_5 C_3 \ddot{v}_{(2)} + \left(C_3 + \frac{C_5}{R_2}\right) \dot{v}_{(2)} - \frac{1}{R_4} v_{(2)} = \frac{C_5}{R_2} \frac{d\hat{v}_1}{dt}$$

In order to solve the two second-order differential equations (14.20) and (14.21) we need the initial conditions $v_2(0)$, $\dot{v}_2(0)$, $v_3(0)$, and $\dot{v}_3(0)$. These initial conditions can be determined from the *initial condition network*. The initial condition network is obtained from the original network when each capacitor is replaced by a dc voltage source whose value equals the initial capacitor voltage, each inductor is replaced by a current source whose value equals the initial inductor current, and each time-varying source is replaced by an equivalent dc source whose value equals the source value at time $t = 0^+$. The initial condition network associated with the network under consideration is shown in Fig. 14-5b. Clearly

$$v_{(2)}(0) = v_3(0)$$

and

$$v_{(3)}(0) = v_5(0)$$

In order to determine $\dot{v}_{\textcircled{2}}(0)$ and $\dot{v}_{\textcircled{3}}(0)$ we recognize that since

$$i_3(t) = C_3 \dot{v}_3(t)$$

and since $v_3(t) = v_{\textcircled{2}}(t)$, then

$$i_3(t) = C_3 \dot{v}_2(t)$$

In particular, for $t = 0$, we have

$$\dot{v}_{\textcircled{2}}(0) = \frac{1}{C_3} i_3(0)$$

By similar reasoning

$$\dot{v}_{\textcircled{3}}(0) = \frac{1}{C_5} i_5(0)$$

The quantities $i_3(0)$ and $i_5(0)$ are determined from the initial condition network:

$$i_3(0) = \frac{1}{R_2} \hat{v}_1(0^+) - \left(\frac{1}{R_2} + \frac{1}{R_4}\right) v_3(0) + \frac{1}{R_4} v_5(0)$$

and

$$i_5(0) = \frac{1}{R_4} v_3(0) - \frac{1}{R_4} v_5(0)$$

Once the initial conditions have been determined, solution of the second-order differential equations for $v_2(t)$ and $v_3(t)$ proceeds as indicated above. □

## Example 14.4

The nodal equilibrium equations for the network of Fig. 14-6a are

$$\left(\frac{1}{R_2} + \frac{1}{L_3 D}\right) v_{\textcircled{1}} - \frac{1}{L_3 D} v_{\textcircled{2}} + i_3(0) = \hat{i}_1(t)$$

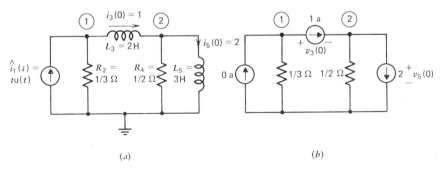

(a)                                    (b)

**Fig. 14-6** (a) Second-order network analyzed in Example 14.4; (b) the initial condition network.

and

$$-\frac{1}{L_3 D} v_① + \left(\frac{1}{L_3 D} + \frac{1}{R_4} + \frac{1}{L_5 D}\right) v_② - i_3(0) + i_5(0) = 0$$

or, after substitution of the element values

$$\left(3 + \frac{1}{2D}\right) v_① - \frac{1}{2D} v_② = -i_3(0) + tu(t)$$

and

$$-\frac{1}{2D} v_① + \left(2 + \frac{5}{6D}\right) v_② = i_3(0) - i_5(0)$$

To put these equations into a form suitable for solution, we first differentiate them:

$$\left(3D + \frac{1}{2}\right) v_① - \frac{1}{2} v_② = u(t)$$

and

$$-\frac{1}{2} v_① + \left(2D + \frac{5}{6}\right) v_② = 0$$

(Note that $d/dt[tu(t)] = t\delta(t) + u(t) = u(t)$.)

These equations may be manipulated in a manner similar to that used in the previous example to obtain the two second-order differential equations

$$\left(12D + 7D + \frac{1}{3}\right) v_① = 4\delta(t) + \frac{5}{3} u(t)$$

and

$$\left(12D^2 + 7D + \frac{1}{3}\right) v_② = u(t)$$

The initial conditions needed to solve these equations are $v_1(0)$, $\dot{v}_1(0)$, $v_2(0)$, and $\dot{v}_2(0)$. The initial condition network is shown in Fig. 14-6$b$. The terms $v_①(0)$ and $v_②(0)$ are determined by writing the nodal equilibrium equations:

$$3v_①(0) = -1$$

and

$$2v_②(0) = -1$$

that is, $v_①(0) = -\frac{1}{3}$ and $v_②(0) = -\frac{1}{2}$. From the inductor branch relationships

$$v_3(t) = L_3 \frac{di_3(t)}{dt}$$

and

$$v_5(t) = L_5 \frac{di_5(t)}{dt}$$

we recognize that

$$\frac{di_3(0)}{dt} = \frac{1}{L_3} v_3(0)$$

and

$$\frac{di_5(0)}{dt} = \frac{1}{L_5} v_5(0)$$

Since

$$v_①(t) = R_2[\hat{\imath}_1(t) - i_3(t)]$$

then

$$\frac{dv_①(t)}{dt} = R_2\left[\frac{d\hat{\imath}_1(t)}{dt} - \frac{di_3(t)}{dt}\right]$$

so that

$$\frac{dv_①(0)}{dt} = R_2\left[\frac{d\hat{\imath}_1(0^+)}{dt} - \frac{di_3(0)}{dt}\right]$$

$$= R_2\left[\frac{d\hat{\imath}_1(0^+)}{dt} - \frac{1}{L_3} v_3(0)\right]$$

$$= R_2\left[\frac{d\hat{\imath}_1(0^+)}{dt} - \frac{1}{L_3}(v_1(0) - v_2(0))\right]$$

$$= \frac{5}{18}$$

Similarly, since

$$v_②(t) = R_4[i_3(t) - i_5(t)]$$

then

$$\frac{dv_②(t)}{dt} = R_4\left[\frac{di_3(t)}{dt} - \frac{di_5(t)}{dt}\right]$$

and

$$\frac{dv_②(0)}{dt} = R_4\left[\frac{di_3(0)}{dt} - \frac{di_5(0)}{dt}\right]$$

$$= R_4\left[\frac{1}{L_3} v_3(0) - \frac{1}{L_5} v_5(0)\right]$$

$$= R_4\left[\frac{1}{L_3} v_1(0) - \left(\frac{1}{L_3} + \frac{1}{L_5}\right) v_2(0)\right]$$

$$= \frac{1}{8} \quad \square$$

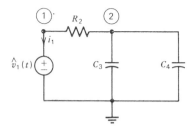

**Fig. 14-7   A network that contains two energy storage elements but which is characterized by a first-order differential equation.**

The following example illustrates that although a network may contain two energy storage elements, it need not necessarily be characterized by a second-order differential equation.

**Example 14.5**

The nodal equilibrium equations for the network of Fig. 14-7 are

$$\frac{1}{R_2} v_{②} + i_1 = -\frac{1}{R_2} \hat{v}_1$$

and

$$\left(\frac{1}{R_2} + DC_3 + DC_4\right) v_{②} = \frac{1}{R_2} \hat{v}_1$$

The second of these equations can be rewritten as

$$\dot{v}_{②}(t) = -\frac{1}{R_2(C_3 + C_4)}(v_{②}(t) - \hat{v}_1(t))$$

which is a first-order differential equation that can be solved using the methods of the previous section.   □

Although we will not pursue this discussion further, it can be shown that whenever a network contains loops of capacitors and voltage sources or cutsets of inductors and current sources, the order of the differential equations that characterize the network is less than the number of energy storage elements in the network.

## 14-3   NODAL EQUILIBRIUM EQUATIONS OF GENERAL *RLC* NETWORKS

The goal of this section is to write the nodal equilibrium equations of a general *RLC* network by inspection.

Recall the procedure first used to obtain the nodal equilibrium equations. For dc networks all the independent *KCL* node equations and branch relationships were written. The node equation *KVL* was expressed by writing

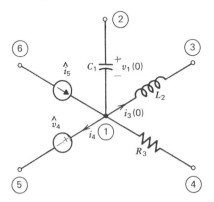

**Fig. 14-8    An arbitrary node in an *RLC* circuit that is used to illustrate writing a nodal equilibrium equation by inspection.**

the branch voltages in terms of the node voltages. The reduction of these equations to a set of equations in terms of the node voltage basis yields the nodal equilibrium equations. We then found that the nodal equilibrium equations could be written by inspection if the branch relationships were used to express the branch currents in the *KCL* node equations directly in terms of the node voltages. This technique can be employed to write the nodal equilibrium equations for *RLC* networks by inspection.

Consider an arbitrary node in an *RLC* circuit as shown in Fig. 14-8. Using the branch relationships to express the *KCL* node equation directly in terms of the node voltages, we have

$$C_1 D[v_①(t) - v_②(t)] + i_2(0) + \frac{1}{L_3 D}[v_①(t) - v_③(t)] + \frac{1}{R_3}[v_①(t) - v_④(t)]$$
$$= -i_4(t) + \hat{i}_5(t)$$

or

$$\left[C_1 D + \frac{1}{L_3 D} + \frac{1}{R_3}\right]v_①(t) - C_1 Dv_②(t) - \frac{1}{L_3 D}v_③(t) - \frac{1}{R_3}v_④(t)$$
$$= -i_2(0) - i_4(t) + \hat{i}_5(t)$$

Define $C_1 D$ as the *capacitance admittance operator*, $1/L_3 D$ as the *inductance admittance operator*, and $1/R_3$ the *resistance admittance operator*. Then the coefficient of $v_①(t)$ is the sum of all admittance operators connected to node ① and the coefficient of $v_k(t)$, $k = 2, 3, \ldots$, is the negative sum of all the admittance operators directly connected between nodes ① and ⓚ. (*Note.* The admittance operators associated with independent voltage and current sources are zero.) The right hand side of this equation is the sum of all independent current source currents, independent voltage source currents (unknowns), and initial inductance currents *entering* node ①.

In general, an $n_n$ node $RLC$ network will have nodal equilibrium equations of the form

$$y_{11} v_{①}(t) + y_{12} v_{②}(t) + \cdots + y_{1m} v_{⑩}(t) = i_{①}(t)$$

$$y_{21} v_{①}(t) + y_{22} v_{②}(t) + \cdots + y_{2m} v_{⑩}(t) = i_{②}(t)$$

$$\cdots\cdots\cdots\cdots\cdots\cdots\cdots$$

$$y_{m1} v_{①}(t) + y_{m2} v_{②}(t) + \cdots + y_{mm} v_{⑩}(t) = i_{⑩}(t)$$

where $m = n_n - 1$ and

$y_{kk}$ = sum of all admittance operators connected between node $⑥$ and all other nodes

$y_{jk}, j \neq k$ = negative sum of all admittance operators directly connected between nodes $ⓙ$ and $⑥$

$i_{⑥}(t)$ = the sum of all independent current source currents, independent voltage source currents, and initial inductance currents entering node $⑥$

### Example 14.6

The nodal equilibrium equations for the network of Fig. 14-9 are

$$\left(\frac{1}{R_1} + C_4 D\right) v_{①}(t) - \frac{1}{R_1} v_{②}(t) = 0$$

$$-\frac{1}{R_1} v_{①}(t) + \left(\frac{1}{R_1} + C_5 D\right) v_{②}(t) - C_5 D v_{⑤}(t) = -i_2(t)$$

$$\left(\frac{1}{L_6 D} + \frac{1}{L_3 D}\right) v_{③}(t) - \frac{1}{L_3 D} v_{④}(t) - \frac{1}{L_6 D} v_{⑤}(t) = -i_6(0) - i_7(0) + i_2(t)$$

$$-\frac{1}{L_3 D} v_{③}(t) + \left(\frac{1}{L_3 D} + \frac{1}{R_7}\right) v_{④}(t) = i_3(0) + i_8(t)$$

$$-C_5 D v_{②}(t) - \frac{1}{L_6 D} v_{③}(t) + \left(C_5 D + \frac{1}{L_6 D} + \frac{1}{R_9}\right) v_{⑤}(t) = i_6(0)$$

We also have the voltage source branch relationship

$$v_{②}(t) - v_{③}(t) = \hat{v}_2(t)$$

Finally, the initial capacitor voltages impose the following initial conditions on the node voltages:

$$v_{①}(0) = v_4(0)$$

and

$$v_{②}(0) - v_{⑤}(0) = v_5(0) \quad \square$$

The nodal equilibrium equations for an $RLC$ network are a set of simultaneous integrodifferential equations. As seen from the examples of the previous section, in order to find explicit analytic expressions for each node

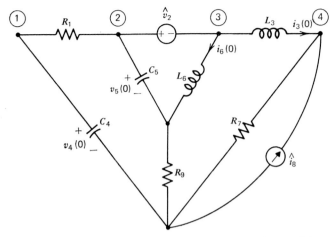

**Fig. 14-9  The *RLC* network of Example 14.6 for which nodal equilibrium equations are written.**

voltage we must reduce this set to an integrodifferential equation in terms of each node voltage.

## Example 14.7

Consider the second-order *RLC* network of Fig. 14-10a. The nodal equilibrium equations are

$$\left(\frac{1}{L_3 D} + C_4 D\right) v_{①}(t) - \frac{1}{L_3 D} v_{②}(t) = \hat{i}_1(t) = i_3(0)$$

$$-\frac{1}{L_3 D} v_{①}(t) + \left(\frac{1}{L_3 D} + \frac{1}{R_2}\right) = i_3(0)$$

We wish to manipulate these equations so as to find an equation solely in terms of $v_{②}(t)$. During our manipulations, we must bear in mind that $D$ is an operator and *not* a number. We can avoid much confusion if we only perform differentiation during our manipulations. If we can make the coefficient of $v_{①}(t)$ in the first equation the same as the coefficient of $v_{①}(t)$ in the second equation, we can eliminate $v_{①}(t)$ by subtraction and have an equation solely in terms of $v_{②}(t)$. Differentiating the second equation, that is, multiplying through by $D$, and then multiplying by $L_3$, we have

$$-v_{①}(t) + \left(1 + \frac{L_3 D}{R_2}\right) v_{②}(t) = 0$$

(*Note.* The equation $Di_3(0) = 0$, since $i_3(0)$ is a constant.) On differentiation and multiplication by $C_4$, this equation becomes:

$$-C_4 D v_①(t) + \left(C_4 D + \frac{L_3 C_4 D^2}{R_2}\right) v_②(t) = 0$$

adding the second nodal equilibrium equation and this expression yields

$$-\left(\frac{1}{L_3 D} + C_4 D\right) v_①(t) + \left(C_4 D + \frac{L_3 C_4}{R_2} D^2 + \frac{1}{L_3 D} + \frac{1}{R_2}\right) v_②(t) = i_3(0)$$

Finally, adding this expression to the first nodal equilibrium equation yields the second order differential equation

$$\left(C_4 D + \frac{L_3 C_4}{R_2} D^2 + \frac{1}{R_2}\right) v_②(t) = \hat{\imath}_1(t)$$

or

$$\frac{d^2 v_②(t)}{dt^2} + \frac{R_2}{L_3} \frac{d v_②(t)}{dt} + \frac{1}{C_4 L_3} v_②(t) = \frac{R_2}{L_3 C_4} \hat{\imath}_1(t)$$

We leave it as an exercise for the student to show that the differential equation involving only $v_①(t)$ is

$$\frac{d^2 v_①(t)}{dt^2} + \frac{R_2}{L_3} \frac{d v_①(t)}{dt} + \frac{1}{C_4 L_3} v_①(t) = \frac{R_2}{L_3 C_4} \hat{\imath}_1(t) + \frac{1}{C_4} \frac{d\hat{\imath}_1}{dt}$$

These differential equations may be solved using the techniques of the previous section. The solution requires knowledge of the initial conditions

$$v_①(0), \quad \frac{d v_①}{dt}\bigg|_{t=0}, \quad v_②(0) \quad \text{and} \quad \frac{d v_②}{dt}\bigg|_{t=0}$$

These initial conditions must be determined from knowledge of the initial capacitance voltage and initial inductance currents, and the fact that

$$\frac{d v_4}{dt}\bigg|_{t=0} = \frac{1}{C_4} i_4(0)$$

and

$$\frac{d i_3}{dt}\bigg|_{t=0} = \frac{1}{L_3} v_3(0)$$

The easiest way to determine the required initial conditions is to analyze the initial condition network of Fig. 13-3b. It is left as an exercise to show that

$$v_①(0) = v_4(0)$$

$$v_②(0) = R_2 i_3(0)$$

$$\dot{v}_①(0) = \frac{1}{C_4} [\hat{\imath}_1(0) - i_3(0)]$$

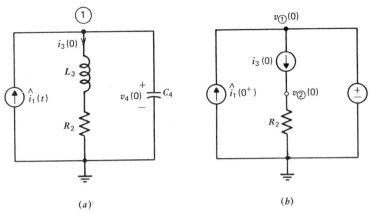

(a)
(b)

**Fig. 14.10** (a) Network studied in Example 14.7; (b) the initial condition network.

and

$$\dot{v}_{②}(0) = \frac{R_2}{L_3} v_{①}(0) - \frac{1}{L_3} i_3(0) \quad \square$$

For the simple network of Fig. 14-10a we ended up having to solve two second-order differential equations. The more energy storage elements in a network the higher the order of the differential equations to be solved. The higher the order of the differential equation the more effort required for its solution (compare the effort required to solve first- and second-order differential equations).

It is beyond the scope of this text to discuss the analytic solution of higher-order differential equations with arbitrary forcing functions. In a later chapter we will show that if limited classes of inputs are allowed much simpler methods may be employed to solve higher-order differential equations. The other alternative is to employ numerical methods, which is the topic of the next chapter.

## 14-4 SUMMARY

In this chapter we considered the time domain analysis of networks that contain arbitrary numbers of energy storage elements. The solution of second-order differential equations that govern second-order networks was discussed. As in the case of first-order differential equations, the solution of a second-order differential equation consists of two parts: the *homogeneous solution* and the *particular solution*.

The *characteristic* equation results from substituting $e^{st}$ into the homogeneous equation. The roots of the characteristic equation are referred to as the *natural frequencies* of the network.

In order to determine the unique solution of a second-order differential equation two initial conditions are required. It was shown that these initial conditions can be found from knowledge of the initial capacitance voltages and inductance currents.

Finally, it was shown how the nodal equilibrium equations of a general *RLC* network may be written by inspection in a manner similar to purely resistive networks by making use of the *capacitance admittance operator*, *inductance admittance operator*, and *resistance admittance operator*.

## PROBLEMS

**14.1**  Find the solutions of the following homogeneous differential equations.

(a) $(D^2 + 3D + 2)v(t) = 0; \ v(0) = 1, \ \dot{v}(0) = 0$

(b) $\ddot{v}(t) + 6\dot{v}(t) + 8v(t) = 0; \ v(0) = 0, \ \dot{v}(0) = 2$

(c) $\ddot{v}(t) + 4\dot{v}(t) + v(t) = 0; \ v(0) = 1, \ \dot{v}(0) = 1$

(d) $(2D^2 + 2D + 2.5)i(t) = 0; \ i(0) = 1, \ \dfrac{di}{dt}(0) = 2$

(e) $\dfrac{d^2i(t)}{dt} + \dfrac{4di(t)}{dt} + 20i(t) = 0; \ i(0) = 1, \ \dfrac{di(0)}{dt} = 1$

**14.2**  Write the loop equation for the network of Fig. P14.2. Sketch the response $i(t)$ for the following situations:

(a) $R_2 = 0$

(b) $R_2{}^2 > \dfrac{4L_1}{C_3}$

(c) $R_2{}^2 < \dfrac{4L_1}{C_3}$

(d) $R_2{}^2 = \dfrac{4L_1}{C_3}$

**14.3**  Find $v_3(t)$ in the network of Fig. P14.3 for the following sets of initial conditions:

(a) $i_1(0) = 1$     and     $v_3(0) = 2$

(b) $i_1(0) = -2$     and     $v_3(0) = 2$

**14.4**  After being closed for a long time, the switch in the network of Fig. P14.4 is opened. Sketch the response $v_1(t)$.

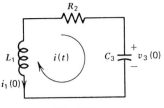

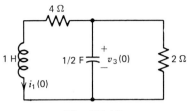

Fig. P14.2                                        Fig. P14.3

**14.5**  Find $i(t)$ for $t \geq 0$ in the network of Fig. P14.5.

**14.6**  The switch in the network of Fig. P14.6 has been closed for a long time. At time $t = 0$ the switch is suddenly opened. Find, and sketch, $v_6(t)$ for $t > 0$.

**14.7**  Find the complete solutions of the following differential equations:

(a) $(D^2 + 5D + 4)i(t) = e^t$, $i(0) = 0$, $\dfrac{di(0)}{dt} = -2$

(b) $(D^2 + 4D + 3)v(t) = tu(t)$, $v(0) = -\dfrac{4}{9}$, $\dot{v}(0) = \dfrac{3}{9}$

(c) $\dfrac{d^2i}{dt^2} + 4\dfrac{di}{dt} + 3i = u(t)$, $i(0) = 0$, $\dfrac{di(0)}{dt} = 0$

(d) $\ddot{v}(t) + 4\dot{v}(t) + 3v(t) = 5$, $v(0) = 1$, $\dot{v}(0) = 0$

**14.8**  Find expressions for the voltage or current (as indicated) as functions of time for each of the networks of Fig. P14.8. (Assume that all initial conditions are zero.)

**14.9**  Find a differential equation for the network of Fig. P14.9 solely in terms of $v_①$ and another solely in terms of $v_②$. Find the initial conditions that are needed to solve these equations, and find the solutions.

**14.10**  Repeat Prob. 14.9 for the network of Fig. P14.10.

**14.11**  Draw an oriented graph for the network of Fig. P14.11. Choose a tree such that all voltage sources and capacitors are tree branches,

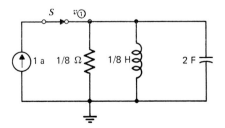

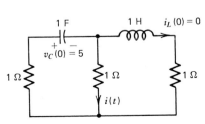

Fig. P14.4                                 Fig. P14.5

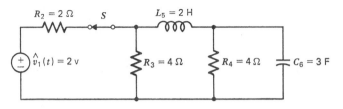

Fig. P14.6

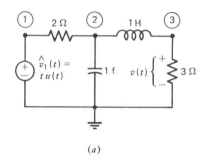

(a)

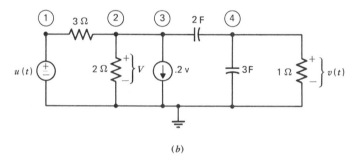

(b)

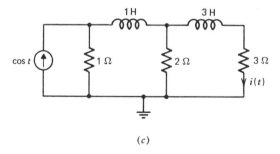

(c)

Fig. P14.8

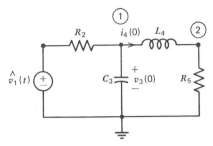

**Fig. P14.9**

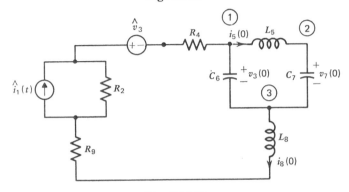

**Fig. P14.10**

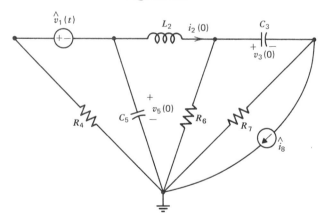

**Fig. P14.11**

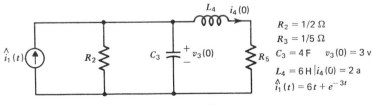

$$R_2 = 1/2 \ \Omega$$
$$R_3 = 1/5 \ \Omega$$
$$C_3 = 4 \, \text{F} \qquad v_3(0) = 3 \, \text{v}$$
$$L_4 = 6 \, \text{H} \mid i_4(0) = 2 \, \text{a}$$
$$\hat{i}_1(t) = 6t + e^{-3t}$$

**Fig. P14.12**

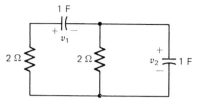

**Fig. P14.13**

and current sources and inductors are links. Resistors may be either tree branches or links. Write the *KCL* fundamental cutset equations, *KVL* fundamental loop equations, and relationships. Manipulate these equations so as to reduce them to a set of equations solely in terms of the fundamental loop currents. Then find a set of equations solely in terms of tree branch voltages. Why is it a good idea to put voltage sources and capacitors in tree branches (if possible) and current sources and inductors in links (if possible)?

**14.12**    Find the node voltages as a function of time for the network of Fig. P14.12.

**14.13**    Find the differential equations for $v_1(t)$ and $v_2(t)$ in the circuit of Fig. P14.13. The total energy stored in the capacitors is

$$e(t) = \frac{1}{2} v_1^2(t) + \frac{1}{2} v_2^2(t)$$

Show that $de(t)/dt \leq 0$ for all $t$.

**14.14**    In the circuit of Fig. P14.14 the input is $\hat{v}_1(t)$ and the response is $v_3(t)$. Find the impulse response and step response.

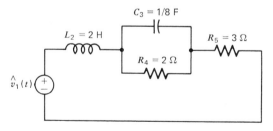

**Fig. P14.14**

**14.15**    Calculate the total energy stored in the capacitor and inductor in the circuit of Fig. P14.2 for the situations given in Prob. 14.2.

## CHAPTER 15

# Numerical Methods of Linear Time Domain Simulation

In Chapter 13 we described two numerical integration methods and showed how they may be used to determine the solution of a first-order network numerically. We developed the companion model for inductors and capacitors and demonstrated that the numerical solution corresponded to repeated dc analysis. In this chapter we extend these ideas to the analysis of higher-order networks as introduced in the previous chapter. We also introduce more accurate numerical integration methods.

## 15-1  NUMERICAL SOLUTION OF GENERAL *RLC* NETWORKS

In this section we extend the companion model concept to general *RLC* networks. For motivational purposes we start with an example that illustrates the technique to be used.

**461**

## Example 15.1

The nodal equilibrium equations, in matrix form, for the network of Fig. 15-1a are

$$(15.1) \quad \begin{bmatrix} 1 & -\dfrac{1}{R_2} & 0 \\[2ex] 0 & \dfrac{1}{R_2} + C_3 D & -C_3 D \\[2ex] 0 & -C_3 D & C_3 D + C_4 D + \dfrac{1}{R_5} \end{bmatrix} \begin{bmatrix} i_1(t) \\[2ex] v_{\textcircled{2}} \\[2ex] v_{\textcircled{3}} \end{bmatrix} = \begin{bmatrix} -\dfrac{1}{R_2} \hat{v}_1(t) \\[2ex] \dfrac{1}{R_2} \hat{v}_1(t) \\[2ex] \hat{i}_6 \end{bmatrix}$$

where we have included the constraint imposed by the voltage source

$$v_{\textcircled{1}}(t) = \hat{v}_1(t)$$

The capacitors impose the initial condition constraints

$$v_{\textcircled{2}}(0) - v_{\textcircled{3}}(0) = v_3(0) \quad \text{and} \quad v_{\textcircled{3}}(0) = v_4(0)$$

If we use the backward Euler integration rule on each capacitance branch relationship then at time $t_{k+1}$, the companion model associated with each capacitor is a resistor in parallel with a voltage source. In particular, the branch relationships at time $t_{k+1}$ are

$$i_3(t_{k+1}) = \frac{C_3}{h} v_3(t_{k+1}) - \frac{C_3}{h} v_3(t_k)$$

for capacitor $C_3$ and

$$i_4(t_{k+1}) = \frac{C_4}{h} v_4(t_{k+1}) - \frac{C_4}{h} v_4(t_k)$$

for capacitor $C_4$. The companion network is shown in Fig. 15-1b. The nodal equilibrium equations of the companion network, in matrix form, are

$$(15.2)$$

$$\begin{bmatrix} 1 & -\dfrac{1}{R_2} & 0 \\[2ex] 0 & \left(\dfrac{1}{R_2} + \dfrac{C_3}{h}\right) & 0 \\[2ex] 0 & -\dfrac{C_3}{h} & \left(\dfrac{C_3}{h} + \dfrac{C_4}{h} + \dfrac{1}{R_5}\right) \end{bmatrix} \begin{bmatrix} i_1(t_{k+1}) \\[2ex] v_{\textcircled{2}}(t_{k+1}) \\[2ex] v_{\textcircled{3}}(t_{k+1}) \end{bmatrix}$$

$$= \begin{bmatrix} -\dfrac{1}{R_2} \hat{v}_1(t_{k+1}) \\[2ex] \dfrac{1}{R_2} \hat{v}_1(t_{k+1}) + \dfrac{C_3}{h} v_3(t_k) \\[2ex] \dfrac{C_4}{h} v_4(t_k) + \hat{i}_6(t_{k+1}) - \dfrac{C_3}{h} v_3(t_k) \end{bmatrix}$$

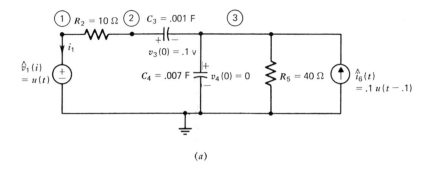

$(a)$

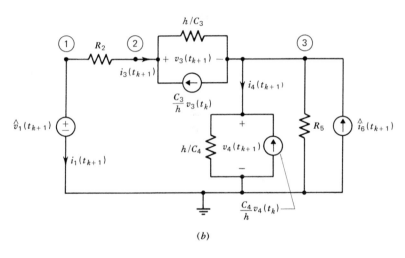

$(b)$

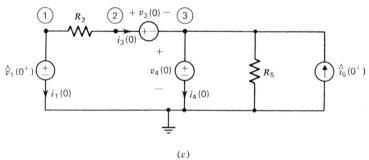

$(c)$

**Fig. 15-1** $(a)$ **The second-order networks used in Example 15.1 to illustrate the numerical solution technique;** $(b)$ **the companion network for the backward Euler integration method;** $(c)$ **the initial condition network.**

Observe that at time $t_{k+1}$ all quantities on the right-hand side of this equation are known. Therefore, the equations may be solved for the node voltages at time $t_{k+1}: v_{\circled{2}}(t_{k+1})$, $v_{\circled{3}}(t_{k+1})$, and the voltage source current $i_1(t_{k+1})$. The companion models are then updated, which in effect means that the right-hand side of (15.2) is altered, and then the equations can be solved again for voltages at the next time point.

Before iteratively solving this network we must determine the behavior at time $t = 0$. This is found by analyzing the initial condition network, shown in Fig. 15-1c. The pertinent equations are

$$
\begin{bmatrix} 1 & 0 & 0 \\ 0 & 1 & 0 \\ 0 & -1 & +1 \end{bmatrix}
\begin{bmatrix} i_1(0) \\ i_3(0) \\ i_4(0) \end{bmatrix} =
\begin{bmatrix} \dfrac{1}{R_2}(v_3(0) + v_4(0) - \hat{v}_1(0^+)) \\ \dfrac{1}{R_2}\hat{v}_1(0^+) \\ \hat{i}_6(0^+) \end{bmatrix}
$$

where we have used the fact that

$$
v_{\circled{1}}(0) = \hat{v}_1(0)
$$
$$
v_{\circled{2}}(0) = v_3(0) + v_4(0)
$$

and

$$
v_{\circled{3}}(0) = v_4(0)
$$

A computer program that may now be used to iteratively solve equation (15-2) is shown in Fig. 15-2a. The results are shown in Fig. 15-2b and 15-2c. □

The procedure for the numerical analysis of general *RLC* networks should be clear. First the initial condition network should be solved to determine the behavior of the network at the initial time $t = 0$. Then each capacitor and inductor is replaced by an appropriate companion model. The nodal equilibrium equations are written for the companion network and arranged so that all knowns, which include capacitor voltages and inductor currents at time $t_k$, appear on the right-hand side and unknowns, the node voltages, voltage source currents, etc., at time $t_{k+1}$, appear on the left-hand side. These equations are recursively solved using GAUSS for $k = 0, 1, 2, \ldots$, at each step updating the right-hand side.

If the step size $h$ is kept constant, then only the coefficients on the right-hand side of the nodal equilibrium equations change for each new $k$. To see that this is true, observe that in the companion model associated with capacitors and inductors, the resistor is a function only of the time step $h$ while the source is dependent on both the time step $h$ as well as the time point $t_k$. Therefore, we are interested in solving the same set of simultaneous algebraic equations for many different values of right-hand sides. The Gaussian elimination procedure is not as efficient for this purpose as the method described in the next section.

```
C      CHAPTER 15 EXAMPLE 15.1
C
C      DIMENSIONS
C
       IMPLICIT REAL*8 (A-H,O-Z)
       DIMENSION Y(5,5),CIN(5),VOUT(2,50),TIME(50)
C
C      READ IN ELEMENT VALUES,INITIAL CAPACITANCE VOLTAGES,
C      TIME STEP AND FINAL TIME
C
     4 CALL READ(R2,C3,C4,R5,V3I,V4I,H,TF)
       T=0.
       NPTS=0
     1 NPTS=NPTS+1
C
C      SET UP AND SOLVE NODE EQUATIONS. PRINT SOLUTION
C
       CALL NODEQN(R2,C3,C4,R5,V3I,V4I,V3,V4,H,T,Y,CIN,N)
       CALL GAUSS(Y,CIN,N,5)
       V3=CIN(1)-CIN(2)
       V4=CIN(2)
       CALL PRINT(V3,V4,H,T)
C
C      STORE SOLUTION FOR PLOTTING
C
       VOUT(1,NPTS)=V3
       VOUT(2,NPTS)=V4
       TIME(NPTS)=T
C
C      IF T IS LESS THAN FINAL TIME (TF) INCREMENT T
C
       IF(E.GT.TF) GO TO 2
       T=T+H
       GO TO 1
C
C      PRINT GRAPH HEADING AND PLOT SOLUTION
C
     2 PRINT 3,H
       CALL PLOT(TIME,VOUT,NPTS,2,0)
     3 FORMAT('1',/38X,'PLOT OF V3 AND V4 VS TIME FOR TIME STEP=',G12.5//
      +53X,'V3= X',10X,'V4= O')
       GO TO 4
       END



       SUBROUTINE READ(R2,C3,C4,R5,V3I,V4I,H,TF)
       IMPLICIT REAL*8 (A-H,O-Z)
       READ(5,1) R2,C3,C4,R5,V3I,V4I,H,TF
     1 FORMAT(8G10.0)
       RETURN
       END
```

**Fig. 15-2a   Program listing for solution of network in Fig. 15-1.**

```
      SUBROUTINE NODEQN(R2,C3,C4,R5,V3I,V4I,V3,V4,H,T,Y,CIN,N)
      IMPLICIT REAL*8 (A-H,O-Z)
      DIMENSION Y(5,5),CIN(5)
C
C     DETERMINE SOURCE VALUES
C
      V1=1.
      CI6=0.
      IF(T.GE..1) CI6=10.D-3
C
C     INITIALIZE COEFFICIENT ARRAYS
C
      DO 1 J=1,5
      CIN(J)=0.DO
      DO 1 K=1,5
    1 Y(J,K)=0.DO
C
C     AT TIME T=0 SOLVE INITIAL CONDITION NETWORK
C
      IF(T.GT.O.) GO TO 2
C
C     INITIAL CONDITION NETWORK
C
      N=5
      GC3=0.DO
      GC4=0.DO
      CI3=0.DO
      CI4=0.DO
      Y(2,4)=1.
      Y(3,4)=-1.
      Y(3,5)=1.
      Y(4,2)=1.
      CIN(4)=V4I
      Y(5,1)=1.
      Y(5,2)=-1.
      CIN(5)=V3I
      GO TO 3
C
C     COMPANION NETWORK
C
    2 N=3
      GC3=C3/H
      GC4=C4/H
      CI3=GC3*V3
      CI4=GC4*V4
    3 Y(1,1)=-1./R2
      Y(1,3)=1.
      CIN(1)=(-1./R2)*V1
      Y(2,1)=(1./R2)+GC3
      Y(2,2)=-GC3
      CIN(2)=(1./R2)*V1+CI3
      Y(3,1)=-GC3
      Y(3,2)=GC3+GC4+(1./R5)
      CIN(3)=CI6+CI4-CI3
      RETURN
      END
```

**Fig. 15-2a** *(Continued)*

466

```
      SUBROUTINE PRINT(V3,V4,H,T)
      IMPLICIT REAL*8 (A-H,O-Z)
      IF(T.NE.O.) GO TO 1
      PRINT 2,H
    1 PRINT 3,T,V3,V4
    2 FORMAT(' RESULTS FOR EXAMPLE 15.1 USING A TIME STEP OF',G12.5///
     +12X,'TIME',18X,'V3',18X,'V4')
    3 FORMAT(10X,G12.5,8X,G15.5,10X,G15.5)
      RETURN
      END
```

----- DATA FOR PROGRAM OF EXAMPLE 15.1 -----

   10.     .001     .007    40.     .1     0.    .025     1.

### Fig. 15-2a (*Continued*)

RESULTS FOR EXAMPLE 15.1 USING A TIME STEP OF 0.25000D-01

| TIME | V3 | V4 |
|---|---|---|
| 0.0 | 0.10000D 00 | 0.0 |
| 0.25000D-01 | 0.68779 | 0.77088D-01 |
| 0.50000D-01 | 0.84548 | 0.91449D-01 |
| 0.75000D-01 | 0.89157 | 0.89998D-01 |
| 0.10000D 00 | 0.88701 | 0.11481 |
| 0.12500 | 0.87055 | 0.13603 |
| 0.15000 | 0.85212 | 0.15525 |
| 0.17500 | 0.83421 | 0.17296 |
| 0.20000 | 0.81737 | 0.18936 |
| 0.22500 | 0.80170 | 0.20457 |
| 0.25000 | 0.78714 | 0.21868 |
| 0.27500 | 0.77363 | 0.23177 |
| 0.30000 | 0.76110 | 0.24392 |
| 0.32500 | 0.74947 | 0.25519 |
| 0.35000 | 0.73868 | 0.26564 |
| 0.37500 | 0.72866 | 0.27534 |
| 0.40000 | 0.71938 | 0.28434 |
| 0.42500 | 0.71076 | 0.29269 |
| 0.45000 | 0.70276 | 0.30044 |
| 0.47500 | 0.69534 | 0.30763 |
| 0.50000 | 0.68846 | 0.31429 |
| 0.52500 | 0.68207 | 0.32048 |
| 0.55000 | 0.67615 | 0.32622 |
| 0.57500 | 0.67065 | 0.33155 |
| 0.60000 | 0.66555 | 0.33649 |
| 0.62500 | 0.66082 | 0.34108 |
| 0.65000 | 0.65643 | 0.34533 |
| 0.67500 | 0.65235 | 0.34928 |
| 0.70000 | 0.64857 | 0.35294 |
| 0.72500 | 0.64507 | 0.35634 |
| 0.75000 | 0.64181 | 0.35949 |
| 0.77500 | 0.63879 | 0.36241 |
| 0.80000 | 0.63599 | 0.36513 |
| 0.82500 | 0.63339 | 0.36764 |
| 0.85000 | 0.63098 | 0.36998 |
| 0.87500 | 0.62875 | 0.37215 |
| 0.90000 | 0.62667 | 0.37416 |
| 0.92500 | 0.62475 | 0.37602 |
| 0.95000 | 0.62296 | 0.37776 |
| 0.97500 | 0.62130 | 0.37936 |
| 1.0000 | 0.61976 | 0.38085 |
| 1.0250 | 0.61834 | 0.38223 |

### Fig. 15-2b  Output of program of Fig. 15-2a.

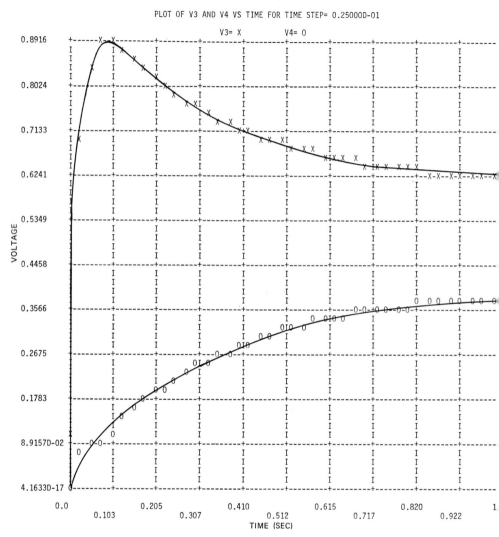

Fig. 15-2c Plot produced by program of Fig. 15-2a.

## 15-2 LU FACTORIZATION

In this section we will introduce a modification of the Gaussian elimination procedure that allows the efficient solution of a set of simultaneous algebraic equations, represented in matrix form by

(15.3)
$$\mathbf{Ax} = \mathbf{b}$$

for many different vectors **b**. Recall that (15.3) is shorthand notation for the set of equations

$$a_{11}x_1 + a_{12}x_2 + \quad \cdots \quad + a_{1m}x_m = b_1$$
$$a_{21}x_1 + a_{22}x_2 + \quad \cdots \quad + a_{2m}x_m = b_2$$
$$\dots\dots\dots\dots\dots\dots\dots\dots\dots\dots\dots$$
$$a_{m1}x_1 + a_{m2}x_2 + \quad \cdots \quad + a_{mm}x_m = b_m$$

where

$$\mathbf{A} = \begin{bmatrix} a_{11} & a_{12} & \cdots & a_{1m} \\ a_{21} & a_{22} & \cdots & a_{2m} \\ \cdot\cdot & \cdots\cdots\cdots\cdots \\ a_{m1} & a_{m2} & \cdots & a_{mm} \end{bmatrix}$$

$$\mathbf{x} = \begin{bmatrix} x_1 \\ x_2 \\ \cdot \\ \cdot \\ \cdot \\ x_m \end{bmatrix} \quad \text{and} \quad \mathbf{b} = \begin{bmatrix} b_1 \\ b_2 \\ \cdot \\ \cdot \\ \cdot \\ b_m \end{bmatrix}$$

Consider the possibility of factoring the matrix **A** into a product of two matrices **L** and **U**:

$$\mathbf{A} = \mathbf{LU}$$

where **L** is a *lower triangular matrix*; that is, it is a matrix with all elements above the main diagonal equal to zero:

$$\mathbf{L} = \begin{bmatrix} l_{11} & 0 & 0 & 0 & \cdots & 0 \\ l_{21} & l_{22} & 0 & 0 & \cdots & 0 \\ l_{31} & l_{32} & l_{33} & 0 & \cdots & 0 \\ \multicolumn{6}{c}{\dots\dots\dots\dots\dots\dots\dots} \\ l_{m1} & l_{m2} & l_{m3} & & \cdots & l_{mm} \end{bmatrix}$$

and **U** is a unit upper triangular matrix; that is, it is a matrix with all elements below the main diagonal equal to zero and all elements on the main diagonal equal to unity:

$$\mathbf{U} = \begin{bmatrix} 1 & u_{12} & u_{13} & \cdots & u_{1m} \\ 0 & 1 & u_{23} & \cdots & u_{2m} \\ 0 & 0 & 1 & \cdots & u_{3m} \\ \multicolumn{5}{c}{\dots\dots\dots\dots\dots\dots} \\ 0 & 0 & 0 & \cdots & 1 \end{bmatrix}$$

If such a factorization is possible, solution of the matrix equation (15.3) is simplified considerably. Observe that given

$$\mathbf{A} = \mathbf{LU}$$

then (15.3) becomes

(15.4) $$\mathbf{LUx} = \mathbf{b}$$

We may define the vector $\hat{\mathbf{b}}$ by

(15.5) $$\mathbf{Ux} = \hat{\mathbf{b}}$$

so that (15.4) becomes

(15.6) $$\mathbf{L}\hat{\mathbf{b}} = \mathbf{b}$$

Thus we can view the solution of (15.3) for $\mathbf{x}$ as a two step procedure: first solve (15.6) for $\hat{\mathbf{b}}$ and then solve (15.5) for $\mathbf{x}$. The advantage to this approach is the simple operations involved to obtain $\hat{\mathbf{b}}$ from $\mathbf{b}$ and $\mathbf{x}$ from $\hat{\mathbf{b}}$. Consider (15.6) first:

$$\mathbf{L}\hat{\mathbf{b}} = \mathbf{b}$$

or

$$
\begin{bmatrix}
l_{11} & 0 & 0 & \cdots & 0 \\
l_{21} & l_{22} & 0 & \cdots & 0 \\
& & & \cdots & \\
l_{m1} & l_{m2} & l_{m3} & \cdots & l_{mm}
\end{bmatrix}
\begin{bmatrix}
\hat{b}_1 \\
\hat{b}_2 \\
\cdot \\
\cdot \\
\cdot \\
\hat{b}_m
\end{bmatrix}
=
\begin{bmatrix}
b_1 \\
b_2 \\
\cdot \\
\cdot \\
\cdot \\
b_m
\end{bmatrix}
$$

or

$$
\begin{aligned}
l_{11}\hat{b}_1 &= b_1 \\
l_{21}\hat{b}_1 + l_{22}\hat{b}_2 &= b_2 \\
&\cdots \\
l_{m1}\hat{b}_1 + l_{m2}\hat{b}_2 + \cdots \ l_{mm}\hat{b}_m &= b_m
\end{aligned}
$$

The first of these equations can be solved for $\hat{b}_1$:

$$\hat{b}_1 = l_{11}^{-1}b_1$$

then the second one for $\hat{b}_2$:

$$\hat{b}_2 = l_{22}^{-1}(b_2 - l_{21}\hat{b}_1)$$

and so on, until the last equation is used to solve for $\hat{b}_m$:

$$\hat{b}_m = l_{mm}^{-1}(b_m - l_{m1}\hat{b}_1 - l_{m2}\hat{b}_2 \cdots - l_{m,m-1}\hat{b}_{m-1})$$

In general

$$(15.7) \qquad \hat{b}_k = l_{kk}^{-1}\left(b_k - \sum_{j=1}^{k-1} l_{kj}\hat{b}_j\right), \qquad k = 1, 2, \ldots, m$$

The process of determining $\hat{b}$ from $b$ is called *forward substitution*.

Now consider using $\hat{b}$ to find $x$ from (15.5):

$$Ux = \hat{b}$$

or

$$\begin{bmatrix} 1 & u_{12} & u_{13} & \cdots & u_{1m} \\ 0 & 1 & u_{23} & \cdots & u_{2m} \\ & & \cdots & & \\ 0 & 0 & 0 & \cdots & 1 \end{bmatrix} \begin{bmatrix} x_1 \\ x_2 \\ \cdot \\ \cdot \\ \cdot \\ x_m \end{bmatrix} = \begin{bmatrix} \hat{b}_1 \\ \hat{b}_2 \\ \cdot \\ \cdot \\ \cdot \\ \hat{b}_m \end{bmatrix}$$

or

$$x_1 + u_{21}x_2 + u_{23}x_3 + \cdots + u_{1m}x_m = \hat{b}_1$$
$$x_2 + u_{23}x_3 + \cdots + u_{2m}x_m = \hat{b}_2$$
$$\cdots\cdots\cdots\cdots\cdots\cdots\cdots\cdots\cdots\cdots\cdots\cdots$$
$$x_m = \hat{b}_m$$

Working from the last of these equations to the first, we have

$$x_m = \hat{b}_m$$
$$x_{m-1} = \hat{b}_{m-1} - u_{m-1,m}x_m$$
$$\cdots\cdots\cdots\cdots\cdots\cdots\cdots$$
$$x_2 = \hat{b}_2 - u_{23}x_3 - u_{24}x_4 - \cdots - u_{2m}x_m$$
$$x_1 = \hat{b}_1 - u_{12}x_2 - u_{13}x_3 - \cdots - u_{1m}x_m$$

or, in general

$$(15.8) \qquad x_k = b_k - \sum_{j=k+1}^{m} u_{kj}x_j, \qquad j = m, m-1, m-2, \ldots, 1$$

This process is called *backward substitution*.

Observe that if the vector $b$ in (15.3) changes, all that is required to determine a new solution $x$ is to perform the forward substitution as indicated by (15.7) to find the new $\hat{b}$ and then a backward substitution as indicated by (15.8). The $L$ and $U$ matrices do not have to be recomputed.

We now turn to the computation of $L$ and $U$ from $A$. We desire that

$$A = LU$$

To see this suppose $\mathbf{A}$ is $(4 \times 4)$ so that the set of equations we are solving is

$$a_{11}x_1 + a_{12}x_2 + a_{13}x_3 + a_{14}x_4 = b_1$$

$$a_{21}x_1 + a_{22}x_2 + a_{23}x_3 + a_{24}x_4 = b_2$$

$$a_{31}x_1 + a_{32}x_2 + a_{33}x_3 + a_{34}x_4 = b_3$$

$$a_{41}x_1 + a_{42}x_2 + a_{43}x_3 + a_{44}x_4 = b_4$$

or in matrix form

$$
\begin{bmatrix}
a_{11} & a_{12} & a_{13} & a_{14} \\
a_{21} & a_{22} & a_{23} & a_{24} \\
a_{31} & a_{32} & a_{33} & a_{34} \\
a_{41} & a_{42} & a_{43} & a_{44}
\end{bmatrix}
\begin{bmatrix}
x_1 \\
x_2 \\
x_3 \\
x_4
\end{bmatrix}
=
\begin{bmatrix}
b_1 \\
b_2 \\
b_3 \\
b_4
\end{bmatrix}
$$

The first pass in Gaussian elimination (see Sections 3–4) is to remove the variable $x_1$ from all equations but the first. The equations that result are (in matrix form)

$$
\begin{bmatrix}
1 & a_{12}^{(1)} & a_{13}^{(1)} & a_{14}^{(1)} \\
0 & a_{22}^{(1)} & a_{23}^{(1)} & a_{24}^{(1)} \\
0 & a_{32}^{(1)} & a_{33}^{(1)} & a_{34}^{(1)} \\
0 & a_{42}^{(1)} & a_{43}^{(1)} & a_{44}^{(1)}
\end{bmatrix}
\begin{bmatrix}
x_1 \\
x_2 \\
x_3 \\
x_4
\end{bmatrix}
=
\begin{bmatrix}
b_1^{(1)} \\
b_2^{(1)} \\
b_3^{(1)} \\
b_4^{(1)}
\end{bmatrix}
$$

where

$$a_{1k}^{(1)} = \frac{a_{1k}}{a_{11}}, \qquad k = 1, 2, 3, 4$$

$$a_{jk}^{(1)} = a_{jk} - a_{j1}a_{1k}^{(1)}, \qquad \begin{matrix} j = 2, 3, 4 \\ k = 2, 3, 4 \end{matrix}$$

$$b_1^{(1)} = \frac{b_1}{a_{11}}$$

and

$$b_j^{(1)} = b_j - a_{j1}b_1^{(1)} \qquad j = 2, 3, 4$$

After the second pass of Gaussian elimination the equations become

$$
\begin{bmatrix}
1 & a_{12}^{(1)} & a_{13}^{(1)} & a_{14}^{(1)} \\
0 & 1 & a_{23}^{(2)} & a_{24}^{(2)} \\
0 & 0 & a_{33}^{(2)} & a_{34}^{(2)} \\
0 & 0 & a_{34}^{(2)} & a_{44}^{(2)}
\end{bmatrix}
\begin{bmatrix}
x_1 \\
x_2 \\
x_3 \\
x_4
\end{bmatrix}
=
\begin{bmatrix}
b_1^{(1)} \\
b_2^{(2)} \\
b_3^{(2)} \\
b_4^{(2)}
\end{bmatrix}
$$

where

$$a_{2k}^{(2)} = \frac{a_{2k}^{(1)}}{a_{22}^{(1)}} \qquad k = 2, 3, 4$$

$$a_{jk}^{(2)} = a_{jk}^{(1)} - a_{j2}^{(1)}a_{2k}^{(2)} \qquad \begin{matrix} j = 3, 4 \\ k = 2, 4 \end{matrix}$$

$$b_2^{(2)} = \frac{b_2}{a_{22}^{(1)}}$$

and

$$b_j^{(2)} = b_j^{(1)} - a_{j2}^{(1)}b_2^{(2)} \qquad j = 3, 4$$

After the third pass we have

$$\begin{bmatrix} 1 & a_{12}^{(1)} & a_{13}^{(1)} & a_{14}^{(1)} \\ 0 & 1 & a_{23}^{(2)} & a_{24}^{(2)} \\ 0 & 0 & 1 & a_{34}^{(3)} \\ 0 & 0 & 0 & a_{44}^{(4)} \end{bmatrix} \begin{bmatrix} x_1 \\ x_2 \\ x_3 \\ x_4 \end{bmatrix} = \begin{bmatrix} b_1^{(1)} \\ b_2^{(2)} \\ b_3^{(3)} \\ b_4^{(3)} \end{bmatrix}$$

and after the fourth, and final, pass we have

$$\begin{bmatrix} 1 & a_{12}^{(1)} & a_{13}^{(1)} & a_{14}^{(1)} \\ 0 & 1 & a_{23}^{(2)} & a_{24}^{(2)} \\ 0 & 0 & 1 & a_{34}^{(3)} \\ 0 & 0 & 0 & 1 \end{bmatrix} \begin{bmatrix} x_1 \\ x_2 \\ x_3 \\ x_4 \end{bmatrix} = \begin{bmatrix} b_1^{(1)} \\ b_2^{(2)} \\ b_3^{(3)} \\ b_4^{(4)} \end{bmatrix}$$

where, in general,

$$(15.9a) \quad a_{jk}^{(l)} = \begin{cases} a_{jk}^{(l-1)}/a_{jj}^{(l-1)} & j = l, k = l, l+1, \ldots \\ a_{jk}^{(l-1)} - a_{jl}^{(l-1)}a_{lk}^{(l)} \\ & j = l+1, l+2, \ldots, k = l+1, l+2, \ldots \end{cases}$$

and

$$(15.9b) \quad b_j^{(l)} = \begin{cases} b_j^{(l-1)}/a_{jj}^{(l-1)} & j = l \\ b_j^{(l-1)} - a_{jl}^{(l-1)}b_l^{(l)} & j = l+1, l+2, \ldots. \end{cases}$$

Observe that the result of Gaussian elimination, before back substitution, is a set of equations of the form

$$\mathbf{Ux} = \hat{\mathbf{b}}$$

where

$$\hat{\mathbf{b}} = [b_1^{(1)}, b_2^{(2)}, b_3^{(3)}, b_4^{(4)}]'$$

In other words, Gaussian elimination transforms $\mathbf{A}$ into $\mathbf{U}$ and $\mathbf{b}$ into $\hat{\mathbf{b}}$. To find $\mathbf{L}$ we need to reconstruct the steps involved in transforming $\mathbf{b}$ into $\hat{\mathbf{b}}$ since

$$\mathbf{L}\hat{\mathbf{b}} = \mathbf{b}$$

In other words, we wish to find a set of equations that can be used to find $\hat{\mathbf{b}}' = (b_1^{(1)}, b_2^{(2)}, b_3^{(3)}, b_4^{(4)})$ from $\mathbf{b}$. To this end we consider equations (15.9b) For $l = 1$ we have [note $a_{jk}^{(0)} = a_{jk}$ and $b_j^{(0)} = b_j$]:

$$b_1^{(1)} = \frac{b_1}{a_{11}}$$

$$b_2^{(1)} = b_2 - a_{21}b_1^{(1)}$$

$$b_3^{(1)} = b_3 - a_{31}b_1^{(1)}$$

and

$$b_4^{(1)} = b_4 - a_{41}b_1^{(1)}$$

If we rearrange these expressions so that the knowns, $b_1$, $b_2$, $b_3$, and $b_4$ appear on the right-hand side, and the unknowns $b_1^{(1)}$, $b_2^{(1)}$, $b_3^{(1)}$, $b_4^{(1)}$ on the left-hand side, we have

$$\begin{bmatrix} a_{11} & 0 & 0 & 0 \\ a_{21} & 1 & 0 & 0 \\ a_{31} & 0 & 1 & 0 \\ a_{41} & 0 & 0 & 1 \end{bmatrix} \begin{bmatrix} b_1^{(1)} \\ b_2^{(1)} \\ b_3^{(1)} \\ b_4^{(1)} \end{bmatrix} = \begin{bmatrix} b_1 \\ b_2 \\ b_3 \\ b_4 \end{bmatrix}$$

For $l = 2$, (15.9b) becomes

$$b_2^{(2)} = b_2^{(1)}/a_{22}^{(1)}$$

$$b_3^{(2)} = b_3^{(1)} - a_{32}^{(1)}b_2^{(2)}$$

and

$$b_4^{(2)} = b_4^{(1)} - a_{42}^{(1)}b_2^{(2)}$$

or after rearranging and putting in matrix form, we have

(15.10b)

$$\begin{bmatrix} 1 & 0 & 0 & 0 \\ 0 & a_{22}^{(1)} & 0 & 0 \\ 0 & a_{32}^{(1)} & 1 & 0 \\ 0 & a_{42}^{(1)} & 0 & 1 \end{bmatrix} \begin{bmatrix} b_1^{(1)} \\ b_2^{(2)} \\ b_3^{(2)} \\ b_4^{(2)} \end{bmatrix} = \begin{bmatrix} b_1^{(1)} \\ b_2^{(1)} \\ b_3^{(1)} \\ b_4^{(1)} \end{bmatrix}$$

where we have added the equation $b_1^{(1)} = b_1^{(1)}$. By substituting (15.10b) into (15.10a), we can obtain $\mathbf{b}^{(2)}$ directly from $\mathbf{b}$ without having to find $\mathbf{b}^{(1)}$ first.

We have

$$\begin{bmatrix} a_{11} & 0 & 0 & 0 \\ a_{21} & 1 & 0 & 0 \\ a_{31} & 0 & 1 & 0 \\ a_{41} & 0 & 0 & 1 \end{bmatrix} \begin{bmatrix} 1 & 0 & 0 & 0 \\ 0 & a_{22}^{(1)} & 0 & 0 \\ 0 & a_{32}^{(1)} & 1 & 0 \\ 0 & a_{42}^{(1)} & 0 & 1 \end{bmatrix} \begin{bmatrix} b_1^{(1)} \\ b_2^{(2)} \\ b_3^{(2)} \\ b_4^{(2)} \end{bmatrix} = \begin{bmatrix} b_1 \\ b_2 \\ b_3 \\ b_4 \end{bmatrix}$$

or

(15.10c)
$$\begin{bmatrix} a_{11} & 0 & 0 & 0 \\ a_{21} & a_{22}^{(1)} & 0 & 0 \\ a_{31} & a_{32}^{(1)} & 1 & 0 \\ a_{41} & a_{42}^{(1)} & 0 & 1 \end{bmatrix} \begin{bmatrix} b_1^{(1)} \\ b_2^{(2)} \\ b_3^{(2)} \\ b_4^{(2)} \end{bmatrix} = \begin{bmatrix} b_1 \\ b_2 \\ b_3 \\ b_4 \end{bmatrix}$$

We now evaluate (15.9b) for $l = 3$:

$$b_3^{(3)} = b_3^{(2)}/a_{33}^{(2)}$$

and

$$b_4^{(3)} = b_4^{(2)} - a_{43}^{(2)} b_3^{(3)}$$

On rearrangement of terms, we have

(15.10d)
$$\begin{bmatrix} 1 & 0 & 0 & 0 \\ 0 & 1 & 0 & 0 \\ 0 & 0 & a_{33}^{(2)} & 0 \\ 0 & 0 & a_{43}^{(2)} & 1 \end{bmatrix} \begin{bmatrix} b_1^{(1)} \\ b_2^{(2)} \\ b_3^{(3)} \\ b_4^{(3)} \end{bmatrix} = \begin{bmatrix} b_1^{(1)} \\ b_2^{(2)} \\ b_3^{(2)} \\ b_4^{(2)} \end{bmatrix}$$

Substituting (15.10c) into (15.10d) yields

(15.10e)
$$\begin{bmatrix} a_{11} & 0 & 0 & 0 \\ a_{21} & a_{22}^{(1)} & 0 & 0 \\ a_{31} & a_{32}^{(1)} & a_{33}^{(2)} & 0 \\ a_{41} & a_{42}^{(1)} & a_{43}^{(2)} & 1 \end{bmatrix} \begin{bmatrix} b_1^{(1)} \\ b_2^{(2)} \\ b_3^{(3)} \\ b_4^{(3)} \end{bmatrix} = \begin{bmatrix} b_1 \\ b_2 \\ b_3 \\ b_4 \end{bmatrix}$$

Finally, for $l = 4$, (15.9b) becomes

$$b_4^{(4)} = b_4^{(3)}/a_{44}^{(3)}$$

so that we have

(15.10f)
$$\begin{bmatrix} 1 & 0 & 0 & 0 \\ 0 & 1 & 0 & 0 \\ 0 & 0 & 1 & 0 \\ 0 & 0 & 0 & a_{44}^{(3)} \end{bmatrix} \begin{bmatrix} b_1^{(1)} \\ b_2^{(2)} \\ b_3^{(3)} \\ b_4^{(4)} \end{bmatrix} = \begin{bmatrix} b_1^{(1)} \\ b_2^{(2)} \\ b_3^{(3)} \\ b_4^{(4)} \end{bmatrix}$$

On substitution of (15.10e) into (15.10e), we have

$$\begin{bmatrix} a_{11} & 0 & 0 & 0 \\ a_{21} & a_{22}^{(1)} & 0 & 0 \\ a_{31} & a_{32}^{(1)} & a_{33}^{(2)} & 0 \\ a_{41} & a_{42}^{(1)} & a_{43}^{(2)} & a_{44}^{(3)} \end{bmatrix} \begin{bmatrix} b_1^{(1)} \\ b_2^{(2)} \\ b_3^{(3)} \\ b_4^{(4)} \end{bmatrix} = \begin{bmatrix} b_1 \\ b_2 \\ b_3 \\ b_4 \end{bmatrix}$$

so that

$$\mathbf{L} = \begin{bmatrix} a_{11} & 0 & 0 & 0 \\ a_{21} & a_{22}^{(1)} & 0 & 0 \\ a_{31} & a_{32}^{(1)} & a_{33}^{(2)} & 0 \\ a_{41} & a_{42}^{(1)} & a_{43}^{(2)} & a_{44}^{(3)} \end{bmatrix}$$

Observe now that the elements of $\mathbf{L}$ are precisely the same as those elements of $\mathbf{A}$ that ultimately are transformed into either 0 or 1 in $\mathbf{U}$ and, moreover, the element of $\mathbf{L}$ in location $(j, k)$ is equal to the element in location $(j, k)$ of $\mathbf{A}^{(l-1)}$ $j = 1, 2, 3, 4$, $k = 1, 2, \ldots, l$ where $\mathbf{A}^{(l-1)}$ represents the matrix $\mathbf{A}$ after the $(l - 1)$st step of Gaussian elimination.

One final note. We know that the Gaussian elimination procedure can be implemented such that the matrix $\mathbf{A}$ is modified at each step and ultimately is replaced by $\mathbf{U}$. In other words, $\mathbf{A}$ is ultimately destroyed and no additional core storage is required. However since we know, a priori, that the lower triangular part of $\mathbf{U}$ contains zeros, and the diagonal of $\mathbf{U}$ contains ones, we do not need to store these values, all that we need do is adjust the back substitution phase to recognize this situation. Now observe that the lower part of the lower triangular matrix $\mathbf{L}$ can be stored in the unneeded space of the lower part of $\mathbf{U}$, that is, the matrices $\mathbf{L}$ and $\mathbf{U}$ can be stored in the same amount of space as that required by $\mathbf{A}$. Moreover, if the Gaussian elimination procedure is implemented such that at the $j$th step the operations that set the coefficients of $x_j$ to zero are avoided, (as had been done in GAUSS) the end result is to replace the upper triangular part of $\mathbf{A}$ with $\mathbf{U}$ and the lower triangular part and diagonal of $\mathbf{A}$ with $\mathbf{L}$.

## Example 15.2

Consider again the set of equations that we solved in Example 3.6:

$$2x_1 + x_2 + 3x_3 = 2$$
$$3x_1 - 2x_2 - x_3 = 1$$
$$x_1 - x_2 + x_3 = -1$$

by Gaussian elimination. In matrix terms these equations can be written as

$$\begin{bmatrix} 2 & 1 & 3 \\ 3 & -2 & -1 \\ 1 & -1 & 1 \end{bmatrix} \begin{bmatrix} x_1 \\ x_2 \\ x_3 \end{bmatrix} = \begin{bmatrix} 2 \\ 1 \\ -1 \end{bmatrix}$$

The first column of **L** is found after the "zeroth" pass of Gaussian elimination on the **A** matrix

$$\begin{bmatrix} 2 & 1 & 3 \\ 3 & -2 & -1 \\ 1 & -1 & 1 \end{bmatrix}$$

where elements of **L** are denoted by dotted squares. After the first pass of Gaussian elimination we have

$$\begin{bmatrix} 2 & 1/2 & 3/2 \\ 3 & -7/2 & -11/2 \\ 1 & -3/2 & -3/2 \end{bmatrix}$$

which yields the first row of **U** (except for $u_{11} = 1$) where we have indicated the elements of **U** by dotted circles. (*Note.* The array shown above is a composite of **L** and **U** and **A** and is *not* a matrix strictly related to **A** since it was not derived by elementary row and column operations.) After the second pass we have

$$\begin{bmatrix} 2 & 1/2 & 3/2 \\ 3 & -7/2 & 11/7 \\ 1 & -3/2 & 13/7 \end{bmatrix}$$

which yields the second column of **L** and the second row of **U** (except for $u_{22} = 1$). The last step would be to divide by $a_{33}^{(2)} = 13/7$ that we avoid since $l_{33} = a_{33}^{(2)}$.

Thus

$$\mathbf{L} = \begin{bmatrix} 2 & 0 & 0 \\ 3 & \dfrac{-7}{2} & 0 \\ 1 & \dfrac{-3}{2} & \dfrac{13}{7} \end{bmatrix}$$

and

$$\mathbf{U} = \begin{bmatrix} 1 & \dfrac{1}{2} & \dfrac{3}{2} \\ 0 & 1 & \dfrac{11}{7} \\ 0 & 0 & 1 \end{bmatrix}$$

The student should verify that

$$\mathbf{A} = \mathbf{LU}$$

The forward substitution proceeds as follows:

$$L\hat{b} = b$$

or

$$\begin{bmatrix} 2 & 0 & 0 \\ 3 & \dfrac{-7}{2} & 0 \\ 1 & \dfrac{-3}{2} & \dfrac{13}{7} \end{bmatrix} \begin{bmatrix} b_1 \\ b_2 \\ b_3 \end{bmatrix} = \begin{bmatrix} 2 \\ 1 \\ -1 \end{bmatrix}$$

Therefore

$$\hat{b}_1 = 1$$

$$\hat{b}_2 = -\frac{2}{7}(1 - 3) = \frac{4}{7}$$

and

$$\hat{b}_3 = \frac{7}{13}\left[-1 - 1 + \left(\frac{3}{2}\right)\left(\frac{4}{7}\right)\right] = \frac{-8}{13}$$

The solution **x** emerges on back substitution:

$$U x = \hat{b}$$

or

$$\begin{bmatrix} 1 & \dfrac{1}{2} & \dfrac{3}{2} \\ 0 & 1 & \dfrac{11}{7} \\ 0 & 0 & 1 \end{bmatrix} \begin{bmatrix} x_1 \\ x_2 \\ x_3 \end{bmatrix} = \begin{bmatrix} 1 \\ \dfrac{4}{7} \\ \dfrac{-8}{13} \end{bmatrix}$$

Thus

$$x_3 = -\frac{8}{13}$$

$$x_2 = \frac{4}{7} + \left(\frac{11}{7}\right)\left(\frac{8}{13}\right) = \frac{20}{13}$$

and

$$x_1 = 1 - \left(\frac{1}{2}\right)\left(\frac{20}{13}\right) + \frac{3}{2}\left(\frac{8}{13}\right) = \frac{15}{13}$$

which agrees with our previous results.

Suppose we now wished to solve the algebraic equations

$$\begin{bmatrix} 2 & 1 & 3 \\ 3 & -2 & -1 \\ 1 & -1 & 1 \end{bmatrix} \begin{bmatrix} x_1 \\ x_2 \\ x_3 \end{bmatrix} = \begin{bmatrix} 1 \\ 2 \\ 0 \end{bmatrix}$$

which except for the right-hand side is the same set of equations we have just solved. Clearly we need not perform another factorization. The **L** and **U** matrices remain unchanged. Only additional forward and backward substitutions are required. In particular, the forward substitution is

$$\begin{bmatrix} 2 & 0 & 0 \\ 3 & \dfrac{-7}{2} & 0 \\ 1 & \dfrac{-3}{2} & \dfrac{13}{7} \end{bmatrix} \begin{bmatrix} \hat{b}_1 \\ \hat{b}_2 \\ \hat{b}_3 \end{bmatrix} = \begin{bmatrix} 1 \\ 2 \\ 0 \end{bmatrix}$$

from which we find that $\hat{b}_1 = \frac{1}{2}$, $\hat{b}_2 = -\frac{1}{7}$, and $\hat{b}_3 = \frac{-5}{13}$. The backward substitution is

$$\begin{bmatrix} 1 & \dfrac{1}{2} & \dfrac{3}{2} \\ 0 & 1 & \dfrac{11}{7} \\ 0 & 0 & 1 \end{bmatrix} \begin{bmatrix} x_1 \\ x_2 \\ x_3 \end{bmatrix} = \begin{bmatrix} \dfrac{1}{2} \\ -\dfrac{1}{7} \\ \dfrac{-5}{13} \end{bmatrix}$$

from which we find the solution

$$x_3 = \frac{5}{13}, \qquad x_2 = \frac{6}{13}, \qquad \text{and} \qquad x_1 = \frac{11}{13} \quad \square$$

## 15-3  THE SUBROUTINE LUSOLV

The FORTRAN program LUSOLV shown in Fig. 15-3 is a subroutine that will solve the simultaneous algebraic equations

$$\mathbf{Ax = b}$$

by first performing an $LU$ factorization on **A**, then performing the forward substitution

$$\mathbf{U\hat{b} = b}$$

```
C
C                      SUBROUTINE 'LUSOLV'
C
C
C
C      LUSOLV IS A SUBROUTINE FOR SOLVING SIMULTANEOUS
C      LINEAR ALGEBRAIC EQUATIONS WITH REAL COEFFICIENTS.
C      THE PROGRAM USES L-U FACTORIZATION AND FORWARD-BACKWARD
C      SUBSTITUTION WITH ROW PIVOTING. LUSOLV IS CALLED
C      FROM ANOTHER PROGRAM BY USING THE STATEMENT
C
C
C                      CALL LUSOLV(A,B,N,M,JPIV)
C
C
C      'A' IS A DOUBLE-SUBSCRIPTED REAL*8 ARRAY DIMENSIONED
C      MXM IN THE CALLING PROGRAM. 'B' IS A SINGLE-SUBSCRIPTED
C      REAL*8 ARRAY AND 'JPIV' IS A SINGLE-SUBSCRIPTED
C      INTEGER ARRAY. 'B' AND 'JPIV' ARE DIMENSIONED M IN THE
C      CALLING PROGRAM AND N IS THE NUMBER OF EQUATIONS
C      TO BE SOLVED. 'A' CONTAINS THE COEFFICIENTS OF THE
C      UNKNOWNS OF THE ALGEBRAIC EQUATIONS. 'B' CONTAINS
C      THE COEFFICIENTS ON THE RIGHT HAND SIDE OF THE
C      ALGEBRAIC EQUATIONS. UPON RETURN FROM LUSOLV 'A'
C      CONTAINS THE ELEMENTS OF THE LOWER AND UPPER TRIANGULAR
C      MATRICES OF 'A' AND 'B' CONTAINS THE SOLUTION.
C
C      LUSOLV ALLOWS THE USER TO ALTER THE RIGHT-HAND SIDE
C      OF THE EQUATIONS AND OBTAIN A NEW SOLUTION WITHOUT
C      HAVING TO DO ANOTHER FACTORIZATION OF 'A'.
C      'JPIV' RECORDS THE PIVOT ORDER USED IN THE
C      FACTORIZATION OF 'A'. THIS IS NECESSARY SO
C      THAT UPON ENTERING FWBW THE ROWS OF 'B' WILL HAVE
C      THE SAME PIVOT ORDER.
C      THE STATEMENT
C
C                      CALL FWBW(A,B,N,M,JPIV)
C
C      YIELDS A NEW SOLUTION WHEN 'B' HAS BEEN ALTERED.
C      USE OF THE TWO INTEGERS M AND N ALLOWS THE
C      DIMENSIONS OF 'A' AND 'B' TO BE SET ONCE IN THE
C      MAIN PROGRAM AND THE ACTUAL NUMBERS OF EQUATIONS TO
C      VARY(NOTE: M≥N), DEPENDING ON THE PROBLEM.
C
       SUBROUTINE LUSOLV(A,B,N,M,JPIV)
       IMPLICIT REAL*8(A-H,O-Z)
       DIMENSION A(M,M),B(M),JPIV(M)
       DO 4 I=1,N
       JPIV(I)=I
       I1=I+1
       IF(DABS(A(I,I)).LE.1.D-10) GO TO 1
       GO TO 15
   1   CONTINUE
       IF(I.EQ.N) GO TO 20

       DO 14 J=I1,N
       IF(DABS(A(J,I)).LE.1.D-10) GO TO 14
       JPIV(I)=J
       GO TO 16
  14   CONTINUE
       GO TO 20
  16   DO 2 K=1,N
```

**Fig. 15-3  Listing of *LU* factorization program.**

480

```
      DO 14 J=I1,N
      IF(DABS(A(J,I)).LE.1.D-10) GO TO 14
      JPIV(I)=J
      GO TO 16
  14  CONTINUE
     ⌐GO TO 20
  16  DO 2 K=1,N
      IPIV=JPIV(I)
      PIV=A(IPIV,K)
      A(IPIV,K)=A(I,K)
   2  A(I,K)=PIV
  15  IF(I.EQ.N) GO TO 3
      DO 8 JI=I1,N
   8  A(I,JI)=A(I,JI)/A(I,I)
      DO 4 J=I1,N
      DO 4 K=I1,N
   4  A(J,K)=A(J,K)-(A(J,I)*A(I,K))
   3  CONTINUE
C
C     ENTER FORWARD-BACKWARD SUBSTITUTION
C
      ENTRY FWBW(A,B,N,M,JPIV)
      DO 61 I=1,N
      IPIV=JPIV(I)
      IF(IPIV.LE.I) GO TO 61
      J=I
      PIVA=B(I)
  62  PIV=B(IPIV)
      B(IPIV)=PIVA
      JPIV(J)=-IPIV
      J=IPIV
      IPIV=JPIV(J)
      PIVA=PIV
      IF(IPIV.GT.0) GO TO 62
  61  CONTINUE
C     FORWARD SUBSTITUTION
C
      DO 31 K=1,N
      SUM=0.0
      IF(K.EQ.1.) GO TO 41
      MM=K-1
      DO 51 J=1,MM
  51  SUM=SUM+A(K,J)*B(J)
  41  B(K)=(1./A(K,K))*(B(K)-SUM)
  31  CONTINUE
C
C     BACKWARD SUBSTITUTION
C
      DO 91 LL=1,N
      K=(N+1)-LL
      SUM=0.0
      IF(K.EQ.N) GO TO 81
      KK=K+1
      DO 71 J=KK,N
  71  SUM=SUM+A(K,J)*B(J)
  81  B(K)=B(K)-SUM
  91  CONTINUE
```

**Fig. 15-3 (*Continued*)**

to determine **b** and then performing the backward substitution

$$Lx = \hat{b}$$

to find the solution **x**. The program listing is virtually self-explanatory. This program is called from another program with the statement

CALL LUSOLV (A, B, N, M, JPIV)

where A is a double precision array, dimensioned (M, M) in the calling program. Before entry to LUSOLV, A contains the coefficients of the unknown. On exit from LUSOLV, A contains the lower triangular part of **L** and the elements of **U** above the main diagonal; that is, the following elements are stored in A:

$$\begin{bmatrix} l_{11} & u_{12} & u_{13} & \cdots & u_{1n} \\ l_{21} & l_{22} & u_{23} & \cdots & u_{2m} \\ l_{31} & l_{32} & l_{33} & \cdots & u_{3m} \\ \cdots\cdots\cdots\cdots\cdots\cdots\cdots \\ l_{m1} & l_{m2} & l_{m3} & & l_{mm} \end{bmatrix}$$

B is a double precision array dimensioned (M) in the calling program. Before entry to LUSOLV, B contains the right-hand side of the equations. On exit from LUSOLV, B contains the solution $x$. JPIV is a full word integer array dimensioned (M) in the calling program. This array stores the pivot order determined by LUSOLV to avoid dividing by zero. N, an integer, is the number of equations to be solved; M, an integer is the dimensions of A and B used in the calling program. Observe that the entry into LUSOLV can be made just after the *LU* factorization but prior to the forward-backward substitution by using the statement

CALL FWBW (A, B, N, M, JPIV)

Thus if the set of equations

$$Ax = b$$

is to be solved for several different values of **b**, that is,

$$Ax = b_1$$

$$Ax = b_2$$

etc., then LUSOLV need only be called once and FWBW called each time a new **b** is used.

### Example 15.3

We wish to determine the node voltages for the network of Fig. 15-4*a* using the backward Euler method. The associated companion network is shown in

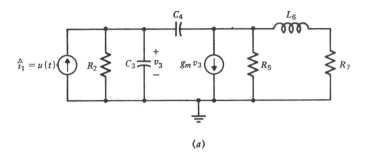

(a)

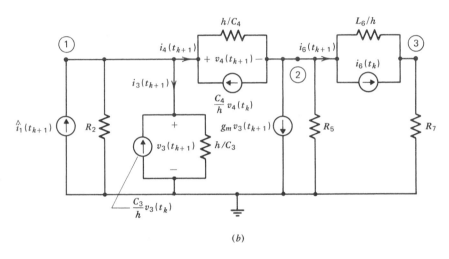

(b)

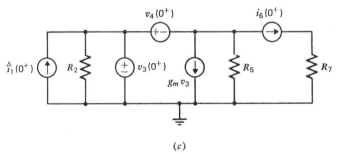

(c)

Fig. 15-4 (a) Network analyzed in Example 15.3; (b) the associated companion network; (c) the initial condition network.

Fig. 15-4$b$. The nodal equilibrium equations for the companion network are

$$
\begin{bmatrix}
\left(\dfrac{1}{R_2} + \dfrac{h}{C_3} + \dfrac{h}{C_4}\right) & -\dfrac{h}{C_4} & 0 \\[2ex]
\left(g_m - \dfrac{h}{C_4}\right) & \left(\dfrac{h}{C_4} + \dfrac{1}{R_5} + \dfrac{L_6}{h}\right) & -\dfrac{L_6}{h} \\[2ex]
0 & -\dfrac{L_6}{h} & \left(\dfrac{L_6}{h} + \dfrac{1}{R_7}\right)
\end{bmatrix}
\begin{bmatrix}
v_{①}(t_{k+1}) \\[2ex]
v_{②}(t_{k+1}) \\[2ex]
v_{③}(t_{k+1})
\end{bmatrix}
$$

$$
=
\begin{bmatrix}
\hat{\imath}_1(t_k) + \dfrac{C_3}{h} v_3(t_k) + \dfrac{C_4}{h} v_4(t_k) \\[2ex]
-\dfrac{C_4}{h} v_4(t_k) + i_6(t_k) \\[2ex]
0
\end{bmatrix}
$$

Observe that the coefficient matrix on the left-hand side remains contant for a given step size $h$, and only the right-hand side changes from one time point to the next. Therefore, the same set of equations are being solved for many different right-hand sides, a situation in which we should employ $LU$ factorization. To solve these equations iteratively, we must determine the initial conditions from the initial condition network of Fig. 15-4$c$. The equations to be solved are

$$
i_3(0^+) + i_4(0^+) = \hat{\imath}_1(0^+) - \frac{1}{R_2} v_3(0^+)
$$

$$
-i_4(0^+) + \frac{1}{R_5} v_{②}(0^+) = i_6(0^+) - g_m v_3(0^+)
$$

$$
\frac{1}{R_7} v_{③}(0^+) = i_6(0^+)
$$

and

$$
v_{②}(0^+) = v_3(0^+) - v_4(0^+)
$$

A computer program that can be used to determine the response of the network of Fig. 15-4$a$ is given in Fig. 15-5$a$. The results are given in Figs. 15-5$b$ and 15-5$c$.    □

## 15-4  TRAPEZOIDAL RULE INTEGRATION

To this point we have only introduced the forward and backward Euler integration methods. Both these techniques assume a piecewise constant, or zeroth-order, approximation to the function being integrated between

```
C      CHAPTER 15 EXAMPLE 15.3
C
C      DIMENSIONS
C
       IMPLICIT REAL*8 (A-H,O-Z)
       DIMENSION Y(5,5),CIN(5),JPIV(5),OUT(1,50),TIME(50)
C
C      READ IN ELEMENT VALUES,INITIAL CAPACITANCE VOLTAGES (V3I,V4I),
C      INITIAL INDUCTANCE CURRENT (CI6I), TIME STEP AND FINAL TIME
C
       CALL READ(R2,C3,C4,R5,EL6,R7,GM,V3I,V4I,CI6I,H,TF)
       T=0.
       NPTS=0
     1 NPTS=NPTS+1
C
C      SET UP AND SOLVE NODE EQUATIONS
C
       CALL NODEQN(R2,C3,C4,R5,EL6,R7,GM,V3I,V4I,CI6I,GC4,CI4,V3,V4,V6,
      +H,T,Y,CIN,N,NPTS)
C
C      CALL FWBW IF NOT SOLVING THE INITIAL CONDITION
C      NETWORK OR THE COMPANION NETWORK AT THE
C      FIRST TIME POINT
C
     6 IF(NPTS.GT.2) GO TO 4
       CALL LUSOLV(Y,CIN,N,5,JPIV)
       GO TO 5
     4 CALL FWBW(Y,CIN,N,5,JPIV)
     5 V3=CIN(1)
       V4=CIN(1)-CIN(2)
       V6=CIN(2)-CIN(3)
C
C      SOLVE FOR THE CURRENT THROUGH R2
C
       OUT1=CIN(1)/R2
       CALL PRINT(OUT1,H,T)
C
C      STORE SOLUTION FOR PLOTTING
C
       OUT(1,NPTS)=OUT1
       TIME(NPTS)=T
C
C      IF T IS LESS THAN FINAL TIME (TF) INCREMENT T
C
       IF(T.GT.TF) GO TO 2
       T=T+H
       GO TO 1
C
C      PRINT GRAPH HEADING AND PLOT SOLUTION
C
     2 PRINT 3,H
       CALL PLOT(TIME,OUT,NPTS,1,0)
     3 FORMAT('1',/42X,'PLOT OF I3 VS TIME FOR TIME STEP=',G12.5//)
       END
```

**Fig. 15-5a  Program listing for Example 15.3.**

```
      SUBROUTINE READ(R2,C3,C4,R5,EL6,R7,GM,V3I,V4I,CI6I,H,TF)
      IMPLICIT REAL*8 (A-H,O-Z)
      READ(5,1) R2,C3,C4,R5,EL6,R7,GM,V3I,V4I,CI6I,H,TF
    1 FORMAT(8G10.0)
      RETURN
      END



      SUBROUTINE PRINT(OUT1,H,T)
      IMPLICIT REAL*8 (A-H,O-Z)
      IF(T.NE.0.) GO TO 1
      PRINT 2,H
    1 PRINT 3,T,OUT1
    2 FORMAT(' RESULTS FOR EXAMPLE 15.3 USING A TIME STEP OF',G12.5///
     +12X,'TIME',18X,'I3')
    3 FORMAT(10X,G12.5,8X,G12.5)
      RETURN
      END



      SUBROUTINE NODEQN(R2,C3,C4,R5,EL6,R7,GM,V3I,V4I,CI6I,GC4,CI4,V3,
     +V4,V6,H,T,Y,CIN,N,NPTS)
      IMPLICIT REAL*8 (A-H,O-Z)
      DIMENSION Y(5,5),CIN(5)
      CI1=.1
C
C     AT TIME T=0 SOLVE INITIAL CONDITION NETWORK
C
      IF(T.GT.0.) GO TO 1
C
C     INITIAL CONDITION NETWORK
C
      N=5
      GC3=0.D0
      GC4=0.D0
      GL6=0.D0
      CI3=0.D0
      CI4=0.D0
      CI6=CI6I
      DO 4 J=1,5
      CIN(J)=0.D0
      DO 4 K=1,5
    4 Y(J,K)=0.D0
      Y(1,4)=1.
      Y(1,5)=1.
      Y(2,5)=-1.
      Y(4,1)=1.
      CIN(4)=V3I
      Y(5,1)=1.
      Y(5,2)=-1.
      CIN(5)=V4I
      GO TO 2
```

**Fig. 15-5a** (*Continued*)

486

```
C
C      COMPANION NETWORK
C
    1 N=3
C
C      UPDATE COMPANION MODELS
C
       GC3=C3/H
       GC4=C4/H
       GL6=H/EL6
       CI3=GC3*V3
       CI4=GC4*V4
       CI6=CI6+(V6/GL6)
C
C      WHEN SOLVING THE COMPANION NETWORK AFTER THE
C      FIRST TIME POINT, ONLY THE RIGHT HAND SIDE
C      OF THE EQUATIONS IS ALTERED
C
       IF(NPTS.GT.2) GO TO 3
       DO 5 J=1,3
       CIN(J)=0.DO
       DO 5 K=1,3
    5 Y(J,K)=0.DO
    2 Y(1,1)=(1./R2)+GC3+GC4
       Y(1,2)=-GC4
       Y(2,1)=GM-GC4
       Y(2,2)=(1./R5)+GC4+GL6
       Y(2,3)=-GL6
       Y(3,2)=-GL6
       Y(3,3)=GL6+(1./R7)
    3 CIN(1)=CI1+CI3+CI4
       CIN(2)=-CI4-CI6
       CIN(3)=CI6
       RETURN
       END
```

----- DATA FOR PROGRAM OF EXAMPLE 15.3 -----

| 1.E2 | 50.E-4 | 20.E-5 | 10. | 10.E-3 | 10. | .01 | .1 |
|------|--------|--------|-----|--------|-----|-----|-----|
| 0.   | 5.E-3  | .05    | 2.0 |        |     |     |    |

(a)

**Fig. 15-5a (Continued)**

sample points. The *trapezoidal rule method* emerges on assuming a first-order approximating curve to the actual function being integrated between sample points. Such an approximation is shown in Fig. 15-6. Under this approximation the integral equation

$$(15.11) \qquad y^{k+1} = y^k + \int_{t_k}^{t_{k+1}} f(y(\tau), \tau) \, d\tau$$

becomes

$$(15.12) \qquad y^{k+1} = y^k + \frac{1}{2} h(f^{k+1} + f^k)$$

RESULTS FOR EXAMPLE 15.3 USING A TIME STEP OF 0.50000D-01

| TIME | I3 |
|---|---|
| 0.0 | 0.10000D-02 |
| 0.50000D-01 | 0.96386D-02 |
| 0.10000D 00 | 0.17550D-01 |
| 0.15000 | 0.24769D-01 |
| 0.20000 | 0.31356D-01 |
| 0.25000 | 0.37366D-01 |
| 0.30000 | 0.42851D-01 |
| 0.35000 | 0.47855D-01 |
| 0.40000 | 0.52420D-01 |
| 0.45000 | 0.56586D-01 |
| 0.50000 | 0.60388D-01 |
| 0.55000 | 0.63856D-01 |
| 0.60000 | 0.67021D-01 |
| 0.65000 | 0.69908D-01 |
| 0.70000 | 0.72543D-01 |
| 0.75000 | 0.74947D-01 |
| 0.80000 | 0.77141D-01 |
| 0.85000 | 0.79142D-01 |
| 0.90000 | 0.80969D-01 |
| 0.95000 | 0.82635D-01 |
| 1.0000 | 0.84156D-01 |
| 1.0500 | 0.85543D-01 |
| 1.1000 | 0.86809D-01 |
| 1.1500 | 0.87964D-01 |
| 1.2000 | 0.89018D-01 |
| 1.2500 | 0.89979D-01 |
| 1.3000 | 0.90857D-01 |
| 1.3500 | 0.91657D-01 |
| 1.4000 | 0.92388D-01 |
| 1.4500 | 0.93054D-01 |
| 1.5000 | 0.93663D-01 |
| 1.5500 | 0.94218D-01 |
| 1.6000 | 0.94724D-01 |
| 1.6500 | 0.95186D-01 |
| 1.7000 | 0.95607D-01 |
| 1.7500 | 0.95992D-01 |
| 1.8000 | 0.96343D-01 |
| 1.8500 | 0.96663D-01 |
| 1.9000 | 0.96955D-01 |
| 1.9500 | 0.97222D-01 |
| 2.0000 | 0.97465D-01 |
| 2.0500 | 0.97687D-01 |

**Fig. 15-5b**   **Output of program in Fig. 15-5a.**

In other words, we are approximating the area under the function $f(y, t)$ between the sample points $t_k$ and $t_{k+1}$ with the area under a trapezoid whose height is $h$ and bases are $f^k = f[y(t_k), t_k]$ and $f^{k+1} = f[y(t_{k+1}), t_{k+1}]$.

**Example 15.4**

We illustrate the use of the trapezoidal rule method of integration by once again solving the differential equation

$$\dot{y}(t) = t$$

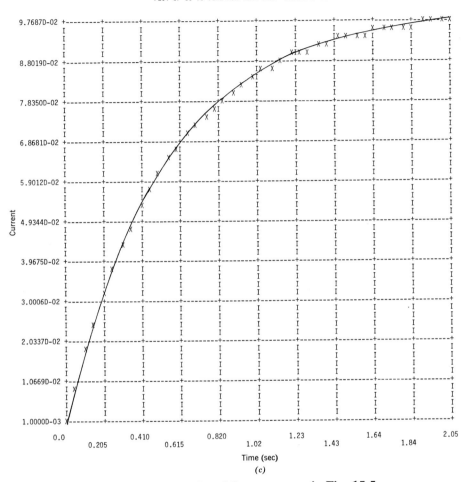

Fig. 15-5c    Plot produced from program in Fig. 15-5a.

which was solved in Example 13.1 using the forward Euler method and in Example 13.3 using the backward Euler method. The recursion relationship (15.12) becomes

$$y^{k+1} = y^k + \frac{1}{2} h[t_{k+1} + t_k]$$

$$= y^k + \frac{1}{2} h[(k+1)\,\Delta t + k\,\Delta t]$$

$$= y^k + \left(k + \frac{1}{2}\right)(h)^2$$

The results of using this recursion relation with a step size of $h = .1$ are summarized in Table 15-1. Observe that the computed solution agrees exactly

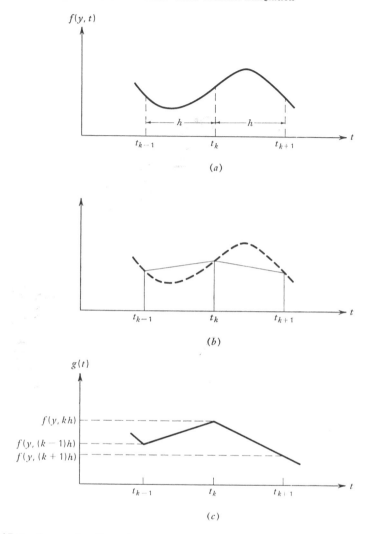

**Fig. 15-6    Approximation of a curve using a first-order curve between sample points.**

with the actual solution since a first-order curve is being approximated by a first-order curve.    □

Recall that the companion model concept arises from a network interpretation of the difference equation that results from application of a numerical integration method to the capacitor or inductor branch relationship. Trapezoidal rule applied to the capacitor branch relationship

$$v_C(t_{k+1}) = v_C(t_k) + \int_{t_k}^{t_{k+1}} \frac{1}{C} i_C(\tau) \, d\tau$$

**Table 15-1**  Comparison of the actual and the trapezoidal rule solutions of the differential equations $\dot{y}(t) = t$ for the step size of $h = .1$

| $t$ | Computed Solution | Actual Solution $t^2/2$ |
|-----|-------------------|-------------------------|
| .0  | .0    | .0    |
| .1  | .005  | .005  |
| .2  | .02   | .02   |
| .3  | .045  | .045  |
| .4  | .08   | .08   |
| .5  | .125  | .125  |
| .6  | .180  | .180  |
| .7  | .240  | .240  |
| .8  | .320  | .320  |
| .9  | .405  | .405  |
| 1.0 | .500  | .500  |

yields

$$(15.13) \qquad v_C(t_{k+1}) = \frac{h}{2C} i_C(t_{k+1}) + \left[ v_C(t_k) + \frac{h}{2C} i_C(t_k) \right]$$

which may be interpreted as a resistor in series with a voltage source as shown in Fig. 15-7a. Alternatively we may rewrite this expression as

$$(15.14) \qquad i_C(t_{k+1}) = \frac{2C}{h} v_C(t_{k+1}) - \left[ i_C(t_k) + \frac{2C}{h} v_C(t_k) \right]$$

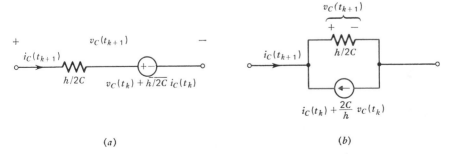

(a)                                        (b)

**Fig. 15-7**  Two equivalent companion models for the capacitor using trapezoidal rule integration.

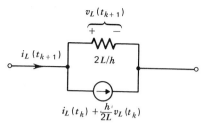

**Fig. 15-8   Companion model associated with an inductor for trapezoidal rule integration.**

which may be interpreted as a current source in parallel with a resistor as shown in Fig. 15-7b. Trapezoidal rule applied to the inductor branch relationship

$$i_L(t_{k+1}) = i_L(t_k) + \int_{t_k}^{t_{k+1}} \frac{1}{L} v_L(\tau)\, d\tau$$

yields

(15.15)  $$i_L(t_{k+1}) = \frac{h}{2L} v_L(t_{k+1}) + \left[ i_L(t_k) + \frac{h}{2L} v_L(t_k) \right]$$

which may be interpreted as a resistor in parallel with a current source as shown in Fig. 15-8. These companion models can be used in a manner similar to the companion models for the forward and backward Euler methods.

**Example 15.5**

Consider the numerical solution of the network of Fig. 15-9a using the trapezoidal rule integration scheme. The capacitor branch relationship is

$$\dot{v}_4 = 5 i_4$$

so that the companion model is described by

$$i_4^{k+1} = \frac{2}{5h} v_4^{k+1} - \left( i_4^k + \frac{2}{5h} v_4^k \right)$$

The companion network (the original network with the capacitor replaced by its companion model) is shown in Fig. 15-9b. The nodal equilibrium equations for the companion network are easily written:

$$\left( 2 + \frac{2}{5h} \right) v_①(t_{k+1}) - \frac{2}{5h} v_②(t_{k+1}) = e^{-t_{k+1}} + \frac{2}{5h} v_4(t_k) + i_4(t_k)$$

and

$$-\frac{2}{5h} v_①(t_{k+1}) + \left( 3 + \frac{2}{5h} \right) v_②(t_{k+1}) = -\frac{2}{5h} v_4(t_k) - i_4(t_k)$$

To solve these equations we need the values for $v_4$ and $i_4$ at time $t_0 = 0$. The initial capacitor voltage $v_4(0)$ is specified while the initial capacitor

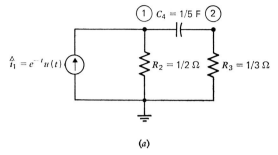

$\hat{i}_1 = e^{-t}u(t)$

$C_4 = 1/5$ F

$R_2 = 1/2\ \Omega$    $R_3 = 1/3\ \Omega$

(a)

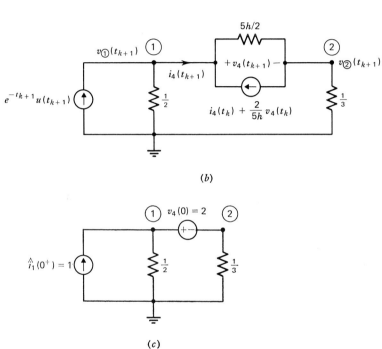

$v_{①}(t_{k+1})$

$5h/2$

$v_{②}(t_{k+1})$

$i_4(t_{k+1})$

$+v_4(t_{k+1})-$

$e^{-t_{k+1}}u(t_{k+1})$

$\dfrac{1}{2}$

$i_4(t_k) + \dfrac{2}{5h}v_4(t_k)$

$\dfrac{1}{3}$

(b)

$v_4(0) = 2$

$\hat{i}_1(0^+) = 1$

$\dfrac{1}{2}$    $\dfrac{1}{3}$

(c)

**Fig. 15-9** **(a)** Network used in Example 15.5; **(b)** the companion network for trapezoidal rule; **(c)** the initial condition network.

current $i_4$ is determined from the initial condition network in Fig. 15-9c. Recall that the initial condition network is obtained from the original network by replacing the capacitor by a voltage source whose value equals the initial capacitance voltage and setting all sources to the values they have at time $t = 0^+$. The initial values are found by solving the nodal equilibrium equations of the initial condition network:

$$2v_{①}(0) + i_4(0) = 1$$

$$\frac{1}{3}v_{②}(0) - i_4(0) = 0$$

and the voltage source branch relation

$$v_①(0) - v_②(0) = 2$$

The student should verify that $i_4(0) = -.43$, $v_①(0) = .71$, and $v_②(0) = -1.29$.

A computer program that can be used to solve the nodal equilibrium equations of the companion network is shown in Fig. 15-10a. In this program we have a means for automatically choosing a good step size. We know from the examples we have worked in this chapter and Chapter 13 that the step size has affected the solution. The simple scheme used in this program for adjusting the step size is as follows.

We start by choosing some initial step size $h_1$. We then use this step size to compute the numerical solution over the time interval of interest. Then, using a step size of one-half the original, that is, $h_2 = \frac{1}{2}h_1$, we recompute the numerical solution. (*Note*. Twice as many time points are required for a step size of $h_2$ than were required for a step size of $h_1$.) Next, the numerical solutions obtained using step sizes of $h_1$ and $h_2$ are compared. If the two solutions are in close agreement (say within 5%) then a step size of $h_1$ suffices. If the two solutions are appreciably different then a step size of $h_3 = \frac{1}{2}h_2 = \frac{1}{4}h_1$ is used to recompute the numerical solution which is then compared with the numerical solution obtained for a step size of $h_2$. If agreement is good then $h_2$ is an appropriate step size, otherwise the step size is halved again, $h_4 = \frac{1}{2}h_3$, and the process repeated.

For our example observe that the initial step size read by the program was .1, which was ultimately reduced to .0125. The results of the analysis are shown in Fig. 15-10b.   □

### Example 15.6

Suppose the trapezoidal rule integration scheme is to be employed to determine the node voltage of the network in Fig. 15-11a as a function of time. The discretized form of the inductor branch relationship under the trapezoidal rule integration method is

$$i_3(t_{k+1}) = \left[ i_3(t_k) + \frac{h}{4} v_3(t_k) \right] + \frac{h}{4} v_3(t_{k+1})$$

Therefore, the companion model is a current source of value $i_3(t_k) + (h/4)v_3(t_k)$ in parallel with a resistor of value $4/h$; the companion network is shown in Fig. 15-11b. The nodal equilibrium equation for the companion network is

$$\left(5 + \frac{h}{4}\right) v_①(t_{k+1}) = -u(t_{k+1}) + i_3(t_k) + \frac{h}{4} v_3(t_k)$$

```
C      CHAPTER 15 EXAMPLE 15.5
C
C      DIMENSIONS
C
       IMPLICIT REAL*8 (A-H,O-Z)
       DIMENSION Y(3,3),CIN(3),JPIV(3),VOUT(2,200),VOTEMP(2,100)
       DIMENSION TIME(200)
C
C       READ IN ELEMENT VALUES, INITIAL CAPACITANCE VOLTAGE,
C          TIME STEP AND FINAL TIME
C
       CALL READ(R2,R3,C4,V4I,H,TF)
       ISTEP=1
   10  T=0.
       NPTS=0
    1  NPTS=NPTS+1
       IF(NPTS.GT.200) STOP 1
C
C      SET UP AND SOLVE NODE EQUATIONS
C
       CALL NODEQN(R2,R3,C4,V4I,V4,H,T,Y,CIN,N,NPTS)
C
C      CALL FWBW IF NOT SOLVING THE INITIAL CONDITION
C          NETWORK OR THE COMPANION NETWORK AT THE
C          FIRST TIME POINT
C
       IF(NPTS.GT.2) GO TO 4
       CALL LUSOLV(Y,CIN,N,3,JPIV)
       GO TO 5
    4  CALL FWBW(Y,CIN,N,3,JPIV)
    5  V1=CIN(1)
       V2=CIN(2)
       V4=CIN(1)-CIN(2)
C
C      STORE SOLUTION FOR PLOTTING
C
       VOUT(1,NPTS)=V1
       VOUT(2,NPTS)=V2
       TIME(NPTS)=T
C
C       IF T IS LESS THAN FINAL TIME (TF) INCREMENT T
C
       IF(T.GT.TF) GO TO 2
       T=T+H
       GO TO 1
C
C      DETERMINE IF STEP SIZE IS APPROPRIATE. IF OKAY PLOT RESPONSE,
C      IF NOT STORE VOUT IN VOTEMP, HALVE STEP SIZE, AND REANALYZE
C      THE NETWORK.
C
    2  IF(ISTEP.EQ.1) GO TO 6
       NN=NPTS/2
       DO 7 I=1,2
       DO 7 J=1,NN
       IF(VOTEMP(I,J).EQ.0.) GO TO 7
       IF(DABS((VOUT(I,J*2-1)-VOTEMP(I,J))/VOTEMP(I,J)).GT..05) GO TO 6
```

**Fig. 15-10a  Program used in Example 15.5 to analyze network of Fig. 15-9 using trapezoidal rule.**

495

```
      7 CONTINUE
        GO TO 9
      6 ISTEP=2
        DO 8 I=1,2
        DO 8 J=1,NPTS
      8 VOTEMP(I,J)=VOUT(I,J)
        H=H/2
        GO TO 10
      9 PRINT 3,H
        CALL PLOT(TIME,VOUT,NPTS,2,0)
      3 FORMAT('1RESULTS FOR EXAMPLE 15.5 USING A TIME STEP OF',G12.5,//,
       +53X,'V1=X',10X,'V2=0')
        END




        SUBROUTINE READ(R2,R3,C4,V4I,H,TF)
        IMPLICIT REAL*8 (A-H,O-Z)
        READ(5,1) R2,R3,C4,V4I,H,TF
      1 FORMAT(6G10.0)
        RETURN
        END




        SUBROUTINE NODEQN(R2,R3,C4,V4I,V4,H,T,Y,CIN,N,NPTS)
        IMPLICIT REAL*8 (A-H,O-Z)
        DIMENSION Y(3,3),CIN(3)
C
C       DETERMINE SOURCE VALUES
C
        CI1=DEXP(-T)
C
C       AT TIME T=0 SOLVE INITIAL CONDITION NETWORK
C       INITIAL CONDITION NETWORK
C
        IF(T.GT.0.) GO TO 1
C
C       INITIAL CONDITION NETWORK
C
        N=3
        GC4=0.D0
        CI4=0.D0
        DO 4 J=1,3
        CIN(J)=0.D0
        DO 4 K=1,3
      4 Y(J,K)=0.D0
        Y(1,3)=1.
        Y(2,3)=-1.
        Y(3,1)=1.
        Y(3,2)=-1.
        CIN(3)=V4I
        GO TO 2
C
C       COMPANION NETWORK
C
      1 N=2
C
C       UPDATE COMPANION MODELS
C
```

**Fig. 15-10a** (*Continued*)

496

```
      GC4=2./(5.*H)
      CC4=(GC4*V4)-CI4
      CI4=CC4+GC4*V4
      IF(NPTS.EQ.2) CI4=CIN(3)+GC4*V4
C
C     WHEN SOLVING THE COMPANION NETWORK AFTER THE
C     FIRST TIME POINT ONLY THE RIGHT HAND SIDE
C     OF THE EQUATIONS IS ALTERED
C
      IF(NPTS.GT.2) GO TO 3
      DO 5 J=1,2
    5 Y(J,K)=0.D0
    2 Y(1,1)=(1./R2)+GC4
      Y(1,2)=-GC4
      Y(2,1)=-GC4
      Y(2,2)=GC4+(1./R3)
    3 CIN(1)=CI1+CI4
      CIN(2)=-CI4
      RETURN
      END
```

----- DATA FOR PROGRAM OF EXAMPLE 15.5 -----

.5          .333          .2          2.          .1          1.5

**Fig. 15-10a (Continued)**

This equation along with the discretized inductor branch relationship

$$i_3(t_{k+1}) = \left[ i_3(t_k) + \frac{h}{4} v_3(t_k) \right] + \frac{h}{4} v_3(t_{k+1})$$

and the relation

$$v_3(t_k) = v_{①}(t_k)$$

can be solved recursively to yield the solution $v_1(t_k)$.  □

## 15-5*  PREDICTOR-CORRECTOR METHODS

We have seen that application of the numerical integration schemes discussed to the differential equation

$$\dot{y}(t) = f(y(t), v(t), t)$$

resulted in a recursion relationship:

$$y^{k+1} = y^k + h f^k$$

for forward Euler;

$$y^{k+1} = y^k + h f^{k+1}$$

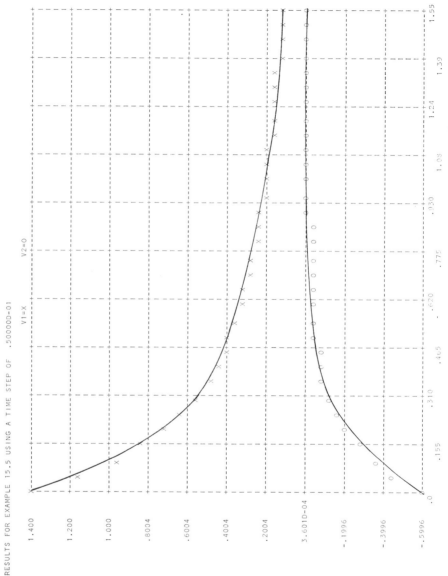

Fig. 15-10b  Output of program in Fig. 15-10a.

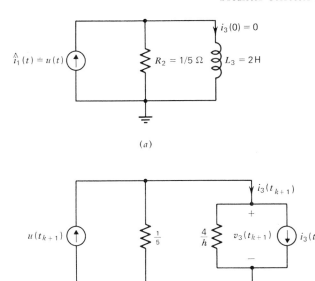

Fig. 15-11   (*a*) Network analyzed in Example 15.6; (*b*) the companion network.

for backward Euler; and

$$y^{k+1} = y^k + \frac{h}{2}(f^{k+1} + f^k)$$

for trapezoidal rule, where

$$y^k \equiv y(t_k)$$
$$f^k \equiv f(y(t_k), v(t_k), t_k)$$

etc. If $f(y, v, t)$ is a linear function of $y$ then the implicit backward Euler and trapezoidal rule recursion relationships can be rearranged to become explicit. An alternative would be to use the explicit forward Euler relationship to *predict* what $y^{k+1}$ is and then substitute this value into the implicit backward Euler or trapezoidal rule relationship to *correct* this value. An example illustrates the technique.

**Example 15.7**

Suppose we wished to solve the differential equation

$$\dot{y}(t) = -y(t), \qquad y(0) = 3$$

using the forward Euler method as the *predictor* and the trapezoidal rule method as the *corrector*. The predictor equation becomes

$$y^{k+1,0} = y^k - hy^k$$
$$= (1 - h)y^k$$

where the added superscript "zero" denotes the predicted value and the corrector becomes

$$y^{k+1,1} = y^k - \frac{h}{2}(y^{k+1,0} + y^k)$$

where the superscript "1" is used to denote the corrected value. For $h = .1$ we have for the first time point $t_1 = .1$:

$$y^{1,0} = (1 - h)y_0$$
$$= 2.7$$

as the predicted value of $y(t_1)$ and

$$y^{1,1} = y^0 - \frac{h}{2}(y^{1,0} + y^0) = 3 - .05(5.7)$$

$$= 2.715$$

as the corrected value of $y(t_1)$. The exact solution is, of course,

$$y(t) = 3e^{-t}$$

so that

$$y(.1) = 2.7145$$

The corrected value is a better estimate of the actual value than was the predicted value. Observe that if trapezoidal rule had been used directly to obtain the explicit recursion relationship

$$y^{k+1} = \frac{\left(1 - \dfrac{h}{2}\right)}{\left(1 + \dfrac{h}{2}\right)} y^k$$

then

$$y^1 = 2.712$$

which is also a poorer estimate of the solution than the corrected value.  □

Observe that the corrector could be used more than once. We would then write, for trapezoidal rule, the corrector as

$$y^{k+1,l+1} = y^k + \frac{h}{2}(f^{k+1,l} + f^k)$$

where $y^{k+1,l+1}$ is the $(l + 1)$st corrected value of $y^{k+1}$, etc.

The motivation behind using predictor-corrector methods is that it is possible to make an estimate of the error between the actual and numerically computed solution at each time point. This error is called the *local truncation error*. If the local truncation error is larger than the error we are willing to except, the step size can be decreased (for example, by half). Alternatively, if the truncation error is much smaller than the maximum allowable error, we might increase the step size (double) and thereby achieve computational savings. In what follows we will indicate how the local truncation error can be estimated from the predicted and corrected values of a solution. The derivation is by no means rigorous.

Consider the differential equation

$$\dot{y} = f(y, t)$$

The forward Euler recursion relationship is

$$\hat{y}^{k+1} = y^k + hf^k = y^k + h\dot{y}^k$$

where $\hat{y}$ is used to denote the computed solution. (We assume $y^k$ and $\dot{y}^k$ are known exactly.) The Taylor series expansion of $y^{k+1}$ about $y^k$ is

$$(15.16) \qquad y^{k+1} = y^k + h\dot{y}^k + \frac{1}{2} h^2 \ddot{y}^k + \frac{1}{3!} h^3 \dddot{y}^k + \cdots$$

because $h = t_{k+1} - t_k$. Therefore, the error involved in the forward Euler method is in the order of

$$\frac{1}{2} h^2 \ddot{y}^k$$

since

$$\varepsilon_p = y^{k+1} - \hat{y}^{k+1} \simeq \frac{1}{2} h^2 \ddot{y}^k$$

Thus if the forward Euler method is used as a predictor, an error of $\varepsilon_p$ is introduced. Suppose we now use the backward Euler method as the corrector. Then

$$\hat{y}^{k+1} = y^k + hf^{k+1} = y^k + h\dot{y}^{k+1}$$

or

$$\hat{y}^{k+1} - h\dot{y}^{k+1} = y^k$$

The Taylor series expansion of $y^{k+1}$ is given in (15.16). The Taylor series expansion of $\dot{y}^{k+1}$ is

$$\dot{y}^{k+1} = \dot{y}^k + h\ddot{y}^k + \frac{h^2}{2} \dddot{y}^k + \cdots$$

Therefore

$$y^{k+1} - h\dot{y}^{k+1} = \left( y^k + h\dot{y}^k + \frac{1}{2} h^2 \ddot{y}^k + \frac{1}{6} h^3 \dddot{y}^k + \cdots \right)$$
$$- h\left( \dot{y}^k + h\ddot{y}^k + \frac{h^2}{2} \dddot{y}^k + \cdots \right)$$
$$= y^k - \frac{1}{2} h^2 \ddot{y}^k + \cdots$$

The error in the backward Euler corrector can now be found:

$$\varepsilon_c = (y^{k+1} - h\dot{y}^{k+1}) - (\hat{y}^{k+1} - h\dot{y}^{k+1}) \simeq -\frac{1}{2} h^2 \ddot{y}^k$$

Now if $y(t_{k+1})$ is the *actual* value of $y$ at time $t_{k+1}$, $y^{k+1,0}$ is the *predicted* value of $y$ at time $t_{k+1}$, and $y^{k+1}$ is the *corrected* value of $y$ at time $t_{k+1}$, then

$$y^{k+1,0} = y(t_{k+1}) - \varepsilon_p$$

and

$$y^{k+1} = y(t_{k+1}) - \varepsilon_c$$

Therefore

$$y^{k+1} - y^{k+1,0} = -\varepsilon_c + \varepsilon_p = -\left(-\frac{1}{2}h^2\ddot{y}^k\right) + \left(\frac{1}{2}h^2\ddot{y}^k\right) = h^2\ddot{y}^k$$

So that an estimate of $\ddot{y}_k$ is

$$\ddot{y}^k = \frac{y^{k+1} - y^{k+1,0}}{h^2}$$

and the truncation error in the corrector is

$$\varepsilon_c = -\frac{1}{2}h^2\ddot{y}^k = -\frac{1}{2}(y^{k+1} - y^{k+1,0})$$

Two observations can be made:

1. This scheme works because the predictor and corrector were of the same order.
2. We can use the estimate of the truncation error to adjust the step size. If $\varepsilon_c$ is larger than an allowable per step error estimate, then $h$ should be decreased, that is, try $h/2$. If $\varepsilon_c$ is smaller than the allowable per step error estimate, then $h$ should be increased, that is, try $2h$.

## 15-6 SUMMARY

In this chapter we returned to the discussion of the numerical techniques for the time domain simulation of linear networks. We began with a discussion of using the companion model approach for networks that contain more than one energy storage element.

It was shown how a matrix **A** could be factored into the product of a *lower triangular* matrix **L** and an *upper triangular matrix* **U**, which has ones on the main diagonal. This *LU factorization* is useful especially when a set of simultaneous algebraic equations are to be solved for more than one right-hand side. The *LU* factorization procedure is implemented in the subroutine LUSOLV.

The *trapezoidal rule* integration scheme was introduced and companion models for capacitors and inductors derived. The trapezoidal rule is a second-order method, and usually more accurate than the first-order backward Euler method.

Finally, a brief discussion on the choice of a suitable step size was given and the usefulness of a predictor-corrector scheme for estimating truncation error was shown.

## PROBLEMS

**15.1**  Consider the networks of Fig. P15.1. Replace each energy storage element by the companion model associated with the trapezoidal integration scheme. Use this network to determine the network's response for a time step equal to half the network's time constant. Compare these results with the actual solution obtained in Prob. 11.5.

**15.2**  Find the **L** and **U** matrices associated with the matrix

$$A = \begin{pmatrix} 1 & -1 & 2 \\ 3 & 2 & 0 \\ 4 & 0 & 5 \end{pmatrix}$$

using Gaussian elimination. Verify that $A = LU$.

**15.3**  Suppose the matrix **B** be the inverse of **A**, that is,

$$AB = 1$$

Let the vectors $b_1, b_2, \ldots b_n$ represent the $n$ columns of **B**, that is

$$\begin{bmatrix} b_1 & b_2 & \cdots & b_n \\ \downarrow & \downarrow & & \downarrow \end{bmatrix}$$

and let $e_k$ denote a vector of all zeros except a one in the $k$th row, that is,

$$e_k = \begin{bmatrix} 0 \\ \cdot \\ \cdot \\ \cdot \\ 0 \\ 1 \\ 0 \\ 0 \\ \cdot \\ \cdot \\ \cdot \\ 0 \end{bmatrix} \leftarrow k\text{th row}$$

then

$$Ab_k = e_k$$

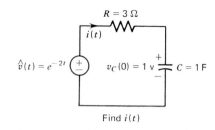

Find $i(t)$

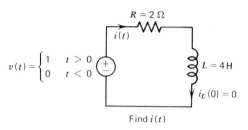

Find $i(t)$

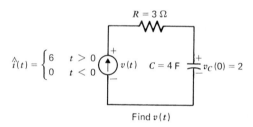

Find $v(t)$

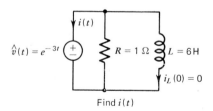

Find $i(t)$

**Fig. P15.1**

Thus once we have performed an *LU* factorization on **A** we can solve for the *k*th column of **B** by performing on forward-backward substitution, or all *n* columns of **B** by performing *n* forward-backward substitutions. Use this scheme to find $\mathbf{A}^{-1}$ for the **A** of Prob. 15.2.

**15.4**    Write a program MATINV that finds the inverse of a matrix by repeatedly calling LUSOLV.

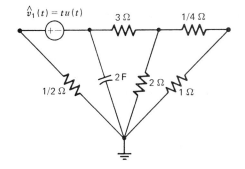

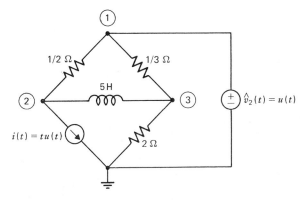

**Fig. P15.5**

**15.5** Use the trapezoidal rule companion model to transform the net-works of Fig. P15.5 into companion networks. Write the nodal equilibrium equations for each of the companion networks. Write a computer program that calls LUSOLV to solve the nodal equili-brium equations numerically. Plot the capacitor voltage or inductor current for various step sizes. Find an appropriate step size in each case.

**15.6** Two numerical integration schemes that we have not considered are

$$y^{k+1} = y^{k-2} + \frac{\Delta t}{3} (\dot{y}^{k-1} + 4\dot{y}^k + \dot{y}^{k+1})$$

and

$$y^{k+1} = y^{k-2} + \frac{\Delta t}{8} (3\dot{y}^{k-2} + 9\dot{y}^{k-1} + 9\dot{y}^k + 3\dot{y}^{k+1})$$

where $\dot{y} = f(y, t)$. Find the companion model associated with capacitors and inductors for each of these integration schemes.

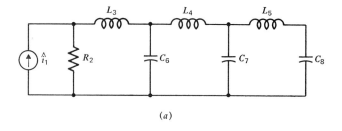

(a)

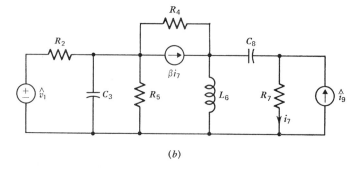

(b)

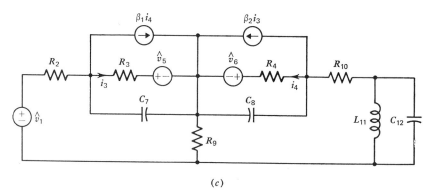

(c)

**Fig. P15.7**

What problems, if any, could be expected using these methods? In particular, consider $k = 0$, 1, and 2. What starting values are needed. How can they be found?

**15.7**    Using the trapezoidal rule write a set of simultaneous algebraic equations in terms of the step size $h$ that can be used iteratively to solve for the time response of each of the networks of Fig. P15.7.

**15.8**    Write a computer program that uses GAUSS to determine the output wave form for the network of Fig.P15.8 using the trapezoidal rule companion model for the capacitors. Use an initial time-step of one nanosecond ($10^{-9}$) and analyze the network from $t = 0$ to $t = 10^{-8}$. Adjust time-step if necessary to guarantee accuracy.

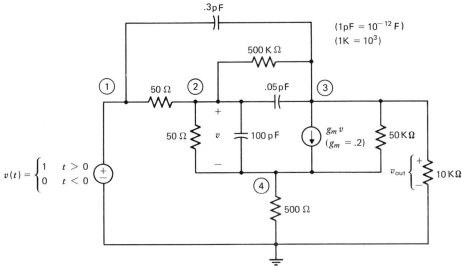

.3pF

$(1pF = 10^{-12} F)$

$(1K = 10^3)$

500 K Ω

50 Ω

.05pF

50 Ω

$v$

100 pF

$g_m v$

$(g_m = .2)$

50 KΩ

$v(t) = \begin{cases} 1 & t > 0 \\ 0 & t < 0 \end{cases}$

$v_{out}$

10 KΩ

500 Ω

**Fig. P15.8**

**15.9**    Show that the truncation error associated with trapezoidal rule is

$$-\frac{1}{12} h^3 \ddot{y}^k$$

**15.10**    Try to determine an estimate of the truncation error in terms of the difference between predictor and corrector values if the forward Euler method is the predictor (a first-order method) and trapezoidal rule is the corrector (a second-order method.)

**15.11**    Consider the second-order predictor

$$y^{k+1} = y^{k-1} + 2h\dot{y}^k$$

Show that the truncation error is

$$\tfrac{1}{3} h^3 \ddot{y}^k$$

by expanding $y^{k+1}$ and $y^{k+1}$ in Taylor series about $y^k$ and subtracting. Discuss Prob. 15.10 with this predictor and a trapezoidal rule corrector.

**15.12**    In Example 15.1 we used the backward Euler method to analyze the network of Fig. 15-1a. Modify the program of Fig. 15-2a so that it uses the trapezoidal rule integration scheme. Use this modified program to analyze the network of Fig. 15-1a and compare your results with those obtained in the text.

**15.13**    Repeat Prob. 15.12 for the program of Fig. 15-5a.

**15.14**     In the program of Fig. 15-10a a proper step size was formed by comparing the results of a complete analysis using a step size $h$ with those of a complete analysis using a step size of $h/2$. If the predicted responses were within 5 percent of each other the step size $h/2$ was accepted. If not, the step size was halved again. An alternative method, and one which is more efficient, would be to allow the step size to halve, or double, *between* each time point. In other words, given the response at time $t = 0$, use a step size $h$ to predict the response at time $t_1 = h$. Then compute the response at time $t_1 = h$ using a step size of $h/2$. (This requires an analysis at time point $h/2$ also.) Compare the two predicted responses at time $h$. If they are within say 5 percent, a step size of $h$ is adequate and we would try a step size of $2h$ for the next time interval. If the predicted responses are not within 5 percent of each other then we would recompute the response at time $h/2$ using a step size of $h/4$. (This requires an analysis at time point $h/4$.) This procedure is repeated until the predicted responses are within 5 percent for two step sizes $\hat{h}$ and $\hat{h}/2$. Then in the next time interval results for step sizes of $\hat{h}$ and $2\hat{h}$ are tried. Observe that this procedure allows the program to take larger step sizes when the response is varying slowly. The price we pay is that the predicted response at each time point requires at least three dc type analyses.

Modify the program of Fig. 15-10a to implement this procedure. Observe that we must redo the $LU$ factorization every time the step size $h$ changes. Use this program to analyze the network of Fig. 15-9 and compare your results with those obtained in the text.

**15.15**     A time varying inductor is characterized by the branch relationship

$$v_L(t) = L(t)\frac{di_L(t)}{dt}$$

Show that if the backward Euler method is used the companion model for this element is a time varying resistor in parallel with a current source.

**15.16**     Derive a companion for the time varying inductor of Prob. 15.15 using the trapezoidal rule integration scheme.

# CHAPTER 16

# Sinusoidal Steady-State Analysis

In the last several chapters we have discussed the time domain analysis of networks under arbitrary excitations. We found it necessary in each instance to solve one or more differential equations, either analytically or numerically. In the next several chapters we will discuss specialized analysis methods that may be employed if the network excitations are restricted to be in certain classes of functions. These techniques avoid the necessity of solving differential equations. The first class of network inputs to be studied is the class of sinusoids.

Study of the behavior of networks under sinusoidal excitation is justified for two reasons. First, many networks are designed to operate under a sinusoidal excitation. For example, hi-fi audio amplifiers are designed to faithfully reproduce but amplify their sine wave inputs. Secondly, we will show that all periodic signals may be represented by a sum of sinusoids so that if the response to a sinusoid is easily found, the response to any periodic signal may be found.

So that we may facilitate discussion, we begin by introducing some useful phraseology and notation. It is assumed that the student is familiar with operations involving complex numbers.

## 16-1  SINUSOIDS AND PHASORS

A general sinusoidal waveform is shown in Fig. 16-1 and is characterized by

$$v(t) = V_m \sin (\omega t)$$

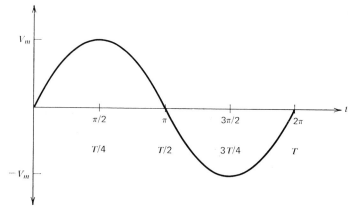

**Fig. 16-1    A sine wave of amplitude $V_m$ and period $T = 2\pi/\omega$.**

where $V_m$ is the maximum value or *amplitude* of the sine wave, $\omega$ is the *frequency* in radians/seconds. The *period* of the sine wave, denoted by $T$, is the time interval required to complete one cycle, that is, the time between $\omega t = 0$ and $\omega t = 2\pi$. Therefore,

$$T = \frac{2\pi}{\omega}$$

A more general form of the sinusoid is

(16.1)     $$v(t) = V_m \sin (\omega t + \phi)$$

which includes the phase angle $\phi$. This sinusoid is shown in Fig. 16-2. The phase angle, measured in radians or degrees, represents a shift to the left of sinusoid with respect to the reference sinusoid. Since corresponding values of the sinusoid $V_m \sin (\omega t + \phi)$ occur $\phi$ radians sooner than those of $V_m \sin (\omega t)$, we say that $V_m \sin (\omega t + \phi)$ *leads* $V_m \sin (\omega t)$ by $\phi$ radians, or $\phi/\omega$ seconds. We can also say that $V_m \sin (\omega t)$ *lags* $V_m \sin (\omega t + \phi)$ by $\phi$ radians.

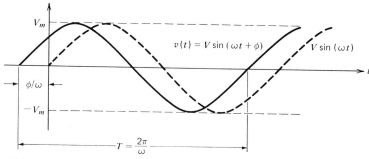

**Fig. 16-2    A sine wave of amplitude $V_m$, phase angle $\phi$ and period $T = 2\pi/\omega$.**

The sine wave (16.1) may also be written in terms of the cosine function

$$v(t) = V_m \cos\left(\omega t + \phi - \frac{1}{2}\pi\right)$$

By defining $\theta \equiv \phi - \pi/2$ we have

$$v(t) = V_m \cos(\omega t + \theta)$$
$$= (V_m \cos\theta)\cos\omega t - (V_m \sin\theta)\sin\omega t$$

In general, we will be studying functions that are sums of sines and cosines. The student should be familiar with Euler's identity

(16.2) $$\hspace{2cm} e^{j\omega t} = \cos\omega t + j\sin\omega t$$

were $j = \sqrt{-1}$. It is easy to show that

$$\cos\omega t = \frac{e^{j\omega t} + e^{-j\omega t}}{2} = \text{Re}\,\{e^{j\omega t}\}$$

where Re $\{\cdot\}$ denotes the *real part of* and

$$\sin\omega t = \frac{e^{j\omega t} - e^{-j\omega t}}{2j} = \text{Im}\,\{e^{j\omega t}\}$$

where Im $\{\cdot\}$ denotes the *imaginary part of.*

The terms $e^{j\omega t}$ and $e^{-j\omega t}$ may be interpreted in terms of *rotating phasors*. In particular, $e^{j\omega t}$ represents a unit phasor rotating in the counter-clockwise direction as shown in Fig. 16-3a since the magnitude of $e^{j\omega t}$ is

$$|e^{j\omega t}| = ((\cos\omega t)^2 + (\sin\omega t)^2)^{1/2} = 1$$

and the phase of $e^{j\omega t}$ is

$$\theta = \tan^{-1}\left[\frac{\text{Im}\,(e^{j\omega t})}{\text{Re}\,(e^{j\omega t})}\right] = \tan^{-1}\left(\frac{\sin\omega t}{\cos\omega t}\right) = \omega t$$

Similarly, $e^{-j\omega t}$ represents a unit phasor rotating in the clockwise direction as shown in Fig. 16-3b. The sine and cosine waves can be interpreted in terms of these rotating phasors. We will use the exponential function $e^{j\omega t}$ in the sequel as a means of determining the responses of networks to sines and cosines.

It will be convenient at times to think of a real-valued signal to be the real part or imaginary part of a complex-valued signal. For example

$$v(t) = 3\cos(\omega t + 2) = \text{Re}\,[3e^{j(\omega t + 2)}]$$

and

$$i(t) = 6\sin(\omega t) = \text{Im}\,(6e^{j\omega t})$$

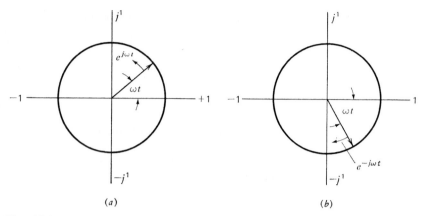

(a)                                    (b)

**Fig. 16-3** (a) **Counter-clockwise rotating phasor of unit amplitude representing** $e^{j\omega t}$; (b) **clockwise rotating phasor representing** $e^{-j\omega t}$.

Any complex sinusoid $v(t)$ may be written as

$$v(t) = Ve^{j\omega t}$$

where $V$ is a complex number called a *phasor*. In particular,

$$V = V_m e^{j\phi}$$

where $V_m$ is the magnitude of $V$:

$$V_m = |V| = \sqrt{\text{Re }\{V\}^2 + \text{Im }\{V\}^2}$$

and $\phi$ is the phase of $V$.

$$\phi = \arg V = \tan^{-1}\left[\frac{\text{Im }(V)}{\text{Re }(V)}\right]$$

Thus we may write

$$v(t) = V_m e^{j\phi} e^{j\omega t}$$
$$= V_m e^{j(\omega t+\phi)}$$
$$= V_m \cos(\omega t + \phi) + jV_m \sin(\omega t + \phi)$$

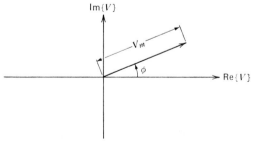

**Fig. 16-4** **Representation of the complex phasor** $V = V_m e^{j\phi}$.

A phasor is easily represented in the complex plane as illustrated in Fig. 16-4. An alternate form for expressing the phasor $V$ is $V = V_m \angle \phi$ where $\angle \phi$ denotes the angle $\phi$.

## 16-2  INTRODUCTION TO THE SINUSOIDAL STEADY-STATE

We introduce the concept of the sinusoidal steady-state by studying the behavior of the second-order network of Fig. 16-5. The nodal equilibrium equation is

$$C\ddot{v}_{①}(t) + \frac{1}{R}\dot{v}_{①}(t) + \frac{1}{L}v_{①}(t) = \dot{i}_1(t)$$

We will let the input be the complex-valued exponential

$$\hat{i}_1(t) = e^{j\omega t}$$

The student should realize that complex-valued signals do not occur in reality however, there is no reason why we cannot manipulate them mathematically. The reason behind allowing a complex-valued source will become clear as we proceed.

The complete solution of the nodal equilibrium equation is the sum of the homogeneous solution and the particular solution. The homogeneous solution is

$$v_{①H}(t) = c_1 e^{s_1 t} + c_2 e^{s_2 t}$$

where

$$s_1, s_2 = -\frac{1}{2RC} \pm \sqrt{\left(\left(\frac{1}{2RC}\right)^2 - \frac{1}{LC}\right)}$$

and $c_1$ and $c_2$ are arbitrary constants. The particular solution is any solution of the original equation. Rather than proceed as was done in Chapter 11, let us assume a particular solution of the form

$$v_{①p}(t) = V_{①}e^{j\omega t}$$

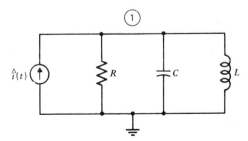

Fig. 16-5  Second-order network used to introduce the sinusoidal steady state.

where $V_{①}$ is a complex phasor:

$$V_{①} = V_m e^{j\phi}$$

Substitution of the particular solution into the original differential equation yields

$$-\omega^2 C V_{①} e^{j\omega t} + j\omega \frac{1}{R} V_{①} e^{j\omega t} + \frac{1}{L} V_{①} e^{j\omega t} = j\omega\, e^{j\omega t}$$

or, on division by $e^{j\omega t}$,

$$\left(-\omega^2 C + j\omega \frac{1}{R} + \frac{1}{L}\right) V_{①} = j\omega$$

which may be rewritten as

$$V_{①} = \frac{j\omega}{\left[\left(\dfrac{1}{L} - \omega^2 C\right) + j\omega \dfrac{1}{R}\right]}$$

$$= \frac{RL}{\sqrt{(R - \omega^2 LCR)^2 + (\omega L)^2}}\, e^{j\phi}$$

where

$$\phi = \tan^{-1}\left(\frac{R - \omega C}{L}\right)$$

Therefore, the particular solution becomes

$$v_{①p}(t) = \frac{RL}{\sqrt{(R - \omega^2 LCR)^2 + (\omega L)^2}}\, e^{j(\omega t + \phi)}$$

We are interested in the case in which the excitation has been applied for a long time and all transients have disappeared. This steady-state response is equal to the particular solution because the homogeneous solution becomes zero as $t$ becomes large.

Observe that from knowledge of the response of this network to $e^{j\omega t}$ the response to the excitations $\cos \omega t$ and $\sin \omega t$ can be determined without resorting again to solving the differential equation. Consider an excitation of $\cos \omega t$ first. All we need to do is recognize that

$$\cos \omega t = \text{Re}\left\{e^{j\omega t}\right\}$$

so that under this excitation the nodal equilibrium equation for this network is

$$C\ddot{v}_{①} + \frac{1}{R}\dot{v}_{①} + \frac{1}{L}v_{①} = \text{Re}\left\{j\omega\, e^{j\omega t}\right\}$$

Now let $v(t)$ be a complex-valued voltage such that

$$v_{①}(t) = \text{Re}\left\{v(t)\right\}$$

Then

$$C \, \text{Re} \, \{\ddot{v}\} + \frac{1}{R} \, \text{Re} \, \{\dot{v}\} + \frac{1}{L} \, \text{Re} \, \{v\} = \text{Re} \, \{j\omega \, e^{j\omega t}\}$$

or

$$C\ddot{v} + \frac{1}{R} \dot{v} + \frac{1}{L} v = j\omega \, e^{j\omega t}$$

However, we found that the steady-state solution of this equation is

$$v(t) = \frac{RL}{\sqrt{(R - \omega^2 LCR)^2 + (\omega L)^2}} \, e^{j(\omega t + \phi)}$$

where

$$\phi = \tan^{-1} \left( \frac{R - \omega C}{L} \right)$$

Therefore,

$$v_{①}(t) = \text{Re} \left\{ \frac{RL}{\sqrt{(R - \omega^2 LCR)^2 + (\omega L)^2}} \, e^{j(\omega t + \phi)} \right\}$$

$$= \frac{RL}{\sqrt{(R - \omega^2 LCR)^2 + (\omega L)^2}} \cos (\omega t + \phi)$$

Similarly, we may show that if the excitation of the network is

$$\sin \omega t = \text{Im} \, \{e^{j\omega t}\}$$

then the response is

$$v_{①}(t) = \text{Im} \left\{ \frac{RL}{\sqrt{(R - \omega^2 LCR)^2 + (\omega L)^2}} \, e^{j(\omega t + \phi)} \right\}$$

$$= \frac{RL}{\sqrt{(R - \omega^2 LCR)^2 + (\omega L)^2}} \sin (\omega t + \phi)$$

We will find that if the excitation of a network is a sinusoid, it is easiest to undertake the solution in terms of the exponential $e^{j\omega t}$ and then take the real or imaginary part, as appropriate.

One of the nice properties of the sinusoidal steady-state response is that it can be found without solving differential equations. This result arises, as seen in the above, from the fact that if the input to a linear network is a sinusoid, then all voltages and currents in the network are sinusoids of the same frequency. Therefore, by assuming all voltages and currents in a network excited by $e^{j\omega t}$ are of the form $Xe^{j\omega t}$, where $X$ is a complex phasor, and dividing through by $e^{j\omega t}$, the differential equations become complex algebraic equations that can be solved for the unknown phasors.

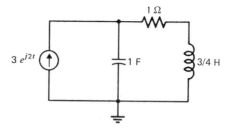

**Fig. 16-6   Network studied in Example 16.1.**

**Example 16.1**

The nodal equilibrium equations of the network in Fig. 16-6 are

$$(1 + D)v_① - 1v_② = 3e^{j2t}$$

$$-1v_① + \left(1 + \frac{4}{3D}\right)v_② = 0$$

Assuming that

$$v_① = V_①e^{j2t} \quad \text{and} \quad v_② = V_②e^{j2t}$$

these equations become

$$(1 + D)V_①e^{j2t} - V_②e^{j2t} = 3e^{j2t}$$

$$-V_①e^{j2t} + \left(1 + \frac{4}{3D}\right)V_②e^{j2t} = 0$$

Recognizing that

$$DV_①e^{j2t} = \frac{d}{dt}[V_①e^{j2t}] = j2V_①e^{j2t}$$

and

$$\frac{1}{D}V_②e^{j2t} = v_②(0) + \int_0^t V_②e^{j2t}\,dt = \frac{1}{j2}V_②e^{j2t}$$

since $v_②(0) = 0$ (all initial conditions are assumed to be zero) the nodal equilibrium equations become

$$(1 + j2)V_① - V_② = 3$$

$$-V_① + \left(1 - j\frac{2}{3}\right)V_② = 0$$

which is a set of simultaneous algebraic equations with complex coefficients. The solution is

$$V_① = \frac{9}{4\sqrt{2}}e^{j\phi_1} \quad \text{and} \quad V_② = \frac{3\sqrt{10}}{8}e^{j\phi_2}$$

where $\phi_1 = -\tan^{-1} 1 = -45°$ and $\phi_2 = \tan^{-1} 3 = 70.5°$. Therefore, the steady state solution to the original node equations is

$$v_①(t) = V_①e^{j2t} \quad \text{and} \quad V_②(t) = v_②e^{j2t} \quad \square$$

It would seem from the previous equation that we should be able to write the complex algebraic "node" equations without first writing the differential equations and then assuming a solution. We pursue this reasoning in the next section.

## 16-3  GENERAL SINUSOIDAL STEADY-STATE ANALYSIS

In the preceding we indicated that all the voltages and currents in a linear network that is excited by a sinusoid of frequency $\omega$ are themselves sinusoids of frequency $\omega$ after all transients have died out. This conjecture can be justified with the following observations.

1. A sinusoid of frequency $\omega$ may be repeatedly differentiated or integrated and it is still a sinusoid of frequency $\omega$.
2. The sum of a number of sinusoids of the same frequency but of different amplitudes is a sinusoid of the same frequency.

Let us study the implications of the above observations. Suppose a $n_b$ branch $n_n$ node network is operating in sinusoidal steady-state. Then we assume all branch voltages are of the form

$$(16.3) \qquad v_l = V_l e^{j\omega t} \qquad l = 1, 2, \ldots, n_b$$

and all branch currents are of the form

$$(16.4) \qquad i_l = I_l e^{j\omega t} \qquad l = 1, 2, \ldots, n_b$$

and all node voltages are of the form

$$(16.5) \qquad v_⑱ = V_⑱ e^{j\omega t} \qquad k = 1, 2, \ldots, n_n - 1$$

where $\omega$ is the frequency of the network's input and $V_l, I_l$, and $V_⑰$ are phasors. (We term $V_l$ a *branch voltage phasor*, $I_l$ a *branch current phasor* and $V_⑰$ a *node voltage phasor*.) Recall that if $\mathbf{A}$ is the nodal incidence matrix of the network and $\mathbf{i}_b$ is the branch current vector, then the KCL node equations may be written as

$$\mathbf{A i}_b = \mathbf{0}$$

or

$$(16.6) \qquad \sum_{k=1}^{n_b} a_{kl} i_l = 0 \qquad l = 1, 2, \ldots, n_n - 1.$$

where the elements of **A** are defined as follows:

$$a_{kl} = \begin{cases} +1 & \text{if branch } l \text{ enters node } \textcircled{k} \\ -1 & \text{if branch } l \text{ leaves node } \textcircled{k} \\ 0 & \text{if branch } l \text{ does not touch node } \textcircled{k} \end{cases}$$

Substituting (16.4) into (16.6) yields

$$\sum_{l=1}^{n_b} a_{kl} I_l e^{j\omega t} = 0$$

or, dividing by $e^{j\omega t}$ we have

(16.7) $$\sum_{l=1}^{n_b} a_{kl} I_l = 0$$

Thus *KCL* node equations can be written directly in terms of the branch current phasors.

If $\mathbf{v}_{\textcircled{n}}$ represents the node voltage vector then we have seen that KVL may be expressed as

$$\mathbf{v}_b = \mathbf{A'v}_{\textcircled{n}}$$

or

(16.8) $$v_l = \sum_{k=1}^{n_b-1} a_{kl} v_{\textcircled{k}} \qquad l = 1, 2, \ldots, n_b$$

On substitution of (16.3) and (16.5) into (16.8), we have

$$V_l e^{j\omega t} = \sum_{k=1}^{n_n-1} a_{kl} V_{\textcircled{k}} \, e^{j\omega t}$$

or after dividing through by $e^{j\omega t}$

(16.9) $$V_l = \sum_{k=1}^{n_n-1} a_{kl} V_{\textcircled{k}}$$

So we see that KVL can be written directly in terms of the branch voltage phasors and node voltage phasors.

Let us now investigate the branch relationships of the linear elements we have studied if the branch voltages and currents have the forms of (16.3) and (16.4), respectively. We have

$$v_R = i_R R$$

for a resistor, or

$$V_R e^{j\omega t} = I_R e^{j\omega t} R$$

so that, in terms of phasors, the resistive branch relationship is

(16.10) $$V_R = I_R R$$

For a capacitor branch

$$i_C = C \frac{dv_C}{dt}$$

or

$$I_C e^{j\omega t} = C \frac{d}{dt}(V_C e^{j\omega t})$$

$$= j\omega C V_C e^{j\omega t}$$

so that

(16.11)     $$I_C = j\omega C V_C$$

Observe that we can interpret this expression as the branch relationship of a complex-valued conductance, or *admittance*, of value

$$Y_C(j\omega) = j\omega C_1$$

so that $I = YV$. Alternatively we may interpret (16.11) as the branch relationship of a complex-valued resistance, or *impedance*, of value

$$Z_C(j\omega) = \frac{1}{j\omega C}$$

so that $V_C = Z_C I_C$.

The inductor branch relationship

$$v_L = L \frac{di_L}{dt}$$

becomes

$$v_L e^{j\omega t} = L \frac{d}{dt}(I_L e^{j\omega t})$$

$$= j\omega L I_L e^{j\omega t}$$

so that

(16.12)     $$V_L = j\omega L I_L$$

Therefore we may interpret the inductor as an impedance of value

$$Z_L(j\omega) = j\omega L$$

or an admittance of value

$$Y_L(j\omega) = \frac{1}{j\omega L}$$

The student will note that we have indicated that impedance and admittance are functions of frequency. This point will be discussed in more detail in the next section. We leave it as an exercise to show that the branch relationships of all the two-port elements introduced in Chapter 8 may also be written directly in terms of phasors.

From this we can conclude that sinusoidal steady-state analysis is much the same as linear dc analysis except that the resistances are complex-valued, that is, they become impedances. The steps involved should be clear. Nodal equilibrium equations may be written by inspection in terms of the node voltage phasors by treating each capacitor as an admittance of value $j\omega C$ and each inductor as impedance of value $j\omega L$ where $\omega$ is the frequency of the input. The sinusoidal steady-state response is then found by calculating the branch voltage phasors or branch current phasors from the node voltage phasors, multiplying by $e^{j\omega t}$, and taking either the real or imaginary part depending on whether the input is a cosine or a sine, respectively.

**Example 16.2**

Let us consider the network of Fig. 16-6. To determine the steady-state response we may view this network in terms of the network of Fig. 16-7 where capacitors and inductors are modelled in terms of impedances. Observe that the input phasor is 3. The term $e^{j2t}$ is dropped temporarily from consideration. The nodal equilibrium equations are

$$(1 + j2)V_{①} - V_{②} = 3$$

$$-V_{①} + \left(1 - j\frac{2}{3}\right)V_{②} = 0$$

These equations are identical to the ones obtained in Example 16.1. Once the phasors $V_{①}$ and $V_{②}$ have been found, the node voltages of the original network may be obtained by multiplying $V_{①}$ and $V_{②}$ by $e^{j2t}$.   □

In general, the nodal equilibrium equations for a network operating in sinusoidal steady-state can be put in a familiar form:

$$Y_{11} V_{①} + Y_{12} V_{②} + \cdots + Y_{1m} V_{⑩} = I_{①}$$
$$Y_{21} V_{①} + Y_{22} V_{②} + \cdots + Y_{2m} V_{⑩} = I_{②}$$
$$\cdots\cdots\cdots\cdots\cdots\cdots\cdots\cdots\cdots\cdots\cdots\cdots\cdots$$
$$Y_{m1} V_{①} + Y_{m2} V_{②} + \cdots + Y_{mm} V_{⑩} = I_{⑩}$$

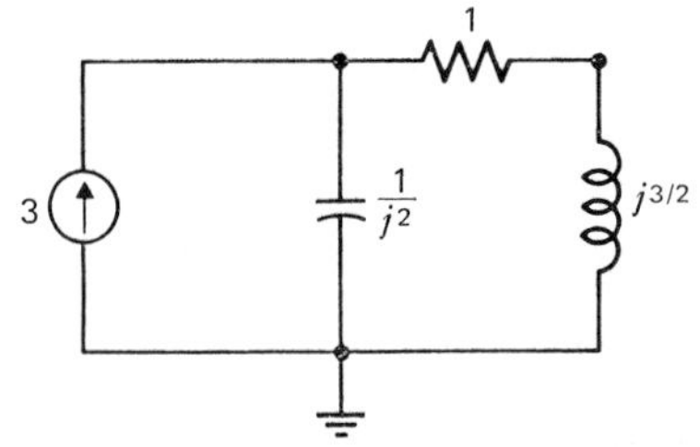

**Fig. 16-7   The network of Fig. 16-6 in terms of phasors and impedances.**

where $m = n_n - 1$,

$Y_{kk}$ = sum of all *admittances* connected between node ⓚ and all other nodes

$Y_{kj}$ = negative sum of all admittances directly connected between nodes ⓚ and ⓙ, $k \neq j$.

and

$I_k$ = sum of all current source current *phasors* and voltage source current *phasors* entering node ⓚ

The student should now be convinced that sinusoidal steady-state analysis is nothing more than a dc analysis using complex numbers. Since all of the techniques and principles presented for a dc analysis still remains valid when complex-valued voltages, currents, and resistances are used, we may employ them for sinusoidal steady-state analysis. One of the most useful of these principles is superposition.

The principle of superposition says that the branch voltage or current of any element in a network containing several independent sources is equal to the sum of the branch voltages or currents due to each source acting alone with all other sources set to zero. The usefulness of superposition is demonstrated by the following example.

## Example 16.3

The network of Fig. 16-8a is excited by two sinusoidal sources of different frequencies. We wish to find the capacitance voltage $v(t)$. It is not clear how to determine $v(t)$ directly using the methods of the previous sections because of the two different frequencies involved. However, by the principle of superposition we can determine the network's response due to each source, with the other source set to zero; the complete response is then the sum of the responses due to each source.

The network of Fig. 16-8b can be used to determine the steady-state response due to current source $\hat{i}_1$ acting alone. Observe that the capacitor is modeled as an impedance of value

$$Z_C = \frac{1}{j\omega C} = \frac{1}{j(2)(1)} = -j.5$$

The capacitor branch voltage phasor under this excitation is

$$V^1 = .47e^{-j83°}$$

and the corresponding time domain response is

$$v^1(t) = \text{Im} \{.47e^{-j83°}e^{j2t}\}$$

$$= .47 \sin(2t - 83°)$$

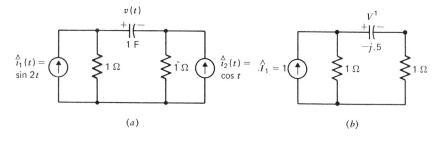

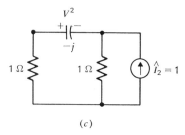

**Fig. 16-8   Sinusoidal analysis of the network in (*a*) is accomplished by invoking superposition. The voltages $V^1$ and $V^2$ are determined in (*b*) and (*c*), then $v(t) = \text{Im}\ (V^1e^{j2t}) + \text{Re}\ (V^2e^{jt})$.**

The network of Fig. 16-11*c* can be used to determine the steady-state response due to current source $\hat{i}_2$ acting alone. For this source, the capacitor is modeled by an impedance of value

$$Z_C = \frac{1}{j\omega C} = \frac{1}{j(1)(1)} = -j$$

The capacitor branch voltage phasor under this excitation is

$$V^2 = .45e^{j116.6°}$$

and the corresponding time domain response is

$$v^2(t) = \text{Re}\ (.45e^{j116.6°}e^{jt})$$

$$= .45 \cos\ (t + 116.6°)$$

The total capacitor voltage is

$$v(t) = v^1(t) + v^2(t)$$

$$= .47 \sin\ (2t - 83°) + .45 \cos\ (t + 116.6°) \quad \square$$

If a network has more than two sources, each of a different frequency, superposition may still be invoked. Of course if the network has more than one source, but they all are of the same frequency, it is not necessary to use the superposition principle.

## 16-4  IMPEDANCE AND ADMITTANCE

In the previous section we found that if a network is operating in the sinu-soidal steady-state, capacitors and inductors behave as complex-valued resistors that are called impedances. In particular, if the input to a network that is composed of $R$'s, $L$'s, and $C$'s is a complex exponential, $e^{j\omega t}$, as illustrated symbolically in Fig. 16-9$a$ the impedance of a capacitor is

$$Z_C(j\omega) = \frac{V_C}{I_C} = \frac{1}{j\omega C}$$

and the impedance of an inductor is

$$Z_L(j\omega) = \frac{V_L}{I_L} = j\omega L$$

These impedances are functions of $j\omega$ because if the value of $j\omega$ changes, so does the value of the impedance. Impedances behave just as resistors. That is to say the total impedance of impedances in series is the sum of the individual impedances. Similarly, the total admittance, defined by

$$Y(j\omega) = \frac{I}{V}$$

of admittances in parallel is the sum of the individual admittances.

Everything that we are able to do with resistors, we may do with imped-ances. For instance, an arbitrary connection of impedances with two acces-sible terminals may be replaced by an equivalent impedance.

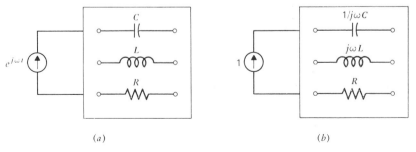

(a)                                    (b)

**Fig. 16-9**  (a) **An arbitrary** *RLC* **network excited by an exponential can be analyzed in terms of (b) a network obtained from the original by replacing all capacitors by impedance of 1/$j\omega C$, all inductors by impedance of $j\omega L$, and the input by a unity-valued source.**

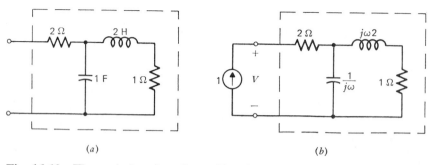

(*a*)                                        (*b*)

**Fig. 16-10**   **The equivalent impedance $Z_{eq}$ of the network in (*a*) may be found by placing a unit-valued current source across the accessible terminals as shown in (*b*) and computing *V*. Then $Z_{eq} = V$.**

## Example 16.4

To compute the value of an equivalent impedance for the network of Fig. 16-10*a* we can connect a unity-valued current source across the terminals, as shown in Fig. 16-10*b*, and compute the voltage phasor across it. Then

$$Z_{eq} = \frac{V}{1}$$

Alternatively, we can try to combine series and parallel combinations of impedances. The second approach is simpler for this example. The inductor and one ohm resistor form a series combination with impedance

$$1 + j\omega 2$$

This impedance, or admittance of value

$$\frac{1}{1 + j\omega 2}$$

is in parallel with the capacitor. The total admittance of this combination is

$$j\omega + \frac{1}{1 + j\omega 2} = \frac{(1 - 2\omega^2) + j\omega}{1 + j\omega 2}$$

or, in terms of impedance

$$\frac{1 + j\omega 2}{(1 - 2\omega^2) + j\omega}$$

This impedance is in series with the two ohm resistor so that

$$\begin{aligned}
Z_{eq} &= 2 + \frac{1 + j\omega 2}{(1 - 2\omega^2) + j\omega} \\
&= \frac{8\omega^4 - 6\omega^2 + 3 + j(-4\omega^3 + \omega)}{(4\omega^4 - 3\omega^2 + 1)}
\end{aligned}$$

Observe that if $\omega = 0$, so that the excitation was a dc source,

$$Z_{eq} = 3$$

This result agrees with our previous reasoning that for a constant signal the capacitor behaves as an open circuit and the inductor behaves as a short circuit. Thus for $\omega = 0$ (a constant or dc signal) the two ohm and one ohm resistors are in series with a total resistance of $3\ \Omega$.  □

It should be clear from this discussion that Thévenin's and Norton's theorems, as well as all the other theorems discussed in Chapter 5, are easily extended to networks operating in the sinusoidal steady state.

## 16-5 FREQUENCY RESPONSE: COMPLEX TWO-PORTS

In most practical steady-state analysis situations we are interested not in the behavior of a network under a sinusoidal excitation at a single frequency, but rather the behavior of the network over a range of frequencies. It is this *frequency domain* or *ac* (for alternating current), response that makes circuits useful for a variety of tasks. For example, a hi-fi amplifier is considered good if it uniformly amplifies sinusoids of frequencies from well below the audio range, about 20 cycles per second or 20 hertz (abbreviated Hz), to well above the audio range, about 20,000 cycles per second or 20 kilohertz (abbreviated KHz). In other applications, it may be desirable that a network amplify only over a very narrow band of frequencies and reject all signals with frequencies outside of this range. Such frequency selective circuits are important in the input stages of a radio receiver.

In order to determine the frequency response of a network, we may proceed as follows. We know that the sinusoidal steady-state analysis of a network under the excitation $e^{j\omega t}$ may be carried out in terms of phasors. If the frequency $\omega$ is allowed to be a variable, then the impedances, and hence the responses, are functions of $\omega$. By analyzing the network for all values of $\omega$ in the range of interest, we obtain the frequency response. Since the network is operating in steady-state, so that all transients have become zero, the time variable $t$ is of no consequence. Of course we can always retrieve the time response due to a cosine or sine wave excitation of a given frequency by multiplying the frequency response phasor by $e^{j\omega t}$ and taking, respectively, the real or imaginary part.

For frequency response studies of a network we are only concerned with the input and the output. (Although only the output is of interest, the entire network must be solved in one way or another to determine the output, and of course, the nodal analysis may be used for this purpose.) Therefore, it is convenient to view the network under consideration to be a two-port where one port is the input and one port is the output. Such a two-port is shown in

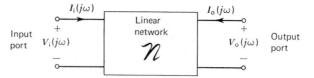

**Fig. 16-11   A two-port network.**

Fig. 16-11. The input port current and voltage phasors, $I_i(j\omega)$ and $V_i(j\omega)$, and the output port current and voltage phasors, $I_o(j\omega)$ and $V_o(j\omega)$ are functions of $\omega$.

From our studies of linear dc two-ports of Chapter 8, we know that the port variables may be related to each other in several ways. The two most common representations are the $z$—parameter representation:

$$\begin{bmatrix} V_i(j\omega) \\ V_o(j\omega) \end{bmatrix} = \begin{bmatrix} Z_{11}(j\omega) & Z_{12}(j\omega) \\ Z_{21}(j\omega) & Z_{22}(j\omega) \end{bmatrix} \begin{bmatrix} I_i(j\omega) \\ I_o(j\omega) \end{bmatrix}$$

and the $y$—parameter representation:

$$\begin{bmatrix} I_i(j\omega) \\ I_o(j\omega) \end{bmatrix} = \begin{bmatrix} Y_{11}(j\omega) & Y_{12}(j\omega) \\ Y_{21}(j\omega) & Y_{22}(j\omega) \end{bmatrix} \begin{bmatrix} V_i(j\omega) \\ V_o(j\omega) \end{bmatrix}$$

The method for calculating the complex-valued $z$ or $y$ parameters of an ac network is identical to the method used to find the real-valued $z$ or $y$ parameters of a dc network.

**Example 16.5**

To calculate the $z$—parameters for the two-port network of Fig. 16-12$a$ we first connect current sources across the input and output ports as shown

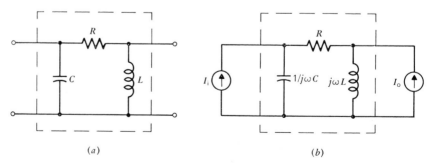

(a)                              (b)

**Fig. 16-12  (a) Network used in Example 16.5. The z-parameters may be found by attaching current sources, as shown in (b), and following the procedure outlined in the text.**

in Fig. 16-12*b*. Because

$$V_i(j\omega) = Z_{11}(j\omega)I_i(j\omega) + Z_{12}(j\omega)I_o(j\omega)$$
$$V_o(j\omega) = Z_{21}(j\omega)I_i(j\omega) + Z_{22}(j\omega)I_o(j\omega)$$

we have

$$Z_{11}(j\omega) = \frac{V_i(j\omega)}{I_i(j\omega)} \quad \text{and} \quad Z_{21}(j\omega) = \frac{V_o(j\omega)}{I_i(j\omega)}$$

if $I_o(j\omega) = 0$; and

$$Z_{12}(j\omega) = \frac{V_i(j\omega)}{I_o(j\omega)} \quad \text{and} \quad Z_{22}(j\omega) = \frac{V_o(j\omega)}{I_o(j\omega)}$$

if $I_i(j\omega) = 0$. Therefore, setting $I_o(j\omega) = 0$ and $I_i(j\omega) = 1$, we find

$$Z_{11}(j\omega) = \frac{R + j\omega L}{1 + j\omega CR - \omega^2 LC}$$

and

$$Z_{21}(j\omega) = \frac{j\omega L}{1 + j\omega CR - \omega^2 LC}$$

Setting $I_i(j\omega) = 0$ and $I_o(j\omega) = 1$, we find

$$Z_{12}(j\omega) = \frac{j\omega L}{1 + j\omega CR - \omega^2 LC}$$

and

$$Z_{22} = \frac{-\omega^2 RLC + j\omega L}{1 + j\omega CR - \omega^2 LC} \quad \square$$

As indicated at the beginning of this section, we are interested in the relationship between the input, or excitation, and output, or response, of a network. We define a *network function*, usually denoted by $H(j\omega)$, as the ratio of the network's output to input. There are four possibilities: $H$ is a ratio of voltages, in which case it is called a *transfer voltage ratio*; $H$ is a ratio of currents, in which case it is called a *transfer current ratio*; $H$ is a ratio of a voltage to a current; or $H$ is a ratio of a current to a voltage. If the network function is a ratio of the response at one port to the excitation at another port, it is sometimes called a *transfer function*. Once $H(j\omega)$ is known we can determine the steady-state response phasor to any sinusoidal input merely by multiplying the input phasor by $H(j\omega)$. The frequency response of a work is found by studying $H(j\omega)$.

**Example 16.6**

Consider the simple *RC* network of Fig. 16-13. Since

$$V_o = \frac{1}{1 + j\omega RC} V_i$$

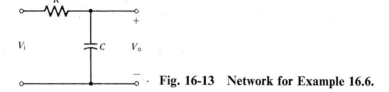

Fig. 16-13    Network for Example 16.6.

the transfer function is

$$H(j\omega) = \frac{1}{1 + j\omega RC}$$

or in polar form

$$H(j\omega) = \left[\frac{1}{1 + (\omega RC)^2}\right]^{1/2} e^{j\phi}$$

where $\phi = -\tan^{-1} \omega RC$. For $\omega = 0$, that is, when the input is a dc signal, $|H| = 1$ and $\phi = 0$. Thus the output equals the input and has the same phase. As $\omega$ increases from zero, $|H|$ decreases and $\phi$ approaches—90°.  □

It is clear from the above example that a pictorial representation of $H(j\omega)$ would be helpful. We discuss a representation in the next section. However, before leaving this section we define two useful quantities. Suppose $H(j\omega)$ is the ratio of a voltage and a current. Therefore, $H(j\omega)$ has units of impedance. In general, $H(j\omega)$ is a complex function of $\omega$ and can be written as

$$H(j\omega) = R(\omega) + jX(\omega)$$

where $R(\omega)$ represents the real part of the impedance and $X(\omega)$ represents the imaginary part of the impedance. The function $X(\omega)$ is called the *reactance*.

Now suppose $H(j\omega)$ is the ratio of a current to a voltage so that it has units of admittance. Then $H(j\omega)$ can be written as

$$H(j\omega) = G(\omega) + jS(\omega)$$

where $G(\omega)$ represents the real part of the admittance and $S(\omega)$ represents the imaginary part of the admittance. The function $S(\omega)$ is called the *susceptance*. Observe that the reactance of a pure inductor is $\omega L$ and the reactance of a pure capacitor is $1/\omega C$; also the susceptance of a pure inductor is $1/\omega L$ and the susceptance of a pure capacitor is $\omega C$.

## 16-6    GRAPHICAL REPRESENTATION OF THE FREQUENCY RESPONSE

A graphical representation of the frequency response of a network is quite useful in evaluating performance. In this section we describe the most commonly used graphical technique—known as the *Bode plot*, which is named after

H. W. Bode. We are interested in displaying pictorially the complex-valued transfer function $H(j\omega)$ as a function of $\omega$. Since it is quite common for the frequency variable $\omega$ to be varied over large ranges of values, say from 1 radian per second to $10^6$ radians per seconds, it is convenient to plot $\log_{10}\omega$ on the abscissa. Because $H(j\omega)$ is complex-valued, we can either plot the real and imaginary parts of $H(j\omega)$ as a function of $\omega$ or magnitude and phase. The latter choice is more meaningful. The magnitude of $H(j\omega)$ is usually plotted in *decibels*, denoted by dB, where

$$|H(j\omega)|_{dB} = 20 \log_{10} |H(j\omega)|$$

and is called the *gain*. (In the sequel we will use log to denote $\log_{10}$ and *ln* to denote $\log_e$.) The phase of $H(j\omega)$ is plotted in either radians or degrees.

Given an analytic expression of a transfer function it is usually a simple matter to sketch the frequency response without actually evaluating $|H(j\omega)|_{dB}$ and $<H(j\omega)$ for many values of $\omega$. The following examples will help illustrate the technique.

**Example 16.7**

The transfer function

$$H(j\omega) = \frac{1}{1 + j\omega T}$$

is of the same form as the transfer function found in Example 16.6 with $T = RC$. Note that $T$ has units of time and represents a time constant. The gain can be found as follows:

$$|H(j\omega)|_{dB} = 20 \log \left| \frac{1}{1 + j\omega T} \right|$$

$$= -20 \log |1 + j\omega T|$$

$$= -20 \log [1 + (\omega T)^2]^{1/2}$$

$$= -10 \log [1 + (\omega T)^2]$$

Now let us determine the approximate *behavior* of the gain based on asymptotes. For values of $\omega$ much smaller than $1/T$ (denoted by $\omega \ll 1/T$) we have that $\omega T \ll 1$ so that the gain can be approximated by

$$|H(j\omega)|_{dB} \simeq -10 \log 1 = 0 \text{ dB}$$

(0 dB indicates unity gain.) For values of $\omega$ much larger than $1/T$ (denoted by $\omega \gg 1/T$) we have that $\omega T \gg 1$ so that the gain can be approximated by

$$|H(j\omega)|_{dB} \simeq -10 \log (\omega T)^2$$

$$= -20 \log \omega T$$

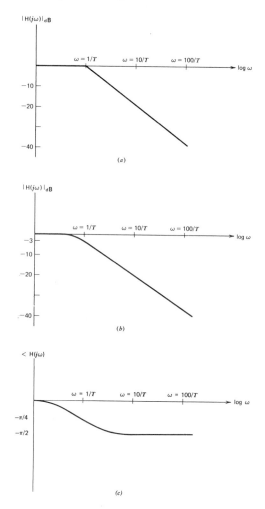

**Fig. 16-14** (a) Approximate behavior of the gain of the transfer function $H(j\omega) = 1/(1 + j\omega T)$ based on asymptotes; (b) a more accurate plot; (c) approximate phase plot.

which indicates that the gain decreases as $\omega$ increases. We can establish the rate of this decrease, that is, the slope, in the following manner. Consider two frequencies $\omega_1$ and $\omega_2$ both of which are much greater than $1/T$. Furthermore, let $\omega_2 = 10\omega_1$, that is, $\omega_2$ is one *decade* higher than $\omega_1$. Observe that

$$|H(j\omega)|_{dB} = -20 \log \omega_2 T$$
$$= -20 \log 10\omega_1 T$$
$$= -20 \log 10 - 20 \log \omega_1 T$$
$$= -20 + |H(j\omega)|_{dB}$$

or that

$$|H(j\omega_2)|_{dB} - |H(j\omega_1)|_{dB} = -20 \text{ dB}$$

Thus the gain decreases at a rate of 20 dB over the decade between $\omega_1$ and $\omega_2$. Alternatively, suppose $\omega_2 = 2\omega_1$, that is, $\omega_2$ is one *octave* higher than $\omega_1$. Then

$$
\begin{aligned}
|H(j\omega_2)|_{dB} &= -20 \log 2\omega_1 T \\
&= -20 \log 2 - 20 \log \omega_1 T \\
&= -6 + |H(j\omega_1)|_{dB}
\end{aligned}
$$

which indicates that the gain decreases at a rate of 6 dB per octave. So we see that a decrease of 20 dB per decade is equivalent to a decrease of 6 dB per octave.

We conclude, therefore, that the approximate behavior of the gain is as shown in Fig. 16-14a. Since this approximation was based on asymptotic behavior of the gain for values of $\omega \ll 1/T$ and $\omega \gg 1/T$, we expect it to be poorest in the neighborhood of the *critical* or *corner frequency* $\omega = 1/T$. Observe that at this frequency

$$
\begin{aligned}
|H(j\omega)|_{dB} &= -10 \log (2) \\
&= -3 \text{ dB}
\end{aligned}
$$

Therefore, our estimate is off by 3 dB and the gain is more accurately represented by the plot shown in Fig. 16-14b. This frequency is also called the *breakfrequency* or 3 dB point. The network whose response is shown in Fig. 16-14b is called a low pass filter because it passes signals whose frequencies are less than $1/T$ while attenuating signals whose frequencies are greater than $1/T$.

Let us now consider the phase plot. We have

$$
\begin{aligned}
<H(j\omega) &= \tan^{-1} \frac{\text{Im} \{H(j\omega)\}}{\text{Re} \{H(j\omega)\}} \\
&= \tan^{-1}(-\omega T) = -\tan^{-1} \omega T
\end{aligned}
$$

We again investigate asymptotic behavior. For $\omega \ll 1/T$, we have $\omega T \ll 1$ and $<H(j\omega) \simeq 0$. For $\omega \gg 1/T$ we have $\omega T \gg 1$ and $< H(j\omega) \simeq -\pi/2$. Finally, for $\omega = 1/T$, $<H(j\omega) = -\tan^{-1} 1 = -\pi/4$. The approximate phase plot is shown in Fig. 16-14c. □

### Example 16.8

Now consider the network of Fig. 16-15. The voltage transfer function is

$$H(j\omega) = \frac{j\omega T}{1 + j\omega T}$$

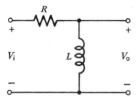

**Fig. 16-15   High-pass filter studied in Example 16.8.**

where $T = L/R$. Let us study the asymptotic behavior of the numerator and denominator separately. We known from the previous example that the denominator tends to remain constant as $\omega$ increases until $\omega = 1/T$, and then it tends to cause the gain to decrease or "roll off" at 6 dB/octance. The numerator of $H(j\omega)$ tends to cause the gain to increase as $\omega$ increases. By following in a manner similar to the one used in Example 16.7 it is easy to show that this rate of increase is 6 dB per octave. Thus we conclude that for $\omega < 1/T$, the gain increases at a rate of 6 dB/octave. For $\omega > 1/T$, since the numerator tends to cause the gain to increase at 6 dB/octave and the denominator tends to cause the gain to decrease at 6 dB/octave, the gain remains constant. For large $\omega$, $|H(j\omega)| \simeq 1$ or 0 dB. This approximation is shown in Fig. 16-16a. We could have arrived at the same approximation by working with $H(j\omega)$ directly. Observe that

$$|H(j\omega)| = \frac{[(\omega T)^4 + (\omega T)^2]^{1/2}}{[1 + (\omega T)^2]}$$

so that

$$|H(j\omega)|_{\text{dB}} = 10 \log [(\omega T)^2((\omega T)^2 + 1)] - 20 \log [1 + (\omega T)^2]$$

$$= 20 \log (\omega T) - 10 \log [1 + (\omega T)^2]$$

For $\omega T \ll 1$,

$$|H(j\omega)|_{\text{dB}} \simeq 20 \log (\omega T)$$

and for $\omega T \gg 1$,

$$|H(j\omega)|_{\text{dB}} \simeq 0$$

This technique of studying the effects of the numerator and denominator on the gain individually and then combining the results is quite useful.

The approximation of the gain shown in Fig. 16-16a is poorest at $\omega = 1/T$. The actual gain at this frequency is

$$|H(j\omega)_{\text{dB}} = 20 \log (1) - 10 \log (2)$$

$$= -3 \text{ dB}$$

An improved sketch of the gain is shown in Fig. 16-16b. A circuit with this type of response is called a high pass filter because all signals whose frequencies are less than $1/T$ are attenuated.

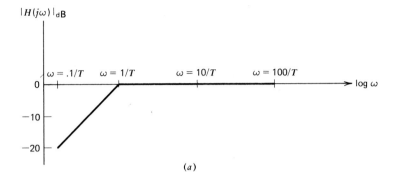

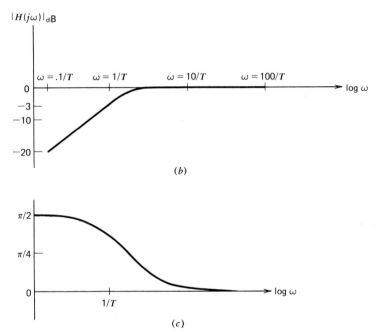

Fig. 16-16(a) Approximation of gain response for network of Fig. 16-15; (b) an improved approximation; (c) the phase plot.

Let us now consider the phase. From the previous example we know that the denominator will cause the phase to decrease from 0 radians to $-\pi/2$ radians as $\omega$ increases, passing through $-\pi/4$ radians at $\omega = 1/T$. The numerator always has a phase of $\pi/2$ radians. Therefore, the phase, as shown in Fig. 16-16c, of $H(j\omega)$ decreases from $\pi/2$ radians to 0 radians as $\omega$ increases, passing through $\pi/4$ radians at $\omega = 1/T$.    □

It is not always possible to estimate frequency response by determining asymptotic behavior as the following example illustrates.

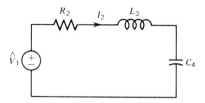

**Fig. 16-17 A simple *RLC* network used to illustrate resonance.**

## Example 16.9

Consider the simple *RLC* network of Fig. 16-17 in which the output is the resistor current $I_2$. The network function is

$$H(j\omega) = \cfrac{1}{R_2 + j\omega L_3 + \cfrac{1}{j\omega C_4}}$$

$$= \cfrac{1}{R_2 + j\left(\omega L_3 - \cfrac{1}{\omega C_4}\right)}$$

so that

$$|H(j\omega)| = \cfrac{1}{\left[R_2^{\,2} + \left(\omega L_3 - \cfrac{1}{\omega C_4}\right)^2\right]^{1/2}}$$

and

$$<H(j\omega) = \tan^{-1}\left[\cfrac{\left(\omega L_3 - \cfrac{1}{\omega C_4}\right)}{R_2}\right]$$

Observe that the magnitude is greatest at a frequency $\omega_0$ for which

$$\omega L_3 - \frac{1}{\omega C_4} = 0$$

that is,

$$\omega_0 = \frac{1}{\sqrt{L_3 C_4}}$$

This frequency is known as the *resonant frequency* and is the frequency at which the reactance of the network function equals zero. Also observe for very large and small values of $\omega$ $|H(j\omega)|$ approaches zero. A plot of the magnitude of $H(j\omega)$ versus frequency is shown in Fig. 16-18a. The phase of $H(j\omega)$ is approximately $\pi/2$ for very small values of $\omega$, then becomes 0 at $\omega = \omega_0$, and finally becomes $-\pi/2$ for very large values of $\omega$. A plot of the phase versus frequency is shown in Fig. 16-18b. □

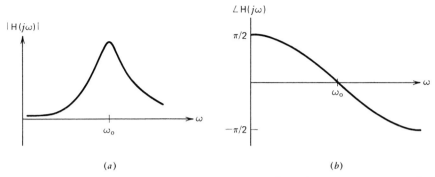

(a)                                             (b)

**Fig. 16-18** (a) **Magnitude and** (b) **phase plots of the frequency response of a series resonant circuit.**

## 16-7 THE SUBROUTINE GAUSSC

For large networks it is now always convenient to find the transfer function explicitly. We may, however, determine the frequency response by analyzing the network for many different values of $\omega$. To perform this task quickly, we require a computer program that is capable of solving the complex-valued nodal equilibrium equations. GAUSSC is such a program.

The FORTRAN subroutine GAUSSC uses Gaussian elimination to solve the set of simultaneous algebraic equations

$$a_{11}x_1 + a_{12}x_2 + \cdots + a_{1n}x_n = b_1$$
$$a_{21}x_1 + a_{22}x_2 + \cdots + a_{2n}x_n = b_2$$
$$\cdots\cdots\cdots\cdots\cdots\cdots\cdots$$
$$a_{n1}x_1 + a_{n2}x_n + \cdots + a_{nn}x_n = b_n$$

where the coefficients $a_{jk}$ and $b_k$ are complex. A listing of this program is given in Fig. 16-19. This program is virtually identical to the subroutine GAUSS described in Chapter 7. The major difference is that the arrays A and B are complex.

GAUSSC is called from another program by using the statement

CALL GAUSSC (A, B, N, M)

A is a double-subscripted, double precision, complex-valued array, dimensioned M × M in the calling program. B is a single subscripted, double precision, complex-valued array dimensioned M in the calling program. As in GAUSS, A contains the complex-valued coefficients of the unknowns of the algebraic equations and B contains the complex-valued coefficients on the right-hand side of the algebraic equations. The integers M and N have the same meaning as they had in GAUSS. M is the dimensions of A and B in the calling program and N is the actual number of equations being solved.

```
C                         SUBROUTINE 'GAUSSC'
C
C
C
C        GAUSSC IS A SUBROUTINE FOR SOLVING SIMULTANEOUS LINEAR
C        ALGEBRAIC EQUATIONS WITH COMPLEX COEFFICIENTS. THE PROGRAM
C        USES GAUSSIAN ELIMINATION WITH ROW PIVOTING.
C
C
C        GAUSSC IS CALLED FROM ANOTHER PROGRAM BY USING THE STATEMENT
C
C                         CALL GAUSSC(A,B,N,M)
C
C        A IS A DOUBLE-SUBSCRIPTED COMPLEX*16 ARRAY,
C        DIMENSIONED MXM IN THE CALLING PROGRAM, B IS A
C        SINGLE-SUBSCRIPTED COMPLEX*16 ARRAY, DIMENSIONED
C        M IN THE CALLING PROGRAM, AND N IS THE NUMBER OF
C        EQUATIONS TO BE SOLVED. A CONTAINS THE COEFFICIENTS OF
C        THE UNKNOWNS OF THE ALGEBRAIC EQUATIONS; B CONTAINS
C        THE COEFFICIENTS ON THE RIGHT HAND SIDE OF THE ALGEBRAIC
C        EQUATIONS; UPON RETURN FROM GAUSSC, B CONTAINS THE
C        SOLUTION AND A CONTAINS THE ELEMENTS OF THE LOWER AND
C        UPPER TRIANGULAR MATRICES OF A. USE OF THE TWO INTEGERS
C        M AND N ALLOWS THE DIMENSIONS OF A AND B TO BE SET
C        ONCE IN THE MAIN PROGRAM AND THE ACTUAL NUMBER OF
C        EQUATIONS TO BE SOLVED TO VARY (NOTE M≥N),DEPENDING
C        ON THE PROBLEM.
C
C
         SUBROUTINE GAUSSC(A,B,N,M)
         IMPLICIT REAL*8(A-H,O-Z)
         COMPLEX*16 A(M,M),B(M),SUM,PIV
         DO 5 I=1,N
         I1=I+1
         IF(CDABS(A(I,I)).LE.1.D-10) GO TO 1
         GO TO 15
   1     CONTINUE
         IF(I.EQ.N) GO TO 10
         DO 14 J=I1,N
         IF(CDABS(A(J,I)).LE.1.D-10) GO TO 14
         IPIV=J
         GO TO 16
  14     CONTINUE
         GO TO 10
```

**Fig. 16-19 Listing of the program GAUSSC.**

The following example serves as an illustration of the use of GAUSSC to determine frequency response.

**Example 16.10**

In Chapter 10 we derived the $\pi$ small signal transistor model (see Example 10.4) that could be used for dc analysis. This model can be modified to portray the small signal sinusoidal steady-state behavior of a transistor by adding the capacitors $C_\pi$ and $C_\mu$ as shown in Fig. 16-20. This model is referred to as the *hybrid-pi* model and is valid only if the transistor is operating in the forward active or normal region. We will use this model to determine the frequency response of the single stage transistor amplifier shown in Fig. 16-21*a*

```
16   DO 2 K=1,N
     PIV=A(IPIV,K)
     A(IPIV,K)=A(I,K)
 2   A(I,K)=PIV
     PIV=B(IPIV)
     B(IPIV)=B(I)
     B(I)=PIV
15   IF(I.EQ.N) GO TO 3
     DO 8 JI=I1,N
 8   A(I,JI)=A(I,JI)/A(I,I)
     B(I)=B(I)/A(I,I)
     DO 5 J=I1,N
     DO 4 K=I1,N
 4   A(J,K)=A(J,K)-(A(J,I)*A(I,K))
 5   B(J)=B(J)-(B(I)*A(J,I))
 3   B(N)=B(N)/A(N,N)
     DO 6 K=2,N
     I=N-K+1
     L=I+1
     SUM=0.D0
     DO 7 J=L,N
 7   SUM=SUM+A(I,J)*B(J)
 6   B(I)=B(I)-SUM
     GO TO 11
10   WRITE(6,9)
 9   FORMAT(' EQUATIONS ARE LINEARLY DEPENDENT')
     STOP
11   RETURN
     END
```

**Fig. 16-19** (*Continued*)

over the frequency range $1 \text{ Hz} \leq \omega \leq 10^8 \text{ Hz}$. Figure 16-21$b$ shows the network with the transistor replaced by the hybrid-pi model.

The nodal equilibrium equations for the network are

$$node \; \textcircled{1} \quad -\frac{1}{R_B} V_{\textcircled{2}} \qquad + I_1 \qquad\qquad\qquad = -\frac{1}{R_B} \hat{V}_1$$

$$node \; \textcircled{2} \quad \left(\frac{1}{R_B} + \frac{1}{r_x}\right) V_{\textcircled{2}} - \frac{1}{r_x} V_{\textcircled{3}} \qquad\qquad = \frac{1}{R_B} \hat{V}_1$$

$$node \; \textcircled{3} \quad \begin{aligned} &-\frac{1}{r_x} V_{\textcircled{2}} + \left(\frac{1}{r_x} + \frac{1}{r_\pi} + j\omega(C_\pi + C_\mu)\right) V_{\textcircled{3}} \\ &\qquad\qquad - j\omega C_\mu V_{\textcircled{4}} - \left(\frac{1}{r_\pi} + j\omega C_\pi\right) V_{\textcircled{5}} = 0 \end{aligned}$$

$$node \; \textcircled{4} \quad \begin{aligned} &(g_m - j\omega C_\mu) V_{\textcircled{3}} + \left(j\omega C_\mu + \frac{1}{r_o} + \frac{1}{R_L}\right) V_{\textcircled{4}} \\ &\qquad\qquad - \left(g_m + \frac{1}{r_o}\right) V_{\textcircled{5}} = 0 \end{aligned}$$

$$node \; \textcircled{5} \quad \begin{aligned} &-\left(\frac{1}{r_\pi} + j\omega C_\pi + g_m\right) V_{\textcircled{3}} - \frac{1}{r_o} V_{\textcircled{4}} \\ &\qquad + \left(g_m + \frac{1}{r_\pi} + \frac{1}{r_o} + \frac{1}{R_E} + j\omega C_\pi\right) V_{\textcircled{5}} = 0 \end{aligned}$$

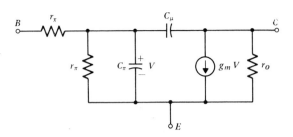

**Fig. 16-20   The hybrid-pi transistor model.**

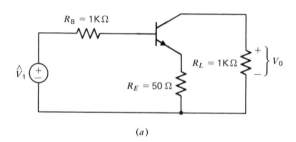

(a)

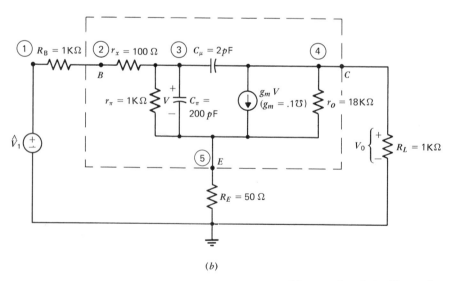

(b)

**Fig. 16-21**   (a) The single stage transistor amplifier analyzed in Example 16.10. (b) The network of (a) with the transistor replaced by the hybrid-pi model.

```
C      CHAPTER 16 EXAMPLE 16.10
C
C      DIMENSIONS
C
       IMPLICIT REAL*8(A-H,O-Z)
       COMPLEX*16 Y(5,5),CIN(5),JW,VO(51),J,V1
       DIMENSION FREQ(51),GAIN(51),PHASE(51),G(51),P(51)
C
C      READ IN ELEMENT VALUES, INITIAL FREQUENCY, FINAL FREQUENCY AND
C      NUMBER OF FREQUENCY POINTS AT WHICH THE NETWORK IS TO BE ANALYZED
C      A LOGARITHMIC PLOT OF THE FREQUENCY RESPONSE IS TO BE GENERATED
C
       CALL READ(RB,RX,RPI,CPI,CMU,GM,RO,RL,RE,V1MAG,V1PHAS,FI,FF,NF)
C
C
C      OBTAIN THE COMPLEX EXPRESSION FOR V1
C
       J=DCMPLX(0.D0,1.D0)
       V1=V1MAG*(DCOS(V1PHAS*3.1412/180.)+(J*DSIN(V1PHAS*3.1412/180.)))
C
C      COMPUTE FREQUENCY MULTIPLIER
C
       DF=10.**((DLOG10(FF)-DLOG10(FI))/(NF-1))
C
C      SET UP AND SOLVE NODE EQUATIONS OVER FREQUENCY RANGE OF INTEREST
C
       DO 1 K=1,NF
       FREQ(K)=FI*DF**(K-1)
       JW=DCMPLX(0.D0,6.2824*FREQ(K))
       CALL NODEQN(RB,RX,RPI,CPI,CMU,GM,RO,RL,RE,V1,JW,Y,CIN)
       CALL GAUSSC(Y,CIN,5,5)
C
C      STORE OUTPUT VOLTAGE FOR PLOTTING
C
       VO(K)=CIN(3)
     1 CONTINUE
       CALL PRINT(VO,FREQ,V1MAG,V1PHAS,NF,G,P)
       PRINT 3
       CALL PLOT(FREQ,G,51,1,1)
       PRINT 4
       CALL PLOT(FREQ,P,51,1,1)
     3 FORMAT('1',//53X,'PLOT OF GAIN(DB) VS FREQUENCY'///)
     4 FORMAT('1',//50X,'PLOT OF PHASE(DEGREES) VS FREQUENCY'///)
       END



       SUBROUTINE READ(RB,RX,RPI,CPI,CMU,GM,RO,RL,RE,V1MAG,V1PHAS,FI,
      +FF,NF)
       IMPLICIT REAL*8(A-H,O-Z)
       READ(5,1) RB,RX,RPI,CPI,CMU,GM,RO,RL,RE,V1MAG,V1PHAS,FI,FF,NF
     1 FORMAT(7G10.0)
       RETURN
       END
```

**Fig. 16-22a  Listing of program which determines the frequency response of the network of Fig. 16-21.**

```
                    SUBROUTINE NODEQN(RB,RX,RPI,CPI,CMU,GM,RO,RL,RE,V1,JW,Y,CIN)
                    IMPLICIT REAL*8(A-H,O-Z)
                    COMPLEX*16 Y(5,5),CIN(5),V1,JW
C
C       INITIALIZE COEFFICIENT ARRAYS
C
                    DO 1 K=1,5
                    CIN(K)=0.D0
                    DO 1 I=1,5
                  1 Y(K,I)=0.D0
C
C       SET UP NODAL EQUILIBRIUM EQUATION COEFFICIENTS
C
                    Y(1,1)=-1./RB
                    Y(1,5)=1.
                    CIN(1)=(-1./RB)*V1
                    Y(2,1)=(1./RB)+(1./RX)
                    Y(2,2)=-1./RX
                    CIN(2)=(1./RB)*V1
                    Y(3,1)=-1./RX
                    Y(3,2)=(1./RX)+(1./RPI)+JW*(CMU+CPI)
                    Y(3,3)=-JW*CMU
                    Y(3,4)=-(1./RPI)-JW*CPI
                    Y(4,2)=GM-JW*CMU
                    Y(4,3)=(JW*CMU)+(1./RO)+(1./RL)
                    Y(4,4)=-(1./RO)-GM
                    Y(5,2)=-(1./RPI)-(JW*CPI)-GM
                    Y(5,3)=-1./RO
                    Y(5,4)=GM+(1./RPI)+(1./RE)+(1./RO)+JW*CPI
                    RETURN
                    END



                    SUBROUTINE PRINT(VO,FREQ,V1MAG,V1PHAS,NF,GAIN,PHASE)
                    IMPLICIT REAL*8(A-H,O-Z)
                    REAL*8 FREQ(51),GAIN(51),PHASE(51)
                    COMPLEX*16 VO(50)
                    PRINT 1
                    DO 2 K=1,NF
C
C       OBTAIN THE MAGNITUDE, REAL AND IMAGINARY PARTS AND
C       THE PHASE OF VO(K)
C
                    VOMAG=CDABS(VO(K))
                    VOIMAG=DAIMAG(VO(K))
                    VOREAL=DREAL(VO(K))
                    VOPHAS=(180./3.1412)*DATAN2(VOIMAG,VOREAL)
C
C       COMPUTE THE GAIN IN DB AND THE PHASE SHIFT IN DEGREES
C
                    GAIN(K)=20.*DLOG10(VOMAG/V1MAG)
                    PHASE(K)=VOPHAS-V1PHAS
                    PRINT 3, FREQ(K),GAIN(K),PHASE(K)
                  2 CONTINUE
                  1 FORMAT(' RESULTS FOR EXAMPLE 16.10'//25X,'GAIN',20X,'PHASE'/
                   +25X,'IN',22X,'IN'/2X,'FREQUENCY',14X,'DB',22X,'DEGREES'/)
                  3 FORMAT(G12.5,12X,G12.5,12X,G12.5)
                    RETURN
                    END
```

**Fig. 16-22a** *(Continued)*

RESULTS FOR EXAMPLE 16.10

| FREQUENCY | GAIN IN DB | PHASE IN DEGREES |
|---|---|---|
| 10.000 | 22.763 | 180.02 |
| 13.804 | 22.763 | 180.02 |
| 19.055 | 22.763 | 180.02 |
| 26.303 | 22.763 | 180.02 |
| 36.308 | 22.763 | 180.02 |
| 50.119 | 22.763 | 180.02 |
| 69.183 | 22.763 | 180.02 |
| 95.499 | 22.763 | 180.02 |
| 131.83 | 22.763 | 180.02 |
| 181.97 | 22.763 | 180.02 |
| 251.19 | 22.763 | 180.02 |
| 346.74 | 22.763 | 180.01 |
| 478.63 | 22.763 | 180.01 |
| 660.69 | 22.763 | 180.01 |
| 912.01 | 22.763 | 180.00 |
| 1258.9 | 22.763 | 179.99 |
| 1737.8 | 22.763 | 179.98 |
| 2398.8 | 22.763 | 179 96 |
| 3311.3 | 22.763 | 179.94 |
| 4570.9 | 22.763 | 179.91 |
| 6309.6 | 22.763 | 179.87 |
| 8709.6 | 22.763 | 179¡81 |
| 12023. | 22.763 | 179.73 |
| 16596. | 22.763 | 179.62 |
| 22909. | 22.762 | 179.46 |
| 31623. | 22.762 | 179.25 |
| 43652. | 22.761 | 178.96 |
| 60256. | 22.760 | 178.56 |
| 83176. | 22.758 | 178.00 |
| 0.11482D 06 | 22.753 | 177.23 |
| 0.15849D 06 | 22.744 | 176.17 |
| 0.21878D 06 | 22.727 | 174.71 |
| 0.30200D 06 | 22.694 | 172.71 |
| 0.41687D 06 | 22.633 | 169.97 |
| 0.57544D 06 | 22.519 | 166.27 |
| 0.79433D 06 | 22.309 | 161.34 |
| 0.10965D 07 | 21.935 | 154.96 |
| 0.15136D 07 | 21.301 | 147.10 |
| 0.20893D 07 | 20.302 | 138.05 |
| 0.28840D 07 | 18.866 | 128.54 |
| 0.39811D 07 | 17.001 | 119.42 |
| 0.54954D 07 | 14.785 | 111.33 |
| 0.75858D 07 | 12.325 | 104.46 |
| 0.10471D 08 | 9.7105 | 98.660 |
| 0.14454D 08 | 7.0044 | 93.587 |
| 0.19953D 08 | 4.2415 | 88.827 |
| 0.27542D 08 | 1.4365 | 83.950 |
| 0.38019D 08 | -1.4115 | 78.516 |
| 0.52481D 08 | -4.3174 | 72.099 |
| 0.72444D 08 | -7.3082 | 64.346 |
| 0.10000D 09 | -10.412 | 55.123 |

(b)

Fig. 16-22b   Output of program in Fig. 16-22a.

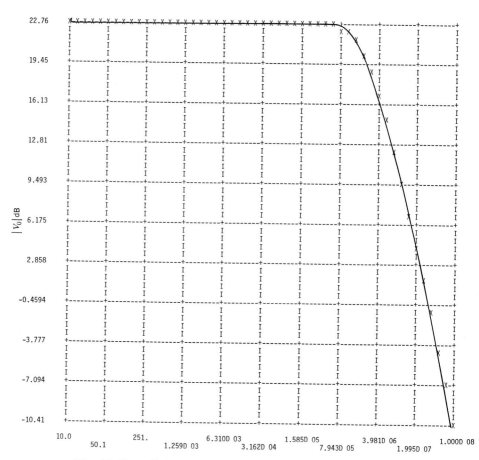

PLOT OF GAIN(DB) VS FREQUENCY

**Fig. 16-22c    Gain plot produced by program in Fig. 16-22a.**

The response, $V_0(j\omega) = V_④(j\omega)$ can be solved over the frequency range $10 \leq \omega \leq 10^8$, Hz, with $\hat{V}_1 = 1$ using the program shown in Fig. 16-22a. The results are shown in Fig. 16-22b and 16-22c.    □

## 16-8    SUMMARY

When we wish to determine the time response of networks under arbitrary excitation we usually were required to solve a differential equation. If however we are only interested in the response of linear time-invariant networks under sinusoidal excitation then solution of differential equations can be avoided. In this chapter we introduced the concept of sinusoidal steady-state analysis.

PLOT OF PHASE(DEGREES) VS FREQUENCY

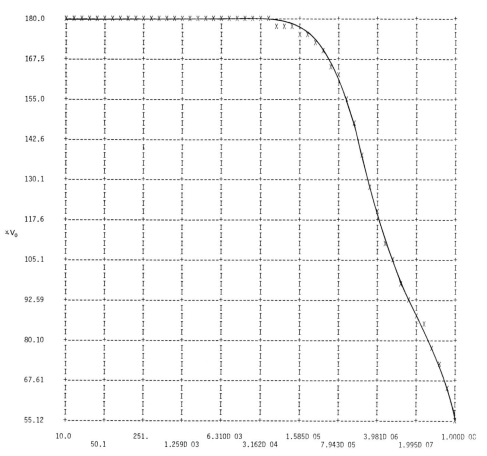

**Fig. 16-22c (Continued)** **Phase plot produced by program in Fig. 16-22a.**

A general *sinusoidal waveform*

$$v(t) = V_m \sin(\omega t + \phi)$$

is a special case of the complex sinusoidal waveform

$$v(t) = V e^{j\omega t}$$

where

$$V = V_m e^{j\phi}$$

is called a *phasor*. If a linear time-invariant network is excited by a sinusoidal input, after some initial transient, all voltages and currents in the networks will be sinusoidal and analysis of the network can be carried out in terms of phasors.

It was shown how sinusoidal steady-state analysis at a single frequency is essentially similar to the dc analysis of linear networks except complex-valued variables are used. In particular resistance and conductance are generalized to *impedance* and *admittance* respectively. The admittance of a capacitor, at frequency $\omega$ is, $j\omega C$ and the impedance of an inductor is $j\omega L$. Nodal equilibrium equations are easily written by inspection using these quantities. These equations are complex-valued simultaneous algebraic equations that can be solved using Gaussian elimination. The subroutine GAUSSC that performs Gaussian elimination on complex-valued sets of equations was discussed.

In many practical situations a network has one input and one output. It is convenient then to characterize the network by a set of complex-valued two-port parameters. The *Bode plot* of a *network function*, which is the ratio of the network's input and output, is quite useful for displaying a network's performance.

## PROBLEMS

**16.1**    Express the following sinusoids in terms of exponentials.

(a) $3 \sin (\omega t + 2)$

(b) $3 \cos (\omega t + 2)$

(c) $3 \cos (\omega t + 2) + j3 \sin (\omega t + 2)$

(d) $3 \sin (\omega t + 2) + j3 \cos (\omega t + 2)$

**16.2**    Express the following exponentials in terms of sines and cosines.

(a) $16e^{j(\pi/4 + \omega t)}$

(b) $\left(\dfrac{16}{\sqrt{2}} + j\dfrac{16}{\sqrt{2}}\right)e^{j\omega t}$

**16.3**    Express the following complex numbers in terms of phasors.

(a) $1 + j\sqrt{4}$

(b) $\dfrac{1}{1 + j}$

(c) $\dfrac{1 - j}{1 + j}$

(d) $\dfrac{1 - 2j}{1 + 3j}$

(e) $\dfrac{(1 + j)^2}{1/j}$

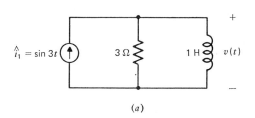

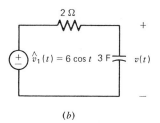

(a)

(b)

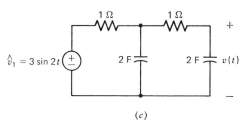

(c)

**Fig. P16.4**

**16.4** Find the sinusoidal steady-state response $v(t)$ in each of the networks of Fig. P16.4 by writing a differential equation and solving it.

**16.5** Find the sinusoidal steady-state response $v(t)$ in each of the networks of Fig. P16.4 using phasors. Compare your results with those obtained in Prob. 16.4.

**16.6** Find the sinusoidal steady-state response $i(t)$ in each of the networks of Fig. P16.6 by writing a differential equation and solving it.

**16.7** Repeat Prob. 16.6 using phasor analysis.

**16.8** Write a set of nodal equilibrium equations in terms of phasors for each of the networks in Fig. P16.8.

**16.9** Show that $n$ impedances with impedance $z_1, z_2, \ldots, z_n$ connected in series behave as a single impedance with a value of $z_1 + z_2 + \cdots + z_n$.

**16.10** Show that $n$ impedances with impedance $z_1, z_2, \ldots, z_n$ connected in parallel behave as a single impedance with a value of

$$\frac{1}{\left(\dfrac{1}{z_1} + \dfrac{1}{z_2} + \cdots + \dfrac{1}{z_n}\right)}$$

**16.11** Find the equivalent impedance and admittance of each of the networks of Fig. P16.11. Use the results of Prob. 16.10.

**16.12** Find the equivalent impedances and admittances of the networks of Fig. P16.12.

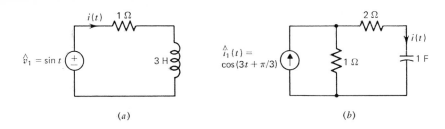

(a)

(b)

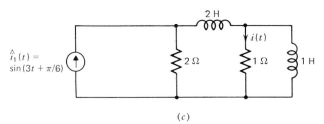

(c)

**Fig. P16.6**

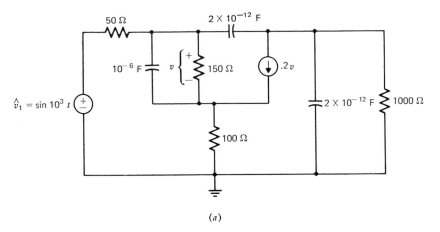

(a)

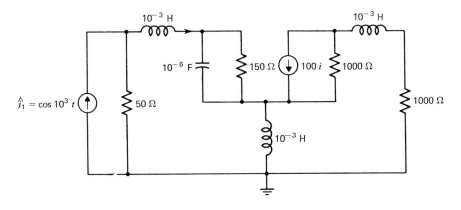

(b)

**Fig. P16.8**

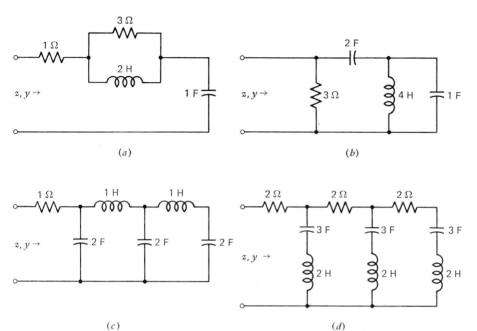

Fig. P16.11

**16.13**   The network of Fig. P16.13 is operating in sinusoidal steady-state. Use phasor analysis to determine $i_2(t)$, $i_3(t)$, and $v_4(t)$.

**16.14**   Find the steady-state solution of the following differential equations

(a) $\dfrac{d^2v}{dt^2} + 5\dfrac{dv}{dt} + 12v = \sin\left(t + \dfrac{\pi}{4}\right)$

(b) $\dfrac{d^3v}{dt^2} + 7\dfrac{d^2v}{dt^3} + 14\dfrac{dv}{dt} + 8v = \cos 3t$

**16.15**   Using the hybrid-pi transistor model (see Fig. 16-20), write nodal equilibrium equations for the networks of Fig. P16.15a and b with $r_x = 150\ \Omega$, $r_\pi = 1\ K$, $C_\pi = 500\ pF$, $C_\mu = 20\ pF$, $g_m = .1\mho$, and $r_o = 10\ \Omega$.

**16.16**   In Chapter 5 we stated and proved Thévenin's and Norton's theorems. Although the proof of these theorems was based upon linear resistive networks, the theorems are valid for any linear network, and in particular, can be employed in phasor analysis. Find the Thévenin and Norton equivalent network as seen by the load resistor, $R_L$, in the small signal common emitter transistor amplifier network of Fig. P16.6.

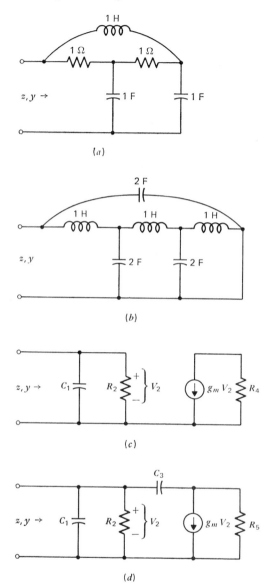

**Fig. P16.12**

**16.17**    The network of Fig. P16.17 is operating in sinusoidal steady-state. Find the node voltages as a function of time. (*Hint*. Use phasor analysis and superposition.)

**16.18**    (a) Find all the branch current phasors and branch voltage phasors in the network of Fig. P16.18 for $\hat{I}_1 = 2$ and $\omega_1 = 2$ and verify

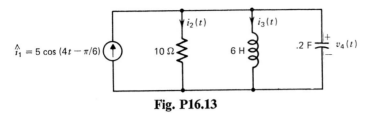

**Fig. P16.13**

that Tellegen's theorem is valid. In other words, show that

$$\sum_B V_B(j\omega_1)I_B(j\omega_1) = 0$$

where the summation is taken over all network branches $B$.

(b) Reanalyze this network for $\hat{I}_1 = 2$ and $\omega_2 = 5$. Let $\hat{V}_B(j\omega_2)$ and $\hat{I}_B(j\omega_2)$ represent the branch voltage and current phasors for the network under this new excitation.

Verify that

$$\sum_B \hat{V}_B(j\omega_2)I_B(j\omega_1) = 0$$

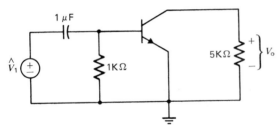

$(a)$

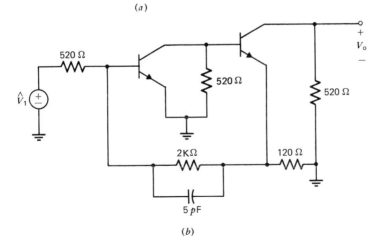

$(b)$

**Fig. P16.15**

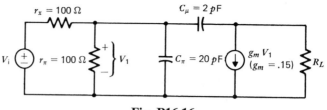

**Fig. P16.16**

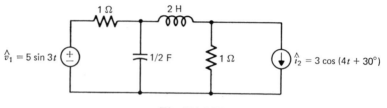

**Fig. P16.17**

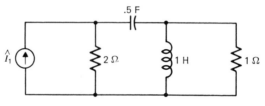

**Fig. P16.18**

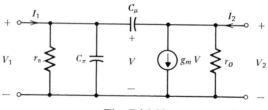

**Fig. P16.19**

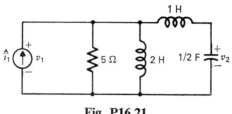

**Fig. P16.21**

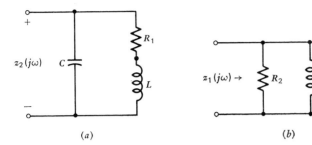

Fig. P16.22

and

$$\sum_B V_B(j\omega_1)\hat{I}_B(j\omega_2) = 0$$

What conclusions can you draw from these results?

**16.19**    Find the $z$-parameters for the two-port network of Fig. P16.19.

**16.20**    Find the $y$-parameters for the two-port network of Fig. P16.19.

**16.21**    Consider the network of Fig. P16.21 with input $\hat{i}_1$.
(a) Calculate the driving point impedance $V_1(j\omega)/I_1(j\omega)$ and the transfer impedance $V_2(j\omega)/I_1(j\omega)$.
(b) Sketch Bode plots of the network functions found in (a).

**16.22**    Calculate the impedances $Z_1(j\omega)$ and $Z_2(j\omega)$ of the networks of Fig. P16.22 where $C = 10^{-9}$ F, $R_1 = 10\ \Omega$, $L = 10^{-3}$ H, and $R_2 = 10^5\ \Omega$. Sketch Bode plots of these impedances over a range of one octave below and one octave above the resonant frequencies.

**16.23**    Use superposition to find the steady-state response $i(t)$ in the network of Fig. P16.23.

**16.24**    The network of Fig. P16.24 is a modification of the network of Fig. 16-21 analyzed in Example 16.10. Modify the program of Fig. 16-22 and use it to plot the frequency plot of this network over

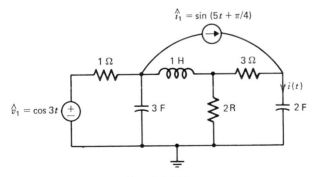

Fig. P16.23

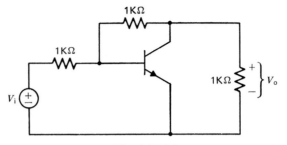

**Fig. P16.24**

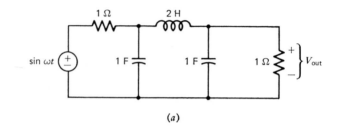

(a)

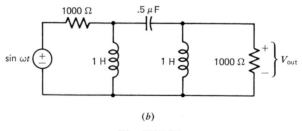

(b)

**Fig. P16.25**

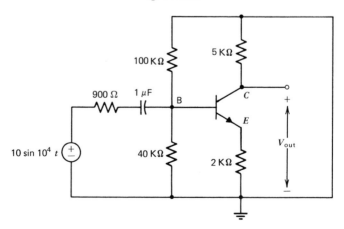

**Fig. P16.26**

the range $10 \leq \omega \leq 10^8$, Hz. (*Hint*. Only the subroutine NODEQN need be modified.)

**16.25**   Find the frequency response of each network in Fig. P16.25 by writing a program similar to the program of Fig. 16-22a. Have the program plot $V_{OUT}$ in dB vs. log $\omega$ for $\omega = .1, 1, 10, 100,$ and 1000 rad/sec. From these plots determine a practical use for these circuits.

**16.26**   Write a computer program similar to the program of Fig. 16-22a to determine the frequency response of the network of Fig. P16.26. Represent the transistor by the hybrid-pi model with parameters $r_x = 100$, $r_\pi = 1000$, $g_m = .1$ mhos, $C_\pi = 200$ pf, $C_\mu = 2$, and $r_o = 5$ K.

## CHAPTER 17

# Generalization of Sinusoidal Steady-State Analysis

We have seen that analysis of a linear network under a sinusoidal excitation can be carried out in the frequency domain avoiding the necessity of solving any differential equations. In this chapter we will generalize the techniques of Chapter 16 so that they may be employed for networks under a somewhat larger class of excitations.

## 17-1 COMPLEX FREQUENCY AND THE EXPONENTIAL FORCING FUNCTION

Consider the waveform $v(t) = Ve^{st}$ where $V$ is a phasor and $s = \sigma + j\omega$ is known as the *complex frequency*. The quantity $\omega$ is radian frequency as before while $\sigma$, which is real and usually negative, is *neper frequency* and has units of *nepers per second*. Observe that special cases of this waveform include the damped sinusoid

$$v(t) = \begin{cases} e^{-\sigma t}\cos \omega t = \text{Re }\{e^{-st}\}, & t > 0 \\ 0, & t < 0 \end{cases}$$

($\sigma$ real) and the constant unity signal

$$v(t) = 1 = e^{st}\big|_{s=0}$$

among other possibilities.

Recognize that because the forcing function $e^{st}$ has the same properties as $e^{j\omega t}$ (see Section 16-3), all voltages and currents in a linear network that is excited by this waveform will themselves be exponentials with complex frequencies. Therefore, we would expect that the same phasor analysis as discussed in Section 16-3 can be extended to the analysis of networks whose inputs are complex exponentials. The only modifications are the branch relationships for capacitors and inductors. In particular, for a capacitor the branch relation

$$i_C = C\frac{dv_C}{dt}$$

becomes

$$I_C e^{st} = C\frac{d}{dt}[V_C e^{st}]$$
$$= sCV_C e^{st}$$

or

$$I_C = sCV_C$$

and for an inductor, the branch relation

$$v_L = L\frac{di_L}{dt}$$

becomes

$$V_L e^{st} = L\frac{d}{dt}[I_L e^{st}]$$
$$= sLI_L e^{st}$$

or

$$V_L = sLI_L$$

The impedances and admittance of capacitors and inductors have thus been generalized to

$$Z_C(s) = \frac{1}{sC} \quad\text{and}\quad Y_C(s) = sC$$

and

$$Z_L(s) = sL \quad\text{and}\quad Y_L(s) = \frac{1}{sL}$$

respectively. Observe that by determining the response of a network to the complex exponential forcing function, we have in effect also determined the response of the network to a dc, exponential, or sinusoidal forcing function as well.

### Example 17.1

In the network of Fig. 17-1 the input is $i_1(t)$ and the output or response is $i_4(t)$. Suppose $i_1(t) = (e^{-3t}\cos 2t)u(t)$. The input can be rewritten as

$$i_1(t) = \text{Re}\,(I_1 e^{st}) \qquad t > 0$$

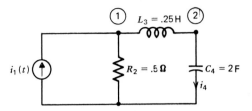

Fig. 17-1   Network studied in Example 17.1.

where $s = -3 + j2$. We can determine the response due to this input by finding the response due to the more general excitation of $i_1(t) = I_1 e^{st}$. Using phasors the node equations are

$$
\begin{bmatrix}
\dfrac{1}{R_2} + \dfrac{1}{sL_3} & -\dfrac{1}{sL_3} \\[2mm]
-\dfrac{1}{sL_3} & \dfrac{1}{sL_3} + sC_4
\end{bmatrix}
\begin{bmatrix}
V_{①} \\[2mm]
V_{②}
\end{bmatrix}
=
\begin{bmatrix}
I_1 \\[2mm]
0
\end{bmatrix}
$$

so that

$$
V_{②} = \frac{I_1}{\dfrac{C_4 L_3}{R_2} s^2 + C_4 s + \dfrac{1}{R_2}}
$$

or, on substitution of element value

$$
V_{②} = \frac{I_1}{s^2 + 2s + 2}
$$

The response $I_4$ is found from

$$
I_4 = C_4 V_{②} = \frac{2s}{s^2 + 2s + 2} I_1
$$

For the input of interest $s = -3 + j2$ and $I_1 = 1$, so that

$$
I_4 = \frac{-6 + j4}{5 - j6} = .92 e^{j.24}
$$

Because $i_1(t) = \mathrm{Re}\,(I_1 e^{st})$ the time response is

$$
\begin{aligned}
i_4(t) &= \mathrm{Re}\,(I_4 e^{st}) \\
&= \mathrm{Re}\,[.92 e^{j.29} e^{(-3+j2)t}] \\
&= .92 e^{-3t} \cos(2t + .29)
\end{aligned}
$$

Observe that we can also quickly determine the response to a constant dc signal. Suppose $i_1(t) = 3$. Then $i_1(t) = 3e^{st}$ with $s = 0$. Therefore, the response of interest is

$$i_4(t) = \text{Re } \{I_4 e^{st}\}\big|_{s=0}$$

$$= 0$$

as we would expect because the capacitor becomes an open circuit under constant excitation.

Finally, observe that the frequency response, as discussed earlier, could be easily formed by setting $s = j\omega$, that is, $\sigma = 0$, and $I_1 = 1$. Then

$$I_4(j\omega) = \frac{2j\omega}{-\omega^2 + 2j\omega + 2}$$

from which a Bode plot can be drawn.   □


## 17-2  POLES AND ZEROS: THE s-PLANE

As mentioned previously, it is the frequency selective properties of linear networks that make them so useful. A convenient means for portraying the frequency response of a network is the Bode plot. For the case under consideration, in which complex frequency is being used, the response is a function of two quantities, $\sigma$ and $\omega$. Thus some alternative form of displaying the response as a function of the complex frequency is needed. In this section we introduce a suitable representation.

In Section 16-4 the concept of a transfer function was introduced. The transfer function, which is the ratio of the network's output to its input, can be also defined in terms of complex frequency. Given a network whose input phasor is $V_i(s)$ and output phasor is $V_o(s)$, then the transfer function is

$$H(s) = \frac{V_o(s)}{V_i(s)}$$

In general $H(s)$ is a ratio of two polynomials of $s$:

$$H(s) = \frac{A(s)}{B(s)} = \frac{a_m s^m + a_{m-1} s^{m-1} + \cdots + a_0}{b_n s^n + b_{n-1} s^{n-1} + \cdots + b_0}$$

(*Note.* The denominator polynomial $B(s)$ is *not* the same as the input phasor $V_i(s)$.) Since a polynomial can be factored into a product of first order polynomials, the transfer function can be written as

$$H(s) = \frac{A(s)}{B(s)} = \frac{K(s - z_1)(s - z_2) \cdots (s - z_m)}{(s - p_1)(s - p_2) \cdots (s - p_n)}$$

where $z_1, z_2, \ldots, z_m$ are the *zeros* or roots of $A(s)$ [i.e., they are the values of $s$ for which $A(s) = 0$] and $p_1, p_2, \ldots, p_n$ are the zeros of $B(s)$. The $z_i$'s are called the *zeros* of $H(s)$ and the $p_i$'s are called the *poles* of $H(s)$. The term $K$ is called the *gain constant*. Observe that knowledge of the poles, zeros, and gain constant of a network is sufficient to predict the response of a network under a large class of excitations. We will see shortly how the poles are also related to the natural or unforced response of a network.

**Example 17.2**

Consider again the network of Fig. 17-1. If $I_1$ is the input and $I_4$ the output, the transfer function, as found in Example 17.1, is

$$H(s) = \frac{2s}{s^2 + 2s + 2}$$

The numerator and denominator of $H(s)$ can be factored as follows

$$H(s) = \frac{2s}{[s + (1 + j1)][s + (1 - j1)]}$$

Thus $H(s)$ has two poles, $p_1 = -1 - j1$ and $p_2 = -1 + j1$, and one zero, $z_1 = 0$. The gain constant is $K = 2$. □

Since poles and zeros are in general complex numbers, it is convenient to represent them by points in the complex plane, called the s-plane. The real axis is labeled $\sigma$ and the imaginary axis is labeled $j\omega$. Poles are usually indicated by the symbol × and zeros by the symbol ○. The pole-zero plot associated with the transfer function discussed in Example 17.2 is shown in Fig. 17-2. Observe that the value of $H(s)$ for a particular value of $s$ can be found graphically. In Example 17.1, we found $H(s)$ for $s = -3 + j2$. Figure 17-3 illustrates the graphical computation. Also observe that the magnitude and phase curve, that is Bode plot, of any transfer function $H(j\omega)$ can be found from the pole-zero diagram of $H(s)$ simply from the magnitude and direction of the vectors drawn from the poles and zeros to the point $j\omega$ as $\omega$ varies between $-\infty$ and $\infty$. □

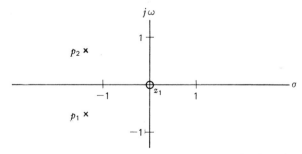

**Fig. 17-2   The pole zero plot for the transfer function**

$$H(s) = \frac{2s}{s^2 + 2s + 2}$$

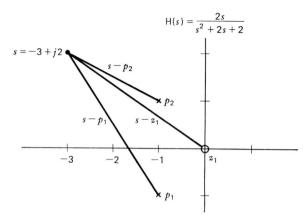

**Fig. 17-3    Graphical illustration of evaluation of $H(s)$ for a particular value of $s$.**

## 17-3    RELATIONSHIP BETWEEN POLES AND THE NATURAL RESPONSE

In Chapter 14 we discussed the time domain solution of networks We showed that solution of a network that contained two energy storage elements usually required solution of a second-order differential equation. In general, a network that contains $n$ energy storage elements requires solution of an $n$th order differential equation, providing that there are no loops of capacitors and voltage sources or cutsets of inductors and current sources. Typically the differential equation that relates the output $y(t)$ to the input $x(t)$ has the form

$$(b_n D^n + b_{n-1} D^{n-1} + \cdots + b_0) y(t) = (a_m D^m + a_{m-1} D^{m-1} + \cdots + a_0) x(t)$$

Let the input be of the form

$$x(t) = X(s) e^{st}$$

The solution $y(t)$ is the sum of the homogeneous solution, or natural response, and the particular solution, or forced response. The homogeneous solution must satisfy the homogeneous equation

$$(b_n D^n + b_{n-1} D^{n-1} + \cdots + b_0) y_H(t) = 0$$

Assume the homogeneous solution is of the form

$$y_H(t) = e^{\lambda t}$$

Substitution of $y_H(t)$ into the homogeneous equation yields

$$(b_n \lambda^n + b_{n-1} \lambda^{n-1} + \cdots + b_0) e^{\lambda t} = 0$$

Thus $y_H(t)$ is a solution if $\lambda$ is a root of the characteristic equation

$$(b_n\lambda^n + b_{n-1}\lambda^{n-1} + \cdots + b_0) = 0$$

If there are $n$ distinct roots $\lambda_1, \lambda_2, \ldots, \lambda_n$, the homogeneous solution is

$$y_H(t) = c_1 e^{\lambda_1 t} + c_2 e^{\lambda_2 t} + \cdots + c_n e^{\lambda_n t}$$

where the constants $c_1, c_2, \ldots, c_n$ depend on the initial conditions. If the $\lambda$'s are not unique the homogeneous solution takes on a slightly different form. Suppose that $\lambda_1$ is of multiplicity $r_1$, $\lambda_2$ is of multiplicity $r_2$, etc. In other words the characteristic equation can be factored as follows

$$(b_n\lambda^n + b_{n-1}\lambda^{n-1} + \cdots + b_0) = (\lambda - \lambda_1)^{r_1}(\lambda - \lambda_2)^{r_2} \cdots (\lambda - \lambda_m)^{r_m}$$

where $r_1 + r_2 + \cdots + r_m = n$. Then the homogeneous solution becomes

$$y_H(t) = \sum_{j=1}^{r_1} c_{1j} t^{(j-1)} e^{\lambda_1 t} + \sum_{j=1}^{r_2} c_{2j} t^{(j-1)} e^{\lambda_2 t} + \cdots + \sum_{j=1}^{r_m} c_{mj} t^{(j-1)} e^{\lambda_m t}$$

Recall that the $\lambda_j$ are the natural frequencies of the network.

Now let us turn to the particular solution. Since the input is a complex exponential, it is reasonable to assume a particular solution of the form

$$y_p(t) = Y(s)e^{st}$$

Substitution of $y_p(t)$ and $x(t)$ into the original differential equation yields

$$(b_n s^n + b_{n-1}s^{n-1} + \cdots + b_0) Y(s) = (a_m s^m + a_{m-1}s^{m-1} + \cdots + a_0)X(s)$$

from which we can determine $Y(s)$:

$$Y(s) = \frac{a_m s^m + a_{m-1}s^{m-1} + \cdots + a_0}{b_n s^n + b_{n-1}s^{n-1} + \cdots + b_0} X(s)$$

The transfer function is

$$H(s) = \frac{Y(s)}{X(s)} = \frac{a_m s^m + a_{m-1}s^{m-1} + \cdots + a_0}{b_n s^n + b_{n-1}s^{n-1} + \cdots + b_0}$$

We now recognize that the poles of $H(s)$, which are the values of $s$ such that

$$b_n s^n + b_{n-1}s^{n-1} + \cdots + b_0 = 0$$

are in fact the natural frequencies of the network. Thus knowledge of the poles and zeros of a network not only yield information concerning the forced response, but also predict the natural response as well.

**Example 17.3**

Consider again the network of Fig. 17-1, redrawn in Fig. 17-4. The transfer function, found in Examples 17.1 and 17.2 is

$$H(s) = \frac{2s}{[s + (1 + j1)][s + (1 - j1)]}$$

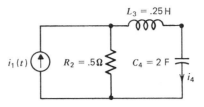

**Fig. 17-4   Network used in Example 17.3 to illustrate calculation of the complete response from knowledge of the poles and zeros.**

Therefore the natural response homogeneous solution is

$$i_{4H}(t) = c_1 e^{-(1+j1)t} + c_2 e^{-(1-j1)t}$$

The complete response then for an input of $i_1(t) = e^{-3t} \cos 2t$ is (see Example 17.1)

$$i_4(t) = c_1 e^{-(1+j1)t} + c_2 e^{-(1-j1)t} + .92 e^{-3t} \cos(2t + .29)$$

The constants $c_1$ and $c_2$ are found by considering the initial conditions $i_4(0$ and $(d/dt)i_4(0)$.   □


## 17-4   POWER AND ENERGY IN SINUSOIDAL STEADY-STATE

Until now we have not discussed in any detail the most important concepts of power and energy. The power absorbed by a two terminal element whose voltage and current is $v(t)$ and $i(t)$, respectively, is given by

$$p(t) = v(t)i(t), \qquad \text{watts} = \text{joules/second}$$

For a resistor, we note that

$$p(t) = v(t)i(t)$$

$$= i^2(t)R = \frac{v^2(t)}{R}$$

Let us now consider a capacitor. The work that is required to assemble a collection of charge on a capacitor is the electrical energy stored in the capacitor. The incremental increase of energy $dw$ and the incremental increase in charge $dq$ are related by

$$dw = v \, dq$$

where $v$ is the voltage across the capacitor. If the total charge on the capacitor is $Q$, the total stored energy is

$$w = \int_0^Q v \, dq$$

For a linear capacitor

$$q = Cv$$

so that the total energy is

$$w = \frac{1}{2} Cv^2 \text{ joules}$$

Power is the rate of change of energy

$$p(t) = \frac{dw}{dt} = v \frac{dq}{dt} = vi$$

In a like manner, we recognize that the incremental increase of energy stored in an inductor and the incremental increase in magnetic flux, $d\phi$, are related by

$$dw = i \, d\phi$$

where $i$ is the inductor current. The energy stored by an inductor's magnetic field that produces a flux $\phi$ is

$$w = \int_0^{\Phi} i \, d\phi$$

For a linear inductor

$$\phi = Li$$

so that

$$w = \frac{1}{2} Li^2$$

In the sinusoidal steady-state we can write, without loss of generality

$$v(t) = \text{Re}\,(Ve^{j\omega t}) \quad \text{and} \quad i(t) = \text{Re}\,(Ie^{j\omega t})$$

where $V$ and $I$ are phasors and $\omega$ is the frequency of the excitation. Power then becomes

$$p(t) = \text{Re}\,(Ve^{j\omega t})\,\text{Re}\,(Ie^{j\omega t})$$

Since $V$ and $I$ are phasors we may write

$$V = |V|\,e^{j\phi_V} \quad \text{and} \quad I = |I|\,e^{j\phi_I}$$

where $\phi_V = \arctan\,[\text{Im}\,(V)/\text{Re}\,(V)]$ and $\phi_I = \arctan\,[\text{Im}\,(I)/\text{Re}\,(I)]$
Then

$$p(t) = |V|\,|I|\,\text{Re}\,[e^{j(\omega t + \phi_V)}]\,\text{Re}\,[e^{j(\omega t + \phi_I)}]$$

$$= |V|\,|I|\,\cos\,(\omega t + \phi_V)\,\cos\,(\omega t + \phi_I)$$

$$= \frac{1}{2}|V|\,|I|\,\cos\,(\phi_V - \phi_I) + \frac{1}{2}|V|\,|I|\,\cos\,(2\omega t + \phi_V + \phi_I)$$

The first term is independent of time and therefore represents an average power that is denoted by $P_{av}$:

$$P_{av} = \frac{1}{2}|V|\,|I|\cos(\phi_V - \phi_I)$$

and the second term is a sinusoidal component with a frequency of twice that of the voltage or current. Observe that when $v$ and $i$ are in phase $\phi_V - \phi_I = 0$ and the average power is at a maximum:

$$\max P_{av} = \frac{1}{2}|V|\,|I| \qquad \phi_V = \phi_I$$

and when $v$ and $i$ are $90°$ out of phase $\phi_V - \phi_I = \pm\pi/2$ and the average power is zero. In the latter situation, the total power $p(t)$ is negative during half the cycle. The term $\cos(\phi_V - \phi_I)$ is called the *power factor*.

### Example 17.4

The network of Fig. 17-5a is operating in sinusoidal steady-state. Analysis of this network is most easily undertaken in terms of the equivalent imped-ance network of Fig. 17-5b where $\hat{I}_1 = -5$, $z_2 = 6 - j$, $z_3 = 1 + j5$, and $z_4 = 3 - j3$. We wish to find the power delivered by the source. To this end we need to find $V_1$:

$$V_1 = \hat{I}_1 Z_T$$

where

$$Z_T = \frac{Z_2(Z_3 + Z_4)}{Z_2 + Z_3 + Z_4}$$
$$= \frac{268 + j54}{101}$$

so that

$$V_1 = 13.3 + j2.67$$
$$= 13.6e^{j11.4°}$$

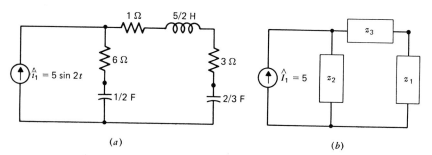

(a)                                       (b)

**Fig. 17-5**  (*a*) **A network operating in sinusoidal steady state, and** (*b*) **the equivalent network in terms of impedances.**

Therefore, the average power delivered by the source is

$$P_{av} = \frac{1}{2} |V_1| |I_1| \cos (\phi_V - \phi_I)$$

$$= 33.3$$

while the total instantaneous power delivered by the source is

$$p(t) = 33.3 + 34 \cos (4t + 11.4°) \quad \square$$

We can express the average power as

$$P_{av} = \frac{1}{2} \text{Re} (VI^*)$$

because

$$\text{Re} (VI^*) = \text{Re} [|V| e^{j\phi_V} |I| e^{j(-\phi_I)}]$$
$$\doteq \text{Re} [|V| |I| e^{j(\phi_V - \phi_I)}] = |V| |I| \cos (\phi_V - \phi_I)$$

It is therefore a simple matter to find the average power for any two-port whose impedance is $Z$. Since $V = ZI$, we have

$$P_{av} = \frac{1}{2} \text{Re} (VI^*)$$

$$= \frac{1}{2} \text{Re} (ZII^*) = \frac{1}{2} |I|^2 \text{Re} \{Z\}$$

Observe that for the case of a pure resistance, $Z = R$ and

$$P_{av} = \frac{1}{2} |I|^2 R$$

for a pure capacitance

$$Z = \frac{1}{j\omega C}$$

and

$$P_{av} = \frac{1}{2} |I|^2 \text{Re} \left(\frac{1}{j\omega C}\right) = 0$$

and for a pure inductance

$$Z = j\omega L$$

so that

$$P_{av} = \frac{1}{2} |I|^2 \text{Re} (j\omega L) = 0$$

Thus only a resistor dissipates power in the steady-state. We often refer to inductors and capacitors as *lossless* elements.

The power factor is very important from the stand point of an electric power company as illustrated in the following example.

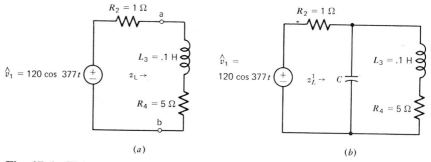

**Fig. 17-6** The networks used in Example 17.5 to illustrate how the power factor affects losses in power transmission cables.

## Example 17.5

An overly simplified view of a power company-customer situation is shown in Fig. 17-6a. The source $\hat{V}_1$ represents the power company's generator, (*Note.* Since $f = 60$ Hz, then $\omega = 2\pi f = 377$ rad/sec) $R_z$ represents the transmission cable loss and $L_3$ and $R_4$ are the customer's load. The impedance of the load is

$$Z_L = 5 + j37.7$$
$$= 38e^{j82.45°}$$

Thus the power factor $\phi$ is

$$\phi = \cos 82.45°$$
$$= .13$$

The current flowing in the circuit is

$$I = \frac{120}{R_2 + Z_L}$$
$$= \frac{120}{6 + j37.7} = .49 - j3.10$$
$$= 3.14e^{-j81.45°}$$

The average power taken by the load is

$$P_{\text{Load}} = \frac{1}{2}|I|^2 \text{ Re } \{Z\}$$
$$= 24.7 \text{ watts}$$

while the average power lossed in the cable is

$$P_{\text{Cable}} = \frac{1}{2}|I|^2 R_2$$
$$= 4.93 \text{ watts}$$

Now suppose we place a capacitance across the load as shown in Fig. 17-6b. The admittance of the new load is

$$Y'_L = j377C + \frac{1}{5 + j37.7} = j377C + \frac{5 - j37.7}{1446}$$

The power factor for the load would be unity, that is, the voltage and current in the load would be in phase, if the imaginary part of $1/L$ were zero, that is, if

$$377C = \frac{37.7}{1446}$$

or

$$C = 69.2\mu F$$

For this situation, the impedance of the load is

$$Z_L = \frac{1446}{5} = 289$$

and the current flowing in the network is

$$I' = \frac{120}{290} = .414$$

The average power taken by the load is

$$P_{\text{Load}} = \frac{1}{2} |I|^2 \operatorname{Re}(Z_L)$$

$$= 24.7 \text{ watts}$$

while the average power lossed in the cable is

$$P_{\text{Cable}} = \frac{1}{2} |I|^2 R_2$$

$$= .086 \text{ watts}$$

Thus we see that by adjusting the power factor to be unity, the loss due to resistance in the cable is reduced while the power delivered to the load remains the same.   ☐

It is convenient at this point to define *complex power* as

$$P = \frac{1}{2} VI^*$$

which has units of volt-amperes, or va and is sometimes called apparent power. Then

$$\operatorname{Re}(P) = \frac{1}{2} \operatorname{Re}(VI^*) = P_{av}$$

We now define the reactive power $Q$ as

$$Q = \text{Im}(P)$$

which has units of *vars* (for volt-ampere reactive). If we let $\phi = \phi_V - \phi_I$ then

$$Q = \frac{|VI|}{2} \sin \phi$$

We can now express the total instantaneous power as

$$p(t) = P_{av} + P \cos 2\omega t - Q \sin 2\omega t$$

Thus the instantaneous power consists of a constant term and two time varying terms that are 90° out of phase with respect to each other.

## 17-5  SUMMARY

The *complex exponential forcing function*

$$v(t) = Ve^{st}$$

where $s = \sigma + j\omega$ is the complex frequency was introduced. It was cited that because this forcing function has the same properties as $e^{j\omega t}$, all voltages and currents in a linear network that is excited by this waveform will themselves be exponentials with complex frequency $s$. Thus the same type of phasor analysis carried out in Chapter 16 is applicable.

When complex exponential forcing functions are present the admittance of a capacitor is generalized to

$$Y_L(s) = sC$$

and the admittance of an inductor is generalized to

$$Y_L(s) = \frac{1}{sL}$$

The concept of a transfer function introduced during the study of the sinusoidal steady-state was extended to cover networks that have complex exponential forcing functions. The transfer function for the latter case can be viewed as a ratio of polynomials in $s$. The roots of the numerator polynomial are called *zeros* while the roots of the denominator polynomial are called *poles*. The relationship between poles and zeros and the natural response was discussed. Finally, the concept of power for networks operating in sinusoidal steady-state was discussed in detail.

## PROBLEMS

**17.1**    Find the equivalent impedances and admittances in terms of the complex frequency variable $s$ for each of the networks of Fig.

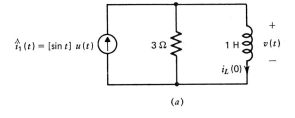

(a)

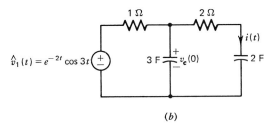

(b)

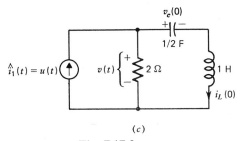

(c)

**Fig. P17.2**

P16.11. Show that when Re $\{s\} = 0$ these impedances and admittances reduce to those found in Prob. 16.11.

**17.2** Find the indicated responses ($v(t)$ or $i(t)$) of the networks of Fig. P17.2 for the excitations shown using the generalized phasor analysis. (Assume all initial conditions are zero.)

**17.3** Use superposition and phasor analysis to find $v(t)$ in the networks of Fig. P17.3.

**17.4** For each of the networks of Fig. P17.4 find the network functions $V_o(s)/V_i(s)$.

**17.5** Find the poles and zeros of the network functions found in Prob. 17.4. Sketch the pole-zero diagram.

**17.6** Find the network function $V(s)/\hat{I}_1(s)$ for the network of Fig. P17.2a. Determine the poles and zeros of the network function and determine the complete response if $i_L(0) = 1a$. (*Hint.* Use the forced response found in Prob. 17.2.)

**17.7** Repeat Prob. (17.6) for the networks of Fig. P17.2b and 17.2c and the network functions $I(s)/\hat{V}_1(s)$ and $V(s)/\hat{I}_1(s)$, respectively.

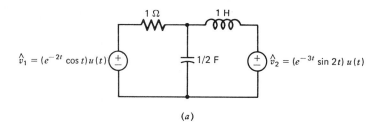

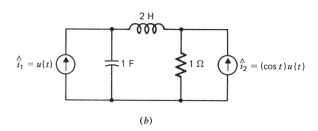

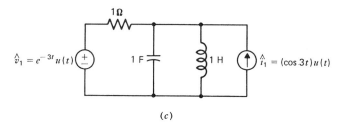

**Fig. P17.3**

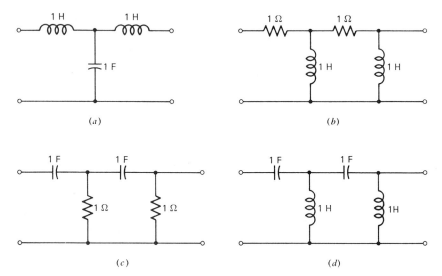

**Fig. P17.4**

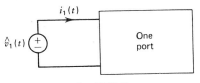

**Fig. P17.8**

**17.8**    In the one-port network of Fig. P17.8 $\hat{v}_1(t) = 75 \cos \omega t$ volts and $i_1(t) = 10 \cos (\omega t + \pi/6)$ amps. Find (a) the instantaneous power, $p(t)$; (b) $P_{av}$ in watts; (c) the reactive power $Q$ in vars; and the power factor.

**17.9**    Write the nodal equilibrium equations that could be used to find the forced response of the network of Fig. P17.9. These equations can be written by inspection by using the admittance $sC$ for capacitors and $1/sL$ for inductors.

**17.10**    Find the values of $R_x$ and $L_x$ in the bridge circuit of Fig. P17.10 in terms of $R_2$, $R_3$, $R_4$, and $C_5$ such that the voltmeter reads zero.

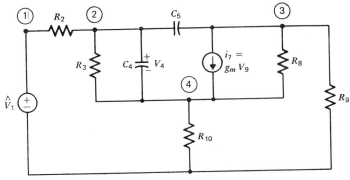

**Fig. P17.9**

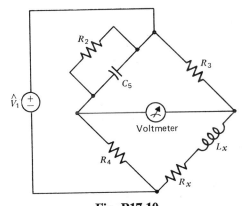

**Fig. P17.10**

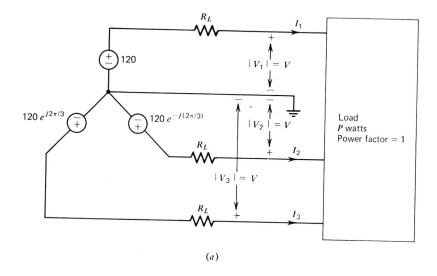

$R_2 = 5\ \Omega$    $L_3 = .05$ H

$\hat{v}_1 = 220 \sin 377t$

$R_4 = 100\ \Omega$

**Fig. P17.11**

Describe how this circuit may be used to measure the resistance that is also present in a physical inductor.

**17.11**    Determine the average power delivered to the load $(L_3 - R_4)$ in the network of Fig. P17.11. What is the power factor seen by the source? What is the power lost in $R_2$?

**17.12**    Suppose a capacitor is placed across terminals $a$ and $b$ in the network of Fig. P17.11. What value must this capacitor have in order

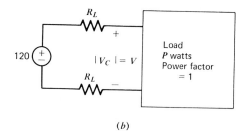

(a)

(b)

**Fig. P17.13**

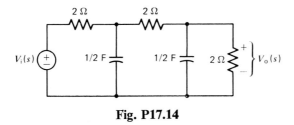

**Fig. P17.14**

to make the power factor of the load unity? For this value of $C$ what is the power lost in $R_2$?

**17.13**    Electric power companies prefer to supply power to their customers using a *three phase* system. The circuit of Fig. P17.13 *a* represents such a system. The three voltage sources all have the same magnitude but are 120° out of phase with respect to each other. Calculate the power lost in the cable, that is, resistors $R_C$, if the load absorbs an average power of $P$ watts with a unity power factor. To see the

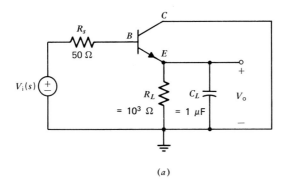

(a)

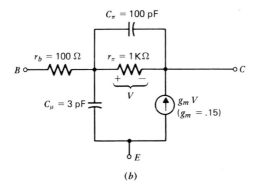

(b)

**Fig. P17.15**

advantage of a three-phase system, consider the single phase situation of Fig. P17.13$b$. Compute the line current for this case as well as the cable loss. Compare these results with those obtained with the three phase system. What conclusions can you draw?

**17.14**     Find the transfer functions $H(s) = V_o(s)/V_i(s)$ for the network of Fig. P17.14. Determine the poles and zeros of $H(s)$ and sketch the form of the response due to initial conditions.

**17.15**     The circuit of Fig. P17.15$a$ is known as an emitter follower. Using the transistor model shown in Fig. P17.15$b$, determine the transfer function $V_o(s)/V_i(s)$. What are the poles and zeros of this transfer function?

## CHAPTER 18

# The Fourier Series and the Fourier Transform

We have seen in the previous two chapters that the steady-state response resulting from the sinusoidal input $e^{j\omega t}$ as well as the forced response resulting from the complex exponential input $e^{st}$ could easily be computed by using phasors and thereby avoiding the necessity of solving differential equations. There are times when it is of interest to study the steady-state behavior of periodic signals that are not sinusoids. In this chapter we first show that any periodic signal may be represented by a sum of sinusoids of harmonically related frequencies. Therefore, by using superposition and the analysis procedure of Chapter 16 we can determine the steady-state response of a linear network to any periodic signal quite easily. We then extend this result to include certain nonperiodic waveforms.

## 18-1 FOURIER SERIES

Let $f(t)$ be a periodic signal with period $T$ so that

$$f(t + T) = f(t)$$

for all $t$. Furthermore, assume that $f(t)$ satisfies the following conditions:

1. The periodic signal $f(t)$ is a single-valued function of $t$.
2. The integral

$$\int_{t_0}^{t_0+T} |f(t)| \, dt$$

exists for all $t_0$.

3. The signal $f(t)$ has a finite number of maxima and minima in any one period.

4. The signal $f(t)$ has a finite number of discontinuities in any one period.

These conditions, known as the *Dirichlet conditions*, are satisfied for most functions with which we are concerned. Then $f(t)$ may be represented by the infinite series

$$(18.1) \qquad f(t) = a_0 + \sum_{n=1}^{\infty} (a_n \cos n\omega_0 t + b_n \sin n\omega_0 t)$$

where the *fundamental frequency* $\omega_0$ is defined by

$$\omega_0 = \frac{2\pi}{T}$$

$n\omega_0$ is called the $n$th harmonic of $\omega_0$, and the *Fourier coefficients* $a_0$, $a_n$, and $b_n$ depend on $n$ and $f(t)$. Expression (18.1) is known as the *trigonometric form of the Fourier series*.

Before discussing the evaluation of the Fourier coefficients, we will introduce some easily proven trigonometric identities. First, observe that

$$\int_0^T \sin n\omega_0 t \, dt = 0$$

for all $n$ and that

$$\int_0^T \cos n\omega_0 t \, dt = 0$$

for all $n \neq 0$, since the average value of a sinusoid over one period is zero. Moreover, we have the following relationships

$$\int_0^T \sin n\omega_0 t \cos m\omega_0 t \, dt = 0 \text{ for all } n \text{ and } m$$

$$\int_0^T \sin n\omega_0 t \sin m\omega_0 t \, dt = 0 \text{ for all } n \neq m$$

$$\int_0^T \cos n\omega_0 t \cos m\omega_0 t \, dt = 0 \text{ for all } n \neq \pm m$$

$$\int_0^T \sin^2 n\omega_0 t \, dt = \frac{T}{2} \text{ for all } n$$

and

$$\int_0^T \cos^2 n\omega_0 t \, dt = \frac{T}{2} \text{ for all } n$$

The coefficients in (18.1) can now be determined. On integration of both sides of (18.1), and substitution of the previous expressions, we have

$$(18.2) \qquad a_0 = \frac{1}{T} \int_0^T f(t) \, dt$$

Hence $a_0$ represents the average value of $f(t)$ over one period, or the *dc value*. The coefficients $a_n$, $n \neq 0$, are found by multiplying both sides of (18.1) by $\cos n\omega_0 t$ and integrating. Then

$$(18.3) \qquad\qquad a_n = \frac{2}{T} \int_0^T f(t) \cos n\omega_0 t \, dt$$

The coefficients $b_n$ are found by multiplying both sides of (18.1) by $\sin n\omega_0 t$ and integrating. Then

$$(18.4) \qquad\qquad b_n = \frac{2}{T} \int_0^T f(t) \sin n\omega_0 t \, dt$$

Expressions (18.2), (18.3), and (18.4) can be used to determine the *Fourier series coefficients* in (18.1).

Because we allow discontinuities we must carefully define the time function at these points. Suppose $t_1$ is a point of discontinuity and

$$f(t) = \begin{cases} f_1(t) & \text{for} \quad t = t_1 - \varepsilon \\ f_2(t) & \text{for} \quad t = t_1 + \varepsilon \end{cases} \qquad \text{where} \qquad 0 < \varepsilon \ll 1$$

then we shall define $f(t_1)$ as the average between $f_1$ and $f_2$ and $t_1$, that is,

$$f(t_1) = \frac{1}{2} [f_1(t_1) + f_2(t_1)] = \frac{1}{2} [f(t_1 + \varepsilon) + f(t_1 - \varepsilon)]$$

### Example 18.1

Figure 18-1 shows a square wave voltage signal for which we wish to find the Fourier series representation. This waveform can be expressed as

$$v(t) = \begin{cases} +1 & 0 < t < \dfrac{T}{2} \\[2mm] -1 & \dfrac{T}{2} < t < T \end{cases}$$

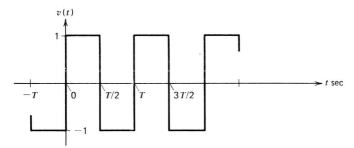

**Fig. 18-1    A square wave of amplitude 1 and period $T$ studied in Example 18.1.**

The average value of this square wave is zero, therefore $a_0 = 0$. To find the coefficients $a_n$ we use (18.3) with $\omega_0 = 2\pi/T$:

$$
\begin{aligned}
a_n &= \frac{2}{T} \int_0^T v(t) \cos \frac{2\pi n t}{T} \, dt \\
&= \frac{2}{T} \int_0^{T/2} \cos \frac{2\pi n t}{T} \, dt - \frac{2}{T} \int_{T/2}^T \cos \frac{2\pi n t}{T} \, dt \\
&= \frac{1}{n\pi} \sin \frac{2\pi n t}{T} \Bigg]_0^{T/2} - \frac{1}{n\pi} \sin \frac{2\pi n t}{T} \Bigg]_{T/2}^T \\
&= \frac{1}{n\pi} [2 \sin n\pi - \sin(0) - \sin 2\pi n] \\
&= 0, \text{ for all } n
\end{aligned}
$$

To find the coefficients $b_n$ we use (18-4)

$$
\begin{aligned}
b_n &= \frac{2}{T} \int_0^T v(t) \sin \frac{2\pi n t}{T} \, dt \\
&= \frac{2}{T} \int_0^T \sin \frac{2\pi n t}{T} \, dt - \frac{2}{T} \int_0^T \sin \frac{2\pi n t}{T} \, dt \\
&= -\frac{1}{n\pi} \cos \frac{2\pi n t}{T} \Bigg]_0^{T/2} + \frac{1}{n\pi} \cos \frac{2\pi n t}{T} \Bigg]_{T/2}^T \\
&= \frac{1}{n\pi} [\cos 2\pi n - 2 \cos n\pi + \cos(0)]
\end{aligned}
$$

Therefore, $b_1 = 4/\pi$, $b_2 = 0$, $b_3 = (\frac{1}{3})(4/\pi)$, $b_4 = 0$, ... or, in general

$$
b_n = \begin{cases} 0 & n \text{ odd } \text{even} \\ \dfrac{4}{n\pi} & n \text{ even } \text{odd} \end{cases}
$$

and the Fourier series representation of $f(t)$ is

$$
f(t) = \frac{4}{\pi} \left( \sin \frac{2\pi t}{T} + \frac{1}{3} \sin \frac{6\pi t}{T} + \frac{1}{5} \sin \frac{10\pi t}{T} + \cdots \right)
$$

Observe that the coefficients of higher harmonic terms are decreasing in magnitude. This will lead us to use a truncated Fourier series representation. The consequences we must pay will be described later.  □

**Example 18.2**

Now consider the triangular waveform of Fig. 18-2. The period is $T = 4$. An analytical expression of this signal is

$$
i(t) = \begin{cases} 1 - t & 0 \le t \le 2 \\ -3 + t & 2 \le t \le 4 \end{cases}
$$

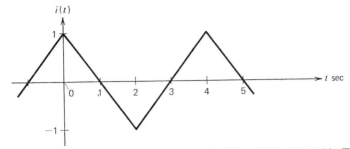

**Fig. 18-2** **Triangular wave, with amplitude 1 and period 4, studied in Example 18.2.**

Since the average value of this waveform is zero, $a_0 = 0$. We use (18.3) to find the coefficients $\omega_n$:

$$a_n = \frac{2}{T} \int_0^T i(t) \cos \frac{2\pi n t}{T}\, dt$$

$$= \frac{1}{2} \int_0^2 (1 - t) \cos \frac{\pi n t}{2}\, dt + \frac{1}{2} \int_2^4 (-3 + t) \cos \frac{\pi n t}{4}\, dt$$

$$= \begin{cases} 0, & n \text{ even} \\ \dfrac{4}{(\pi n)^2}, & n \text{ odd} \end{cases}$$

We leave it as an exercise for the student to show that from (18.4)

$$b_n = 0 \qquad \text{for all } n$$

Thus the Fourier series representation of the triangular wave is

$$i(t) = \frac{4}{\pi^2}\left( \cos \frac{\pi t}{2} + \frac{1}{9} \cos \frac{3\pi t}{2} + \frac{1}{25} \cos \frac{5\pi t}{2} + \cdots \right) \qquad \square$$

Observe that the waveform in Fig. 18-1 could be represented by a Fourier series with only sine terms, and the waveform of Fig. 18-2 could be represented by a Fourier series with only cosine terms. Moreover, only odd harmonic terms were required in each case. These situations arise when a waveform displays certain symmetry. We will study some of the properties of the Fourier series in the problems.

Observe that the coefficients $a_n$ and $b_n$ display the "harmonic content" of a signal. For instance, in Example 18.2 above, we see that the triangular wave of Fig. 18-2 has a frequency component at the fundamental frequency with an amplitude of $4/\pi^2$, a frequency component at the third harmonic with an amplitude of $4/9\pi^2$, etc. The frequency content, or discrete spectrum can be plotted as shown in Fig. 18-3.

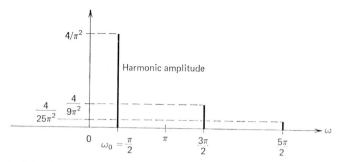

**Fig. 18-3   The frequency spectrum of the triangular waveform of Fig. 18-2.**

Assume that we have our periodic function $f(t)$ stored on a computer as a table of numbers, as might happen after the computer analysis of a circuit. To determine the harmonic content of this function using the expressions (18.2) and (18.3) we would first have to determine an analytic expression for $f(t)$. This procedure is cumbersome. In a later section we discuss an alternative method using the *discrete Fourier series*.

A more convenient form of the Fourier series is the *complex* or *exponentia* form. This form is obtained from the trigonometric form (18.1) by expressing the sines and cosines in terms of the exponential $e^{jn\omega_0 t}$. From the Euler identity

$$e^{jn\omega_0 t} = \cos n\omega_0 t + j \sin n\omega_0 t$$

we have

$$\cos n\omega_0 t = \frac{1}{2} [e^{jn\omega_0 t} + e^{-jn\omega_0 t}]$$

and

$$\sin n\omega_0 t = \frac{1}{2j} [e^{jn\omega_0 t} - e^{-jn\omega_0 t}]$$

Substitution of these equations into (18.9) yields

$$f(t) = a_0 + \sum_{n=1}^{\infty} \left( a_n \frac{e^{jn\omega_0 t} + e^{-jn\omega_0 t}}{2} + b_n \frac{e^{jn\omega_0 t} - e^{-jn\omega_0 t}}{2j} \right)$$

$$= a_0 + \sum_{n=1}^{\infty} \left[ \left( \frac{a_n - jb_n}{2} \right) e^{jn\omega_0 t} + \left( \frac{a_n + jb_n}{2} \right) e^{-jn\omega_0 t} \right]$$

We can simplify the expression by defining the coefficients

$$c_n = \frac{a_n - jb_n}{2}$$

(18.5)

$$c_{-n} = \frac{a_n + jb_n}{2} \qquad n = 1, 2, \ldots$$

and $c_0 = a_0$. Then we have

(18.6) $$f(t) = c_0 + \sum_{n=1}^{\infty} (c_n e^{jn\omega_0 t} + c_{-n} e^{-jn\omega_0 t})$$

or

(18.7) $$f(t) = \sum_{n=-\infty}^{\infty} c_n e^{jn\omega_0 t}$$

Observe that

$$c_{-n} = \frac{(a_n + jb_n)}{2} = \frac{(a_n - jb_n)^*}{2}$$

where * denotes complex conjugate. Therefore

$$c_{-n} = c_n{}^*$$

and (18.6) can be written as

(18.8) $$f(t) = c_0 + \sum_{n=1}^{\infty} [c_n e^{jn\omega_0 t} + (c_n e^{jn\omega_0 t})^*]$$
$$= c_0 + \sum_{n=1}^{\infty} 2 \operatorname{Re} \{c_n e^{jn\omega_0 t}\}$$

An expression for $c_n$ is obtained by substituting (18.3) and (18.4) into (18.5):

$$c_n = \frac{1}{T} \int_0^T f(t) \cos n\omega_0 t \, dt - \frac{j}{T} \int_0^T f(t) \sin n\omega_0 t \, dt$$
$$= \frac{1}{T} \int_0^T f(t)(\cos n\omega_0 t - j \sin n\omega_0 t) \, dt$$

or

(18.9) $$c_n = \frac{1}{T} \int_0^T f(t) e^{-jn\omega_0 t} \, dt$$

Observe that if we let $n = 0$ in (18.9),

$$c_0 = \frac{1}{T} \int_0^T f(t) \, dt$$

which agrees with the definition that $c_0 = a_0$. The student should verify that (18.9) is also valid for negative $n$.

### Example 18.3

Let us find the complex Fourier series for the square wave shown in Fig. 18-1:

$$v(t) = \begin{cases} +1 & 0 < t < \dfrac{T}{2} \\[2mm] -1 & \dfrac{T}{2} < t < T \end{cases}$$

To find the coefficients of the complex Fourier series we use (18.9)

$$c_n = \frac{1}{T} \int_0^T f(t) e^{-jn\omega_0 t}\, dt$$

Since the average value of $f(t)$ is zero,

$$c_0 = 0$$

For $n \neq 0$ we have

$$c_n = \frac{1}{T}\left[ \int_0^{T/2} (1) e^{-jn\omega_0 t}\, dt + \int_{T/2}^T (-1) e^{-jn\omega_0 t}\, dt \right]$$

$$= \frac{1}{T}\left\{ \left[ \frac{1}{-jn\omega_0} e^{-jn\omega_0 t} \right]_0^{T/2} - \left[ \frac{1}{-jn\omega_0} e^{-jn\omega_0 t} \right]_{T/2}^T \right\}$$

$$= \frac{j}{\pi n} (e^{-jn\pi} - 1)$$

where we have used the fact that $\omega_0 = 2\pi/T$. Observe that

$$e^{-jn\pi} = \cos n\pi + j \sin n\pi = \begin{cases} -1 & \text{for} \quad n \text{ odd} \\ 1 & \text{for} \quad n \text{ even} \end{cases}$$

Hence

$$c_n = \begin{cases} \dfrac{-j2}{n^\pi} & \text{for} \quad n \text{ odd} \\ 0 & \text{for} \quad n \text{ even} \end{cases}$$

The complex Fourier series

$$f(t) = \sum_{n=-\infty}^{\infty} c_n e^{jn\omega_0 t}$$

becomes

$$f(t) = \cdots + \frac{j2}{3\pi} e^{-j3\omega_0 t} + \frac{j2}{\pi} e^{-j\omega_0 t} - \frac{j2}{\pi} e^{j\omega_0 t} - \frac{j2}{3\pi} e^{j3\omega_0 t} \cdots$$

Notice that we may rewrite this expression as

$$f(t) = \sum_{n=1,\, n \text{ odd}}^{\infty} 2 \operatorname{Re}\left[ \frac{-j2}{n\pi} e^{jn\omega_0 t} \right]$$

$$= \sum_{n=1,\, n \text{ odd}}^{\infty} 2 \operatorname{Re}\left[ \frac{-j2}{n\pi} (\cos n\omega_0 t + j \sin n\omega_0 t) \right]$$

$$= \sum_{n=1,\, n \text{ odd}}^{\infty} \frac{4}{n\pi} \sin n\omega_0 t$$

$$= \frac{4}{\pi}\left( \sin \frac{2\pi t}{T} + \frac{1}{3} \sin \frac{6\pi t}{T} + \frac{1}{5} \sin \frac{10\pi t}{T} \cdots \right)$$

which is identical with the Fourier series representation found in Example 18.1.  □

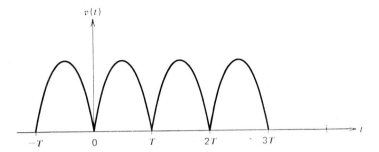

$v(t)$

$-T$    $0$    $T$    $2T$    $3T$    $t$

**Fig. 18-4    Full-wave rectifier waveform of Example 18.4.**

**Example 18.4**

The full-wave rectifier waveform shown in Fig. 18-4 is the magnitude of a sine function:

$$v(t) = V_m \sin \frac{\pi}{T} t \qquad 0 \leq t \leq T$$

To find the coefficients of the complex Fourier series we use (18.9) with $\omega_0 = 2\pi/T$;

$$c_n = \frac{1}{T} \int_0^T f(t) e^{-j2\pi nt/T} \, dt$$

$$= \frac{1}{T} \int_0^T V_m \sin \frac{\pi}{T} t \, e^{-j2\pi nt/T} \, dt$$

$$= \frac{V_m}{j2T} \left\{ \left[ \frac{T}{j\pi(1-2n)} (e^{j\pi/T(1-2n)t}) \right]_0^T - \left[ \frac{T}{j\pi(1+2n)} (e^{-j\pi/T(1+2n)t}) \right]_0^T \right\}$$

$$= -jV_m \left[ \frac{j2}{\pi(1+4n^2)} \right]$$

$$= \frac{2V_m}{\pi(1+4n^2)}$$

Therefore, the complex Fourier series is

$$v(t) = \frac{2V_m}{\pi} \left( \cdots + \frac{1}{37} e^{-j3\omega_0 t} + \frac{1}{17} e^{-j2\omega_0 t} + \frac{1}{5} e^{-j\omega_0 t} \right.$$

$$\left. + 1 + \frac{1}{5} e^{j\omega_0 t} + \frac{1}{17} e^{-j2\omega_0 t} + \frac{1}{37} e^{j3\omega_0 t} + \cdots \right) \qquad \square$$

## 18-2  STEADY-STATE ANALYSIS UNDER PERIODIC EXCITATIONS

In the last section we found that any periodic signal that satisfies the Dirichlet conditions may be represented in terms of a sum of exponentials. Therefore, the steady-state analysis of linear networks under a periodic excitation may be simplified by expressing the periodic signal in terms of a Fourier series and invoking superposition. To elaborate, consider the situation illustrated in Fig. 18-5$a$. The excitation $v(t)$ is periodic with period $T$ and, therefore, can be represented by a Fourier series (18.8) with $\omega_0 = 2\pi/T$:

$$v(t) = c_0 + \sum_{n=1}^{\infty} 2\text{Re}\,[c_n e^{jn\omega_0 t}]$$

Assume that after $N$ terms the magnitudes of the coefficients $c_n$ become negligible so that

$$v(t) \simeq c_0 + \sum_{n=1}^{N} 2\text{Re}\,[c_n e^{jn\omega_0 t}]$$

$$= c_0 + 2\text{Re}\,[c_1 e^{j\omega_0 t}] - 2\text{Re}\,[c_2 e^{j2\omega_0 t}]$$

$$+ \cdots + 2\text{Re}\,[c_N e^{jN\omega_0 t}]$$

Thus the voltage source $v(t)$ can be thought of as being composed of a series connection of voltage sources, as shown in Fig. 18-5$b$. By superposition then,

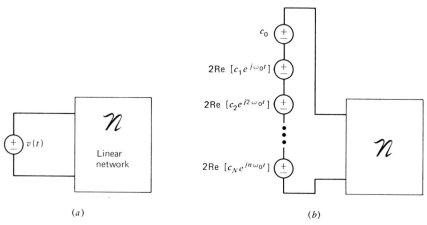

$(a)$                              $(b)$

**Fig. 18-5**  ($a$) A linear network with a periodic input. ($b$) The same network where the input voltage source is separated into a series connection of voltage sources. Each source represents another term in the Fourier series.

the steady-state response of the network to $v(t)$ is equal to the sum of the steady-state responses due to each of the sources in Fig. 18-5$b$ acting alone with all other sources set to zero. However, the steady-state response to each of the sources in Fig. 18-5$b$ is easy to compute using the phasor concept of Chapter 16.

## Example 18.5

We desire the steady-state response $v_o$ of the network of Fig. 18-6$a$ when $\hat{v}_1(t)$ is the square-wave of Fig. 18-1:

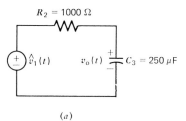

$$\hat{v}_1(t) = \begin{cases} +1 & 0 \leq t \leq \dfrac{T}{2} \\[2ex] -1 & \dfrac{T}{2} \leq t \leq T \end{cases}$$

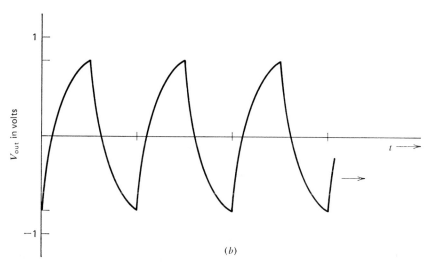

$(a)$

$(b)$

**Fig. 18-6**  $(a)$ **Network studied in Example 18.5.** $(b)$ **Response of the network due to the square wave of Fig. 18-1.**

The complex Fourier series of this square-wave was found in Example 18.3:

$$\hat{v}_1(t) = \cdots \frac{+j2}{5\pi} e^{-j5\omega_0 t} + \frac{j2}{3\pi} e^{-j3\omega_0 t} + \frac{j2}{\pi} e^{-j\omega_0 t}$$

$$- \frac{j2}{\pi} e^{j\omega_0 t} - \frac{j2}{3\pi} e^{j3\omega_0 t} - \frac{j2}{5\pi} e^{j5\omega_0 t} - \cdots$$

$$= 2 \operatorname{Re} \left[ \frac{-j2}{\pi} e^{j\omega_0 t} - \frac{j2}{3\pi} e^{j3\omega_0 t} - \frac{j2}{5\omega} e^{j5\omega_0 t} - \cdots \right]$$

where $\omega_0 = 2\pi/T = 2\pi$ rad/sec ($T = 1$ sec). Assume that the first three terms of this series adequately represent $\hat{v}_1(t)$. We can then use superposition to determine the response of the network to the square-wave, as follows. First, we recognize that the input $\hat{v}_1(t)$ can be written as

$$\hat{v}_1(t) = \hat{v}_1^1(t) + \hat{v}_1^2(t) + \hat{v}_1^3(t)$$

where

$$\hat{v}_1^1(t) = \operatorname{Re} \left[ \frac{-j4}{\pi} e^{j\omega_0 t} \right]$$

$$\hat{v}_1^2(t) = \operatorname{Re} \left[ \frac{-j4}{3\pi} e^{j3\omega_0 t} \right]$$

and

$$\hat{v}_1^3(t) = \operatorname{Re} \left[ \frac{-j4}{5\pi} e^{j5\omega_0 t} \right]$$

The response due to $\hat{v}_1(t)$ is equal to the sum of the responses to the $\hat{v}_1^1(t)$, $\hat{v}_1^2(t)$, and $\hat{v}_1^3(t)$. Now observe that if the input to the network is the phasor $\hat{V}_1$, the response phasor is

$$V_0 = \frac{1}{1 + j\omega R_2 C_2} \hat{V}_1$$

or on substitution of element values

(18.11)
$$V_0 = \frac{1}{1 + j\omega.25} \hat{V}_1$$

where $\omega$ is the radian frequency of the excitation. We can now easily find the time response due to $\hat{v}_1^1(t)$. Let $\omega = \omega_0 = 2\pi$ and $\hat{V}_1 = \hat{V}_1^1 = -j4/\pi$ in (18.11), then

$$V_0^1 = \frac{1}{1 + j.5\pi} \left( \frac{-j4}{\pi} \right) = .684 e^{-j2.6}$$

and the time response due to $\hat{v}_1^1(t)$ is

$$v_0^1(t) = \operatorname{Re} (V_0^1 e^{j\omega_0 t})$$

$$= .684 \cos (\omega_0 t - 2.6)$$

Next we can find the time response due to $\hat{v}_1^2(t)$. Let $\omega = 3\omega_0 = 6\pi$ and $\hat{V}_1 = \hat{V}_1^2 - (j^4/3\pi)$ in (18.11), then

$$V_o^2 = \frac{1}{1 + j.5\pi}\left(\frac{-j4}{3\pi}\right) = .088e^{-j2.9}$$

and the time response due to $\hat{v}_1^2(t)$ is

$$v_o^2(t) = \text{Re}\,(V_o{}^2 e^{j3\omega_0 t})$$

$$= .088\cos\,(3\omega_0 t - 2.9)$$

Finally, in an analogous manner we establish the time response due to $\hat{v}_1^3(t)$ as

$$v_1^3(t) = .032\cos\,(5\omega_0 t - 3.0)$$

The time response due to square-wave is

$$v_0(t) = .684\cos\,(\omega_0 t - 2.6) + .088\cos\,(3\omega_0 t - 2.9)$$

$$+ .032\cos\,(5\omega_0 t - 3.0)$$

which is sketched in Fig. 18-6b. Observe that the Fourier series coefficient of $v_0(t)$ are $V_0^1$, $V_0^2$, $V_0^3$, etc.   □

## 18-3  THE DISCRETE FOURIER SERIES: THE SUBROUTINE FOURR

We now consider how Fourier series coefficients can be computed numerically on the computer. The complex form of the Fourier series, given by

$$(18.12) \qquad f(t) = \sum_{n=-\infty}^{\infty} c_n e^{jn\omega_0 t}$$

where

$$(18.13) \qquad c_n = \frac{1}{T}\int_0^T f(t)e^{-jn\omega_0 t}\,dt$$

is a continuous function of time. Suppose now that $K$ equally spaced values of the periodic signal $f(t)$ were stored on the computer. We denote these samples of $f(t)$ as $f_0, f_1, \ldots, f_{K-1}$ where

$$f_k = f(k\Delta T), \qquad k = 0, 1, 2, \ldots, K-1, \qquad \text{and} \qquad \Delta T = \frac{T}{K}$$

Notice that $f(K\Delta T) = f(T) = f(0)$. Furthermore, since we are interested in representing the Fourier series on a computer we can only use a finite number of terms. Using the first $(N - 1)$ harmonics, (18.12) becomes

$$f(k\Delta T) = \sum_{n=-(N-1)}^{N-1} c_n e^{jnk\omega_0 \Delta T}$$

Now recognize that

$$\omega_0 = \frac{2\pi}{T} = \frac{2\pi}{K\Delta T}$$

so that the above expression becomes

(18.14) $$f(k\Delta T) = \sum_{n=-(N-1)}^{N-1} c_n e^{j2\pi nk/K}$$

Expression (18.13) may be discretized using the forward Euler integration scheme:

$$c_n = \frac{1}{T} \sum_{k=0}^{K-1} f(t_k) e^{-jn\omega_0 t_k} \Delta T$$

Since

$$\frac{\Delta T}{T} = \frac{1}{K}$$

and

$$\omega_0 t_k = \left(\frac{2\pi}{K\Delta T}\right)(k\Delta T) = \frac{2\pi k}{K}$$

we have

(18.15) $$c_n = \frac{1}{K} \sum_{k=1}^{K-1} f(k\Delta T) e^{-j2\pi nk/K}$$

Since $f(t)$ is a real-valued function of time,

$$c_{-n} = c_n^*$$

and (18.14) may be written as

(18.16) $$f(k\Delta T) = c_0 + 2 \sum_{n=1}^{N-1} \text{Re}\,(c_n e^{j2\pi nk/K})$$

Using expressions (18.15) and (18.16) a simple computer subroutine can be written to evaluate the Fourier series of a periodic function. The listing of such a program is shown in Fig. 18-7. Because subscripts of arrays in FORTRAN must be greater than zero, the time samples are stored in F(K) where F(1) corresponds to $f_0$; F(2) corresponds to $f_1$, etc. Similarly, the Fourier coefficients are stored in the array C(N) where C(1) corresponds to $c_0$, C(2) corresponds to $c_1$, etc.

We can use the subroutine FOURR to study the effect of using a truncated Fourier series to represent a periodic function. Before doing so however, we state without proof the least squares property of the Fourier series. Let

```
         SUBROUTINE FOURR(F,C,N,K,ITRAN)
C
C        THIS SUBROUTINE COMPUTES THE FOURIER COEFFICIENTS AND THE FOURIER
C        SERIES.  THE ARGUMENTS HAVE THE FOLLOWING MEANINGS:
C               F - A REAL ARRAY DIMENSIONED M IN THE CALLING PROGRAM WHICH
C                   CONTAINS THE VALUES OF THE TIME FUNCTION AT THE EQUALLY
C                   SPACED TIME POINTS 0, DELT, 2*DELT, ... ,(K-1)*DELT.
C               C - THE COMPLEX VALUED ARRAY DIMENSIONED M IN THE CALLING
C                   PROGRAM WHICH CONTAINS THE FOURIER COEFFICIENTS. C(1)
C                   CONTAINS THE COEFFICIENT OF THE DC TERM, C(2) CONTAINS
C                   THE COEFFICIENT OF THE FUNDAMENTAL HARMONIC, ETC.
C               K - THE NUMBER OF EQUALLY SPACED TIME POINTS.
C               N - THE NUMBER OF TERMS IN THE FOURIER SERIES.  THIS NUMBER
C                   MUST BE SUCH THAT N IS ODD AND 2*N .LE. K.
C               ITRAN - IF EQUAL TO 1 FOURIER COEFFICIENTS ARE COMPUTED FROM F
C                       IF EQUAL TO 2 FOURIER SERIES IS COMPUTED FROM C.
C
         IMPLICIT REAL*8 (A-H,O-Z)
         REAL*8 F(1), PI
         COMPLEX*16 C(1),W
         PI=4.*DATAN(1.D0)
         GO TO (5,6), ITRAN
5        DO 1 J=1,N
C
C        J TH COEFFICIENT IS STORED IN C(J+1), J=0,1,2,...,N
C
         C(J)=(0.,0.)
         DO 2 L=1,K
         W=DCMPLX(0.D0,2.*PI*(J-1)*(L-1)/K)
2        C(J)=C(J)+F(L)*CDEXP(-W)
1        C(J)=C(J)/K
         GO TO 7
6        DO 8 L=1,K
C
C        COMPUTE VALUE OF F AT J-TH TIME POINT AND STORE IN F(J+1), J=0,1,2...,N-1
C
         F(L)=C(1)
         DO 9 J=2,N
         W=DCMPLX(0.D0,2.*PI*(J-1)*(L-1)/K)
9        F(L)=F(L)+(C(J)*CDEXP(W)+DCONJG(C(J)*CDEXP(W)))
8        CONTINUE
7        RETURN
         END
```

**Fig. 18-7  Listing of the Subroutine FOURR which computes the Fourier coefficients and the Fourier series of a periodic function.**

$f_N(t)$ denote Fourier series of $f(t)$ truncated after the first $N$ terms. The *truncation error* is the difference between $f(t)$ and $f_N(t)$:

$$\varepsilon_N(t) = f(t) - f_N(t)$$

The *mean-squared error* is defined as

$$E_N = \frac{1}{T} \int_0^T [\varepsilon_N(t)]^2 \, dt$$

The truncated Fourier series has the property that no smaller $E_N$ may be found for another series with the same number of terms.

Given that no series can do better in the least squares sense than the Fourier series, the question that we are concerned with is how many terms in the truncated series are needed to accurately portray $f(t)$. The worst error, as might be expected, arises near a discontinuity. To illustrate this point we will use the subroutine FOURR to find the truncated Fourier series of the square-wave

$$f(t) = \begin{cases} 1 & 0 < t < \dfrac{1}{2} \\ -1 & \dfrac{1}{2} < t < 1 \end{cases}$$

At the points of discontinuity, $f(t) = 0$. A program that uses FOURR to determine the Fourier coefficients and the truncated Fourier series of the function $f(t)$ is shown in Fig. 18-8. Some words of explanation are in order.

We must first recognize that there are two places for errors to creep in: the truncation error, as was stated, and the discretization error associated with the forward Euler integration and sampling of $f(t)$. To minimize discretization error, we use 50 samples of $f(t)$. The sampled values of $f(t)$ are stored in the array FA. [*Note.* The $f(t)$ is defined by the function subprogram F(T).] FOURR is now called to compute the first 25 Fourier coefficients. (The reasoning behind computing only 25 coefficients will be discussed later.) We now compute the truncated Fourier series approximation to $f(t)$ at 200 time points using N terms of the Fourier series (one dc term and $N - 1$ harmonics). This approximation is stored in the array FH. The actual function $f(t)$ is also evaluated at these 200 time points for comparison purposes. The results are plotted using the subroutine PLOT. Some of these plots are shown in Figs. 18-9a to 18-9f. Observe that as the number of terms in the truncated series increases from $N = 3$ to $N = 21$, the better the approximation to the square wave. For the case $N = 21$, note that the approximation is poorest near the discontinuities at time points $t = 0$, .5, and 1. This effect is called the Gibbs phenomena.[1] Finally, note that contrary to what we might first expect the approximation for $N = 25$ is poorer than for $N = 21$. This rather curious result can be explained by the *sampling theorem*, which we state without proof.[2]

**Sampling Theorem**

Let the function $f(t)$ be bandlimited; in other words, $f(t)$ has no harmonic content outside of some frequency range, $\omega > |\omega_C|$. Then $f(t)$ can be uniquely

---

[1] For a more detailed discussion see E. A. Guillemin, *Mathematics of Circuit Analysis*, Wiley, 1949, pp. 482–496.

[2] A proof of this theorem is given in S. W. Director and R. A. Rohrer, *Introduction to System Theory*, McGraw-Hill, 1972, pp. 276–277.

```
C     PROGRAM TO ILLUSTRATE THE GIBBS PHENOMENON
C
      IMPLICIT REAL*8 (A-H,O-Z)
      REAL*8 FA(200),FH(200),TIM(200),Y(2,200)
      COMPLEX*16 C(100),W
C
C     SET NO. OF TIME POINTS K, EVALUATE F AT EACH POINT AND
C     STORE IN FA.
C
      K=50
      DELT=1./K
      DO 1 J=1,K
      TIM(J)=DELT*(J-1)
      FA(J)=F(TIM(J))
1     CONTINUE
      PRINT 5, ( TIM(J), FA(J), J=1,K)
5     FORMAT('1TIME FUNCTION',/,7X,'TIME',14X,'FA'/,(2X,2G14.3))
C
C     COMPUTE THE FIRSE 25 FOURIER COEFFICIENTS.
C
      CALL FOURR(FA,C,25,K,1)
      PRINT 3
3     FORMAT(///' FOURIER COEFFICIENTS',/,' TERM',5X,'REAL',8X,
     1'IMAGINARY',//)
      DO 4 J=1,25
      JM1=J-1
4     PRINT 6, JM1,C(J)
6     FORMAT(' ',I3,3X,2G14.3)
C
C     COMPUTE THE TRUNCATED FOURIER SERIES APPROXIMATION
C     TO F AND STORE IN FH.
C     USE 200 TIME POINTS FOR BETTER PLOTS.
C
      L=200
      DELT=1./L
      DO 11 J=1,L
      TIM(J)=DELT*(J-1)
      Y(1,J)=F(TIM(J))
11    CONTINUE
      DO 99 N=3,25,2
16    CALL FOURR(FH,C,N,L,2)
C
C     PLOT FH AND FA AS A FUNCTION OF TIME
C
      DO 22 J=1,L
22    Y(2,J)=FH(J)
      PRINT 9, K,N
9     FORMAT('1'///,' PLOT OF ACTUAL FUNCTION (X) AND APPROXIMAT FU'
     +,'NCTION (0) VERSUS TIME.',/' NUMBER OF TIME SAMPLE POINTS',
     +' USED TO GENERATE FOURIER COEFFICIENTS =',I4,/' NUMBER OF',
     +' TERMS IN FOURIER SERIES =',I4,//)
      CALL PLOT(TIM,Y,L,2,0)
99    CONTINUE
      END


      REAL FUNCTION F*8 (T)
      REAL*8 T
      IF(O..LT.T.AND.T.LT..5) F=1.
      IF(.5.LT.T.AND.T.LT.1.) F=-1
      IF(T.EQ.0..OR.T.EQ..5)  F=0
      IF(.4998.LT.T.AND..5002.GT.T) F=0.
      RETURN
      END
```

**Fig. 18-8  Program used to compute Fourier coefficients and truncated Fourier series of a square wave.**

PLOT OF ACTUAL FUNCTION (X) AND APPROXIMAT FUNCTION (O) VERSUS TIME.
NUMBER OF TIME SAMPLE POINTS USED TO GENERATE FOURIER COEFFICIENTS = 50
NUMBER OF TERMS IN FOURIER SERIES = 3

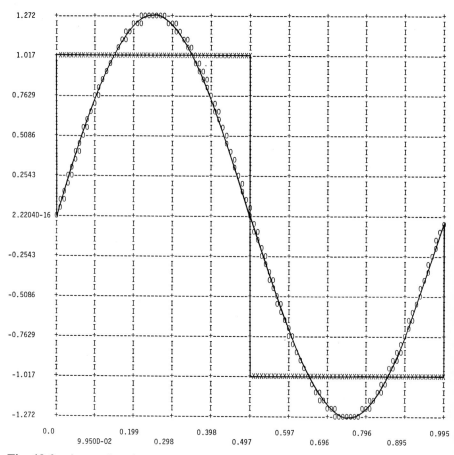

**Fig. 18-9  Approximation to square wave using 3 terms (dc plus two first harmonic terms) in the Fourier series.**

determined by its values

$$f_n = f\left(\frac{n\pi}{\omega_C}\right) \qquad n = 0, \pm1, \pm2, \pm3, \ldots$$

at a sequence of equidistant points that are $\pi/\omega_C$ apart.  □

If we assume that $N$ terms in the Fourier series (one dc term and $(N-1)$ harmonic terms) is adequate representation, then we have essentially assumed that $f(t)$ has no harmonic content for frequencies greater than $\omega_C = 2\pi(N-1)/T$. By the sampling theorem, $f(t)$ must be sampled at the points

$$\frac{n\pi}{\omega_C} = \frac{nT}{2(N-1)} \qquad n = 0, \pm1, \pm2, \pm3$$

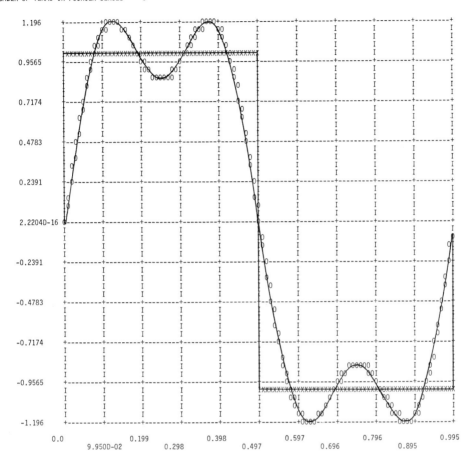

PLOT OF ACTUAL FUNCTION (X) AND APPROXIMAT FUNCTION (0) VERSUS TIME.
NUMBER OF TIME SAMPLE POINTS USED TO GENERATE FOURIER COEFFICIENTS =  50
NUMBER OF TERMS IN FOURIER SERIES =   5

**Fig. 18-9$b$** **Approximation to square wave using 5 terms (dc plus two first harmonic and two third harmonic) of the Fourier series.**

Because the particular function we are interested in is periodic with period $T$ we only need to evaluate $f(t)$ at the sample points

$$0, \frac{T}{2(N-1)}, \frac{2T}{2(N-1)}, \frac{3T}{2(N-1)}, \ldots, \frac{[2(N-1)-1]T}{2(N-1)}$$

since

$$f\left(\frac{2(N-1)T}{2(N-1)}\right) = f(0)$$

$$f\left(\frac{[2(N-1)+1]T}{2(N-1)}\right) = f\left(\frac{T}{2(N-1)}\right)$$

PLOT OF ACTUAL FUNCTION (X) AND APPROXIMAT FUNCTION (O) VERSUS TIME.
NUMBER OF TIME SAMPLE POINTS USED TO GENERATE FOURIER COEFFICIENTS =  50
NUMBER OF TERMS IN FOURIER SERIES =  7

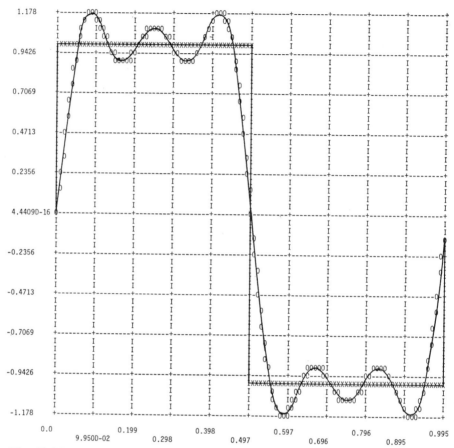

**Fig. 18-9(c)  Approximation to square wave using up to the 5th harmonic of the Fourier series.**

etc. Thus to adequately represent $f(t)$ with $N$ terms of a Fourier series (one dc term and $(N - 1)$ harmonic terms) we must sample $f(t)$ at least at $2(N - 1)$ equidistant points in one period $T$. Alternatively, we can say that if we sample $f(t)$ at $K$ equidistant time points, we can expect to determine accurately all Fourier series coefficients up to the coefficient corresponding to the $K/2$ harmonic. (*Note.* Counting the dc term there would then be $K/2 + 1$ terms in the Fourier series.)

For the example on hand, $K = 50$; so we would expect to at most be able to compute accurately the coefficient corresponding to the 25th harmonic. However as can be seen from Figs. 18-9e and 18-9f, the approximation using 21 terms appears to be superior than the approximation using 25 terms. This

PLOT OF ACTUAL FUNCTION (X) AND APPROXIMAT FUNCTION (O) VERSUS TIME.
NUMBER OF TIME SAMPLE POINTS USED TO GENERATE FOURIER COEFFICIENTS =  50
NUMBER OF TERMS IN FOURIER SERIES =  15

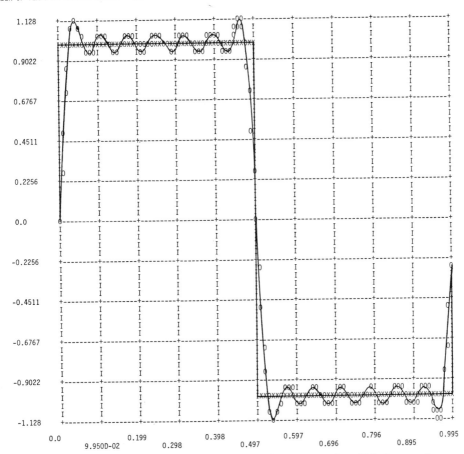

**Fig. 18-9d Approximation to square wave using up to the 13th harmonic term in the Fourier Series.**

situation results because we are "pushing the limit" in trying to predict accurately 25 harmonics using only 50 samples.

## Example 18.6

In this example we wish to illustrate how the subroutine FOURR can be combined with a phasor analysis program to determine the steady-state time domain response of a linear network that is under a nonsinusoidal periodic excitation. The network of Fig. 18-10a was first studied in Example 16-10. The program of Fig. 16-22 was used to determine the frequency response of the network. Now suppose we wished to determine the response of the network to the square wave input shown in Fig. 18-1 for periods of $T = 10^{-4}$ sec,

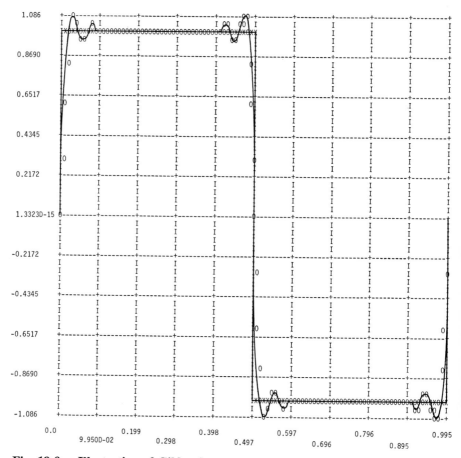

PLOT OF ACTUAL FUNCTION (X) AND APPROXIMAT FUNCTION (0) VERSUS TIME.
NUMBER OF TIME SAMPLE POINTS USED TO GENERATE FOURIER COEFFICIENTS = 50
NUMBER OF TERMS IN FOURIER SERIES = 21

**Fig. 18-9*e*  Illustration of Gibbs phenomenon.**

$10^{-5}$ sec, $10^{-6}$ sec, and $10^{-7}$ sec. The program of Fig. 18-10c is a modified version of the program of Fig. 16-22 that can be used for this purpose. Observe that the input waveform is specified in the function subprogram F(TIM, T).

Let us briefly review the steps followed in the main program. After calling the subroutine READ to read in the element values, the period of the input T, the number of points K used to sample the input, and the number of harmonics N to be used in the Fourier series representation of the input, the function F is called to generate the appropriate sampled value of the input. These values are stored in the array FA. The next step is to call the subroutine FOURR to determine the N Fourier series coefficients of the input. The

PLOT OF ACTUAL FUNCTION (X) AND APPROXIMAT FUNCTION (0) VERSUS TIME.
NUMBER OF TIME SAMPLE POINTS USED TO GENERATE FOURIER COEFFICIENTS = 50
NUMBER OF TERMS IN FOURIER SERIES = 25

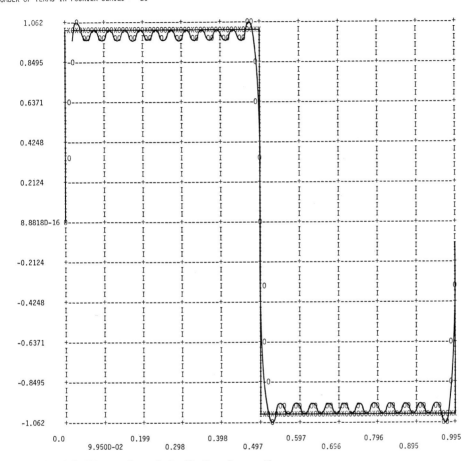

**Fig. 18-9***f*  **Illustration of the limits of sampling.**

network is then analyzed at each of the harmonic frequencies using the appropriate Fourier coefficient as the input voltage phasor V1. The results of these analyses are stored in the array VO. These output phasors are the Fourier series coefficients of the output $v_o(t)$. We evaluate the Fourier series of $v_o(t)$ at 200 time points by calling FOURR again and plotting it.

Plots of the input and response of the network for the four different input frequencies are shown in Fig. 18-11*a* to 18-11*d*. Observe that as the frequency of the input approaches, and exceeds, the bandwidth of the circuit (which is about $10^6$ Hz, see Fig. 16-21*b*), the output deteriorates from a square wave. Once again observe the Gibbs phenomenon in Figs. 18-11*a* and 18-11*b*.  □

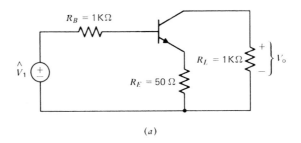

(a)

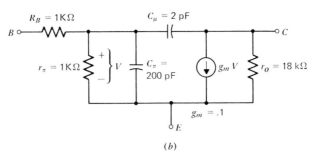

(b)

**Fig. 18-10(a)   The network studied in Example 18.6; (b) the transistor model.**

## 18-4   THE FOURIER INTEGRAL

We now wish to extend what has been done above to certain nonperiodic waveforms. Let $f(t)$ be an arbitrary function of time as shown in Fig. 18-12a. We artificially create a periodic signal $\hat{f}(t)$ by repeating the portion of $f(t)$ for which $0 < t < T$ as shown in Fig. 18-12b. Thus

$$\hat{f}(t) = f(t) \qquad \text{for} \qquad 0 \le t < T$$

and $\hat{f}(t)$ is periodic with period $T$. We can expand $\hat{f}(t)$ in a Fourier series:

(18.17)
$$\hat{f}(t) = \sum_{n=-\infty}^{\infty} c_n e^{jn\Delta\omega t}$$

where we have defined $\Delta\omega = 2\pi/T$ and

$$c_n = \frac{1}{T} \int_{-T/2}^{T/2} \hat{f}(t) e^{-jn\Delta\omega t}\, dt$$

Because $\hat{f}(t)$ is periodic with period $T$ we can rewrite this expression as

(18.18)
$$c_n = \frac{1}{T} \int_{0}^{T} \hat{f}(t) e^{-jn\Delta\omega t}\, dt$$

It is convenient to define

$$\omega \equiv n\Delta\omega$$

```
C       CHAPTER 18 EXAMPLE 18.6
C
C       DIMENSIONS
C
        IMPLICIT REAL*8(A-H,O-Z)
        COMPLEX*16 Y(5,5),CIN(5),JW,VO(51),V1
        COMPLEX*16 C(51)
        REAL*8 FREQ(51),G(51),P(51)
        REAL*8 FA(200),FH(200),TIM(200),YY(2,200)
C
C       READ IN ELEMENT VALUES,THE NUMBER OF HARMONICS TO BE
C       USED IN THE FOURIER SERIES REPRESENTATION OF THE INPUT,
C       THE PERIOD OF THE INPUT, AND THE NUMBER OF TIME POINTS
C       TO BE USED TO SAMPLE THE INPUT.
C
      4 CALL READ(RB,RX,RPI,CPI,CMU,GM,RO,RL,RE,N,T,K)
C
C       EVALUATE INPUT AT EACH TIME POINT AND STORE IN FA
C
        DELT=T/K
        DO 3 J=1,K
        TIM(J)=DELT*(J-1)
        FA(J)=F(TIM(J),T)
      3 CONTINUE
        PRINT 5,T,K,DELT,N
      5 FORMAT('1RESULTS FOR PROBLEM 18.6'/,' PERIOD=',G10.3,5X,
       +'NO. OF SAMPLE POINTS=',I3,5X,'SAMPLING INTERVAL=',G10.3,
       +5X,'NO. OF HARMONICS=',I3,///,' INPUT SIGNAL'/,2X,'TIME',
       +19X,'VOLTAGE INPUT')
        PRINT 6,(TIM(J),FA(J),J=1,K)
      6 FORMAT(' ',G12.5,2X,G12.5)
C
C       COMPUTE THE FIRST N FOURIER COEFFICIENTS OF THE INPUT
C
        CALL FOURR(FA,C,N,K,1)
        PRINT 7
      7 FORMAT(///' FOURIER COEFFICIENTS OF INPUT VOLTAGE'/,2X,
       +'HARMONIC',15X,'REAL PART',15X,'IMAGINARY PART')
        DO 10 J=1,N
        L=J-1
   10   PRINT 8,L,C(J)
      8 FORMAT(I5,19X,G12.5,12X,G12.5)
C
C       ANALYZE THE NETWORK AT EACH OF THE HARMONIC FREQUENCY POINTS
C       USING THE APPROPRIATE FOURIER COEFFICIENT AS THE INPUT PHASOR
C
        DO 1 J=1,N
        V1=C(J)
        FREQ(J)=(J-1)/T
        JW=DCMPLX(0.D0,6.2824*FREQ(J))
        CALL NODEQN(RB,RX,RPI,CPI,CMU,GM,RO,RL,RE,V1,JW,Y,CIN)
        CALL GAUSSC(Y,CIN,5,5)
C
C       STORE OUTPUT VOLTAGE PHASOR
C
        VO(J)=CIN(3)
      1 CONTINUE
```

**Fig. 18-10(c)  A program used to determine the steady-state response of the network of Fig. 18-10(a) for periodic inputs.**

```
          CALL PRINT(VO,FREQ,N,G,P)
C
C         COMPUTE THE TRUNCATED FOURIER SERIES APPROXIMATION
C         TO VO AND STORE IN FH
C         USE 200 TIME POINTS FOR BETTER PLOTS
C
          L=200
          CALL FOURR(FH,VO,N,L,2)
C
C         PLOT THE INPUT AND OUTPUT AS A FUNCTION OF TIME
C
          DELT=T/L
          DO 2 J=1,L
          TIM(J)=DELT*(J-1)
          YY(1,J)=F(TIM(J),T)
     2    YY(2,J)=FH(J)
          PRINT 11,T,K,DELT,N
    11    FORMAT('1RESULTS FOR PROBLEM 18.6'/,' PERIOD=',G10.3,5X,
         +'NO. OF SAMPLE POINTS=',I3,5X,'SAMPLING INTERVAL=',G10.3,
         +5X,'NO. OF HARMONICS=',I3,//)
          PRINT 9,T
     9    FORMAT(' PLOT OF INPUT (X) AND APPROXIMATE OUTPUT (O) '
         +,'VERSUS TIME. PERIOD OF THE INPUT IS',G12.5,' SECONDS..')
          CALL PLOT(TIM,YY,L,2,0)
          GO TO 4
          END




          SUBROUTINE READ(RB,RX,RPI,CPI,CMU,GM,RO,RL,RE,N,T,K)
          IMPLICIT REAL*8(A-H,O-Z)
          READ(5,1) RB,RX,RPI,CPI,CMU,GM,RO,RL,RE,N,T,K
     1    FORMAT(8G10.0)
          RETURN
          END




          SUBROUTINE NODEQN(RB,RX,RPI,CPI,CMU,GM,RO,RL,RE,V1,JW,Y,CIN)
          IMPLICIT REAL*8(A-H,O-Z)
          COMPLEX*16 Y(5,5),CIN(5),V1,JW
C
C         INITIALIZE COEFFICIENT ARRAYS
C
          DO 1 K=1,5
          CIN(K)=0.DO
          DO 1 I=1,5
     1    Y(K,I)=0.DO
C
C         SET UP NODAL EQUILIBRIUM EQUATION COEFFICIENTS
C
          Y(1,1)=-1./RB
          Y(1,5)=1.
          CIN(1)=(-1./RB)*V1
          Y(2,1)=(1./RB)+(1./RX)
          Y(2,2)=-1./RX
          CIN(2)=(1./RB)*V1
          Y(3,1)=-1./RX
```

**Fig. 18-10c (Continued)**

600

```
      Y(3,2)=(1./RX)+(1./RPI)+JW*(CMU+CPI)
      Y(3,3)=-JW*CMU
      Y(3,4)=-(1./RPI)-JW*CPI
      Y(4,2)=GM-JW*CMU
      Y(4,3)=(JW*CMU)+(1./RO)+(1./RL)
      Y(4,4)=-(1./RO)-GM
      Y(5,2)=-(1./RPI)-(JW*CPI)-GM
      Y(5,3)=-1./RO
      Y(5,4)=GM+(1./RPI)+(1./RE)+(1./RO)+JW*CPI
      RETURN
      END

      SUBROUTINE PRINT(VO,FREQ,NF,GAIN,PHASE)
      IMPLICIT REAL*8(A-H,O-Z)
      REAL*8 FREQ(51),GAIN(51),PHASE(51)
      COMPLEX*16 VO(50)
      PRINT 1
      DO 2 K=1,NF
C
C     OBTAIN THE MAGNITUDE, REAL AND IMAGINARY PARTS AND
C     THE PHASE OF VO(K)
C
      VOMAG=CDABS(VO(K))
      VOIMAG=DAIMAG(VO(K))
      VOREAL=DREAL(VO(K))
      IF(VOREAL.NE.O.) GO TO 4
      VOPHAS=0.
      GO TO 5
    4 VOPHAS=(180./3.1412)*DATAN2(VOIMAG,VOREAL)
C
C     COMPUTE THE MAGNITUDE AND PHASE OF THE OUTPUT
C
    5 GAIN(K)=VOMAG
      PHASE(K)=VOPHAS
      PRINT 3, FREQ(K),GAIN(K),PHASE(K)
    2 CONTINUE
    1 FORMAT(//' OUTPUT'//,25X,'MAGNITUDE',15X,'PHASE'/
     +49X,'IN'/2X,'FREQUENCY',38X,'DEGREES'/)
    3 FORMAT(G12.5,12X,G12.5,12X,G12.5)
      RETURN
      END
      REAL FUNCTION F*8 (TIM,T)
      REAL*8 TIM,T
      IF(0..LT.TIM.AND.TIM.LT.T/2.) F=1.
      IF(T/2..LT.TIM.AND.TIM.LT.T) F=-1.
      IF(TIM.EQ.O..OR.TIM.EQ.T/2.)  F=0.
      IF(.4998*T.LT.TIM.AND..5002*T.GT.TIM) F=0.
      RETURN
      END
```

----- DATA FOR PROGRAM OF EXAMPLE 18.6 -----

|      | 1000. | 100. | 1000. | 200.E-12 | 2.E-12 | .1 | 18.E3 | 1000. |
|------|-------|------|-------|----------|--------|-----|-------|-------|
| 50.  |       | 21   | 1.E-4 | 50       |        |     |       |       |
|      | 1000. | 100. | 1000. | 200.E-12 | 2.E-12 | .1 | 18.E3 | 1000. |
| 50.  |       | 21   | 1.E-5 | 50       |        |     |       |       |
|      | 1000. | 100. | 1000. | 200.E-12 | 2.E-12 | .1 | 18.E3 | 1000. |
| 50.  |       | 21   | 1.E-6 | 50       |        |     |       |       |
|      | 1000. | 100. | 1000. | 200.E-12 | 2.E-12 | .1 | 18.E3 | 1000. |
| 50.  |       | 21   | 1.E-7 | 50       |        |     |       |       |

**Fig. 18-10c (Continued)**

and rewrite (18.17) and (18.18) as

(18.19)
$$\hat{f}(t) = \sum_{\substack{\omega=n\Delta\omega \\ n=-\infty}}^{\infty} c_{\omega} e^{j\omega t}$$

and

(18.20)
$$c_{\omega} = \frac{1}{T} \int_0^T f(t) e^{-j\omega t} \, dt$$

We now define the function $F(j\omega)$ such that

$$F(j\omega) = Tc_{\omega}$$

so that (18.19) and (18.20) become

(18.21)
$$\hat{f}(t) = \sum_{\substack{\omega=n\omega \\ n=-\infty}}^{\infty} \frac{1}{2\pi} F(j\omega) e^{j\omega t} \Delta\omega$$

and

(18.22)
$$F(j\omega) = \int_0^T f(t) e^{-j\omega t} \, dt$$

If we now let $T \to \infty$ so that $f(t) = \hat{f}(t)$ for all time, $\Delta\omega$ will become infinitesimally small and the summation in (18.21) becomes an integral

(18.23)
$$f(t) = \frac{1}{2\pi} \int_{-\infty}^{\infty} F(j\omega) e^{j\omega t} \, d\omega$$

and (18.22) becomes

(18.24)
$$F(j\omega) = \int_{-\infty}^{\infty} f(t) e^{-j\omega t} \, dt$$

Expression (18.23) is known as the *Fourier integral* and $F(j\omega)$ is called the *Fourier transform* of $f(t)$. Expression (18.24) is known as the *inverse Fourier transform*. It is convenient to designate this relationship as

$$F(j\omega) = \mathscr{F}[f(t)]$$

and

$$f(t) = \mathscr{F}^{-1}[F(j\omega)]$$

RESULTS FOR PROBLEM 18.6
PERIOD= 0.100D-03      NO. OF SAMPLE POINTS= 50      SAMPLING INTERVAL= 0.500D-06      NO. OF HARMONICS= 21

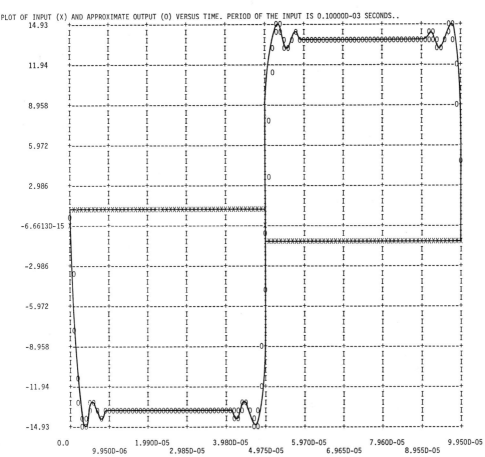

**Fig. 18-11a   The response of the network of Fig 18-10a to a square wave input with a period of $10^{-4}$ sec.**

What we have done above has not in any way been rigorous. Rigorous justification of the existence of the Fourier transform is beyond the scope of this text. However, suffice it to say that most any physical waveform of interest has a Fourier transform. A *sufficient* condition for a function $f(t)$ to have a Fourier transform is

$$\int_{-\infty}^{\infty} |f(t)| \, dt < \infty$$

However, this condition is not necessary. For example, we will show that the unit step function has a Fourier transform. We state, without proof, that

RESULTS FOR PROBLEM 18.6
PERIOD= 0.100D-04    NO. OF SAMPLE POINTS= 50    SAMPLING INTERVAL= 0.500D-07    NO. OF HARMONICS= 21

PLOT OF INPUT (X) AND APPROXIMATE OUTPUT (0) VERSUS TIME. PERIOD OF THE INPUT IS 0.10000D-04 SECONDS..

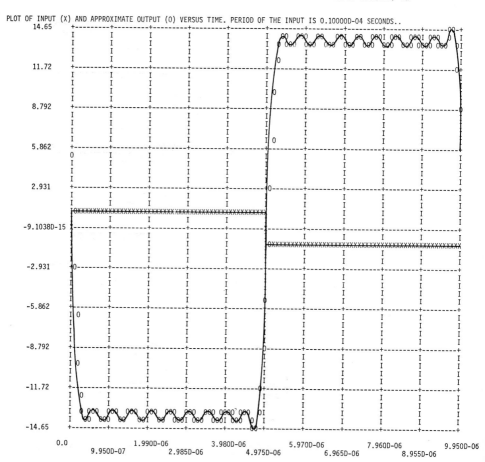

**Fig. 18-11*b*** **The response of the network of Fig. 18-10*a* to a square wave input with a period of $10^{-5}$ sec.**

the Fourier transform pair is unique. In other words, for a given $f(t)$ there is a unique $F(j\omega)$ and vice versa.

Evaluation of the Fourier transform for some common functions is shown below.

**Example 18.7**

Consider first the time function $f(t) = e^{-at}u(t)$. Observe that

$$f(t) = \begin{cases} e^{-at} & t \geq 0 \\ 0 & t < 0 \end{cases}$$

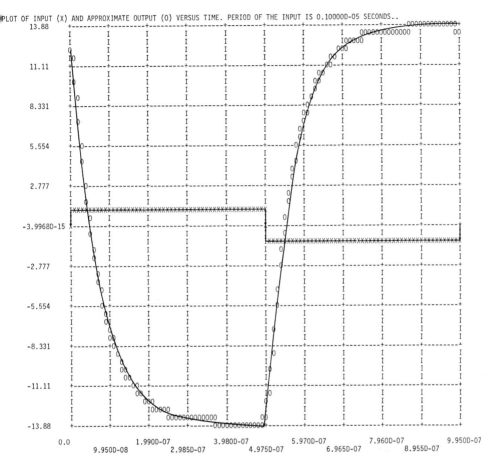

RESULTS FOR PROBLEM 18.6
PERIOD= 0.100D-05    NO. OF SAMPLE POINTS= 50    SAMPLING INTERVAL= 0.500D-08    NO. OF HARMONICS= 21

PLOT OF INPUT (X) AND APPROXIMATE OUTPUT (O) VERSUS TIME. PERIOD OF THE INPUT IS 0.10000D-05 SECONDS..

**Fig. 18-11c** The response of the network of Fig 18-10a to a square wave input with a peroid of $10^{-6}$ sec.

and that

$$\int_{-\infty}^{\infty} |f(t)|\ dt = \int_{0}^{\infty} e^{-at}\ dt$$

$$= -\frac{1}{a}\ e^{-at} \bigg]_{0}^{\infty}$$

$$= \frac{1}{a}$$

RESULTS FOR PROBLEM 18.6
PERIOD= 0.100D-06    NO. OF SAMPLE POINTS= 50    SAMPLING INTERVAL= 0.500D-09    NO. OF HARMONICS= 21

PLOT OF INPUT (X) AND APPROXIMATE OUTPUT (O) VERSUS TIME. PERIOD OF THE INPUT IS 0.10000D-06 SECONDS..

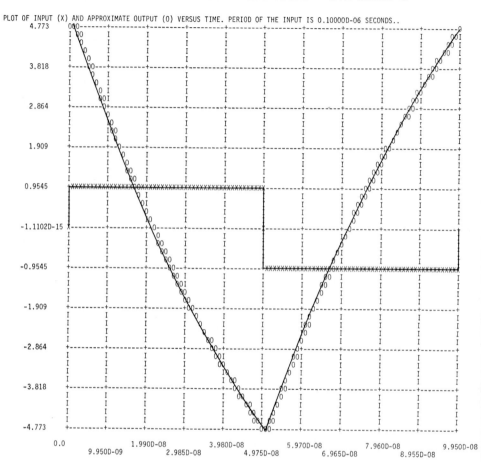

Fig. 18-11*d* **The response of the network of Fig. 18-10*a* to a square wave input with a period of $10^{-7}$ sec.**

so that the Fourier transform does exist. From (18.24) we have

$$F(j\omega) = \mathscr{F}[e^{-at}u(t)]$$

$$= \int_0^\infty e^{-at}e^{-j\omega t}\, dt$$

$$= \int_0^\infty e^{-(a+j\omega)t}\, dt$$

$$= \frac{1}{a + j\omega} \quad \square$$

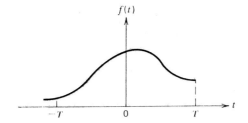

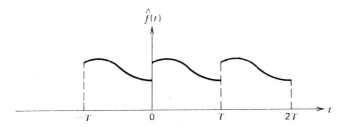

**Fig. 18-12** (*a*) **An arbitrary signal** $f(t)$. (*b*) **Periodic representation of this signal.**

### Example 18.8

Now consider the function $f(t) = (e^{-at} \cos \omega_0 t)u(t)$. It is easy to verify that

$$\int_{-\infty}^{\infty} |f(t)|\, dt < \infty$$

so that the Fourier transform does exist. From (18.24) we have

$$F(j\omega) = \int_{-\infty}^{\infty} (e^{-at} \cos \omega_0 t)u(t)e^{-j\omega t}\, dt$$

$$= \int_{0}^{\infty} \frac{1}{2} e^{-(a+j\omega)t}(e^{j\omega_0 t} + e^{-j\omega_0 t})\, dt$$

$$= \frac{1}{2}\int_{0}^{\infty} e^{-[a+j(\omega-\omega_0)]t}\, dt + \frac{1}{2}\int_{0}^{\infty} e^{-[a+j(\omega+\omega_0)]t}\, dt$$

$$= \frac{1}{2[a+j(\omega-\omega_0)]} + \frac{1}{2[a+j(\omega+\omega_0)]}$$

$$= \frac{a+j\omega}{(a+j\omega)^2 + \omega_0^2} \qquad \square$$

## Example 18.9

Consider the unit pulse signal of width $T$, defined by

$$f(t) = \begin{cases} 1 & -\dfrac{T}{2} \le t \le \dfrac{T}{2} \\ 0 & \text{elsewhere} \end{cases}$$

The Fourier transform exists because

$$\int_{-\infty}^{\infty} |f(t)|\, dt < \infty$$

We have

$$F(j\omega) = \int_{-\infty}^{\infty} f(t) e^{-j\omega t}\, dt$$

$$= \int_{-T/2}^{T/2} e^{-j\omega t}\, dt$$

$$= \frac{e^{j\omega T/2} - e^{-j\omega T/2}}{j\omega} = T\,\frac{\sin(\omega T/2)}{(\omega T/2)} \qquad \square$$

Recall that the Fourier series coefficients associated with a periodic signal have physical significance. In particular, a plot of these coefficients as a function of frequency represented the discrete frequency content or spectrum of the original signal. It is therefore reasonable to assume that the Fourier transform, which is a continuous function of the frequency variable $\omega$, has physical meaning. Suppose that $f(t)$ is either the voltage across or the current through a one ohm resistor. Then $f^2(t)$ represents the instantaneous power delivered to the resistor. The total energy delivered to this resistor is obtained by integrating the power over all time:

$$W = \int_{-\infty}^{\infty} f^2(t)\, dt$$

Assuming that $f(t)$ is Fourier transformable we can rewrite this expression as

$$W = \int_{-\infty}^{\infty} f(t) \left[ \frac{1}{2\pi} \int_{-\infty}^{\infty} F(j\omega) e^{j\omega t}\, d\omega \right] dt$$

$$= \frac{1}{2\pi} \int_{-\infty}^{\infty} F(j\omega) \left[ \int_{-\infty}^{\infty} f(t) e^{j\omega t}\, dt \right] d\omega$$

after an interchange of the order of integration. But

$$\int_{-\infty}^{\infty} f(t) e^{j\omega t}\, dt = F(-j\omega)$$

so that

$$W = \frac{1}{2\pi} \int_{-\infty}^{\infty} F(j\omega)F(-j\omega)\, d\omega$$

$$= \frac{1}{2\pi} \int_{-\infty}^{\infty} |F(j\omega)|^2\, d\omega$$

because since $f(t)$ is real, $F(-j\omega) = F^*(j\omega)$. Therefore, we have shown that

$$\int_{-\infty}^{\infty} f^2(t)\, dt = \frac{1}{2\pi} \int_{-\infty}^{\infty} |F(j\omega)|^2\, d\omega$$

This result is known as *Parseval's theorem*. Thus we see that the energy associated with $f(t)$ is obtained by integrating over all time in the time domain or $1/2\pi$ times an integration over all frequencies in the frequency domain.

To continue, we recognize that $|F(j\omega)|^2$ is an even function[3] of $\omega$ so that

$$W = \frac{1}{2\pi} \int_{-\infty}^{\infty} |F(j\omega)|^2\, d\omega = \frac{1}{\pi} \int_{0}^{\infty} |F(j\omega)|^2\, d\omega$$

since $\omega = 2\pi f$ we have

$$W = \int_{-\infty}^{\infty} |F(j2\pi f)|^2\, df = 2\int_{0}^{\infty} |F(j2\pi f)|^2\, df$$

If we divide the entire frequency range into vanishingly small increments of width $df$ then the area under $|F(j2\pi f)|^2$ over one increment is $|F(j2\pi f)|^2\, df$. The sum of all such areas as $f$ ranging over the range $-\infty$ to $\infty$ is the total one ohm energy contained in $f(t)$. Thus $|F(j2\pi f)|^2$ is the $(1 - \Omega)$ *energy density* or energy per unit bandwidth of $f(t)$. We can determine that portion of the total energy lying within any interval by integrating $|F(j2\pi f)|^2$ over that interval.

## 18-5   PROPERTIES OF THE FOURIER TRANSFORM

Let us now investigate some of the properties of the Fourier transform that make it so useful for network analysis. Additional properties are discussed in the problems. In the following we assume that $F(j\omega)$ and $G(j\omega)$ are the Fourier transforms of the real-valued functions $f(t)$ and $g(t)$, respectively.

1. *Linearity*

(18.25) $$\mathscr{F}[\alpha f(t) + \beta g(t)] = \alpha F(j\omega) + \beta G(j\omega)$$

[3] $|F(-j\omega)|^2 = |F(j\omega)|^2$

Proof:

$$\mathcal{F}[\alpha f(t) + \beta g(t)] = \int_{-\infty}^{\infty} [\alpha f(t) + \beta g(t)] e^{-j\omega t} \, dt$$

$$= \alpha \int_{-\infty}^{\infty} f(t) e^{-j\omega t} \, dt + \beta \int_{-\infty}^{\infty} g(t) e^{-j\omega t} \, dt$$

$$= \alpha F(j\omega) + \beta G(j\omega)$$

2. *Scale change*

(18.26)     $$\mathcal{F}[f(\alpha t)] = \frac{1}{|\alpha|} F\left(\frac{j\omega}{\alpha}\right), \qquad \text{where } \alpha \text{ is a constant.}$$

Proof:

$$\mathcal{F}[f(\alpha t)] = \int_{-\infty}^{\infty} f(\alpha t) e^{-j\omega t} \, dt$$

Let $\tau = \alpha t$ so that $t = \tau/\alpha$ and $dt = (1/\alpha) \, d\tau$

Then

$$\int_{-\infty}^{\infty} f(\alpha t) e^{-j\omega t} \, dt = \begin{cases} \dfrac{1}{\alpha} \displaystyle\int_{-\infty}^{\infty} f(\tau) e^{-j(\omega/\alpha)\tau} \, d\tau & \text{if} \quad \alpha > 0 \\[2em] -\dfrac{1}{\alpha} \displaystyle\int_{-\infty}^{\infty} f(\tau) e^{-j(\omega/\alpha)\tau} \, d\tau & \text{if} \quad \alpha < 0 \end{cases}$$

Therefore

$$\mathcal{F}F[f(\alpha t)] = \frac{1}{|\alpha|} F\left(\frac{j\omega}{\alpha}\right)$$

3. *Time shift*

(18.27)     $$\mathcal{F}[f(t - \tau)] = e^{-j\omega\tau} F(j\omega)$$

Proof:

$$\mathcal{F}[f(t - \tau)] = \int_{-\infty}^{\infty} f(t - \tau) e^{-j\omega t} \, dt$$

Let $\sigma = t - \tau$, so that $t = \sigma + \tau$ and $dt = d\sigma$. Then

$$\int_{-\infty}^{\infty} f(t - \tau) e^{-j\omega t} \, dt = \int_{-\infty}^{\infty} f(\sigma) e^{-j\omega(\sigma+\tau)} \, d\sigma$$

$$= e^{-j\omega\tau} \int_{-\infty}^{\infty} f(\sigma) e^{-j\omega\sigma} \, d\sigma$$

$$= e^{-j\omega\tau} F(j\omega)$$

4. *Frequency shifting*

(18.28)     $$\mathcal{F}[e^{j\omega_0 t} f(t)] = F[j(\omega - \omega_0)]$$

Proof:

$$\mathscr{F}[e^{j\omega_0 t}f(t)] = \int_{-\infty}^{\infty} e^{j\omega_0 t}f(t)e^{-j\omega t}\,dt$$

$$= \int_{-\infty}^{\infty} f(t)e^{-j(\omega-\omega_0)t}\,dt$$

$$= F[j(\omega - \omega_0)]$$

5. *Time differentiation*

(18.29)
$$\mathscr{F}\left[\frac{df(t)}{dt}\right] = j\omega F(j\omega)$$

and

$$\mathscr{F}\left[\frac{d^n f(t)}{dt^n}\right] = (j\omega)^n F(j\omega)$$

Proof:

$$f(t) = \frac{1}{2\pi}\int_{-\infty}^{\infty} F(j\omega)e^{j\omega t}\,d\omega$$

Differentiating both sides with respect to $t$ yields

$$\frac{df(t)}{dt} = \frac{1}{2\pi}\int_{-\infty}^{\infty} j\omega F(j\omega)e^{j\omega t}\,d\omega$$

which establishes the result. Generalization to the $n$th derivative follows immediately.

6. *Miscellaneous*

(18.30)         (a) $F^*(j\omega) = F(-j\omega)$ for real $f(t)$

(18.31)         (b) $f(t)$ even $\Rightarrow F(j\omega)$ real

(18.32)         (c) $f(t)$ odd $\Rightarrow F(j\omega)$ imaginary

Proof of the above is left to the problems.

(18.33)         (d) $F(t) = 2\pi\mathscr{F}^{-1}[f(-\omega)]$

Proof

Observe that $t$ and $\omega$ may be interchanged in the expression

$$F(\omega) = \int_{-\infty}^{\infty} f(t)e^{-j\omega t}\,dt$$

to yield

$$F(t) = \int_{-\infty}^{\infty} f(\omega)e^{-jt\omega}\,d\omega$$

Now let $\omega \to -\omega$ so that $d\omega \to -d\omega$ then

$$F(t) = \int_{-\infty}^{\infty} f(-\omega)e^{j\omega t}\, d\omega$$

$$= 2\pi \left[ \frac{1}{2\pi} \int_{-\infty}^{\infty} f(-\omega)e^{j\omega t}\, d\omega \right]$$

$$= 2\pi \mathscr{F}^{-1}[f(-\omega)]$$

We observed earlier that the Fourier transform does not exist in the ordinary sense for the unit impulse and unit step. Since these two functions are quite useful for network studies, it is quite convenient to extend the Fourier transform to these functions as well. Recall from Chapter 12 that the unit impulse $\delta(t)$ is defined by its behavior on integration. Given an ordinary function $f(t)$ the unit impulse is "defined" by

$$\int_{-\infty}^{\infty} f(t)\, \delta(t)\, dt = f(0)$$

Using this definition we can formally find the Fourier transform of $\delta(t)$:

(18.34a) $$\mathscr{F}[\delta(t)] = \int_{-\infty}^{\infty} \delta(t)e^{-j\omega t}\, dt = 1$$

Furthermore

(18.34b) $$\mathscr{F}[\delta(t - \tau)] = \int_{-\infty}^{\infty} \delta(t - \tau)e^{-j\omega t}\, dt = e^{-j\omega \tau}$$

Now consider the function $f(t) = 1$. Observe that

$$\int_{-\infty}^{\infty} |f(t)|\, dt = \infty$$

However we may employ (18.33) to find $F(\omega)$. Since we have

$$F[\delta(t)] = 1$$

then

$$1 = 2\pi \mathscr{F}^{-1}[\delta(-\omega)]$$

or

(18.35) $$\mathscr{F}(1) = 2\pi\delta(-\omega)$$

$$= 2\pi\delta(\omega)$$

since the unit impulse is an even function.

Another useful singularity function is the *signum function*, sgn($t$), defined by

$$\text{sgn}\,(t) = \begin{cases} -1 & t < 0 \\ 1 & t > 0 \end{cases}$$

In order to find the Fourier transform of sgn $(t)$ we observe that

$$\text{sgn}\,(t) = \lim_{\substack{\alpha \to 0 \\ \alpha > 0}} [e^{-\alpha t}u(t) - e^{\alpha t}u(-t)]$$

Then

$$\mathscr{F}[\text{sgn}\,(t)] = \lim_{\alpha \to 0} \left( \int_0^\infty e^{-\alpha t}e^{-j\omega t}\,dt - \int_{-\infty}^0 e^{\alpha t}e^{-j\omega t}\,dt \right)$$

(18.36)

$$= \lim_{\alpha \to 0} \left( \frac{-2j\omega}{\alpha^2 + \omega^2} \right) = \frac{2}{j\omega}$$

The Fourier transform of the unit step involves some additional complication. Since

$$\delta(t) = \frac{d}{dt}\,u(t)$$

then from (18.29)

$$\mathscr{F}\left[\frac{d}{dt}\,u(t)\right] = j\omega\mathscr{F}[u(t)]$$

or

(18.37) $$j\omega\mathscr{F}[u(t)] = 1$$

We must be careful at this point because we are not dealing with ordinary functions. Observe that (18.37) is identical to the expression

$$j\omega\mathscr{F}[u(t)] = 1 + j\omega k\delta(\omega)$$

in a general sense ($k$ is a constant) because if both sides were multiplied by some function $g(\omega)$ and integrated

$$\int_{-\infty}^\infty j\omega k\,\delta(\omega)g(\omega)\,d\omega = 0$$

Therefore, we conclude that the Fourier transform of the unit step is

$$\mathscr{F}[u(t)] = \frac{1}{j\omega} + k\,\delta(\omega)$$

In order to find the appropriate value of $k$ we recognize that the unit step may be written as

$$u(t) = \frac{1}{2} + \frac{1}{2}\,\text{sgn}\,(t)$$

so that

$$\mathscr{F}[u(t)] = \frac{1}{2}\,\mathscr{F}(1) + \frac{1}{2}\,\mathscr{F}[\text{sgn}\,(t)]$$

(18.38)

$$= \pi\,\delta(\omega) + \frac{1}{j\omega}$$

from (18.35) and (18.36).

These Fourier transform pairs can be used to easily evaluate the Fourier transform of the sine and cosine functions.

### Example 18.10

Consider the Fourier transform of function $e^{j\omega_0 t}$

$$\mathcal{F}(e^{j\omega_0 t}) = \int_{-\infty}^{\infty} e^{j\omega_0 t} e^{-j\omega t} dt$$

$$= \int_{-\infty}^{\infty} e^{-j(\omega - \omega_0) t} dt$$

$$= 2\pi \, \delta(\omega - \omega_0)$$

which follows directly from (18.28) and (18.35).    □

### Example 18.11

Now consider the sinusoidal functions $\cos \omega_0 t$ and $\sin \omega_0 t$. We have

$$\mathcal{F}(\cos \omega_0 t) = \int_{-\infty}^{\infty} \cos \omega_0 t \, e^{-j\omega t} dt$$

$$= \frac{1}{2} \int_{-\infty}^{\infty} (e^{j\omega_0 t} + e^{-j\omega_0 t}) e^{-j\omega t} dt$$

$$= \frac{1}{2} \left[ \int_{-\infty}^{\infty} e^{-j(\omega - \omega_0) t} dt + \int_{-\infty}^{\infty} e^{-j(\omega + \omega_0) t} dt \right]$$

$$= \pi[\delta(\omega - \omega_0) + \delta(\omega + \omega_0)]$$

and

$$\mathcal{F}(\sin \omega_0 t) = \int_{-\infty}^{\infty} \sin \omega_0 t \, e^{-j\omega t} dt$$

$$= \frac{1}{2j} \int_{-\infty}^{\infty} (e^{j\omega_0 t} - e^{-j\omega_0 t}) e^{-j\omega t} dt$$

$$= \frac{1}{2j} \left[ \int_{-\infty}^{\infty} e^{-j(\omega - \omega_0) t} dt - \int_{-\infty}^{\infty} e^{-j(\omega + \omega_0) t} dt \right]$$

$$= j\pi[\delta(\omega + \omega_0) - \delta(\omega - \omega_0)]$$

These transforms agree with physical reasoning since we expect that pure sinusoids should have all their energy concentrated about a single frequency.

□

Table 18.1 lists some common time functions and their Fourier transforms.

**Table 18-1**    Fourier transforms of some common functions

| $f(t)$ | $F(j\omega)$ |
|---|---|
| $e^{-at}u(t) \ \text{Re} \ [a] > 0$ | $\dfrac{1}{a + j\omega}$ |
| $(e^{-at} \cos \omega_0 t)u(t)$ | $\dfrac{a + j\omega}{(a + j\omega)^2 + \omega_0^2}$ |
| $(e^{-at} \sin \omega_0 t)u(t)$ | $\dfrac{\omega_0}{(a + j\omega)^2 + \omega_0^2}$ |
| $p_T(t) = \begin{cases} 1 & -T/2 \le t \le T/2 \\ 0 & \text{elsewhere} \end{cases}$ | $\dfrac{T \sin (\omega T/2)}{\omega T/2}$ |
| $e^{-a\lvert t \rvert}, \ a > 0$ | $\dfrac{2a}{a^2 + \omega^2}$ |
| $1$ | $2\pi\delta(\omega)$ |
| $e^{j\omega_0 t}$ | $2\pi\delta(\omega - \omega_0)$ |
| $\text{sgn} \ t$ | $\dfrac{2}{j\omega}$ |
| $u(t)$ | $\dfrac{1}{j\omega} + \pi\delta(\omega)$ |
| $\cos \omega_0 t$ | $\pi[\delta(\omega - \omega_0) + \delta(\omega + \omega_0)]$ |
| $\sin \omega_0 t$ | $j\pi[\delta(\omega + \omega_0) - \delta(\omega - \omega_0)]$ |
| $\delta(t)$ | $1$ |

## 18-6    THE FOURIER TRANSFORM IN NETWORK ANALYSIS

Consider the $n$th order linear time-invariant network $\mathcal{N}$ that has a single input, $x(t)$, and a single output $y(t)$. The output can be any voltage or current in the network. Suppose we wish to determine the output under the assumption that all initial conditions are zero. The output can be found from knowledge of $x(t)$ in one of several ways. First, of course, is to use the standard nodal analysis, or equivalent, method in which appropriate differential and algebraic equations as written and solved. For high-order networks an analytic solution is usually difficult to obtain but numerical techniques can be employed. If, however, the input is of the form $x(t) = Xe^{j\omega t}$, and we only desire the steady-state response, we can employ phasor analysis so that only complex algebraic equations need be solved, avoiding the necessity of solving differential equations. Even if the input has the somewhat more general form, $x(t) = Xe^{st}$, where $s = \sigma + j\omega$, we can still employ phasor analysis to determine the forced response and thereby avoid direct solution of differential equations. In this section we show that if the input $x(t)$ is Fourier transformable we can also determine the output $y(t)$ without having to solve differential equations directly.

Recall from Chapters 14 and 17 that, in general, a network that contains $n$ energy storage elements requires solution of an $n$th order-differential equation providing that there are no loops of capacitors and voltage sources or cutsets of inductors and current sources. Typically the differential equation that relates the output $y(t)$ to the input $x(t)$ has the form

(18.39)

$$(b_n D^n + b_{n-1} D^{n-1} + \cdots + b_0) y(t) = (a_m D^m + a_{m-1} D^{m-1} + \cdots + a_0) x(t)$$

or

$$b_n \frac{d^n y(t)}{dt^n} + b_{n-1} \frac{d^{n-1} y(t)}{dt^{n-1}} + \cdots + b_0 y(t)$$

$$= a_m \frac{d^m y(t)}{dt^m} + a_m \frac{d^{m-1} y(t)}{dt^m} + \cdots + a_0 y(t)$$

We may take the Fourier transform of this expression, and employing linearity and (18.29) we have

(18.40)   $$[b_n (j\omega)^n + b_{n-1}(j\omega)^{n-1} + \cdots + b_0] Y(\omega)$$

$$= [a_m (j\omega)^m + a_{m-1}(j\omega)^{m-1} + \cdots + a_0] X(\omega)$$

where

$$Y(j\omega) = \mathcal{F}[y(t)] \quad \text{and} \quad X(j\omega) = \mathcal{F}[x(t)]$$

So the Fourier transform of the output is easily determined from (18.40)

(18.41)   $$Y(j\omega) = \frac{[b_n (j\omega)^n + b_{n-1}(j\omega)^{n-1} + \cdots + b_0]}{[a_m (j\omega)^m + a_{m-1}(j\omega)^{m-1} + \cdots + a_0]} X(j\omega)$$

Thus we see that if we know the differential equation that relates the input and output, and if the input is Fourier transformable, we can easily find the Fourier transform of the output. To determine the output $y(t)$ we may take the inverse Fourier transform of $Y(\omega)$.

**Example 18.12**

Consider the $RLC$ network of Fig. 18-13. The input $i_1(t)$ is Fourier transformable, and the output is $i_4(t)$. The differential equation that relates $i_4(t)$ and $i_1(t)$ is

$$\left( R_4 C_3 D^2 + D + \frac{R_4}{L_2} \right) i_4(t) = D i_1(t)$$

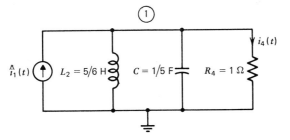

**Fig. 18-13  Network studied in Example 18.12.**

or on substitution of element values

$$\left[\frac{1}{5}(j\omega)^2 + j\omega + \frac{6}{5}\right]I_4(\omega) = j\omega\hat{I}_1(\omega)$$

or

$$I_4(j\omega) = \frac{j\omega}{(j\omega)^2 + 5j\omega + 6}\hat{I}_1(\omega)$$

Suppose $\hat{i}_1(t) = e^{-t}u(t)$. From Table 18.1 we know that

$$\hat{I}_1(j\omega) = \frac{1}{(1 + j\omega)}$$

so

$$I_4(j\omega) = \left[\frac{j\omega}{(j\omega)^2 + 5j\omega + 6}\right]\left[\frac{1}{(1 + j\omega)}\right] \qquad \square$$

Observe that we can rewrite (18.41) as

$$Y(j\omega) = H(j\omega)X(j\omega)$$

where

$$H(j\omega) = \frac{[b_n(j\omega)^n + b_{n-1}(j\omega)^{n-1} + \cdots + b_0]}{[a_m(j\omega)^m + a_{m-1}(j\omega)^{m-1} + \cdots + a_0]}$$

is the network's transfer function. Moreover, if the input is the unit impulse, $x(t) = \delta(t)$, then $X(j\omega) = 1$, and $Y(j\omega) = H(j\omega)$. Therefore, the transfer function is the Fourier transform of the impulse response.

Often we are satisfied with the knowledge of the frequency domain behavior of a circuit. If, however, we need the time domain behavior we have to determine the inverse Fourier transform. To find the inverse Fourier transform of a function $F(j\omega)$ we may employ (18.23) directly. However, this integral is usually difficult to evaluate. An alternative is to separate $F(j\omega)$ into common functional forms and use a table such as Table 18.1 to find the inverse transform.

**Example 18.13**

Consider again Example 18.12. We found

$$I_4(j\omega) = \frac{j\omega}{[(j\omega)^2 + 5j\omega + 6](1 + j\omega)^2}$$

$$= \frac{j\omega}{(j\omega + 3)(j\omega + 2)(1 + j\omega)}$$

We can effect a partial fraction expression as follows:

$$I_4(j\omega) = \frac{A}{(j\omega + 3)} + \frac{B}{(j\omega + 2)} + \frac{C}{(j\omega + 1)}$$

$$= \frac{A(j\omega + 2)(j\omega + 1) + B(j\omega + 3)(j\omega + 1) + C(j\omega + 3)(j\omega + 2)}{(j\omega + 3)(j\omega + 2)(j\omega + 1)}$$

Equating powers of $j\omega$ we find that

$$A + B + C = 0$$

$$3A + 4B + 5C = 1$$

$$2A + 3B + 6C = 0$$

so that $A = -3/2$, $B = 2$, and $C = -1/2$ and

$$I_4(j\omega) = \frac{-3}{2(j\omega + 3)} + \frac{2}{(j\omega + 2)} - \frac{1}{2(j\omega + 1)}$$

Using Table 18.1 we find that

$$i_4(t) = \left(-\frac{3}{2}e^{-3t} + 2e^{-2t} - \frac{1}{2}e^{-t}\right)u(t) \quad \square$$

**PROBLEMS**

18.1   Find the trigonometric Fourier series for each of the functions shown in Fig. P18.1. Sketch the discrete frequency spectrum.

18.2   Show that if a periodic function $f(t)$ is an even function, that is, $f(-t) = f(t)$, then only cosine terms and a constant are present in the trigonometric Fourier series representation of $f(t)$.

18.3   Show that if a periodic function $f(t)$ is an odd function, that is, $f(-t) = -f(t)$, then only sine terms are present in the trigonometric Fourier series representation of $f(t)$.

18.4   Show that if a periodic function $f(t)$ displays half-wave symmetry, that is, $f(t) = -f(t \pm T/2)$, only odd harmonic terms are present in the trigonometric Fourier series representation of $f(t)$.

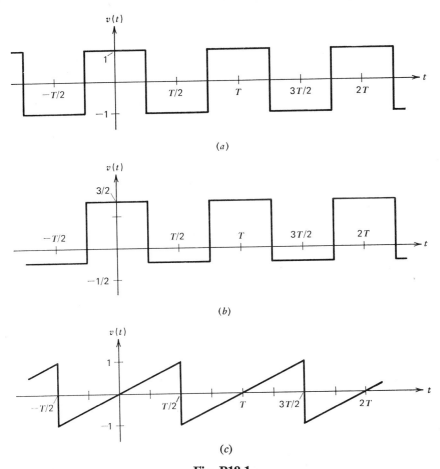

(a)

(b)

(c)

**Fig. P18.1**

**18.5**   Find the complex Fourier series for each of the waveforms shown in Fig. P18.1.

**18.6**   Determine the response of the network shown in Fig. P18.6 due to each of the wave forms of Fig. P18.1. (*Hint.* Use a truncated Fourier series.)

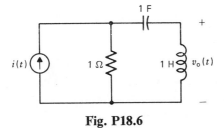

**Fig. P18.6**

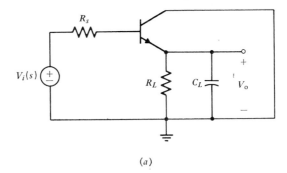

(a)

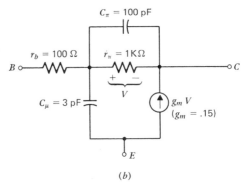

(b)

**Fig. P18.11**

**18.7**    Modify the program of Fig. 18-8 and study the effects of using a truncated Fourier series to represent the waveforms of Figs. P18.1(b) and P18.1(c) for $T = 1$. (*Note.* Only the function subprogram F(T) need be modified.)

**18.8**    Using the program of Fig. 18-10, study the effect of the number of time sample points $K$ and the number of terms in the Fourier series used to represent the input $N$ on the predicted output. In particular compare the results for the following cases:

(a) $K = 50, N = 3$
(b) $K = 50, N = 5$
(c) $K = 50, N = 25$
(d) $K = 22, N = 3$
(e) $K = 22, N = 5$
(f) $K = 22, N = 11$

**18.9**    Modify the subprogram function F in the program of Fig. 18-10, and use the program to find the steady-state response of the network of Fig. 18-9 under a triangular wave input, similar to the one of Fig. 18-2, whose period is $10^{-4}$ sec, $10^{-5}$ sec, $10^{-6}$ sec, and $10^{-7}$ sec.

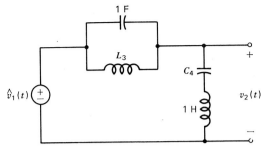

**Fig. P18.12**

**18.10**  Repeat Prob. 18.9 for the waveforms of Figs. P18.1*b* and P18.1*c*.
**18.11**  Modify the program of Fig. 18-10 and use it to determine the steady state response of the emitter follower circuit of Fig. P18.11 to a square input whose period is $10^{-6}$ seconds.
**18.12**  In the circuit of Fig. P18.12 suppose the voltage source $\hat{v}_1$ represents the output of another circuit that has unwanted harmonic components at $\omega = 3$ and 5 rad/sec. Find the values of $L_3$ and $C_4$ such that $v_2(t)$ does not contain these frequency components.
**18.13**  Find the Fourier transforms of the following functions:

(a) $f(t) = [e^{-t} - e^{-3t}]u(t)$

(b) $f(t) = \delta(t) - e^{-2t}u(t)$

(c) $f(t) = e^{-t}\cos\left(3t + \dfrac{\pi}{4}\right)u(t)$

**18.14**  Suppose a time function $f(t)$ has a Fourier transform of

$$F(j\omega) = \frac{1 + j\omega}{8 - \omega^2 + j6\omega}$$

Find the Fourier transform of

(a) $f(2t)$

(b) $f(t - 3)$

(c) $f(2t - 3)$

(d) $3f\left(\dfrac{t}{2}\right)$

**18.15**  Determine the $(1-\Omega)$ energy density contained in the function $te^{-2t}u(t)$. What fraction of this energy lies in the frequency band $-\omega_0 < \omega < \omega_0$?
**18.16**  Repeat Prob. 18.13 for the time functions $3e^{-2t}u(t)$ and $3e^{-2t}$.
**18.17**  Let $\hat{v}_1(t) = u(t)$ in the initially relaxed circuit of Fig. P18.15 and find $v_2(t)$ using Fourier transforms. Show that if $\hat{v}_1(t) = \delta(t)$, the response $v_2(t)$ is the derivative of what it was before.

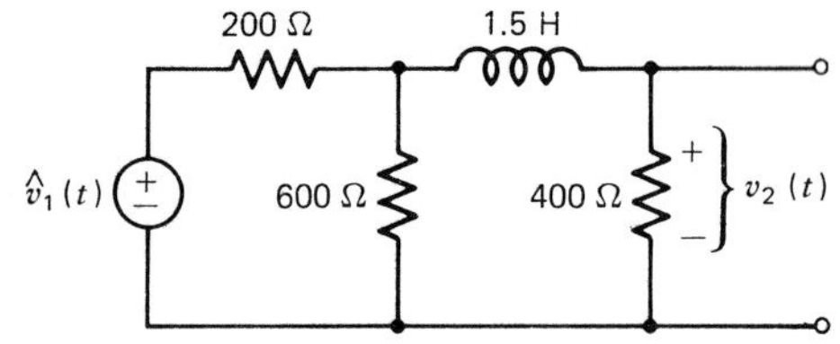

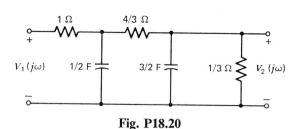

**Fig. P18.15**

**Fig. P18.20**

**18.18**     Sketch the time functions whose Fourier transforms are

(a) $\dfrac{j\omega}{(1+\omega^2)}$

(b) $\dfrac{1}{(2+j\omega)(1+j\omega)}$

(c) $\dfrac{1}{1-\omega^2+j\omega}$

**18.19**     Find the Fourier transforms of the periodic waves of Fig. 18-1 and 18-2. Can you draw any conclusions about the frequency spectrum of the Fourier transform of a periodic wave?

**18.20**     Find the transfer functions $H(j\omega) = V_1(j\omega)/V_2(j\omega)$ for the network of Fig. P18.20.

## CHAPTER 19

# Introduction to Computer-Aided Design

The majority of this text has been concerned with the analysis of circuits. In this chapter we will introduce some elementary circuit design considerations, and show how the computer may be employed in the design process. So that we can keep the discussion on an introductory level we will consider only the frequency domain design of networks that contain $R$, $L$, $C$'s and voltage controlled current sources. In particular, we will be able to design simple active networks.

## 19-1 THE PERFORMANCE FUNCTION

Whereas the designer is capable of making a qualitative judgment as to how well a circuit is performing, the computer can only make a quantitative judgment. In a circuit design program, the *performance function* is used to assign a numerical value to circuit behavior. This function is a measure of the error between the actual network response and the desired network response. The object of a computer-aided design program is to adjust the elements of a network so as to reduce, that is, minimize, the value of the performance function. In effect the original network design problem is recast into a *mathematical optimization problem* as will be seen below.

Suppose we wish to design an *RLC* network to have a specified voltage transfer function $H(j\omega)$. In other words, we wish to find a network $\mathcal{N}$, shown

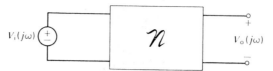

**Fig. 19-1    A network $\mathcal{N}$ is to be found so that**

$$V_0(j\omega)/V_i(j\omega) = H(j\omega)$$

in Fig. 19-1, such that

$$\frac{V_o(j\omega)}{V_i(j\omega)} = H(j\omega)$$

for all values of $\omega$. Since we are working on the computer, we can evaluate $V_o(j\omega)$ at only a finite number of discrete frequency points. The frequency points are denoted by $\omega_1, \omega_2, \ldots, \omega_N$. Therefore, we desire that

$$\frac{V_o(j\omega_k)}{V_i(j\omega_k)} = H(j\omega_k)$$

for $k = 1, 2, \ldots, N$.

The previous design problem can be recast into a mathematical optimization problem that can be solved by the computer as follows. First, recognize that the desired design will have been realized if we apply a unity excitation to $\mathcal{N}$, that is, $V_i(j\omega_k) = 1$, and the output is $H(j\omega_k)$. Thus the desired output phasor, under unity excitation is $H(j\omega_k)$, which we will denote by $\hat{V}_o(j\omega_k)$. For any network under unit excitation, let $V_o(j\omega_k)$ represent the actual output phasor at frequency $\omega_k$. Then a measure of the deviation of a network performance from the desired performance is the squared error function

$$(19.1a) \qquad \varepsilon = \frac{1}{2}\sum_{k=1}^{N}|V_o(j\omega_k) - \hat{V}_o(j\omega_k)|^2$$

or

$$(19.1b) \qquad \varepsilon = \frac{1}{2}\sum_{k=1}^{N}[|V_o(j\omega_k)| - |\hat{V}_o(j\omega_k)|]^2$$

if only the magnitude of the output is to be specified. If $\mathbf{P}$ denotes the vector of adjustable elements in $\mathcal{N}$ (i.e., $\mathbf{p}$ contains $R$'s, $L$'s, and $C$'s) then

$$\varepsilon = \varepsilon(\mathbf{p})$$

because the output $V_o(j\omega)$ depends upon the network elements. The network design problem has thus been transformed into a mathematical optimization problem: adjust the network elements $\mathbf{p}$ in such a way so as to minimize $\varepsilon(\mathbf{p})$.

**Example 19.1**

As a specific example of a performance function suppose that we wished to design a low pass filter network to have the frequency response shown

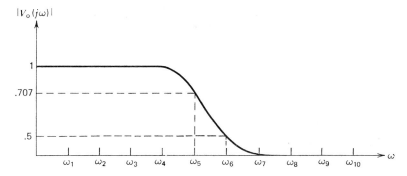

**Fig. 19-2   A desired low pass output.**

in Fig. 19-2. Then performance function (19.1b) could be used with $N = 10$ and

$$|V_o(j\omega_k)| = \begin{cases} 1 & \text{for} \quad k = 1, 2, 3, 4 \\ .707 & \text{for} \quad k = 5 \\ .5 & \text{for} \quad k = 6 \\ 0 & \text{for} \quad k = 7, 8, 9, 10 \end{cases} \quad \square$$

In the next section we will discuss an algorithm that may be used to minimize (19.1). However, before leaving this section a slight generalization of (19.1) is introduced. In practice it usually turns out that we do not really want to give equal attention to the errors at each frequency point $\omega_k$. Rather, we desire to minimize the errors at some frequencies more than others. To allow for this situation we rewrite (19.1a) and (19.1b) as

(19.2a)
$$\varepsilon = \frac{1}{2} \sum_{k=1}^{N} W_k |V_o(j\omega_k) - \hat{V}_o(j\omega_k)|^2$$

and

(19.2b)
$$\varepsilon = \frac{1}{2} \sum_{k=1}^{N} W_k [|V_o(j\omega_k)| - |\hat{V}_o(j\omega_k)|]^2$$

where the $W_k$ are real, nonnegative weighting functions that may be arbitrarily specified.

## 19-2   THE METHOD OF STEEPEST DESCENT

By way of introduction to the minimization of the performance function (19.2) consider the following situation. Suppose we are given the scalar function $f(x)$ and we wished to find the value of $x$ denoted by $x^*$ that minimizes $f(x)$. Then

(19.3a)
$$f(x^*) \leq f(x) \text{ for all } x,$$

(19.3b)
$$\left.\frac{df}{dx}\right|_{x=x^*} = 0$$

and

(19.3c)
$$\left.\frac{d^2f}{dx^2}\right|_{x=x^*} > 0$$

If $f(x)$ was a simple explicit function of $x$ we could simply employ equations (19.3b) and (19.3c) to determine $x^*$. However, if $f(x)$ is an implicit function of $x$, as in the network design problem we are considering, an alternative approach is needed.

Let $x^j$ and $x^{j+1}$ denote the $j$th and $(j + 1)$st estimate of $x^*$, respectively, and define $\Delta x^j$ as

$$\Delta x^j \equiv x^{j+1} - x^j$$

We desire a value of $\Delta x^j$ such that

$$f(x^{j+1}) = f(x^j + \Delta x^j) < f(x^j)$$

that is, the value of $f$ at the $(j + 1)$st estimate is less than the value of $f$ at the $j$th estimate. Expanding $f(x^j + \Delta x^j)$ in a Taylor series about $x^j$ yields

$$f(x^j + \Delta x^j) = f(x^j) + \left.\frac{df}{dx}\right|_{x=x^j} \Delta x^j + \cdots$$

Using only the first two terms of this expansion, we have

$$f(x^j) + \left.\frac{df}{dx}\right|_{x=x^j} \Delta x^j < f(x^j)$$

Therefore, for a decrease in $f$ we must have

$$\left.\frac{df}{dx}\right|_{x=x^j} \Delta x^j < 0$$

Thus if the derivative of $f$ evaluated at $x^j$ is positive, $\Delta x^j$ should be negative, that is,

$$x^{j+1} < x^j$$

and if the derivative of $f$ evaluated at $x^{j+1}$ is negative, $\Delta x^j$ should be positive, that is,

$$x^{j+1} > x^j$$

Thus the sign of the derivative of $f$ indicates how $x$ should be changed to reduce $f$.

Let us now return to the real problem of interest. We wish to adjust the parameter vector $\mathbf{p}$ so as to minimize the performance function $\varepsilon(\mathbf{p})$. Following the same procedure as has been discussed, let $\mathbf{p}^j$ and $\mathbf{p}^{j+1}$ denote the $j$th and $(j+1)$st values of $\mathbf{p}$, and define

(19.4)
$$\Delta \mathbf{p} = \mathbf{p}^{j+1} + \mathbf{p}^j$$

Since $\mathbf{p}$ is a vector we can think of $\mathbf{p}^j$ and $\mathbf{p}^{j+1}$ as points in the *parameter space*. For example, if there are only two designable parameters, $p_1$ and $p_2$, then

$$\mathbf{p}^j = \begin{pmatrix} p_1^j \\ p_2^j \end{pmatrix} \quad \text{and} \quad \mathbf{p}^{j+1} = \begin{pmatrix} p_1^{j+1} \\ p_2^{j+1} \end{pmatrix}$$

represents a point in the two dimensional plane defined by plotting $p_1$ on the $x$-axis and $p_2$ on the $y$-axis, as illustrated in Fig. 19-3. The *change vector* $\Delta \mathbf{p}^j$ then indicates the direction and magnitude of the difference between $\mathbf{p}^j$ and $\mathbf{p}^{j+1}$.

We desire a $\Delta \mathbf{p}^j$ such that

$$\varepsilon(\mathbf{p}^{j+1}) = \varepsilon(\mathbf{p}^j + \Delta \mathbf{p}^j) < \varepsilon(\mathbf{p}^j)$$

Taking a multidimensional Taylor series expansion of $\varepsilon(\mathbf{p}^j + \Delta \mathbf{p}^j)$ about $\mathbf{p}^j$ and retaining only the first two terms, yields

$$\varepsilon(\mathbf{p}^j + \Delta \mathbf{p}^j) = \varepsilon(\mathbf{p}^j) + [\nabla \varepsilon(\mathbf{p}^j)]\Delta \mathbf{p}^j < \varepsilon(\mathbf{p}^j)$$

where the *gradient vector* $\nabla \varepsilon(\mathbf{p}^j)$ is a row vector given by

$$\nabla \varepsilon(\mathbf{p}^j) = \left( \frac{\partial \varepsilon}{\partial p_1} \bigg|_{\mathbf{p}=\mathbf{p}^j}, \frac{\partial \varepsilon}{\partial p_2} \bigg|_{\mathbf{p}=\mathbf{p}^j}, \ldots, \frac{\partial \varepsilon}{\partial p_n} \bigg|_{\mathbf{p}=\mathbf{p}^j} \right)$$

Thus

(19.5)
$$(\nabla \varepsilon(\mathbf{p}^j)) \, \Delta \mathbf{p}^j < 0$$

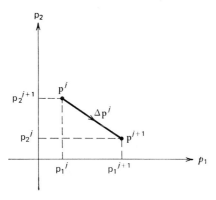

**Fig. 19-3** The two-dimensional parameter space defined by $p_1$ and $p_2$ and the change vector $\Delta p^j$.

Let **s** be a unit vector in the same direction as $\Delta\mathbf{p}^j$.

(19.6)
$$\mathbf{s}^i = \frac{\Delta\mathbf{p}^j}{\|\Delta\mathbf{p}^j\|}$$

where

$$\|\Delta\mathbf{p}^j\| = \sqrt{(p_1^j)^2 + (p_2^j)^2 + \cdots + (p_n^j)^2}$$

is the magnitude of the vector $\Delta\mathbf{p}^j$. Then

(19.7)
$$\Delta\mathbf{p}^j = (\mathbf{p}^{j+1} - \mathbf{p}^j) = \alpha^j\mathbf{s}^i$$

where $\alpha^j$ is a real number that represents the magnitude of $\Delta\mathbf{p}^j$ and is called the *step size*. We wish to determine $\alpha^j$ and $\mathbf{s}^j$ such that

(19.8)
$$\alpha^j[\nabla\varepsilon(\mathbf{p}^j)]\mathbf{s}^j < 0$$

or

(19.9)
$$[\nabla\varepsilon(\mathbf{p}^j)]\mathbf{s}^j < 0$$

for $\alpha^j > 0$. Recognize that (19.9) represents the "dot" product of two vectors:

$$[\nabla\varepsilon(\mathbf{p}^j)]\mathbf{s}^j = \|\nabla\varepsilon(\mathbf{p}^j)\| \, \|\mathbf{s}^j\| \cos\theta$$

where $\theta$ is the angle between $\nabla\varepsilon(\mathbf{p}^j)$ and $\mathbf{s}^j$. This product takes on its most negative value when $\theta = 180°$, that is, when

$$\mathbf{s}^j = - \frac{[\nabla\varepsilon(\mathbf{p}^j)]^T}{\|\nabla\varepsilon(\mathbf{p}^j)\|}$$

In other words, to ensure that $\varepsilon$ decreases when $\varepsilon$ is changed from $\mathbf{p}^j$ to $\mathbf{p}^{j+1}$, we wish to change $\mathbf{p}^j$ along a direction that is opposite to the direction of the gradient of $\varepsilon$ evaluated at $\mathbf{p}^j$. Since the gradient represents the direction of greatest increase in a function, the parameter vector is adjusted along the direction of greatest decrease that is called the direction of *steepest descent*. Until now nothing has been said about the distance along this direction that $\mathbf{p}^j$ should be changed. We will return to this point after an example.

### Example 19.1

In order to illustrate the method of steepest decent, consider the performance function

$$\varepsilon(\mathbf{p}) = p_1{}^2 + p_1 p_2 + p_2{}^2$$

where $\mathbf{p} = (p_1, p_2)^T$ are the designable parameters (we are using this performance function for illustrative purposes only—it is not meant to be useful for circuit design applications). Since $\varepsilon$ is an explicit function of $\mathbf{p}$, we can determine the minimum analytically by computing the gradient, setting it equal to zero and solving for $p_1$ and $p_2$:

$$\frac{\partial \varepsilon}{\partial p_1} = 2p_1 + p_2 = 0$$

$$\frac{\partial \varepsilon}{\partial p_2} = p_1 + 2p_2 = 0$$

so that $p_1 = p_2 = 0$ is the point at which $\varepsilon(\mathbf{p})$ is a minimum. Observe that

$$\frac{\partial^2 \varepsilon}{\partial p_1{}^2} = \frac{\partial^2 \varepsilon}{\partial p_2{}^2} = 2 > 0$$

and

$$\frac{\partial^2 \varepsilon}{\partial p_1 \partial p_2} = \frac{\partial^2 \varepsilon}{\partial p_2 \partial p_1} = 1 > 0$$

We now solve this same problem using the method of steepest descent. Let us initially guess that the minimum occurs at the point $p_1^0 = 2$, $p_2^0 = -3$, where the superscript zero denotes the initial guess. Then

$$\varepsilon(\mathbf{p}^0) = (2)^2 + (2)(-3) + (-3)^2 = 7$$

and

(19.10) $$\nabla \varepsilon(\mathbf{p}) = (2p_1 + p_2 \quad , \quad p_1 + 2p_2)$$

so that

$$\nabla \varepsilon(\mathbf{p}^0) = (1 \quad -4)$$

The direction in which we should move $\mathbf{p}$ to achieve the greatest reduction in $\varepsilon$ is

$$\mathbf{s}^0 = -\frac{[\nabla \varepsilon(\mathbf{p}^0)]^T}{\|\nabla \varepsilon(\mathbf{p}^0)\|} = \frac{-1}{\sqrt{(1)^2 + (-4)^2}} \begin{pmatrix} 1 \\ -4 \end{pmatrix}$$

$$= \begin{pmatrix} -\dfrac{1}{\sqrt{17}} \\ \dfrac{4}{\sqrt{17}} \end{pmatrix}$$

The next estimate of $\mathbf{p}$ (from (19.7)) is

$$\mathbf{p}^1 = \mathbf{p}^0 + \alpha^0 \mathbf{s}^0$$

or

$$\mathbf{p}^1 = \begin{pmatrix} 2 - \dfrac{\alpha^0}{\sqrt{17}} \\ -3 + \dfrac{4\alpha^0}{\sqrt{17}} \end{pmatrix}$$

**Table 19.1**

| $\alpha^0$ | $p_1^1$ | $p_2^1$ | $\varepsilon(\mathbf{p}^1)$ |
|---|---|---|---|
| 0.5 | 1.88 | $-2.51$ | 5.13 |
| 1.0 | 1.76 | $-2.03$ | 3.64 |
| 2.0 | 1.51 | $-1.06$ | 1.81 |
| 2.5 | 1.40 | $-\ .575$ | 1.47 |
| 2.6 | 1.37 | $-\ .478$ | 1.45 |
| 2.7 | 1.35 | $-\ .381$ | 1.44 |
| 2.8 | 1.32 | $-\ .284$ | 1.45 |
| 3.0 | 1.27 | $-\ .090$ | 1.51 |

where $\alpha^0 > 0$. Table 19.1 lists the values of $\mathbf{p}^1$ and $\varepsilon(\mathbf{p}^1)$ for several different values of $\alpha^0$. Observe that there is a continuous decrease in $\varepsilon$ as $\alpha^0$ increases to 2.7. When $\alpha^0$ increases past 2.7 $\varepsilon$ begins increasing again. The best choice for $\alpha^0$ is 2.7 so that

$$\mathbf{p}^1 = \begin{pmatrix} 1.35 \\ -.381 \end{pmatrix} \quad \text{and} \quad f(\mathbf{p}^1) = 1.44$$

(Although the search for $\alpha$ may seem to be haphazard there are systematic methods available, one of which is to be discussed in the next section.)

We now reevaluate the gradient (19.10) at $\mathbf{p}^1$ and compute another estimate of $\mathbf{p}$:

$$\nabla\varepsilon(\mathbf{p}^1) = (2.32 \quad\quad .588)$$

and

$$\mathbf{s}^1 = \frac{-1}{2.39}\begin{pmatrix} 2.32 \\ .588 \end{pmatrix}$$

The next estimate of $\mathbf{p}$, from (21.7) is

$$\mathbf{p}^2 = \mathbf{p}^1 + \alpha^1 \mathbf{s}^1$$

or

$$\mathbf{p}^2 = \begin{pmatrix} 1.35 - \alpha^1(.971) \\ -.381 - \alpha^1(.246) \end{pmatrix}$$

where $\alpha^1 > 0$. Table 19.2 lists the values of $\mathbf{p}^1$ and $\varepsilon(\mathbf{p}^1)$ for several values of $\alpha^1$. A minimum along this *search direction* is reached for a value of $\alpha^1 = 1.0$. This procedure would be repeated again and again until a suitably small value of $\varepsilon$ is obtained. $\square$

The previous example illustrates how the steepest descent method transforms the multidimensional search problem of finding the $n$-dimensional

**Table 19.2**

| $\alpha^1$ | $p_1^2$ | $p_2^2$ | $\varepsilon(\mathbf{p}^2)$ |
|:---:|:---:|:---:|:---:|
| .8 | .573 | −.578 | .331 |
| .9 | .476 | −.602 | .303 |
| 1.0 | .379 | −.627 | .299 |
| 1.1 | .282 | −.651 | .320 |
| 1.2 | .185 | −.676 | .366 |

vector $\mathbf{p}$, which minimizes $\varepsilon(\mathbf{p})$ into a series of *one-dimensional*, or *linear*, *search* problems of finding the scalar $\alpha^j$, which minimizes $\varepsilon(\mathbf{p}^j + \alpha^j \mathbf{s}^j)$ where $\mathbf{p}^j$ and $\mathbf{s}^j$ are known. In order, for this procedure to work, $\varepsilon$ must be a *unimodal*. Figure 19-4 illustrates some unimodal functions. In general $\varepsilon$ does not have this property, however we may employ this algorithm anyway realizing that the minimum we find is a *local minimum* that may, or may not be a *global minimum*. At present there is no real means of determining whether a minimum is local or global.

It is instructive to illustrate the steepest descent method graphically. To this end, consider a performance or "error" function of two variables.

$$\varepsilon(\mathbf{p}) = \varepsilon(p_1, p_2)$$

that is shown in Fig. 19-5. The curves labeled $\varepsilon_1$, $\varepsilon_2$, $\varepsilon_3$, and $\varepsilon_4$ represent lines of constant error. These "equierror" contours can be projected onto the parameter space as shown in Fig. 19-6. Consider the initial estimate of the minimum, $\mathbf{p}°$. The gradient at $\mathbf{p}°$ is perpendicular (normal) to the tangent at $\mathbf{p}°$. The point $\mathbf{p}^1$ is found by adjusting $\alpha°$ so as to minimize

$$\varepsilon(\mathbf{p}° + \alpha \mathbf{s}°)$$

that is

$$\mathbf{p}^1 = \mathbf{p}° + \alpha° \mathbf{s}°$$

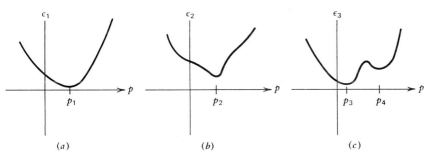

$(a)$         $(b)$         $(c)$

**Fig. 19-4** $(a)$ and $(b)$ show unimodal functions of $p$; $(c)$ is not unimodal. Points $p_1$, $p_2$, and $p_3$ are global minima; $p_1$, $p_2$, $p_3$, and $p_4$ are local minima.

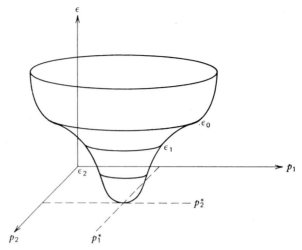

**Fig. 19-5   A 3-dimensional plot of a unimodal performance function of two variables.**

and

$$\varepsilon^1 \equiv \varepsilon(\mathbf{p}^1)$$

Since there are other values of $\mathbf{p}$ for which $\varepsilon(\mathbf{p}) = \varepsilon^1$, $\varepsilon^1$ is another equierror contour as shown in the figure. Observe that the line $\mathbf{p}^\circ + \alpha\mathbf{s}^\circ$ is tangent to the contour $\varepsilon^1$ at point $\mathbf{p}^1$. Since the new search direction is normal to the tangent at point $\mathbf{p}^1$, it is also normal to the previous search direction $\mathbf{s}^\circ$.

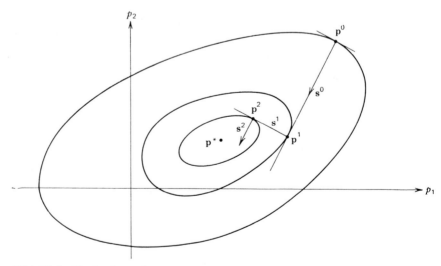

**Fig. 19-6   Projection of the equierror contours shown in Fig. 19-5 onto the parameter space and illustration of the steepest descent method.**

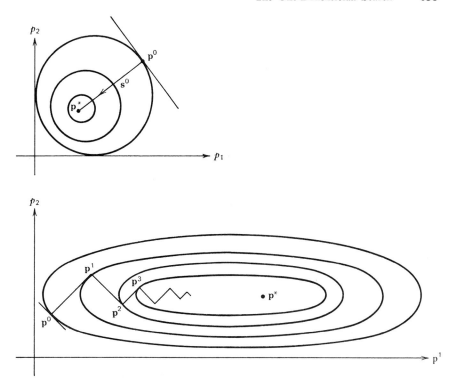

**Fig. 19-7** (*a*) Circular equierror contours require only one iteration to reach the minimum while (*b*) ellongated ellipses require very many iterations to reach the minimum.

There are two situations that deserve special mention. If the equierror contours are perfect circles as shown in Fig. 19-7*a*, only one iteration is required to reach the minimum because the negative gradient direction points directly at the minimum. If on the other hand, the contours are elongated as shown in Fig. 19-7*b*, the gradient direction can almost parallel the path that leads directly to the minimum and a very large number of iterations are required to reach the minimum. There are methods that can be employed which avoid this problem, one of which is discussed in Section 19-4.

## 19-3   THE ONE-DIMENSIONAL SEARCH

The method of steepest descent reduces the search for the minimum of the performance function $\varepsilon(\mathbf{p})$ to a series of searches for the minimum of $\varepsilon$ along a given direction $\mathbf{s}^j$. This search is one dimensional since only the step size $\alpha^j$ for which $\varepsilon(\mathbf{p}^j + \alpha^j \mathbf{s}^j)$ is minimized must be found. Although there are a number of methods for efficiently carrying out a one dimensional search in this section, just one of them will be briefly described.

We wish to find the value of $\alpha^j$ for which $\varepsilon(\mathbf{p}^j + \alpha^j \mathbf{s}^j)$ is minimized. For convenience we will drop the superscript $j$ and observe that $\varepsilon(\mathbf{p} + \alpha \mathbf{s})$ is a function of $\alpha$:

$$\varepsilon(\mathbf{p} + \alpha \mathbf{s}) = f(\alpha)$$

The $\alpha$ we seek is such that

$$\frac{df}{d\alpha} = 0$$

We observe that

$$f(0) = \varepsilon(p)$$

$$\frac{df}{d\alpha} = \nabla \varepsilon(\mathbf{p} + \alpha \mathbf{s})\mathbf{s}$$

and

$$\frac{df(0)}{d\alpha} = \nabla \varepsilon(\mathbf{p})\mathbf{s} < 0$$

for steepest descent.

The first step in the procedure is referred to as *extrapolation* and its purpose is to bound the minimum; in other words, we must determine the limits $\alpha_l$ and $\alpha_u$ such that

$$f(\alpha_l) \leq f(\alpha) \leq f(\alpha_u)$$

for all values of $\alpha$ such that $\alpha_l \leq \alpha < \alpha_u$. These limits can be found by initially choosing a small step size, say $\alpha_0$, $f(\alpha_0)$ then doubling the step size and evaluating $f(2\alpha_0)$, then doubling the step size again and evaluating $f(4\alpha_0)$, etc., until there is an increase in error, that is, $f(2^j\alpha_0) < f(2^{j+1}\alpha_0)$. We then know that the correct value of $\alpha$ lies between $2^{(j-1)}\alpha_0 \leq \alpha \leq 2^{(j+1)}\alpha_0$, that is, $\alpha_l = 2^{j-1}\alpha_0$ and $\alpha_u = 2^{j+1}\alpha_0$. Figure 9-8 illustrates this reasoning. The student

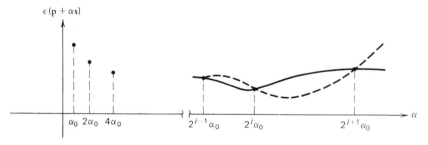

**Fig. 19-8    Illustration of the method of bounding $\alpha$.**

$$f(\alpha_0) > f(2\alpha_0) > f(4\alpha_0) > \cdots f(2^{j-1}\alpha_0) > f(2^j\alpha_0),$$

but $f(2^j\alpha_0) < f(2^{j+1}\alpha_0)$ so that a suitable lower bound on the desired value of $\alpha$ is $\alpha_1 = 2^{j-1}\alpha_0$ and a suitable upper bound is $\alpha_u = 2^{j+1}\alpha_0$. The solid and broken curves illustrate two possible situations and show why the value $2^j\alpha_0$ cannot be a boundary.

should notice that this procedure is valid because we have *assumed* that the performance function is unimodal. The interval $(\alpha_l, \alpha_u)$ is termed the *interval of uncertainty*.

Once $\alpha_l$ and $\alpha_u$ have been found, we can observe that the values $f(\alpha_l)$, $f(\alpha_u)$ and the derivatives $df/d\alpha|_{\alpha_l}$ and $df/d\alpha|_{\alpha u}$ define a cubic curve. The value of $\alpha$ minimizes the cubic curve. In particular, it can be shown[1] that this value of $\alpha$ is given by

$$\alpha^* = \alpha_u - \left( \frac{\left.\dfrac{df}{d\alpha}\right|_{\alpha_u} + w - z}{\left.\dfrac{df}{d\alpha}\right|_{\alpha_u} - \left.\dfrac{df}{d\alpha}\right|_{\alpha l} + 2w} \right)(\alpha_u - \alpha_l)$$

where

$$z = 3\frac{f(\alpha_l) - f(\alpha_u^u)}{\alpha_u - \alpha_l} + \left.\frac{df}{d\alpha}\right|_{\alpha l} + \left.\frac{df}{d\alpha}\right|_{\alpha u}$$

and

$$w = \left( z^2 - \left.\frac{df}{d\alpha}\right|_{\alpha l} \left.\frac{df}{d\alpha}\right|_{\alpha u} \right)^{1/2}$$

This scheme is implemented in the minimization program described in the next section.

## 19-4  THE SUBROUTINE CNJGRD

As indicated previously steepest descent is not suitable for minimizing a function if its error contours are enlongated. There are several methods that try to avoid this problem. One such method is the method of conjugate gradients.[2] In this method the search direction is partially determined by the previous search direction as well as the gradient. It is beyond the scope of this text to discuss in detail the conjugate gradient method, however, because it is generally superior to steepest descent we will employ it in our design program. The subroutine CNJGRD shown in Fig. 19-9 is an implementation of the conjugate gradient minimization algorithm. Observe that cubic interpolation is employed. The subrouting CNJGRD is called from another program by the statement

CALL CNJGRD (N, P, F, G, S, EST, EPS, LIMIT, IER, $PR)

[1] R. Fletcher and M. J. D. Powell, "A Rapidly Convergent Descent Method for Minimization," *Computer Journal*, Vol. 6, pp. 163–168, 1963.

R. Fletcher and C. M. Reeves, "Function Minimization by Conjugate Gradients,' *Computer J.*, Vol. 7, pp. 149–154, 1964.

```
        SUBROUTINE CNJGRD(N,P,F,G,S,EST,EPS,LIMIT,IER,$PR)
        IMPLICIT REAL*8 (A-H,O-Z)
        DIMENSION G(1),S(1),P(1)
        LOGICAL*1 $PR
C
C       THIS SUBROUTINE MINIMIZES A FUNCTION OF N PARMATERS USING
C       THE CONJUGATE GRADIENT METHOD. FOR A FULL DESCRIPTION OF
C       THE USE OF THIS ROUTINE SEE SECTION 19-4 OF THE TEXT.
C·
C
C       INITIALIZE VARIABLES
C
        ICOUNT=0
        IER=0
        N1=N+1
C
C       CALCULATE INITIAL ERROR AND GRADIENT
C
        CALL FUNCT(N,P,F,G,ICOUNT)
        IF($PR) PRINT 105,F
105     FORMAT(' INITIAL ERROR= ',G12.5)
C
C       BEGIN ITERATION LOOP
C
100     DO 99 I=1,N1
        ICOUNT = ICOUNT+1
        OLDF=F
C
C       COMPUTE SQUARE OF GRADIENT AND CHECK FOR ZERO VALUE
C
        GG=0.
        DO 1 J=1,N
1       GG=GG+G(J)*G(J)
        IF(GG.EQ.0) GO TO 50
C
C       USE STEEPEST DESCENT AT BEGINNING OF EVERY N+1 ITERATIONS
C
        IF(I.GT.1) GO TO 3
        DO 2 J=1,N
2       S(J)=-G(J)
        GO TO 4
3       BETA=GG/OLDGG
        DO 5 J=1,N
5       S(J)=-G(J)+BETA*S(J)
C
C       BEGIN LINEAR SEARCH
C
4       YB=F
        VB=0
        SS=0
        DO 6 J=1,N
C
C       SAVE OLD PARAMETER VALUES AND COMPUTE DIRECTIONAL DERIVATES
C
        S(J+N)=P(J)
        VB=VB+G(J)*S(J)
6       SS=SS+S(J)*S(J)
C
C       CHECK THAT DIRECTION OF SEARCH WILL DECREASE ERROR
C
        IF(VB.GE.0) GO TO 20
```

**Fig. 19-9    The subroutine CNJGRD.**

```
C
C      ESTIMATE INITIAL STEP SIZE
C
       ALFA=2*(EST-F)/VB
       AMBDA=1./DSQRT(SS)
       IF(ALFA.LE.0) GO TO 7
8      IF(ALFA**2*SS.GE.1) GO TO 7
       AMBDA=ALFA
7      ALFA=0.
C
C      DOUBLE STEP SIZE UNTIL MINIMUM IS BOUNDED
C
15     YA=YB
       VA=VB
C
C      STEP PARAMETERS ALONG SEARCH DIRECTION
C
       CALL PRMCHG(P,AMBDA,S,N)
C
C      CALCULATE FUNCTION AND GRADIENT FOR NEW PARAMETERS
C
       CALL FUNCT(N,P,F,G,ICOUNT)
       IF($PR) PRINT 101,ICOUNT,F
101    FORMAT(' EXTRAPOLATION --- ITERATION NO. ',I4,'  ERROR=',G12.5)
       YB=F
C
C      CALCUALTE NEW DIRECTIONAL DERIVATIVES -- IF POSITIVE MINIMUM
C      HAS BEEN PASSED -- IF ZERO MINIMUM HAS BEEN FOUND
C
       VB=0
       DO 10 J=1,N
10     VB=VB+G(J)*S(J)
       IF(VB) 11,12,21
C
C      CHECK TO SEE IF ERROR HAS INCREASED
C
11     IF(YB.GT.YA) GO TO 21
C
C      DOUBLE STEP SIZE AND REPEAT ABOVE PROCEDURE
C
       AMBDA=AMBDA+ALFA
       ALFA=AMBDA
C
C      TERMINATE IF STEP SIZE IS TOO LARGE
C
       IF(SS*AMBDA.LT.1.E10) GO TO 15
23     IER=3
       RETURN
C
C      PERFORM CUBIC INTERPOLATION
C
21     T=0.
13     IF(AMBDA.EQ.0) GO TO 12
       Z=3*(YA-YB)/AMBDA+VA+VB
       W=DSQRT(Z**2-VA*VB)
       ALFA=AMBDA*(VB+W-Z)/(VB+2.*W-VA)
       CALL PRMCHG(P,T-ALFA,S,N)
C
C      TERMINATE IF F IS LESS THEN VALUES AT EACH END OTHERWISE
C      REDUCE INTERVAL
C
```

**Fig. 19-9 (*Continued*)**

```
      CALL FUNCT(N,P,F,G,ICOUNT)
      IF($PR) PRINT 102,ICOUNT,F
102   FORMAT(10X,'INTERPOLATION --- ITERATION NO.',I4,'  ERROR=',G12.5)
      IF(F.LE.YA.AND.F.LE.YB) GO TO 12
      VC=0.
      DO 19 J=1,N
19    VC=VC+G(J)*S(J)
      IF(VC.GE.0) GO TO 18
      YA=F
      VA=VC
      AMBDA=ALFA
      T=AMBDA
      GO TO 13
18    YB=F
      VB=VC
      AMBDA=AMBDA-ALFA
      GO TO 21
C
C     COMPUTE LENTGH OF CHANGE VECTOR
C
12    T=0.
      DO 27 J=1,N
      K=J+N
      S(K)=P(J)-S(K)
27    T=T+DABS(S(K))
C
C     CHECK TO SEE IF LENGTH OF CHANGE VECTOR IS INSIGNIFICANT
C     IF AT LEAST N+1 ITERATIONS HAVE BEEN TAKEN
C
      IF(ICOUNT.LT.N1) GO TO 22
      IF(T.LT.EPS) GO TO 29
C
C     TERMINATE IF F HAS NOT DECREASED DURING LAST ITERATION
22    IF(F.GT.(OLDF+EPS)) GO TO 23
20    OLDGG=GG
      IF(ICOUNT.GT.LIMIT) GO TO 53
99    CONTINUE
      GO TO 100
C
C     CHECK FOR SUFFICIENTLY SMALL GRADIENT
C
29    IF(GG-EPS) 50,50,23
53    IER=2
      RETURN
50    IER=1
      RETURN
      END
```

**Fig. 19-9 (*Continued*)**

where

    N  the number of adjustable parameters

    P  a Real* 8 array dimensional P(N) in the calling program that contains the values of the parameters

    F  a Real* 8 variable that contains the final value of the performance function on return to the calling program

G a Real* 8 array dimensioned G(N) in the calling program that stores the gradient of the performance function

S a Real* 8 array dimensioned S(2*N) in the calling program

EST an estimate of the mininum value of F

EPS an estimate of how many significant digits of accuracy are required for convergence, a good value is $10^{-9}$

LIMIT the maximum number of iterations

IER is set to a value between 1 and 3 on return to the calling program to indicate the following conditions:

IER = 1, convergence has occurred

IER = 2, Maximum number of iterations exceeded

IER = 3, error in gradient calculation

$PR a Logical* 1 variable that when set to true will cause the value of F to be printed each time the performance function is evaluated

In order to use CNJGRD, two additional subroutines FUNCT and PRMCHG must be supplied. The subroutine FUNCT returns the values of the performance function and its gradient for each set of parameters. This subroutine has the arguments

SUBROUTINE FUNCT (N, P, F, G, ICOUNT)

where ICOUNT denotes the number of iterations and can be used for the purpose of printing as will be seen in the examples. The subroutine PRMCHG has the arguments

SUBROUTINE PRMCHG (P, AMBDA, S, N)

and is used to alter the parameter vector P along the direction S using a step size of AMBDA. A suitable version of this subroutine is shown in Fig. 19-10. The reason for allowing the user the option of altering the manner in which the parameters are changed will become clear later when we discuss logarithmic scaling of the parameter vector.

**Example 19.2**

To illustrate the use of the subroutine CNJGRD consider the performance function

$$\varepsilon(\mathbf{p}) = p_1^2 + p_1 p_2 + p_2^2$$

```
                    SUBROUTINE PRMCHG(P,AMBDA,S,N)
                    IMPLICIT REAL*8 (A-H,O-Z)
                    DIMENSION P(1),S(1)
          C
                    DO 1 J=1,N
          1         P(J)=P(J)+AMBDA*S(J)
                    RETURN
                    END
```

**Fig. 19-10   The subroutine PRMCHG which is called by CNJGRD to alter the parameters linearly along the search direction.**

originally considered in Example 19.1. The program of Fig. 19-11$a$ calls CNJGRD to minimize $\varepsilon(\mathbf{p})$. The subroutine FUNCT evaluates $\varepsilon(\mathbf{p})$ and $\nabla\varepsilon(\mathbf{p})$ for each value of $\mathbf{p}$. The results of the program are shown in Fig. 19-11$b$. Observe that the minimum was found in two iterations. If steepest descent had been employed it would have required a significantly greater number of iterations to find the minimum.   □

```
          C
          C         CHAPTER 19 EAMPLE 19.2
          C
                    IMPLICIT REAL*8 (A-H,O-Z)
                    DIMENSION P(2),G(2),S(4)
                    LOGICAL*1 $PR
          C
          C         INITIALIZE VARIABLES AND CALL CNJGRD
          C
                    N=2
                    $PR=.TRUE.
                    P(1)=2
                    P(2)=-3
                    CALL CNJGRD(N,P,F,G,S,10.D-5,10.D-9,50,IER,$PR)
                    PRINT 1,IER
          1         FORMAT('1IER=',I3)
          C
          C         PRINT FINAL VALUES
          C
                    PRINT 2,F,(P(J),J=1,2)
          2         FORMAT('OF=',G12.5,'P(1)=',G12.5,'P(2)=',G12.5)
                    END


                    SUBROUTINE FUNCT(N,P,F,G,ICOUNT)
                    IMPLICIT REAL*8 (A-H,O-Z)
                    DIMENSION P(1),G(1)
                    F=P(1)**2+P(1)*P(2)+P(2)**2
                    G(1)=2*P(1)+P(2)
                    G(2)=P(1)+2*P(2)
                    RETURN
                    END
```

**Fig. 19-11$a$   Program of Example 19.2 to illustrate the use of CNJGRD.**

```
INITIAL ERROR=    7.0000
EXTRAPOLATION --- ITERATION NO.    1  ERROR=  3.6416
EXTRAPOLATION --- ITERATION NO.    1  ERROR=  1.8126
EXTRAPOLATION --- ITERATION NO.    1  ERROR=  2.7429
           INTERPOLATION --- ITERATION NO.   1  ERROR=  1.4423
EXTRAPOLATION --- ITERATION NO.    2  ERROR= 0.11775
EXTRAPOLATION --- ITERATION NO.    2  ERROR= 0.26489
           INTERPOLATION --- ITERATION NO.   2  ERROR= 0.14059D-31
EXTRAPOLATION --- ITERATION NO.    3  ERROR=  1.4370
           INTERPOLATION --- ITERATION NO.   3  ERROR= 0.25037D-32
```

**Fig. 19-11*b*   Output of program of 19-11*a*.**

## 19-5   CALCULATION OF THE GRADIENT: THE ADJOINT NETWORK

In the case of network design the performance function is not usually an explicit function of the designable parameter, but rather implicitly depends on them through the output voltage or current, as in (19.1) and (19.2). Therefore, it is not practical to find an analytic expression for the gradient. In this section we will discuss a numerical technique that can be used to compute the gradient accurately and efficiently.

Consider the $j$th component of the gradient vector $\partial \varepsilon / \partial p_j$. If the performance function is that given by (19.2a) we have

(19.11a)
$$\frac{\partial \varepsilon}{\partial p_j} = \frac{1}{2} \sum_{k=1}^{N} W_k \frac{\partial}{\partial p_j} \left( [V_o(j\omega_k) - \hat{V}_o(j\omega_k)][V_o^*(j\omega_k) - \hat{V}_o^*(j\omega_k)] \right)$$
$$= \sum_{k=1}^{N} \mathrm{Re} \left\{ W_k [V_o^*(j\omega_k) - \hat{V}_o^*(j\omega_k)] \frac{\partial V_o(j\omega_k)}{\partial p_j} \right\}$$

On the other hand, if the performance function is that given by 19.26 we have

(19.11b)
$$\frac{\partial \varepsilon}{\partial p_j} = \frac{1}{2} \sum_{k=1}^{N} W_k \frac{\partial}{\partial p_j} \left\{ [V_o(j\omega_k)V_o^*(j\omega_k)]^{1/2} - |\hat{V}_o(j\omega_k)| \right\}^2$$
$$= \sum_{k=1}^{N} \mathrm{Re} \left\{ W_k \frac{V_o^*(j\omega_k)}{|V_o(j\omega_k)|} [|V_o(j\omega_k)| - |\hat{V}_o(j\omega_k)|] \frac{\partial V_o(j\omega_k)}{\partial p_j} \right\}$$

Thus we see that the gradient of the performance function $\partial \varepsilon / \partial \mathbf{p}$ is related to the sensitivity of the output voltage with respect to the designable parameters at each frequency point, $\partial V_o(j\omega_k) / \partial \mathbf{p}$. Efficient computation of network sensitivities was first discussed in Chapter 6 for the dc case. In this section we will extend these results for networks that contain $R$, $L$, $C$'s and voltage controlled current sources.

Before proceeding we will briefly digress to reconsider Tellegen's theorem which was first introduced in Chapter 5. An alternative form of this theorem is as follows. Let $\mathbf{V}_b(j\omega_k)$ and $\mathbf{I}_b(j\omega_k)$ denote the vector of branch voltage phasors and the vector of branch current phasors, respectively for a given network; then

$$V_b{}^T I_b = \sum_{j=1}^{n_b} V_j I_j = 0$$

This result is easily proven. Observe that if $A$ is the nodal incidence matrix of the network, and $V_n$ the vector of node voltage phasors,

$$V_b{}^T I_b = V_n{}^T A I_b = 0$$

because of *KCL*

$$A I_b = 0$$

Using the same reasoning, we can show that if $V_b$ is the branch voltage vector of one network and $\Phi_b$ is the branch current vector of another network which has the same topology as the first (i.e., both networks have the same nodal incidence matrix), then

$$V_b{}^T \Phi_b = 0$$

This form of Téllegen's theorem is used below.

We now return to our goal of calculating the sensitivity of the output voltage $V_o$ with respect to all the parameters $\mathbf{p}$. Let $\mathcal{N}$ denote the network under consideration. We assume that $\mathcal{N}$ has a single input source $V_i$ and that the output $V_o$ is measured across a zero valued current source $I_o = 0$. The remaining elements in $\mathcal{N}$ are resistors, capacitors, inductors, and voltage controlled current sources. For convenience of notation resistor, branch voltages, and currents will be indicated by $V_R$ and $I_R$, respectively; capacitor branch voltages and currents will be indicated by $V_C$ and $I_C$, respectively; and inductor branch voltages and currents will be indicated by $V_L$ and $I_L$, respectively. The voltage controlled current sources are assumed to consist of two branches, as shown in Fig. 19-12. The voltage and current of the controlling branch will be designated by $V_{g_m C}$ and $I_{g_m C}$ while the voltage and current of the controlled, or dependent branch, will be designated by $V_{g_m D}$ and $I_{g_m D}$.

We will also be concerned with the network $\hat{\mathcal{N}}$. At this point the only requirement we impose on $\hat{\mathcal{N}}$ is that it have the same topology as $\mathcal{N}$. The branch voltages and currents in $\hat{\mathcal{N}}$ will be denoted by $v$ and $\Phi$, respectively. Applying Téllegen's theorem between $\mathcal{N}$ and $\hat{\mathcal{N}}$ yields.

$$V_b{}^T \Phi_b = 0$$

and

$$I_b{}^T N_b = 0$$

These expressions may be subtracted to yield

(19.12) $$V_b{}^T \Phi_b - I_b{}^T N_b = 0$$

If the parameters in $\mathcal{N}$ experience a small perturbation; that is, $\mathbf{p}$ becomes $\mathbf{p} + \Delta\mathbf{p}$, then the voltages and currents in $\mathcal{N}$ are also perturbed; they become

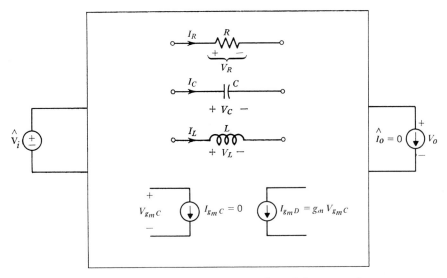

**Fig. 19-12    Symbolic representation of the network.**

$V + \Delta V$ and $I + \Delta I$, respectively. Expression (19.12) becomes

$$(\mathbf{V}_b^T + \Delta \mathbf{V}_b^T)\mathbf{\Phi}_b - (\mathbf{I}_b^T + \Delta \mathbf{I}_b^T)\mathbf{\nu}_b = 0$$

which means that

(19.13)
$$\Delta \mathbf{V}_b^T \mathbf{\Phi}_b - \Delta \mathbf{I}_b^T \mathbf{\nu}_b = 0$$

which can be rewritten as

$$\sum_{k=1}^{n_b} (\Delta V_k \Phi_k - \Delta I_k \nu_k) = 0$$

This summation can be expanded as follows

$$\Delta V_i \Phi_i - \Delta I_i \nu_i + \Delta V_o \Phi_o - \Delta I_o \nu_o + \sum_R [\Delta V_R \Phi_R - \Delta I_R \nu_R]$$

$$+ \sum_C (\Delta V_C \Phi_C - \Delta I_C \nu_C) + \sum_L (\Delta V_L \Phi_L - \Delta I_L \nu_L)$$

(19.14)
$$+ \sum_{g_m} (\Delta V_{g_m C} \Phi_{g_m C} - \Delta I_{g_m C} \nu_{g_m C} + \Delta V_{g_m D} \Phi_{g_m D} - \Delta I_{g_m D} \nu_{g_m D})$$

$$= 0$$

where $\sum_R$ denotes summation over all resistive branches, $\sum_C$ denotes summation overall capacitance branches, etc. Observe that the subscripts used on $\Phi$ and $\nu$ do not necessarily indicate the types of elements that appear in $\hat{\mathcal{N}}$, rather they are merely used to indicate the correspondence between the branches in $\mathcal{N}$ and $\hat{\mathcal{N}}$.

Let us now consider the branch relationships in $\mathcal{N}$. The input voltage source is described by

$$V_i = \hat{V}_i$$

so that $\Delta V_i = 0$, independent of changes in the element values. Similarly, the output current source is described by

$$I_o = \hat{I}_o = 0$$

so that $\Delta I_o = 0$. Resistance branches are characterized by

$$V_R = R I_R$$

so that after a perturbation of the element values, we have

$$(V_R + \Delta V_R) = (R + \Delta R)(I_R + \Delta I_R)$$

or

$$\Delta V_R = R\Delta I_R + I_R\Delta R$$

if second order terms are neglected. In a similar manner we arrive at the relations

$$\Delta I_C = j\omega C\,\Delta V_C + j\omega V_C\,\Delta C$$

for capacitor branches and

$$\Delta V_L = j\omega L\,\Delta I_L + j\omega I_L\,\Delta L$$

for inductor branches. Finally, consider voltage controlled current sources. The controlling branch is characterized by

$$I_{g_mC} = 0$$

so that

$$\Delta I_{g_mC} = 0$$

while the controlled branch is characterized by

$$I_{g_mD} = g_m V_{g_mC}$$

so that

$$\Delta I_{g_mD} = g_m\,\Delta V_{g_mC} + V_{g_mC}\,\Delta g_m$$

On substitution of these relationships, (19.14) becomes

$$-\Delta I_i \nu_i + \Delta V_o \Phi_o + \sum_R [(R\Phi_R - \nu_R)\Delta I_R + I_R\Phi_R \Delta R]$$

$$+ \sum_C [(\Phi_C - j\omega C\nu_C)\Delta V_C - j\omega V_{CC}\,\nu\Delta C]$$

$$(19.15) \qquad + \sum_L [(j\omega L\Phi_L - \nu_L)\Delta I_L + j\omega I_L\Phi_L \Delta L]$$

$$+ \sum_{g_m} [\Phi_{g_mD}\Delta V_{g_mD} + (\Phi_{g_mC} - g_m\nu_{g_mD})\Delta V_{g_mC} - V_{g_mC}\nu_{g_mD}\Delta g_m]$$

$$= 0$$

What we desire is an expression that relates the change in the output, $\Delta V_o$, to changes in the parameters, $\Delta R$, $\Delta C$, $\Delta L$, and $\Delta g_m$. Observe that (19.15) would be in the desired form if we can eliminate terms involving $\Delta I_i$, $\Delta I_R$, $\Delta V_C$, $\Delta I_L$, $\Delta V_{g_m D}$, and $\Delta V_{g_m C}$. We can eliminate $\Delta I_i$ if we can set $v_i = 0$, which can be done if branch $R$ in $\hat{\mathcal{N}}$ is chosen to be a zero-valued voltage source. Similarly $\Delta I_R$ can be eliminated if branch $R$ in $\hat{\mathcal{N}}$ is chosen to be a resistor so that

$$v_R = R\Phi_R \quad \text{or} \quad R\Phi_R - v_R = 0$$

Likewise, $\Delta V_C$ and $\Delta I_L$ can be eliminated if branches $C$ and $L$ in $\hat{\mathcal{N}}$ are chosen to be capacitors and inductors. Finally, $\Delta V_{g_m D}$ and $\Delta V_{g_m C}$ can be eliminated if we allow

$$\Phi_{g_m D} = 0$$

and

$$\Phi_{g_m C} = g_m v_{g_m D}$$

in $\hat{\mathcal{N}}$. These branch relationships correspond to a voltage controlled current source; however, observe that the controlling branch in $\mathcal{N}$ corresponds to the controlled branch in $\hat{\mathcal{N}}$. If we now choose the 0th branch in $\hat{\mathcal{N}}$ to be a unity valued current source, so that

$$\Phi_0 = 1$$

expression (19.15) can be written as

$$\Delta V_o = \sum_R (-I_R\Phi_R)\Delta R + \sum_C (j\omega V_C v_C)\Delta C$$
$$+ \sum_L (-j\omega I_L\Phi_L)\Delta L + \sum_{g_m} (V_{g_m C} v_{g_m D})\Delta g_m$$

Observe that the change in $V_o$ is also related to changes in individual parameters by

$$\Delta V_o = \sum_R \frac{\partial V_o}{\partial R}\Delta R + \sum_C \frac{\partial V_o}{\partial C}\Delta C + \sum_L \frac{\partial V_o}{\partial L}\Delta L + \sum_{g_m} \frac{\partial V_o}{\partial g_m}\Delta g_m$$

we conclude that

(19.16)
$$\frac{\partial V_o}{\partial R} = -I_R\Phi_R$$

(19.17)
$$\frac{\partial V_o}{\partial C} = j\omega V_C v_C$$

(19.18)
$$\frac{\partial V_o}{\partial L} = -j\omega I_L\Phi_L$$

and

(19.19)
$$\frac{\partial V_o}{\partial g_m} = V_{g_m C} v_{g_m D}$$

In summary, what we have shown is that the sensitivity of the output voltage $V_o$, with respect to all the parameters in the network at one frequency can be obtained by:

1. Analyzing the original network and computing certain branch currents and voltages.
2. Forming the adjoint network, exciting it with a unity valued current source at the branch which corresponds to the output of the original network, and computing certain branch currents and voltages.
3. Forming the weighted products of voltages and currents in $\mathcal{N}$ with the corresponding voltages and currents in $\hat{\mathcal{N}}$, as indicated by (19.16) to (19.19).

**Example 19.3**

To illustrate that the adjoint network approach does yield the correct sensitivity expressions consider the network of Fig. 19-13a. It is easy to show that

$$V_o = \frac{-g_m R_2}{j\omega C_5 (R_2 + j\omega L_1)} V_i$$

The sensitivity of $V_o$ with respect to the elements can be obtained by direct differentiation:

$$\frac{\partial V_o}{\partial L_1} = \frac{g_m R_2 V_i}{C_5 (R_2 + j\omega L_1)^2}$$

$$\frac{\partial V_o}{\partial R_2} = \frac{-j\omega g_m L_1 V_i}{j\omega C_5 (R_2 + j\omega L_1)^2}$$

$$\frac{\partial V_o}{\partial C_5} = \frac{-g_m R_2 V_i}{j\omega C_5^2 (R_2 + j\omega L_1)}$$

and

$$\frac{\partial V_o}{\partial g_m} = \frac{-R_2 V_i}{j\omega C_5 (R_2 + j\omega L_1)}$$

The adjoint network that corresponds to this network is shown in Fig. 19-13b. We need the currents in $L_1$ and $R_2$ and the voltages across $C_5$ and the controlling branch of the controlled source in both $\mathcal{N}$ and $\hat{\mathcal{N}}$. (*Note.* The controlling branch in $\mathcal{N}$ is #3 and the controlling branch in $\hat{\mathcal{N}}$ is #4):

$$I_1 = \frac{V_i}{R_2 + j\omega L_1}$$

$$I_2 = \frac{V_i}{R_2 + j\omega L_1}$$

$$V_3 = \frac{R_2 V_i}{R_2 + j\omega L_1}$$

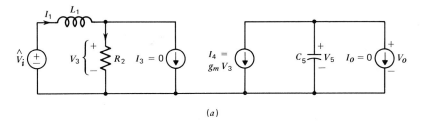

$(a)$

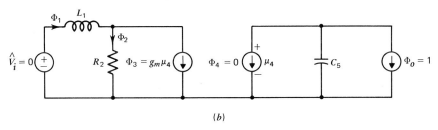

$(b)$

**Fig. 19-13** *(a)* **Network used to illustrate sensitivity computation in Example 19.3;** *(b)* **the corresponding adjoint network.**

and

$$V_5 = \frac{g_m R_2 V_i}{j\omega C_5 (R_2 + j\omega L_1)}$$

for the original network; and

$$\Phi_1 = \frac{-g_m R_2}{j\omega C_5 (R_2 + j\omega L_1)}$$

$$\Phi_2 = \frac{j\omega g_m L_1}{j\omega C_5 (R_2 + j\omega L_1)}$$

$$\nu_4 = \frac{-1}{j\omega C_5}$$

and

$$\nu_5 = \frac{-1}{j\omega C_5}$$

for the adjoint network. From (19.16)–(19.19):

$$\frac{\partial V_o}{\partial L_1} = -j\omega I_1 \Phi_1 = \frac{g_m R_2 V_i}{C_5 (R_2 + j\omega L_1)^2}$$

$$\frac{\partial V_o}{\partial R_2} = -I_2 \Phi_2 = \frac{-j\omega g_m L_1 V_i}{j\omega C_5 (R_2 + j\omega L_1)^2}$$

$$\frac{\partial V_o}{\partial C_5} = j\omega V_5 \nu_5 = \frac{-g_m R_2 V_i}{j\omega C_5^{\ 2} (R_2 + j\omega L_1)}$$

which agrees with the result for direct differentiation.  □

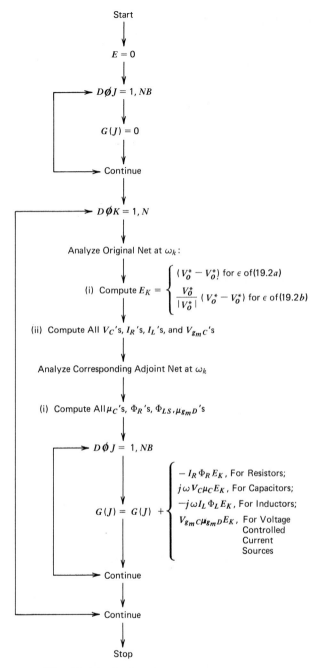

Fig. 19-14 Evaluation of the gradients.

Now that we have an efficient means for computing sensitivities, the gradient expressions (19.11a) and (19.11b) can be evaluated using the steps outlined in the flow diagram of Fig. 19-14.

Before presenting an example of computer-aided network design we will demonstrate a special relationship that exists between the original and adjoin networks which will greatly reduce the computational effort needed to evaluate the sensitivities.

## 19-6 RELATION BETWEEN THE ORIGINAL AND ADJOINT NETWORKS

Recall the branch relationships for the nonsource elements in $\mathcal{N}$:

$$V_R = RI_R$$

for resistors,

$$I_C = j\omega C V_C$$

for capacitors,

$$V_L = j\omega L I_L$$

for inductors, and

$$I_{g_m C} = 0$$

$$I_{g_m D} = g_m V_{g_m C}$$

for voltage controlled current sources. If $\mathbf{V}_R$ and $\mathbf{I}_R$ denote the vectors of resistor branch voltages and currents, respectively, with similar interpretations for $\mathbf{I}_C, \mathbf{V}_C, \mathbf{V}_L, \mathbf{I}_L, \mathbf{I}_{g_m C}, \mathbf{I}_{g_m D}, \mathbf{V}_{g_m C}$, and $\mathbf{V}_{g_m D}$, then by proper numbering of the elements we can write

$$\mathbf{V}_b = \begin{bmatrix} \mathbf{V}_R \\ \mathbf{V}_C \\ \mathbf{V}_L \\ \mathbf{V}_{g_m C} \\ \mathbf{V}_{g_m D} \end{bmatrix} \quad \text{and} \quad \mathbf{I}_b = \begin{bmatrix} \mathbf{I}_R \\ \mathbf{I}_C \\ \mathbf{I}_L \\ \mathbf{I}_{g_m C} \\ \mathbf{I}_{g_m D} \end{bmatrix}$$

The branch relations can be expressed in matrix form:

(19.20) $$\mathbf{I}_b = \mathcal{Y} \mathbf{V}_b$$

where

(19.21) $$\mathcal{Y} = \begin{bmatrix} \mathbf{R}^{-1} & 0 & 0 & 0 & 0 \\ 0 & j\omega \mathbf{C} & 0 & 0 & 0 \\ 0 & 0 & \dfrac{1}{j\omega} \mathbf{L}^{-1} & 0 & 0 \\ 0 & 0 & 0 & 0 & 0 \\ 0 & 0 & 0 & \mathbf{G}_m & 0 \end{bmatrix}$$

where $\mathbf{R}$, $\mathbf{C}$, $\mathbf{L}$, and $\mathbf{G}_m$ are appropriate diagonal matrices of element values.

Now consider the branch relationships of nonsource elements in the adjoint network $\hat{\mathcal{N}}$:

$$v_R = R\Phi_R$$

for resistors,

$$\Phi_C = j\omega C v_C$$

for capacitors,

$$v_L = j\omega L\Phi_L$$

for inductors, and

$$\Phi_{g_m C} = g_m v_{g_m D}$$

$$\Phi_{g_m D} = 0$$

for voltage controlled current sources (the subscripts here indicate branch correspondence with $\mathcal{N}$,) or in matrix form

$$\mathbf{\Phi}_b = \hat{\mathscr{Y}}\mathbf{v}_b$$

where

$$\hat{\mathscr{Y}} = \begin{bmatrix} \mathbf{R}^{-1} & 0 & 0 & 0 & 0 \\ 0 & j\omega\mathbf{C} & 0 & 0 & 0 \\ 0 & 0 & \dfrac{1}{j\omega}\mathbf{L}^{-1} & 0 & 0 \\ 0 & 0 & 0 & 0 & \mathbf{G}_m \\ 0 & 0 & 0 & 0 & 0 \end{bmatrix}$$

Observe that the branch admittance matrix $\hat{\mathscr{Y}}$ of the adjoint network the transpose of the branch admittance matrix $\mathscr{Y}$ of the original network

$$\hat{\mathscr{Y}} = \mathscr{Y}^T$$

We conclude, therefore, that the nodal admittance matrix of the adjoint network, which is related to $\hat{\mathscr{Y}}$ by

$$\hat{\mathbf{Y}} = \mathbf{A}\hat{\mathscr{Y}}\mathbf{A}^T$$

is the transpose of the nodal admittance matrix of $\mathcal{N}$:

$$\hat{\mathbf{Y}} = \mathbf{A}\hat{\mathscr{Y}}\mathbf{A}^T = \mathbf{A}\mathscr{Y}^T\mathbf{A}^T = \mathbf{Y}^T$$

The fact that $\hat{\mathbf{Y}} = \mathbf{Y}^T$ can be used to reduce the computational effort in determining the network sensitivities. The nodal equilibrium equations of $\hat{\mathcal{N}}$ are

$$\mathbf{Y}\mathbf{V}_n = \mathbf{I}$$

Recall that solving these equations using *LU* factorization involves three steps:

1. The factorization of $\mathbf{Y}$ into the lower triangular matrix $\mathbf{L}$ and the upper triangular matrix $\mathbf{U}$ so that $\mathbf{Y} = \mathbf{LU}$.

2. Forward substitution, that is, solving $\mathbf{Lb} = \mathbf{I}$.
3. Backward substitution, that is, solving $\mathbf{UV}_n = \mathbf{b}$.

The nodal equilibrium equations of $\mathcal{N}$ are

$$\hat{\mathbf{Y}}\mathbf{v}_n = \mathbf{\Phi}$$

Observe that

$$\hat{\mathbf{Y}} = \mathbf{Y}^T = (\mathbf{LU})^T = \mathbf{U}^T\mathbf{L}^T$$

and that $\mathbf{U}^T$ is lower triangular and $\mathbf{L}^T$ is upper triangular. Therefore, if the $\mathbf{L}$ and $\mathbf{U}$ factors are known for $\mathbf{Y}$, the first step in the $LU$ factorization solution of the adjoint network nodal equilibrium equations can be skipped. The two steps that remain are

1. Forward substitution, that is, solving

$$\mathbf{U}^T\hat{\mathbf{b}} = \mathbf{\Phi}$$

2. Backward substitution, that is, solving

$$\mathbf{L}^T\mathbf{v}_n = \hat{\mathbf{b}}$$

## 19-7  A NETWORK DESIGN EXAMPLE

In this section we will discuss the optimization of the transistor amplifier circuit of Fig. 19-15$a$, the equivalent circuit of which is shown in Fig. 19-15$b$. We will write a special purpose optimization program for this circuit. Our purpose is to illustrate network optimization without getting bogged down in the intricacies of writing a general purpose design program.

Suppose that we desire that the amplifier have a flat gain of 100, or 40 dB over as large a range as possible. Since we are unconcerned with the phase of the output, we can state this specification on gain as follows:

For $\hat{V}_1 = 1$ design the network to yield $|V_o| = 100$ for all $\omega$. A suitable performance function is (19.2$b$)

$$\varepsilon = \frac{1}{2}\sum_{k=1}^{N} W_k(|V_o(j\omega_k)| - |\hat{V}_o(j\omega_k)|)^2$$

where

$$|\hat{V}_o(j\omega)| = 100 \qquad \text{for all} \qquad \omega_k \qquad k = 1, 2, \ldots, N$$

$\omega_k$ are the sample frequency points
$N$ is the number of frequency points
and
$W_k$ are a set of weights

We will choose $R_B$, $r_\pi$, $R_E$, $C_\pi$, $g_m$, and $R_L$ to be designable parameters with the initial values as shown. (*Note.* This circuit has been considered previously in Example 16.10.)

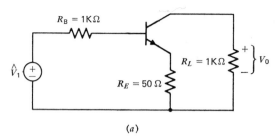

(a)

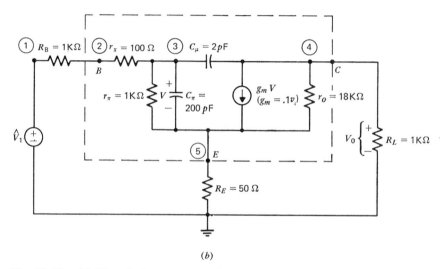

(b)

**Fig. 19-15**  (a) **The single stage transistor amplifier to be designed.** (b) **The network of** (a) **with the transistor replaced by its hybrid-pi model.**

Before proceeding with a description of the program, we will introduce the concept of parameter scaling. Recall from our earlier discussions on steepest descent that the more circular the error contours are the quicker the minimum could be found. Consider now that in our present example the resistances are many orders of magnitudes larger in values than the capacitor. In order to "equalize" this difference and "round out" the parameter space we can consider the natural logarithm of the parameters to be designable rather than the parameters themselves. In other words, if $p$ is a parameter of the network then $\ln p$ is the designable parameter. The gradient of the performance function with respect to $\ln p$ is:

$$\frac{\partial \varepsilon}{\partial \ln p} = \frac{\partial \varepsilon}{\partial p} \frac{\partial p}{\partial \ln p}$$

$$= p \frac{\partial \varepsilon}{\partial p}$$

```
C      CHAPTER 19 NETWORK DESIGN EXAMPLE
C
C      DIMENSIONS
C
       IMPLICIT REAL*8 (A-H,O-Z)
       COMMON/ANAL/V1(50),FREQ(50),DESOUT(50),WEIGHT(50),NF,NPARM
       DIMENSION P(9),G(9),S(18)
       EQUIVALENCE (P(1),RB),(P(2),RPI),(P(3),RE),(P(4),CPI),
      +(P(5),GM),(P(6),RL),(P(7),RX),(P(8),CMU),(P(9),RO)
       COMPLEX*16 V1
       LOGICAL*1 $PR
C
C      READ INITIAL ELEMENT VALUES
C
       READ 1,RB,RX,RPI,CPI,CMU,GM,RO,RL,RE
1      FORMAT(8G10.0)
C
C      SET NO. OF ADUSTABLE PARAMETERS
C
       NPARM=6
       N=NPARM
C
C      READ NUMBER OF FREQUENCY POINTS, THE FREQ,V1,DESIRED OUTPUT,
C      AND WEIGHT DATA AT EACH POINT
C
C      PRINT HEADINGS
C
       PRINT 8
8      FORMAT(' SPECIFICATIONS',/' FREQUENCY',12X,'INPUT',17X,
      +'DESIRED',6X,'WEIGHT',//)
C
C      READ FREQUENCY POINTS, INPUTS, AND DESIRED OUTPUTS
C
       READ 2,NF
       DO 3 K=1,NF
       READ 4, FREQ(K),V1(K),DESOUT(K),WEIGHT(K)
3      PRINT 9,FREQ(K),V1(K),DESOUT(K),WEIGHT(K)
9      FORMAT(' ',5(G12.5,2X))
2      FORMAT(I2)
4      FORMAT(5G10.0)
       $PR=.TRUE.
       CALL CNJGRD(N,P,F,G,S,10.D-9,10.D-5,100,IER,$PR)
       PRINT 7,IER
7      FORMAT('1IER=',I3)
C
C      PRINT FINAL NETWORK DESCRIPTION AND RESPONSE
C
       PRINT 6,(P(J),J=1,9)
6      FORMAT(' FINAL NETWORK VALUES',/,' RB=',G12.5,2X,'RPI=',G12.5,
      +2X,'RE=',G12.5,2X,'CPI=',G12.5,2X,'GM=',G12.5,2X,'RL=',G12.5,2X,
      +/' RX=',G12.5,2X,'CMU=',G12.5,2X,'RO=',G12.5)
C
       CALL FUNCT(N,P,F,G,-1)
C
       END
```

**Fig. 19-16   Listing of program which designs the amplifier of Fig. 19-15.**

```
      SUBROUTINE FUNCT(N,P,F,G,ICOUNT)
      IMPLICIT REAL*8 (A-H,O-Z)
      COMMON /ANAL/ V1(50),FREQ(50),DESOUT(50),WEIGHT(50),NF,NPARM
      DIMENSION P(1),G(1),JPIV(6),PY(2,50)
      COMPLEX*16 Y(6,6),CIN(6),JW,V1,V(6),T*
      COMPLEX*16 CRB,CRPI,CRE,VCPI,VGM,CRL
      EQUIVALENCE (V(1),CIN(1))
C
C     INITIALIZE ELEMENT VALUES
C
      RB=P(1)
      RPI=P(2)
      RE=P(3)
      CPI=P(4)
      GM=P(5)
      RL=P(6)
      RX=P(7)
      CMU=P(8)
      RO=P(9)
C
C     INITIALIZE GRADIENT AND ERROR
C
      DO 1 J=1,NPARM
1     G(J)=0.
      F=0.
C
C     CALCULATE PERFORMANCE FUNCTION AND GRADIENT
C
C     RUN OVER EACH FREQUENCY POINT
C
      DO 2 K=1,NF
      JW=DCMPLX(0.D0,6.2824*FREQ(K))
C
C     SOLVE ORIGINAL NETWORK
C
      CALL NODEQN(RB,RX,RPI,CPI,CMU,GM,RO,RL,RE,V1(K),JW,Y,CIN)
      CALL LUSLVC(Y,CIN,6,6,JPIV)
C
C     STORE OUTPUT FOR PLOTTING
C
      PY(1,K)=CDABS(V(4))
C
C     CALUCULATE BRANCH VOLTAGES AND CURRENTS NEEDED FOR
C     GRADIENT COMPUTATION
C
      CRB=(V(2)-V(1))/RB
      CRPI=(V(3)-V(5))/RPI
      CRE=V(5)/RE
      VCPI=V(3)-V(5)
      VGM=V(3)-V(5)
      CRL=V(4)/RL
      ERR=(CDABS(V(4))-DESOUT(K))
      T=ERR*DCONJG(V(4))/CDABS(V(4))
```

**Fig. 19-16 (*Continued*)**

```
C
C      SET UP ADJOINT NETWORK EXCITATION
C
       DO 6 L=1,6
6      CIN(L)=(0.,0.)
       CIN(4)=(-1.,0.)
C
C      COMPUTE ADJOINT NETWORK VOLTAGES AND CURRENTS
C
       CALL FWBWCT(Y,CIN,6,6,JPIV)
C
C      COMPUTE GRADIENTS
C
       G(1)=G(1)-P(1)*DREAL(T*CRB*(V(2)-V(1))/RB)
       G(2)=G(2)-P(2)*DREAL(T*CRPI*(V(3)-V(5))/RPI)
       G(3)=G(3)-P(3)*DREAL(T*CRE*V(5)/RE)
       G(4)=G(4)+P(4)*DREAL(T*JW*VCPI*(V(3)-V(5)))
       G(5)=G(5)+P(5)*DREAL(T*VGM*(V(4)-V(5)))
       G(6)=G(6)-P(6)*DREAL(T*CRL*V(4)/RL)
       F=F+.5*ERR**2
2      CONTINUE
C
C      PLOT RESPONSE OF INIEIAL AND FINAL NETWORKS
C
       IF (ICOUNT.EQ.-1) GO TO 9
       IF(ICOUNT.NE.0) GO TO 3
       PRINT 4,ICOUNT
4      FORMAT('1PLOT OF ACTUAL RESPONSE (X) VERSUS DESIRED RESPONSE(0)',
      +' IN DB VERSUS FREQUENCY IN HZ',//' ITERATION NO.',I3,///)
       GO TO 10
9      PRINT 11
11     FORMAT('1FINAL NETWORK RESPONSE -- ACTUAL (X) AND DESIRED (0)',
      +' RESPONSE IN DB VERSUS FREQUENCY IN HZ')
10     DO 5 K=1,NF
5      PY(2,K)=DESOUT(K)
       CALL PLOT(FREQ,PY,NF,2,1)
       PRINT 8,F
8      FORMAT('1ERROR = ',G12.5//,6X,'FREQUENCY',10X,'ACTUAL',11X,
      +'DESIRED')
       PRINT 7, (FREQ(K),PY(1,K),PY(2,K),K=1,NF)
7      FORMAT(3(5X,G12.5))
3      RETURN
       END


       SUBROUTINE PRMCHG(P,AMBDA,S,N)
       IMPLICIT REAL*8(A-H,O-Z)
       DIMENSION P(1),S(1)
C
C      CHANGE PARAMETERS -- ASSUMING LOGARITHMIC SCALING
C
       DO 1 J=1,N
       P(J)=P(J)*DEXP(AMBDA*S(J))
1      CONTINUE
       RETURN
       END
```

**Fig. 19-16** (*Continued*)

```
        SUBROUTINE NODEQN(RB,RX,RPI,CPI,CMU,GM,RO,RL,RE,V1,JW,Y,CIN)
        IMPLICIT REAL*8(A-H,O-Z)
        COMPLEX*16 Y(6,6),CIN(6),V1,JW
C
C       INITIALIZE COEFFICIENT ARRAYS
C
        DO 1 K=1,6
        CIN(K)=(0.,0.)
        DO 1 I=1,6
      1 Y(K,I)=(0.,0.)
C
C       SET UP NODAL EQUILIBRIUM EQUATION COEFFICIENTS
C
        Y(1,1)=1./RB
        Y(1,2)=-1./RB
        Y(1,6)=1.
        Y(2,1)=-1./RB
        Y(2,2)=(1./RB)+(1./RX)
        Y(2,3)=-1./RX
        Y(3,2)=-1./RX
        Y(3,3)=(1./RX)+(1./RPI)+JW*(CMU+CPI)
        Y(3,4)=-JW*CMU
        Y(3,5)=-(1./RPI)-JW*CPI
        Y(4,3)=GM-JW*CMU
        Y(4,4)=(JW*CMU)+(1./RO)+(1./RL)
        Y(4,5)=-(1./RO)-GM
        Y(5,3)=-(1./RPI)-(JW*CPI)-GM
        Y(5,4)=-1./RO
        Y(5,5)=GM+(1./RPI)+(1./RE)+(1./RO)+JW*CPI
        Y(6,1)=1.
        CIN(6)=1.
        RETURN
        END
```

**Fig. 19-16 (*Continued*)**

where $\partial\varepsilon/\partial p$ is the gradient we can compute using the adjoint network. This method of parameter scaling will be incorporated into our program.

We will now discuss the program of Fig. 19-16a which optimizes the amplifier under consideration by adjusting the designable parameters to minimize the performance function. The main program reads in the initial element values, the frequency sample points, FREQ, (in Hz, not radians/second) the input, V1, (real and imaginary parts,) the magnitude of the desired output DESOUT, and the weights, WEIGHT. The subroutine CNJGRD is then called to iteratively adjust the parameters and find the minimum of $\varepsilon$. On return from CNJGRD, the final network description is printed and its response plotted.

The subroutine FUNCT on being called evaluates the performance function and gradient by analyzing the original network and its adjoint at each frequency point. The nodal equilibrium equations for the original network are set up using the most recent set of parameter values by the subroutine NODEQN. These equations are then solved by LUSLVC (shown in Fig. 19-17), which is a complex-valued version of LUSOLV with an additional

```
C
C                        SUBROUTINE 'LUSLVC'
C
C
C
C
C       LUSLVC IS A SUBROUTINE FOR SOLVING SIMULTANEOUS
C       LINEAR ALGEBRAIC EQUATIONS WITH COMPLEX-VALUED COEFFICIENTS.
C       THE PROGRAM USES L-U FACTORIZATION AND FORWARD-BACKWARD
C       SUBSTITUTION WITH ROW PIVOTING. LUSLVC IS CALLED
C       FROM ANOTHER PROGRAM BY USING THE STATEMENT
C
C
C                   CALL LUSLVC(A,B,N,M,JPIV)
C
C
C       'A' IS A DOUBLE-SUBSCRIPTED COMPLEX*16 ARRAY DIMENSIONED
C       MXM IN THE CALLING PROGRAM. 'B' IS A SINGLE-SUBSCRIPTED
C       COMPLEX*16 ARRAY AND 'JPIV' IS A SINGLE-SUBSCRIPTED
C       INTEGER ARRAY. 'B' AND 'JPIV' ARE DIMENSIONED M IN THE
C       CALLING PROGRAM AND N IS THE NUMBER OF EQUATIONS
C       TO BE SOLVED. 'A' CONTAINS THE COEFFICIENTS OF THE
C       UNKNOWNS OF THE ALGEBRAIC EQUATIONS. 'B' CONTAINS
C       THE COEFFICIENTS ON THE RIGHT HAND SIDE OF THE
C       ALGEBRAIC EQUATIONS. UPON RETURN FROM LUSLVC 'A'
C       CONTAINS THE ELEMENTS OF THE LOWER AND UPPER TRIANGULAR
C       MATRICES OF 'A' AND 'B' CONTAINS THE SOLUTION.
C
C       LUSLVC ALLOWS THE USER TO ALTER THE RIGHT-HAND SIDE
C       OF THE EQUATIONS AND OBTAIN A NEW SOLUTION WITHOUT
C       HAVING TO DO ANOTHER FACTORIZATION OF 'A'.
C       'JPIV' RECORDS THE PIVOT ORDER USED IN THE
C       FACTORIZATION OF 'A'. THIS IS NECESSARY SO
C       THAT UPON ENTERING FWBWC THE ROWS OF 'B' WILL HAVE
C       THE SAME PIVOT ORDER.
C       THE SEATEMENT
C
C                   CALL FWBWC(A,B,N,M,JPIV)
C
C       YIELDS A NEW SOLUTION WHEN 'B' HAS BEEN ALTERED.
C       USE OF THE TWO INTEGERS M AND N ALLOWS THE
C       DIMENSIONS OF 'A' AND 'B' TO BE SET ONCE IN THE
C       MAIN PROGRAM AND THE ACTUAL NUMBERS OF EQUATIONS TO
C       VARY(NOTE: M≥N), DEPENDING ON THE PROBLEM.
C
C       SINCE THE MAIN PURPOSE OF THIS PROGRAM IS TO BE
C       USED IN AN OPTIMIZATION PROGRAM THE ENTRY POINT
C       FWBWCT IS INCLUDED WHICH ALLOWS THE FORWARD -
C       BACKWARD SUBSTITUION USING THE TRANSPOSED U AND
C       L MATRICES.  TO USE THIS ENTRY POINT FOR ADJOINT
C       NETWORK COMPUTATIONS USE THE STATEMENT
C
C                   CALL FWBWCT(A,B,N,M,JPIV)
C
        SUBROUTINE LUSLVC(A,B,N,M,JPIV)
        IMPLICIT REAL*8(A-H,O-Z)
        COMPLEX*16 A(M,M),B(M),PIV,PIVA,SUM
        DIMENSION JPIV(M)
```

**Fig. 19-17   Listing of the subroutine LUSLVC.**

```
      DO 4 I=1,N
      JPIV(I)=I
      I1=I+1
      IF(CDABS(A(I,I)).LE.1.D-10) GO TO 1
      GO TO 15
    1 CONTINUE
      IF(I.EQ.N) GO TO 20
      DO 14 J=I1,N
      IF(CDABS(A(J,I)).LE.1.D-10) GO TO 14
      JPIV(I)=J
      GO TO 16
   14 CONTINUE
      GO TO 20
   16 DO 2 K=1,N
      IPIV=JPIV(I)
      PIV=A(IPIV,K)
      A(IPIV,K)=A(I,K)
    2 A(I,K)=PIV
   15 IF(I.EQ.N) GO TO 3
      DO 8 JI=I1,N
    8 A(I,JI)=A(I,JI)/A(I,I)
      DO 4 J=I1,N
      DO 4 K=I1,N
    4 A(J,K)=A(J,K)-(A(J,I)*A(I,K))
    3 CONTINUE
C
C     ENTER FORWARD-BACKWARD SUBSTITUTION
C
      ENTRY FWBWC(A,B,N,M,JPIV)
      IFB=1
   64 DO 61 I=1,N
      IPIV=JPIV(I)
      IF(IPIV.LE.I) GO TO 61
      J=I
      PIVA=B(I)
   62 PIV=B(IPIV)
      B(IPIV)=PIVA
      JPIV(J)=-IPIV
      J=IPIV
      IPIV=JPIV(J)
      PIVA=PIV
      IF(IPIV.GT.0) GO TO 62
   61 CONTINUE
      DO 63 I=1,N
   63 JPIV(I)=IABS(JPIV(I))
      GO TO (65,165),IFB
C
C     FORWARD SUBSTITUTION
C
   65 DO 31 K=1,N
      SUM=(0.,0.)
      IF(K.EQ.1.) GO TO 41
      MM=K-1
      DO 51 J=1,MM
   51 SUM=SUM+A(K,J)*B(J)
```

**Fig. 19-17 (*Continued*)**

```
   41  B(K)=(1./A(K,K))*(B(K)-SUM)
   31  CONTINUE
C
C      BACKWARD SUBSTITUTION
C
       DO 91 LL=1,N
       K=(N+1)-LL
       SUM=(0.,0.)
       IF(K.EQ.N) GO TO 81
       KK=K+1
       DO 71 J=KK,N
   71  SUM=SUM+A(K,J)*B(J)
   81  B(K)=B(K)-SUM
   91  CONTINUE
       GO TO 30
   20  PRINT 21
   21  FORMAT(' EQUATIONS ARE LINEARLY DEPENDENT')
       STOP
   30  CONTINUE
       RETURN
C
C      ENTER FORWARD-BACKWARD SUBSTITUTION USING
C      THE TRANSPOSED U AND L MATRICES
C
       ENTRY FWBWCT(A,B,N,M,JPIV)
       IFB=2
       GO TO 64
C
C      FORWARD SUBSTITUION
C
  165  DO 131 K=2,N
       L=K-1
       SUM=(0.,0.)
       DO 151 J=1,L
  151  SUM=SUM+A(J,K)*B(J)
       B(K)=B(K)-SUM
  131  CONTINUE
C
C      BACKWARD SUBSTITUTION
C
       DO 191 LL=1,N
       K=N-LL+1
       SUM=(0.,0.)
       IF(K.EQ.N) GO TO 181
       KK=K+1
       DO 171 J=KK,N
  171  SUM=SUM+A(J,K)*B(J)
  181  B(K)=(B(K)-SUM)/A(K,K)
  191  CONTINUE
       RETURN
       END
```

**Fig. 19-17** (*Continued*)

| 1000. | 100. | 1000. | 200. | E-12 | 2. | E-12 | .1 | 18000. | 1000. |
|-------|------|-------|------|------|----|------|----|--------|-------|

50.

21

| 100. | 1. | | 40. | 1. |
|------|----|-|-----|----|
| 200. | 1. | | 40. | 1. |
| 400. | 1. | | 40. | 1. |
| 600. | 1. | | 40. | 1. |
| 800. | 1. | | 40. | 1. |
| 1000. | 1. | | 40. | 1. |
| 2000. | 1. | | 40 | 1. |
| 4000. | 1. | | 40. | 1. |
| 6000. | 1. | | 40. | 1. |
| 8000. | 1. | | 40. | 1. |
| 10000. | 1. | | 40. | 1. |
| 20000. | 1. | | 40. | 1. |
| 40000. | 1. | | 40. | 1. |
| 60000. | 1. | | 40. | 1. |
| 80000. | 1. | | 40. | 1. |
| 100000. | 1. | | 40. | 1. |
| 200000. | 1. | | 40. | 1. |
| 400000. | 1. | | 40. | 1. |
| 600000. | 1. | | 40. | 1. |
| 800000. | 1. | | 40. | 1. |
| 1000000. | 1. | | 40. | 1. |

**Fig. 19-18    Data used for program in Fig. 19-16.**

entry point discussed below, and the branch voltages and currents needed for the gradient calculation are computed and stored. The adjoint network is "analyzed" by the subroutine FWBWCT, an entry point into LUSLVC, which performs a forward and backward substitution using the transposed **U** and **L** factors of the nodal admittance matrix of the original network. The appropriate adjoint network branch voltages and currents are then calculated and multiplied by appropriately weighted original network branch voltages and currents and the parameter value to form the gradient. Observe the gradients expressions are evaluated with respect to ln $p$.

The subroutine PRMCHG adjusts the ln $p$ along the direction specified by CNJGRD. Observe that the $(k + 1)$st updated value of the $j$th designable parameter ln $p_j^{k+1}$ is obtained from the previous value of the $j$th designable parameter, ln $p_j^k$ by

$$\ln p_j^{k+1} = \ln p_j^k + \alpha s_j$$

or

$$p_j^{k+1} = p_j^k \exp{(\alpha s_j)}$$

The program was run using the data shown in Fig. 19-18 and the results are shown in Figs. 19-19$a$ and 19-19$b$.

## 19-8  SUMMARY

The purpose of this chapter was to introduce the student to computer-aided circuit design. A detailed investigation of all concepts presented in this chapter is well beyond the scope of the text, however it is hoped that the

SPECIFICATIONS

| FREQUENCY | INPUT | | DESIRED | WEIGHT |
|---|---|---|---|---|
| 100.00 | 1.0000 | 0.0 | 40.000 | 1.0000 |
| 200.00 | 1.0000 | 0.0 | 40.000 | 1.0000 |
| 400.00 | 1.0000 | 0.0 | 40.000 | 1.0000 |
| 600.00 | 1.0000 | 0.0 | 40.000 | 1.0000 |
| 800.00 | 1.0000 | 0.0 | 40.000 | 1.0000 |
| 1000.0 | 1.0000 | 0.0 | 40.000 | 1.0000 |
| 2000.0 | 1.0000 | 0.0 | 40.000 | 1.0000 |
| 4000.0 | 1.0000 | 0.0 | 40.000 | 1.0000 |
| 6000.0 | 1.0000 | 0.0 | 40.000 | 1.0000 |
| 8000.0 | 1.0000 | 0.0 | 40.000 | 1.0000 |
| 10000. | 1.0000 | 0.0 | 40.000 | 1.0000 |
| 20000. | 1.0000 | 0.0 | 40.000 | 1.0000 |
| 40000. | 1.0000 | 0.0 | 40.000 | 1.0000 |
| 60000. | 1.0000 | 0.0 | 40.000 | 1.0000 |
| 80000. | 1.0000 | 0.0 | 40.000 | 1.0000 |
| 100000. | 1.0000 | 0.0 | 40.000 | 1.0000 |
| 0.20000D 06 | 1.0000 | 0.0 | 40.000 | 1.0000 |
| 0.40000D 06 | 1.0000 | 0.0 | 40.000 | 1.0000 |
| 0.60000D 06 | 1.0000 | 0.0 | 40.000 | 1.0000 |
| 0.80000D 06 | 1.0000 | 0.0 | 40.000 | 1.0000 |
| 0.10000D 07 | 1.0000 | 0.0 | 40.000 | 1.0000 |

**Fig. 19-19a** **The design specifications outputed by the program of Fig. 19-17.**

ERROR = 7303.2

| FREQUENCY | ACTUAL | DESIRED |
|---|---|---|
| 100.00 | 13.745 | 40.000 |
| 200.00 | 13.745 | 40.000 |
| 400.00 | 13.745 | 40.000 |
| 600.00 | 13.745 | 40.000 |
| 800.00 | 13.745 | 40.000 |
| 1000.0 | 13.745 | 40.000 |
| 2000.0 | 13.745 | 40.000 |
| 4000.0 | 13.745 | 40.000 |
| 6000.0 | 13.745 | 40.000 |
| 8000.0 | 13.745 | 40.000 |
| 10000. | 13.745 | 40.000 |
| 20000. | 13.744 | 40.000 |
| 40000. | 13.743 | 40.000 |
| 60000. | 13.740 | 40.000 |
| 80000. | 13.737 | 40.000 |
| 100000. | 13.733 | 40.000 |
| 0.20000D 06 | 13.697 | 40.000 |
| 0.40000D 06 | 13.557 | 40.000 |
| 0.60000D 06 | 13.332 | 40.000 |
| 0.80000D 06 | 13.036 | 40.000 |
| 0.10000D 07 | 12.682 | 40.000 |

**Fig. 19-19b** **Response of the network of Fig. 19-15 for initial parameter values.**

661

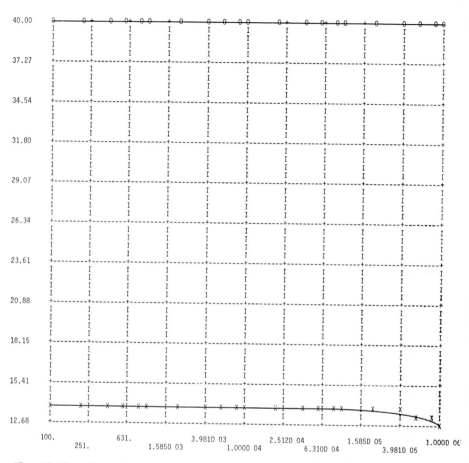

**Fig. 19-19c  Plot of the response of the network of Fig. 19-15 for initial parameter values.**

```
IER=  3
FINAL NETWORK VALUES
RB= 0.24818     RPI=  1056.4     RE= 0.64060     CPI= 0.21547D-11  GM=  14.496      RL=  28.657
RX=  100.00     CMU= 0.20000D-11  RO=  18000.
```

**Fig. 19-19d  The final parameter values determined by the program of Fig. 19-17.**

662

ERROR =   0.19602D-02

| FREQUENCY | ACTUAL | DESIRED |
|---|---|---|
| 100.00 | 40.006 | 40.000 |
| 200.00 | 40.006 | 40.000 |
| 400.00 | 40.006 | 40.000 |
| 600.00 | 40.006 | 40.000 |
| 800.00 | 40.006 | 40.000 |
| 1000.0 | 40.006 | 40.000 |
| 2000.0 | 40.006 | 40.000 |
| 4000.0 | 40.006 | 40.000 |
| 6000.0 | 40.006 | 40.000 |
| 8000.0 | 40.006 | 40.000 |
| 10000. | 40.006 | 40.000 |
| 20000. | 40.006 | 40.000 |
| 40000. | 40.006 | 40.000 |
| 60000. | 40.006 | 40.000 |
| 80000. | 40.005 | 40.000 |
| 100000. | 40.005 | 40.000 |
| 0.20000D 06 | 40.004 | 40.000 |
| 0.40000D 06 | 39.997 | 40.000 |
| 0.60000D 06 | 39.986 | 40.000 |
| 0.80000D 06 | 39.971 | 40.000 |
| 0.10000D 07 | 39.951 | 40.000 |

**Fig. 19-19e   Response of the network of Fig. 19-15 for final parameter values.**

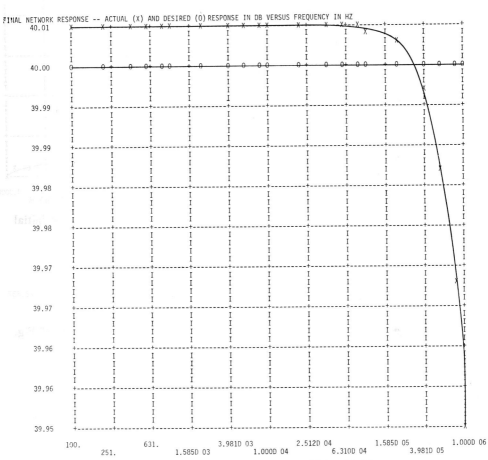

**Fig. 19-19f   Plot of the response of Fig. 19-19e.**

flavor of circuit optimization has been transmitted. The student should bear in mind that network optimization programs are not intended to replace the designer but are to aid him in design. Moreover, a network optimization program is only effective if the designer provides it with a reasonable initial estimate of the network structure and element values.

## PROBLEMS

**19.1**   Sketch the following performance functions in the region $-2 \leq p_1 \leq 2$, $-2 \leq p_2 \leq 2$. Also sketch the equierror contours in the parameter space. Indicate all minima.

(a) $\varepsilon = p_1 p_2$

(b) $\varepsilon = p_1{}^2 + p_2{}^2$

(c) $\varepsilon = p_1{}^1 p_2{}^2$

**19.2**   Use the program of Fig. 19-11 to find the minimum of the functions in Prob. 19.2. (*Hint.* Modify the subroutine FUNCT.) Take as initial estimates of the parameters $p_1 = 2$ and $p_2 = -3$.

**19.3**   Find the adjoint network element that corresponds to gyrators and current controlled current sources. Determine expressions in terms of original and adjoint network voltages and currents that can be used to evaluate the sensitivity of an output voltage to $\alpha$ for a gyrator and $\beta$ for a current controlled current source.

**19.4**   Use the program of Fig. 19-16 to design the network of Fig. 19-15 for a constant gain of 40 dB over a frequency of range of $100-10^{10}$ Hz. Take the initial element values of those given in Fig. 19-19*d*.

**19.5**   It is possible to obtain an even wider bandwidth for the amplifier of Fig. 19-15 by putting a capacitor $C_E$ across $R_E$. Modify the program of Fig. 19-16 by replacing $C_\pi$ as a designable parameter with $C_E$ and repeat Prob. 19.4. Take the initial value of $C_E$ to be $10^{-12}$ F. (*Note.* Be sure to modify the subroutine FUNCT which computes the gradient in addition to NODEQN.)

# Answers to Selected Problems

(Complete solutions to most problems can be found in the Teachers Manual, which accompanies this text.)

**CHAPTER 1**

**1.1** (a) node ②   $i_2 + i_4 + i_5 = 0$
        node ③   $-i_1 + i_3 - i_4 = 0$
      (b) node ②   $-i_3 - i_5 - i_6 = 0$
        node ③   $i_1 - i_2 + i_3 = 0$
      (c) node ②   $i_3 + i_4 + i_5 = 0$
        node ③   $i_1 - i_5 - i_6 = 0$

**1.2** (a) loop ①   $v_1 - v_2 + v_3 + v_4 + v_7 = 0$
        loop ②   $-v_2 + v_5 - v_6 = 0$
      (b) loop ①   $v_2 - v_5 + v_6 = 0$
        loop ②   $-v_3 + v_4 - v_5 = 0$

**1.3** (a) independent
      (b) dependent
      (c) independent
      (d) independent
      (e) dependent
      (f) independent

**1.4** (a) node ①   $-i_1 + i_2 + i_3 = 0$
        node ②   $-i_3 - i_4 = 0$
        node ③   $i_1 - i_2 + i_4 = 0$
      (b) node ①   $i_1 + i_2 - i_4 = 0$
        node ②   $-i_1 - i_3 - i_5 = 0$
        node ③   $-i_2 + i_3 = 0$
        node ④   $i_4 + i_5 = 0$

**1.5** (a) loop ①   $v_1 + v_2 = 0$
        loop ②   $-v_2 + v_3 - v_4 = 0$
        loop ③   $v_1 + v_3 - v_4 = 0$

      Equations are linearly dependent

      (b) loop ①   $-v_1 + v_2 + v_3 = 0$
        loop ②   $-v_1 - v_4 + v_5 = 0$
        loop ③   $v_2 + v_3 + v_4 - v_5 = 0$

      Equations are linearly dependent

**1.6**    Branch 1 has the associated reference directions Branch 2 does not

**1.7**    $p = 800/3$ watts

**1.8**    $-\hat{\imath}_1 + \left(\dfrac{1}{R_2} + \dfrac{1}{R_3}\right) v_2 + \dfrac{1}{R_3} v_4 = 0$

       $-\dfrac{1}{R_3} v_2 - \left(\dfrac{1}{R_4} + \dfrac{1}{R_3}\right) v_4 = 0$

**1.9**    $i(t) = \dfrac{5 \sin 2\pi t}{5000} = 10^{-3} \sin 2\pi t$ amp

       $p(t) = v(t)i(t) = 5 \times 10^{-3} \sin^2 2\pi t$ watts

# CHAPTER 2

**2.1**    (d) $n_n = 4$, $n_b = 6$

       (e) $n_n = 5$, $n_b = 8$

**2.2**    (a) linearly independent

       (b) linearly independent

       (c) linearly dependent

       (d) linearly independent

**2.3**    (b) ①    $-i_1 + i_2 - i_5 - i_6 = 0$

         ②            $i_1 - i_2 = 0$

         ③         $i_3 + i_4 + i_5 = 0$

         ④       $-i_3 - i_4 + i_6 = 0$

            ① $+$ ② $+$ ③ $= -$④

**2.4**    (b) ⊡1    $v_1 - v_2 - v_4 - v_6 - v_9 = 0$

         ⊡2          $-v_4 + v_5 - v_8 = 0$

         ⊡3         $-v_2 + v_3 - v_4 = 0$

       These equations are linearly independent.

       (d) ⊡1           $v_1 + v_2 + v_4 = 0$

          ⊡2          $v_3 - v_4 + v_6 = 0$

          ⊡3          $v_5 - v_6 - v_7 = 0$

          ⊡4    $v_1 + v_2 + v_3 + v_5 - v_7 = 0$

       These equations are linearly dependent

**2.6**    ① closed surface #1          $i_2 + i_7 - i_8 = 0$

       ② closed surface #2        $-i_4 + i_5 - i_7 = 0$

       ③ closed surface #3    $-i_1 + i_3 - i_5 + i_7 = 0$

①′ node #1      $i_1 + i_2 = 0$

②′ node #2      $-i_2 - i_3 + i_4 = 0$

③′ node #3      $-i_4 + i_5 - i_6 = 0$

④′ node #4      $i_6 - i_7 = 0$

⑤′ node #5      $i_3 - i_5 + i_8 = 0$

⑥′ node #6      $-i_1 + i_7 - i_8 = 0$

where ① = ①′ + ⑥′

     ② = ③′ + ④′

     ③ = ⑤′ + ⑥′

**2.8**    Cutset equations for tree 1:

$$i_1 + i_2 + i_3 + i_4 = 0$$
$$-i_2 - i_4 + i_5 + i_{11} + i_{12} = 0$$
$$-i_3 + i_6 - i_{11} - i_{12} = 0$$
$$-i_2 + i_7 + i_8 = 0$$
$$-i_4 - i_8 - i_9 + i_{11} + i_{12} = 0$$
$$-i_8 + i_{10} + i_{11} = 0$$
$$-i_3 - i_{11} + i_{13} = 0$$

loop equations for tree 1:

$$v_1 - v_2 - v_5 - v_7 = 0$$
$$v_1 - v_3 - v_6 - v_{13} = 0$$
$$v_1 - v_4 - v_5 + v_9 = 0$$
$$v_7 - v_8 + v_9 - v_{10} = 0$$
$$v_5 - v_6 - v_9 + v_{10} - v_{11} - v_{13} = 0$$
$$v_5 - v_6 - v_9 - v_{12} = 0$$

**2.9**    $i_1 = -10$, $i_5 = -1$, $i_6 = 11$, $i_7 = 1$, $i_9 = 2$, $i_{10} = -3$, and $i_{13} = 7$

Given $i_2 = 3$, $i_3 = 2$, $i_4 = 5$, $i_8 = 2$, $i_{11} = 5$, $i_{12} = 4$

$$i_1 = -i_2 - i_3 - i_4 = -10$$
$$i_5 = i_2 + i_4 - i_{11} - i_{12} = -1$$
$$i_6 = i_3 + i_{11} + i_{12} = 11$$
$$i_7 = i_2 - i_8 = 1$$
$$i_9 = -i_4 - i_8 + i_{11} + i_{12} = 2$$
$$i_{10} = i_8 - i_{11} = -3$$
$$i_{13} = i_3 + i_{11} = 7$$

for tree #1

Using the fundamental cutset equations for tree #2 and tree #3, we get the same answers (procedure omitted.)

**2.10**    Given $i_2 = 4$, $i_3 = -5$, $i_5 = 2$, $i_7 = 3$, $i_9 = 1$, $i_{12} = 3$ from equation ④ $i_8 = i_2 - i_7 = 1$

The rest of the branch currents are impossible to obtain.

**2.11**    $v_2 = -11$, $v_3 = -18$, $v_4 = 5$, $v_8 = 6$, $v_{11} = -13$, and $v_{12} = -10$

**2.14**    (b) mesh $\boxed{1}$    $v_2 - v_3 - v_6 = 0$

mesh $\boxed{2}$    $v_3 + v_4 - v_5 = 0$

mesh $\boxed{3}$    $v_1 - v_2 - v_4 = 0$

mesh $\boxed{4}$    $v_1 - v_5 - v_6 = 0$

# CHAPTER 3

**3.1**    (a) $5\Omega$, (b) $14/11 \ \mho$

**3.2**    (a) $0.4$a, (b) $16/7$v

**3.3**    $v = .334$v

**3.7**    (a) $x_1 = 1$, $x_2 = 2$, $x_3 = -1$

**3.8**    (b) $x_1 = 1$, $x_2 = 1$, $x_3 = -2$

(c) $x_1 = -53.03$, $x_2 = 50.79$, $x_3 = 69.66$

**3.10**    (b) $v_① = -3.1$, $v_② = 6.96$, $v_③ = 2.7$

**3.12**    (a) 6 watts

(c) $p_1 = 19/3$ watts

$p_2 = 33/2$ watts

**3.17**    (a) $R_{\text{eq}} = .636 \ \Omega$

# CHAPTER 4

**4.2**    (a) $-v_② + i_1 = -3.1$

$3v_② = -1$

$i_7 = 3.1$

**4.3**    (b) $v_① = -7.428$

$v_② = -1.428$

$v_③ = -2.715$

$v_④ = -3.428$

$v_⑤ = -1.715$

$i_1 = -1.571$

$i_2 = -0.286$

**4.6**    $p = 4.42$ watts

**4.8**    $R = 4.914 \ \Omega$

## CHAPTER 5

**5.3** (a) $v = 15/7$, (b) $i = 7/11$

**5.4** $p = .96$ watts

**5.7** (a) $i_3 = 2/13$; (b) $i_3 = .25$ amp

**5.11** (a) $i = i = 2/3\, i_1 + 1/6\, i_2$ and $v = i_1 - 1/2\, i_2$

**5.14** $i_{1\,\Omega} = -4$ amp, $i_{3\,\Omega} = 18/7$ amp

**5.16** $v_{oc} = 1.585$, $R_{\mathrm{TH}} = 2.179$, $i_{sc} = .727$

**5.17** $v_{oc} = 2.62$, $R_{\mathrm{TH}} = 1.31$, $i_{sc} = 2$

## CHAPTER 7

**7.3** $i_{sc} = .619$ amp, $R_{\mathrm{N}} = .4038\ \Omega$

**7.5** (a) $x \simeq .4$

**7.6** $v_{\textcircled{1}} \simeq .667$

**7.11** $p_{R_1} = .0625$ watts, $p_{R_3} = .125$ watts, $p_{\hat{v}_3} = .03177$ watts

**7.12** $i_1 = 1.28$ amps

**7.15** $i = 5 + \cos t/25$ amps

## CHAPTER 8

**8.2** $R_2 = 16\ \Omega$

**8.3** (a) $v_{oc} = 3.2$ volts, $R_{\mathrm{TH}} = 3\ \Omega$

**8.6** $v_0 = -24990$ volts

**8.7** $v_0 = 24010$ volts

**8.9** $v_0 = -490$ volts; input impedance of each circuit $= 100\ \Omega$

**8.13** (a) $z_{11} = z_{22} = 15/4$, $z_{12} = z_{21} = 13/4$

$y_{11} = y_{22} = 15/14$, $y_{21} = y_{12} = -13/14$

(b) $z_{11} = (R_1 + R_3)(R_2 + R_4)/(R_1 + R_2 + R_3 + R_4)$

$z_{12} = z_{21} = (R_1 R_4 - R_2 R_3)/(R_1 + R_2 + R_3 + R_4)$

$z_{22} = (R_1 + R_2)(R_3 + R_4)/(R_1 + R_2 + R_3 + R_4)$

(c) $z_{11} = n_1^2 R_2(R_1 + R_3)/((n_1 - n_2)^2 R_2 + n_2^2(R_1 + R_3))$

$z_{21} = z_{12} = n_1 n_2 R_2(R_1 + R_3)/((n_1 - n_2)^2 R_2 + n_2^2(R_1 + R_3))$

$z_{22} = n_2^2 R_2(R_1 + R_3)/((n_1 - n_2)^2 R_2 + n_2^2(R_1 + R_3))$

**8.15** $y_{11} = 1/R_1$, $y_{12} = 0$, $y_{21} = -r_m R_2/(R_4 R_1(R_2 - r_m))$, $y_{22} = 1/R_4$

**8.20** $n_1/n_2 = \sqrt{\dfrac{R_2}{R_3}}$

**8.21** $y_{11} = y_{22} = 0$, $y_{21} = 1/R$, $y_{12} = -1/R$

## CHAPTER 9

**9.1**    (b) $\begin{bmatrix} 13 & 5 & 1 \\ 39 & 13 & -1 \\ 8 & 2 & -1 \end{bmatrix}$

**9.13**    (a)
$$A = \begin{bmatrix} -1 & 2 & 1 \\ 0 & 1 & 2 \\ 1 & 0 & -1 \end{bmatrix}$$

(b) $\det (A) = 4$

(c)
$$A^{-1} = \begin{bmatrix} -.25 & .5 & -.25 \\ .5 & 0 & .5 \\ .75 & .5 & -.25 \end{bmatrix}$$

**9.15**    $v_① = 55/67$, $v_② = 41/67$, $v_③ = 50/67$

## CHAPTER 11

**11.5**    (b) $i(t) = 1/2(1 - e^{-1/2t})$
(d) $i(t) = -17/18e^{-3t} - 1/18$

**11.6**    (a) $v(t) = 2e^{-2t}$
(c) $v(t) = -2e^{-3t} + 2e^{-(5/2)t}$

**11.9**    (a) $-2/5e^{-(1/3)t} + 2/5e^{-2t}$.

**11.11**    $v(t) = 2/3 + 4/3e^{-(1/2)t}$, $t > 0$

**11.13**    $L_4 = 6.49$H.

**11.17**    Zero-input response is $-.05e^{-(5/18)t}$

## CHAPTER 12

**12.5**    (a) 1; (b) $\infty$; (c) 1

**12.6**    (a) $\delta(t) - \delta(t - 1)$; (b) $\delta(t - 1)$;
(d) $-\alpha e^{-\alpha t}u(t - \tau) + e^{-\alpha t}\delta(t - \tau)$

**12.7**    $v_0(t) = 5u(2 - t) - \dfrac{5L_2}{R_3}\delta(t - 2)$

**12.11**    (a) 0, (b) 1, (c) 4, (d) $\infty$

## CHAPTER 13

**13.1**    (c) Forward Euler
$$x^{k+1} = (1 + 3kh^2)x^k + kh^2$$
Backward Euler
$$x^{k+1} = \frac{1}{1 - 3(k + 1)h^2}x^k + \frac{(k + 1)h^2}{1 - 3(k + 1)h^2}$$

## CHAPTER 14

**14.1**   (e) $i = (1/2 - j3/8)e^{(-2+j4)t} + (1/2 + 3/8j)e^{(-2-j4)t}$

**14.3**   (b) $v_3(t) = 8e^{-2t} - 6e^{-3t}$

**14.5**   $i(t) = -5/2e^{-t}$

**14.7**   (d) $v(t) = 5/2 - 9/4e^{-t} + 9/12e^{-3t}$

**14.8**   (b) $v_2(t) = 1.27(e^{-.16t} - e^{-.95t})$

## CHAPTER 15

**15.2**   $L = \begin{pmatrix} 1 & 0 & 0 \\ 3 & 5 & 0 \\ 4 & 4 & 9/5 \end{pmatrix}$   $U = \begin{pmatrix} 1 & -1 & 2 \\ 0 & 1 & -6/5 \\ 0 & 0 & 1 \end{pmatrix}$

**15.3**   $A^{-1} = \begin{pmatrix} 10/9 & 5/9 & -4/9 \\ -5/3 & -1/3 & 2/3 \\ -8/9 & -4/9 & 5/9 \end{pmatrix}$

## CHAPTER 16

**16.1**   (a) $\dfrac{3}{2j}(e^{j(\omega t+2)} - e^{-j(\omega t+2)})$.

**16.3**   (c) $e^{-j90°} = 1 < -90° = -j$

**16.4**   (a) $v = 9/6(\cos 3t + \sin 3t)$

**16.6**   (b) $i(t) = .3293 \cos(3t + \pi/3) - .0366 \sin(3t + \pi/3)$

**16.13**   $i_2(t) = .085 \cos(4t - \pi/6) + .648 \sin(4t - \pi/6)$
$i_3(t) = -.270 \cos(4t - \pi/6) + .0356 \sin(4t - \pi/6)$

**16.20**   $y_{11} = -j\omega C_\mu - g_\pi + j\omega(C_\pi + C_\mu)$
$y_{21} = -(j\omega C_\mu - g_m)$
$y_{22} = j\omega C_\mu + g_0$
$y_{12} = -j\omega C_\mu$

**16.21**   (a) $V_1(j\omega)/I_1(j\omega) = \dfrac{5j\omega((j\omega)^2 + 2)}{(j\omega)^3 + 15/2(j\omega)^2 + 2j\omega + 5}$

$V_2(j\omega)/I_1(j\omega) = \dfrac{10j\omega}{(j\omega)^3 + 15/2(j\omega)^2 + 2j\omega + 5}$

## CHAPTER 17

**17.1**   (a) $Y = \dfrac{s(3 + 2s)}{8s^2 + 5s + 3}$

**17.4**   (c) $V_o = \dfrac{s^2}{s^2 + 3s + 1}$

**17.6**    $v(t) = -3.9e^{-3t} + .95 \sin (t + 71.6°)$

**17.8**    (b) 325 watts; (c) $-187.5$ vars

**17.11**    $p_{av} = 212.6$ watts, $p_{5\,\Omega} = 10.6$ watts, P.F. $= -.984$

## CHAPTER 18

**18.1**    (b) $f(t) = 1/2 + \displaystyle\sum_{\substack{n=1 \\ \text{nodd}}}^{\infty} (-1)^{(n-1)/2}\, \dfrac{4}{\pi n} \cos \left(\dfrac{2\pi n t}{T}\right)$

**18.6**    $v_0(t) \simeq 4/\pi \cos (t + \pi/2) + \dfrac{1.37}{\pi} \cos (3t - 149°) + \dfrac{\pi}{0.82}$

$$\times \cos (5t + 12°)$$

**18.13**    (b) $F(j\omega) = \dfrac{1 + j\omega}{2 + j\omega}$

**18.14**    (d) $\dfrac{6 + j12\omega}{8 - 4\omega^2 + j12\omega}$

**18.20**    $V_1(j\omega)/V_2(j\omega) = (64 - 2\omega^2 + 27j\omega)/8$

# Bibliography

The following is not meant to be an exhaustive list of references. It is a list of a few books covering several areas discussed in this text that may be helpful for the student who would like either to pursue his study of a given topic or a different presentation of a topic.

## GENERAL CIRCUIT THEORY

1. C. A. Desoer and E. S. Kuh, *Basic Circuit Theory*, McGraw-Hill, 1969.
2. W. H. Hayt, Jr. and J. E. Kemmerly, *Engineering Circuit Analysis*, McGraw-Hill, 1971.
3. M. E. Van Valkenburg, *Network Analysis*, Third Edition, Prentice-Hall, 1974.

## CIRCUIT THEORY WITH A COMPUTATIONAL FLAVOR

4. L. P. Huelsman, *Basic Circuit Theory with Digital Computations*, Prentice-Hall, 1972.
5. S. C. Gupta, J. W. Bayless, and B. Peikari, *Circuit Analysis with Computer Applications to Problem Solving*, Intext Education Publishers, 1972.
6. O. Wing, *Circuit Theory with Computer Methods*, Holt, Rinehart and Winston, 1972.
7. B. Kinariwala, F. F. Kuo, and N. Tsao, *Linear Circuits and Computation*, John Wiley, 1973.
8. R. W. Jensen and B. O. Watkins, *Network Analysis-Theory and Computer Methods*, Prentice-Hall, 1974.

## NUMERICAL TECHNIQUES

9. R. Beckert and J. Hurt, *Numerical Calculations and Algorithms*, McGraw-Hill, 1967.
10. G. Forsythe and C. B. Moler, *Computer Solution of Linear Algebraic Systems*, Prentice-Hall, 1967.
11. S. A. Hovanessian and L. A. Pipes, *Digital Computer Methods in Engineering*, McGraw-Hill, 1969.
12. R. W. Hamming, *Introduction to Applied Numerical Analysis*, McGraw-Hill, 1971.

## COMPUTER-AIDED CIRCUIT DESIGN

13. D. A. Calahan, *Computer-Aided Network Design*, McGraw-Hill, 1972.
14. S. W. Director, *Computer-Aided Circuit Design: Simulation and Optimization*, Dowden, Hutchinson, and Ross, 1974.

# INDEX